EAST ANGLIAN ARCHAEOLOGY

This volume is dedicated to Graham Webster,
who turned me from other paths to archaeology.

Orton Hall Farm: A Roman and Early Anglo-Saxon Farmstead

by D.F. Mackreth

with major contributions from
Dr Helen Bamford, Dr A.S. Esmonde Cleary, K.F. Hartley, Martin Howe, Joan King, Frances Lee, J.R. Perrin, R.J. Spain and Felicity Wild

and contributions from
Dr J.R. Baker, A. Challands, Mary Harman, Dr M. Henig and Dr Graham Webster

principal illustrations by
Edward Curry and Linda Meadows

photographs by
James Brown

East Anglian Archaeology
Report No.76, 1996

Nene Valley Archaeological Trust

EAST ANGLIAN ARCHAEOLOGY
REPORT NO.76

Published by
Nene Valley Archaeological Trust
Department of Art History and Archaeology
The Architecture Building
Victoria University of Manchester
Manchester M13 9PL

in conjunction with
The Scole Archaeological Committee

Editor: Stanley West
EAA Managing Editor: Jenny Glazebrook

Set in Times Roman by Joan Daniells using ® Ventura Publisher
Printed by Derry & Sons Limited, Nottingham

ISBN 0 9528105 0 6

For details of *East Anglian Archaeology*, see inside front cover

This volume is published with the aid of a grant from English Heritage

Cover Illustration
Reconstruction of the interior of Barn 4 as a brewery
(by Edward Curry)

Contents

List of Contents v
List of Plates vi
List of Figures vi
List of Tables viii
Contents of Microfiche ix
Contributors x
Abbreviations x
Preface and Acknowledgements xi
General Introduction xv

Chapter 1. Description of the Site
I. Pre-period 1 activity 1
II. Period 1 1
Summary 1
Phases *a* and *b* 2
Phase *c* 4
Phase *d* 5
Phase *e* 5
Phase *f* 7
Unphased features 8
III. Period 2 8
Summary 8
Phase *a* 9
Phase *b* 11
Phase *c* 12
IV. Period 3 14
Summary 14
The eastern side of the Main Yard 14
The north side of the Yard and the Droveway 15
The enclosures in the Main Yard 17
The Small Yard 17
Barn 3 19
V. Period 4 20
Summary 20
The east side of the yard and Barn 4 20
The Rectangular Building and the yard to its east 22
The north-east corner of the Main Yard 23
The north-west corner of the Main Yard, and the area north 23
Barn 1, walled yard and the House 25
Barn 3 and the south side of the Main Yard 26
VI. Period 5 27
Introduction 27
The eastern side, centre and south 28
The southern enclosure 32
Barn 4, and the north-east corner 34
The Main Yard 35
The central part of the north range 37
The Small Yard and the House 38
Barn 1, Barn 3, and the area between the two 39
Summary 40
VII. Post-period 5 activity 42

Chapter 2. The Roman Buildings
I. The Barns 55
II. The House 70
III. The Rectangular Building 70
IV. The Mill-house, west of the Rectangular Building 72
V. The building south of the Mill-house 74

Chapter 3. Features Associated with Fire
I. The Driers 75
II. Other Features 80

Chapter 4. The Anglo-Saxon Buildings
I. The Eastern Building 85
II. The Sunken-featured Building 86
III. The Western Building 87
IV. The Granary 89
V. Other Buildings 91

Chapter 5. The Finds
I. The Worked Flints and Other Prehistoric Finds, by Dr Helen Bamford [MF7]
II. The Coins, by Dr A.S. Esmonde Cleary 92
III. Roman Finds, by D.F. Mackreth 92
IV. Anglo-Saxon Finds, by Martin Howe and D.F. Mackreth 102
V. Medieval Objects, by D.F. Mackreth [MF8]
VI. The Glass, by John Shepherd [MF8]
VII. The Millstones, by R.J. Spain 105

Chapter 6. The Roman Pottery, by J.R. Perrin, with contributions by F. Wild and K.F. Hartley
I. Introduction 114
II. The layout of the report 114
III. Introduction to the catalogue 115
IV. The wares and fabrics 115
1 Lower Nene Valley 116
2 Shell-gritted wares 119
3 Grey wares 120
4 'Belgic' and grey tempered wares 121
5 London ware type 121
V. Discussion of the dating and catalogue from Periods 1–5 121
VI. Wares and forms: function and status 177
VII. Types of pottery related to areas of the site 181
a Samian ware 181
b Beakers 181
c Mortaria 181
d Large RSG bowls 181
e Anglo-Saxon 181
Interpretation 182
Pottery and population size 182
VIII. Pottery and continuity, residuality or survival 188
IX. The Samian, by Felicity Wild 190
X. The Mortaria, by J.R. Perrin with K.F. Hartley 191
Catalogue 192
A type series for mortaria found in the Lower Nene Valley, by K.F. Hartley 199

Chapter 7. The Anglo-Saxon Pottery
I. Introduction 205
II. Group 1 206
III. Group 2 209
IV. Other stratified pottery 212
V. Unstratified 214
VI. Previous excavations 215

Chapter 8. The Bones
I. The Animal Bones, by Joan King 216
Introduction 216
Method 216
Butchering and use 217
Use shown by disease 218
Conclusions 218
II. The Bird Bones, by Mary Harman 218
III. Human Pathology, by Frances Lee [MF10]

Chapter 9. The Economy and Use of the Site
I. Introduction 220
II. The Land 220
III. The Animals 221
Cattle 221
Sheep 224
Pig 224
Horse 225
Deer 225
IV. Cereals and Mills 225
V. Aspects of the Use of the Site 226
Animals 226
Crops 227
Equipment 228
VI. The Inhabitants 228
Food 228
Drink 229
VII. Conclusion 230

Chapter 10. The Site in its Setting: Romans and Anglo-Saxons
I. Settlement in the Nene and Welland Valleys 232
II. The Fens and Durobrivae in the Second Century 233
III. The Third and Fourth Centuries 235
IV. Roman and Anglo-Saxon (late fourth century-early sixth century) 237

Bibliography 240
Index, by Peter Gunn 250
Microfiche

List of Plates

Plates I–VII are in the wallet inside back cover

Plate I Plan of pre-Period 1 with all section lines, and feature numbers not appearing on part-plans
Plate II Plan of Period 1
Plate III Plan of Period 2
Plate IV Plan of Period 3
Plate V Plan of Period 4
Plate VI Plan of Period 5
Plate VII Plan of post-Period 5, with all archaeological features and furrows
Plate VIII The Intaglio, No. 36. (*Photographer*: James Brown) 96
Plate IX Barred Zoomorphic Comb, No. 251. (*Photographer*: James Brown) 106

List of Figures

Fig. 1 Location map: the Fen edge and settlement around Orton Hall Farm xii
Fig. 2 The site in relation to modern development xiii
Fig. 3 Key to section conventions xvi
Fig. 4 Period 1: specialised pits and possible post-holes in the north-west corner of the site 3
Fig. 5 Period 1: probable gateway system in main route across the site 6
Fig. 6 Period 1: structure and wear hollow in the southern part of the central enclosures 7
Fig. 7 Period 2: Droveway, Period 3 features in broken outline 9
Fig. 8 Period 3: gateway into the eastern central enclosure 14
Fig. 9 Period 3: features in the Droveway, Period 2 features in broken outline 16
Fig. 10 Period 4: gateway on the east side of the Droveway 25
Fig. 11 Period 5: Anglo-Saxon gateway into the main eastern enclosure 28
Fig. 12 Period 5: overall plan of the eastern enclosures showing 'primary' and 'secondary' Anglo-Saxon features 29
Fig. 13 Period 5: adult female inhumation 33

Fig. 14 Sections [1]–[9] 44
Fig. 15 Sections [10]–[13] 44
Fig. 16 Sections [14]–[19] 45
Fig. 17 Sections [20]–[24] 45
Fig. 18 Sections [25]–[28] 46
Fig. 19 Sections [29]–[34] 46
Fig. 20 Sections [35]–[39] 47
Fig. 21 Sections [40]–[45] 47
Fig. 22 Sections [46]–[49] 48
Fig. 23 Sections [50]–[55] 48
Fig. 24 Sections [56]–[63] 49
Fig. 25 Sections [64]–[68] 49
Fig. 26 Sections [69]–[72] 50
Fig. 27 Sections [73]–[76] 50
Fig. 28 Sections [77]–[82] 51
Fig. 29 Sections [83]–[88] 51
Fig. 30 Sections [89]–[96] 52
Fig. 31 Sections [97]–[104] 52
Fig. 32 Sections [105]–[107] 53
Fig. 33 Sections [108]–[109] 53
Fig. 34 Sections [110]–[118] 54
Fig. 35 Plan of Barn 1 56
Fig. 36 Sections of pier bases in Barn 1 57
Fig. 37 Sections of pier bases in Barn 1 57
Fig. 38 Plan of Barn 2, the first north-east barn 59
Fig. 39 Sections of pier bases in Barn 2 60
Fig. 40 Sections of pier bases in Barn 2 60
Fig. 41 Plan of Barn 3, the South Barn 61
Fig. 42 Sections of pier bases in Barn 3 62
Fig. 43 Plan of Barn 4, the second north-east barn 64
Fig. 44 Sections of pier bases in Barn 4 65
Fig. 45 Plan of the rectangular building 71
Fig. 46 Sections A–B, in the rectangular building; C in the Mill-house 72
Fig. 47 Plan of the Mill-house 73
Fig. 48 Plan of the Period 3 drier, F156, in Barn 2 75
Fig. 49 Plans of the Period 4 driers, F168 and F169, in Barn 4 76
Fig. 50 Sections of the Periods 3 and 4 driers, F156, F168 and F169 76
Fig. 51 Plans of the Period 5 drier F159/F160 77
Fig. 52 Sections of the Period 5 drier F159/F160 78
Fig. 53 Hearths *etc.*, Sections 1–5 81
Fig. 54 Hearths *etc.*, Sections 6–10 82
Fig. 55 Plans of the eastern Anglo-Saxon House and the Sunken-featured Building 86
Fig. 56 Plan of the Anglo-Saxon hall and annex superimposed on the Roman House 88
Fig. 57 Plan of the Granary 90

Figures 58–60, 66–73 and 76 are in microfiche

Fig. 58 The flints, Nos 1–10
Fig. 59 The flints, Nos 11–22
Fig. 60 The flints, Nos 23–29; stone axe, No. 30; Bronze Age object, No. 31
Fig. 61 Brooches, Nos 1–15 94
Fig. 62 Belt fittings and buckles, Nos 23–29; finger rings, Nos 30–35; bracelets, Nos 37–57 97
Fig. 63 Bracelets, Nos 58–59; beads *etc.*, Nos 61–63; pins, Nos 64–81 99
Fig. 64 Combs and cosmetic items, Nos 94–99; boot-plates, Nos 101 and 103; counter, No.104 101
Fig. 65 Cult objects: pipe-clay figurine, No.106; copper-alloy animal, No.107 102
Fig. 66 Vessels, Nos 108–110; knives and shears, Nos 111–113, 116, 119–120; knife handles, Nos 122–123
Fig. 67 Styli, Nos 124–130
Fig. 68 Needles, Nos 132–137. Tools, Nos 139–142
Fig. 69 Tools, Nos 143–152
Fig. 70 Tools, Nos 153–163, 168–169. Weaving equipment, Nos 171–176, 178
Fig. 71 Whetstones, Nos 179–180, 185; keys, Nos 190–192; suspension items, Nos 195–202
Fig. 72 Suspension items, Nos 203–204; fittings, No.205, 208– 210, 212–126, 218–220. Fittings, Nos 222–225
Fig. 73 Fittings, Nos 226 and 229; miscellaneous, Nos 232–242
Fig. 74 Anglo-Saxon objects, Nos 243–250. Medieval objects, Nos 253–255 103
Fig. 75 The Barred Zoomorphic Comb, No.251 107
Fig. 76 The Glass (microfiche)
Fig. 77 Millstones, Nos 1–3 109
Fig. 78 Millstones, No. 4 111
Fig. 79 Millstones, Nos 5–7 112
Fig. 80 Roman pottery. Period 1: Nos 1–22 124
Fig. 81 Roman pottery. Period 1: Nos 23–45 125
Fig. 82 Roman pottery. Period 1: Nos 46–73 126
Fig. 83 Roman pottery. Period 1: Nos 74–103 128
Fig. 84 Roman pottery. Period 1: Nos 104–125 129
Fig. 85 Roman pottery. Period 1: Nos 126–146 130
Fig. 86 Roman pottery. Period 1: Nos 147–179 132
Fig. 87 Roman pottery. Period 1: Nos 180–202 133
Fig. 88 Roman pottery. Period 1: Nos 203–225 135
Fig. 89 Roman pottery. Period 2: Nos 226–249 140
Fig. 90 Roman pottery. Period 3: Nos 250–272 144
Fig. 91 Roman pottery. Period 3: Nos 273–292 145
Fig. 92 Roman pottery. Period 3: Nos 293–316 147
Fig. 93 Roman pottery. Period 3: Nos 317–335 148
Fig. 94 Roman pottery. Period 3: Nos 336–359 150
Fig. 95 Roman pottery. Period 3: Nos 360–380 151
Fig. 96 Roman pottery. Period 3: Nos 381–390, Period 4: Nos 391–403 156
Fig. 97 Roman pottery. Period 4: Nos 404–430 158
Fig. 98 Roman pottery. Period 4: Nos 431–451 159
Fig. 99 Roman pottery. Period 4: Nos 452–475 161
Fig. 100 Roman pottery. Period 4: Nos 476–498 162
Fig. 101 Roman pottery. Period 5: Nos 499–532 167
Fig. 102 Roman pottery. Period 5: Nos 533–553 168
Fig. 103 Roman pottery. Period 5: Nos 554–578 170
Fig. 104 Roman pottery. Period 5: Nos 579–597 171
Fig. 105 Roman pottery. Period 5: Nos 598–609 173
Fig. 106 Roman pottery. Period 5: Nos 610–623 175
Fig. 107 Key to Figures 108–112 182

Fig. 108 Period 1, distribution of types of Roman pottery 183
Fig. 109 Period 2, distribution of types of Roman pottery 184
Fig. 110 Period 3, distribution of types of Roman pottery 185
Fig. 111 Period 4, distribution of types of Roman pottery 186
Fig. 112 Period 5, distribution of types of Roman pottery 187
Fig. 113 Samian pottery, Nos 1–8 191
Fig. 114 Roman mortaria. Periods 1, 2 and 3 194
Fig. 115 Roman mortaria. Periods 3 and 4 195
Fig. 116 Roman mortaria. Periods 4 and 5 197
Fig. 117 Roman mortaria. Period 5 198
Fig. 118 Anglo-Saxon pottery. Nos 1–7 208
Fig. 119 Anglo-Saxon pottery. Nos 8–20 209
Fig. 120 Anglo-Saxon pottery. Nos 21–40 211
Fig. 121 Anglo-Saxon pottery. Nos 41–76 213

List of Tables

Table 1 Spacings, ratios and areas of the four Barns 68
Table 2 Spacings, ratios and areas of some other barns 68
Table 3 Floor areas and ratios of the driers 77
Table 4 Flints [MF8]
Table 5 Period 1, main fabrics A% 122
Table 6 Period 1, main vessel forms/fabrics 122
Table 7 F588 system, main vessel forms/fabrics 123
Table 8 F813 system, main vessel forms/fabrics 123
Table 9 F1048 system, main vessel forms/fabrics 127
Table 10 F814 system, main vessel forms/fabrics 131
Table 11 F1025 system, main vessel forms/fabrics 131
Table 12 F525/6, main vessel forms/fabrics 134
Table 13 Period 1, RSG, main vessel forms 134
Table 14 Period 1, grey ware, main vessel forms 136
Table 15 Period 1, LNVGW, main vessel forms 137
Table 16 Period 1, LNVCC, main vessel forms 137
Table 17 Period 1, c/w, main vessel forms 138
Table 18 Period 2, main fabrics A% 139
Table 19 Period 2, main vessel forms/fabrics 139
Table 20 F189 system, main vessel forms/fabrics 139
Table 21 F969, main vessel forms/fabric 139
Table 22 F437 system, main vessel forms/fabric 140
Table 23 Barn 2 ditches, main vessel forms/fabric 141
Table 24 Period 2, main RSG vessel forms 141
Table 25 Period 2, main grey ware vessel forms 141
Table 26 Period 2, main LNVGW vessel forms 141
Table 27 Period 2, main LNVCC vessel forms 142
Table 28 Period 2, main c/w vessel forms 142
Table 29 Period 3, main fabric/feature A% 142
Table 30 Period 3, main vessel forms/fabric 143
Table 31 Barn 2, internal features, main vessel forms/fabric 143
Table 32 F1094 *etc.*, main vessel forms/fabrics 143
Table 33 The Yards, main vessel forms/fabrics 146
Table 34 F1016 well, main vessel forms/fabrics 146
Table 35 F435 system, main vessel forms/fabrics 149
Table 36 F441 system, main vessel forms/fabrics 149
Table 37 F772 system, main vessel forms/fabrics 149
Table 38 F254 well, main vessel forms/fabrics 149
Table 39 'F969' well, main vessel forms/fabrics 152
Table 40 Period 3, main RSG vessel forms 152
Table 41 Period 3, main grey ware vessel forms 152
Table 42 Period 3, main LNVGW vessel forms 153
Table 43 Period 3, main LNVCC vessel forms 153
Table 44 Period 3, main c/w vessel forms 154
Table 45 Period 4, main fabrics/feature A% 155
Table 46 Period 4, main vessel forms/fabric 155
Table 47 F434, main vessel forms/fabrics 155
Table 48 F200 pond, main vessel forms/fabrics 157
Table 49 F1052 well, main vessel forms/fabrics 157
Table 50 F164 system, main vessel forms/fabrics 157
Table 51 F148 system, main vessel forms/fabrics 157
Table 52 F911 yard, main vessel forms/fabrics 160
Table 53 F542 system, main vessel forms/fabrics 160
Table 54 Period 4, main RSG vessel forms 163
Table 55 Period 4, main grey ware vessel forms 163
Table 56 Period 4, main LNVGW vessel forms 163
Table 57 Period 4, main LNVCC vessel forms 164
Table 58 Period 5, main fabric/feature A% 165
Table 59 Period 5, main vessel forms/fabric 165
Table 60 F31 system, main vessel forms/fabric 165
Table 61 F80 ditch, main vessel forms/fabric 166
Table 62 F81 ditch, main vessel forms/fabric 166
Table 63 F200 pond, main vessel forms/fabric 166

Table 64	Rectangular Building, floors, main vessel forms/fabrics	169
Table 65	F1061/F1065 system, main vessel forms/fabric	169
Table 66	F55 ditch, main vessel forms/fabric	169
Table 67	F171 system, main vessel forms/fabric	172
Table 68	F12 gulley, main vessel forms/fabric	172
Table 69	F172 well, main vessel forms/fabric	172
Table 70	AF48 pit, main vessel forms/fabric	172
Table 71	F751 slot, main vessel forms/fabrics	172
Table 72	Period 5, main RSG vessel forms	174
Table 73	Period 5, main grey ware vessel forms	175
Table 74	Period 5, main LNVGW vessel forms	176
Table 75	Period 5, main LNVCC vessel forms	176
Table 76	Periods 1–5, number and percentages of main vessel forms	178
Table 77	Number of vessels and size of population	188
Table 78	Samian vessel types, Periods 1–5	188
Table 79	Mortaria in Periods 1–5, numbers and origins	192
Table 80	Total bones by animals and Periods [MF9]	
Table 81	Number of birds of different species recognised in each Period	219
Table 82	Human pathology: diaphyseal long bone measurements from infant burials [MF10]	
Table 83	Number and percentages of bone per twenty-five-year term within Periods 1–5	224

Contents of Microfiche

MF1 Context groups
MF2 Layer list
MF3 Feature list
MF4 Post-hole list
MF5 List of driers from Orton Hall Farm and other sites, nos 1–83 (see Chapter 3.I)
MF6 Summary of the late G.F. Dakin's excavation
The Buildings
Small finds
Pottery
Building materials
Coins
Human and animal bones
MF7 Chapter 5 I. The worked flints and other prehistoric finds, by Dr Helen Bamford
Figs 58–60 The flints and prehistoric objects
Table 4 Distributions of flints by Period
Flint flakes and scrapers: measurements
MF8 The small finds, and site small finds list
Chapter 5 III. Roman Finds
Figs 66–73 Roman finds
V. Medieval Objects (on Fig. 74)
VI. The Glass, by John Shepherd
Fig. 76 The Glass
Small finds list (including G.F. Dakin's)
MF9 Animal bone
Table 80 Total bones by animals and Periods
Tables for cattle, sheep, pig, horse, dog, cat
MF10 Chapter 8 III. Human Osteology, by Frances Lee
Table 82 Diaphyseal long bone measurements from infant burials
MF11 Pottery: processing system
Pottery Quantification by Period

Contributors

Dr J.R. Baker

Dr Helen Bamford, MA, FSA
Monuments Protection Programme Fieldworker, English Heritage

A. Challands, BA, MIFA
Archaeological Consultant,
The Old School House, Glinton Road, Helpston Road, Peterborough

Dr Simon Esmonde Cleary
Department of Ancient History and Archaeology, University of Birmingham

Edward Curry
Archaeological Draughtsman

Mary Harman
Animal Bone Specialist

K.F. Hartley, FSA
Roman Pottery Consultant

Dr M. Henig, MA, DPhil, FSA
Institute of Archaeology, Oxford

the late Martin Howe
formerly Curator, Peterborough Museum

Joan King
Animal Bone Specialist

Frances Lee

Linda Meadows
Archaeological Illustrator

J.R. Perrin, MLitt., FSA
Central Archaeology Service, English Heritage

John Shepherd

R.J. Spain

Dr Graham Webster, OBE, PhD, D.Litt., FSA, AMA, MIFA
Consultant to Historic Buildings and Monuments Commission

Felicity Wild, MA, FSA
Roman Pottery Consultant

Abbreviations

A%	average percentage (see p.114)
BB1/BB2	Black-burnished ware
CC	colour-coated ware
CCC	Colchester colour-coated ware
CGCC	Central Gaulish colour-coated ware
CR	colour range(s) (pottery)
c/w	cream and white wares
EVE(s)	estimated vessel equivalent(s)
F	Feature
GW	grey ware
HS	hand-stripping
L	Layer
LIASG	Late Iron Age shell-gritted ware
LNVCC	Lower Nene Valley colour-coated ware
LNVC/W	Lower Nene Valley cream and white wares
LNVGW	Lower Nene Valley grey ware
LRCC	Lower Rhineland colour-coated ware
M	mortarium
MS	machine-stripping
OXCC	Oxfordshire colour-coated ware
OXW	Oxfordshire white/cream ware (mortaria)
RSG	Roman shell-gritted ware
SG	shell-gritted
TSG	Late Iron Age to early Roman, transitional shell-gritted ware
UA	unassociated (layer not assigned to a feature or area)
UF	unfeatured
VR	Verulamium region ware
WM	Jar wide-mouthed jar
1971	1971 excavation layers

For other conventions used in the pottery catalogue, see Chapter 6, III; for abbreviations of pottery forms, see Chapter 6, VI.

Preface and Acknowledgements

The site lay at TL 17659555 in the parish of Orton Longueville on what is now the southern boundary of Orton Township (Fig. 1) and immediately west of the junction of the Soke Parkway with the Fletton Parkway, under the main carriageways and slip-roads leading up to the former (Fig. 2). The site was excavated in advance of these road-works. The work reported on here was the third campaign and, where possible, the results of the earlier ones are incorporated.

The Field Section of the Peterborough Museum Society, with the kind permission of the then landowner, Mr John Hunting, carried out weekend excavation from 1964 to 1967. Although the finds recovered survive, none of the records has been located apart from a sketch plan of what is called here Barn 1. The work was directed by the late Mr G.F. Dakin and a summary text prepared by him was issued and is published here (MF6). The site was included in the survey of the archaeology of the Designated Area of Peterborough New Town prepared by the Royal Commission on Historical Monuments in advance of development (RCHM 1969, 30(7)).

In the summer of 1971, before the Nene Valley Research Committee had been able to create a full-time body to deal with rescue work, an excavation was set up to deal with the outstanding problems of the site before construction work on the new Soke Parkway began. The policy followed was to investigate, by trenching, the major anomalies which appeared on a magnetometer survey of the area under threat. In one area, the trenches exposed driers which were ultimately found to lie in Barn 4, and in another the trenches were opened out to reveal part of a ditch system to the east of that. The season's work was directed by A. Challands, W. Hanson and S. Upex. The 1971 results were encouraging and, with buildings, ditches and a pond, pointed to a much larger rural site than Mr Dakin's work had indicated. However, the excavation was discontinued when it was found that the area marked by the Department of the Environment as being scheduled was wrongly plotted and was not under immediate threat.

Excavation began again in 1973 when a year was allowed for its completion before contractors started on the first stage of the Fletton Parkway. In the event, the worsening economic climate slowed the rate of development in the new town areas and the site was finally finished in 1975 just as the roadside ditches and fences for the new road were being put in. The whole site is now covered by road-works or the extensive landscaping and tree-planting which lie next to it on the south. The final work was observation of the earth-stripping for the continuation to the east of the dual carriageway, but, as is common in such cases, the large machines used did not leave a surface clean enough for much to be seen, and the removal of the soil prevented any collection of sherds. Thus, although there is a faint possibility that the site continued to the east in a more definite form than subsidiary ditches belonging to fields, there were certainly no further remains of buildings with stone walling. The greatest loss to knowledge concerned the whereabouts of the eastern limit of the Anglo-Saxon occupation, but the more ephemeral traces which would have marked it stood no real chance of being detected in the circumstances. Trial trenches were cut from the main excavation area both to the north and the west and neither set revealed any major extension to the site. On the south, the limit of the excavation was the hedge which marked the edge of the Designated Area of the New Town and beyond that is a large London Brick Company clay pit. An unsubstantiated rumour refers to the finding of a building in it when it was first opened, but examination of the exposed edge failed to reveal anything, and the digging of an electricity main along the hedge is also said to have produced nothing. This suggests that there are no good grounds for supposing that any significant element belonging to the Roman site has been lost.

It was not until the major stripping of the topsoil began in 1973 that it was appreciated that there had been extensive damage caused by medieval furrows. The discovery of the site had been a result of the destruction of this ridge and furrow in the early 1960s, and its existence had passed out of mind, save for the farmer's, in fewer than ten years. The stripping of the topsoil was done exclusively with a JCB although a trailer system was used to prevent double handling of soil. The JCB proved to be ideal for the purpose, because the operator was closer to the cutting edge than he would have been on a larger machine and was able to take into account the differing levels caused by the furrows and the areas of stonework which marked the positions of the buildings. Most of the furrows were left unemptied because of the economy in time and cost. The stone rubble areas penetrated into the topsoil and were left for detailed hand-stripping at a later time as it was quickly appreciated that to have planed down using a machine would remove most of the evidence present. My thanks go to Mr C. Clapham, the JCB driver, who proved to have a delicate hand and, from previous experience, had an exact appreciation of the needs of archaeologists.

The recording system used on the site was a double serial list, one for features and the other for layers, each one being given a unique number and both tied into plans located on a co-ordinate grid. Although three or four base-lines and grids had been applied to parts or all of the site previously, these were in imperial measurements. The new grid was metric and its point of origin was sufficiently far away for there to be no fear of confusion if the co-ordinates for any given point within the excavation were reversed by accident. Each of the period plans (Pls I–VII) bears the major intersections of this grid to allow the reader to relate the site plans easily to the published interpretation. The period plans also show the direction of the numbering of the grid and this enables the co-ordinates given for features to be applied directly.

The area finally uncovered was 14,725m^2 in extent, excluding the trial trenching. Plate VII shows the density of feature which was visible between the furrows while Plate I shows all the archaeology except the furrows. The exigencies of the original time limit and the real shortage of funds and facilities — no store, or headquarters, no ancillary staff — militated against the complete removal of all deposits. The objective was, therefore, to sample extensively to recover both dating evidence and stratigraphical relationships. This means that lengths of features away from furrows (Pl. VII) were dug, and

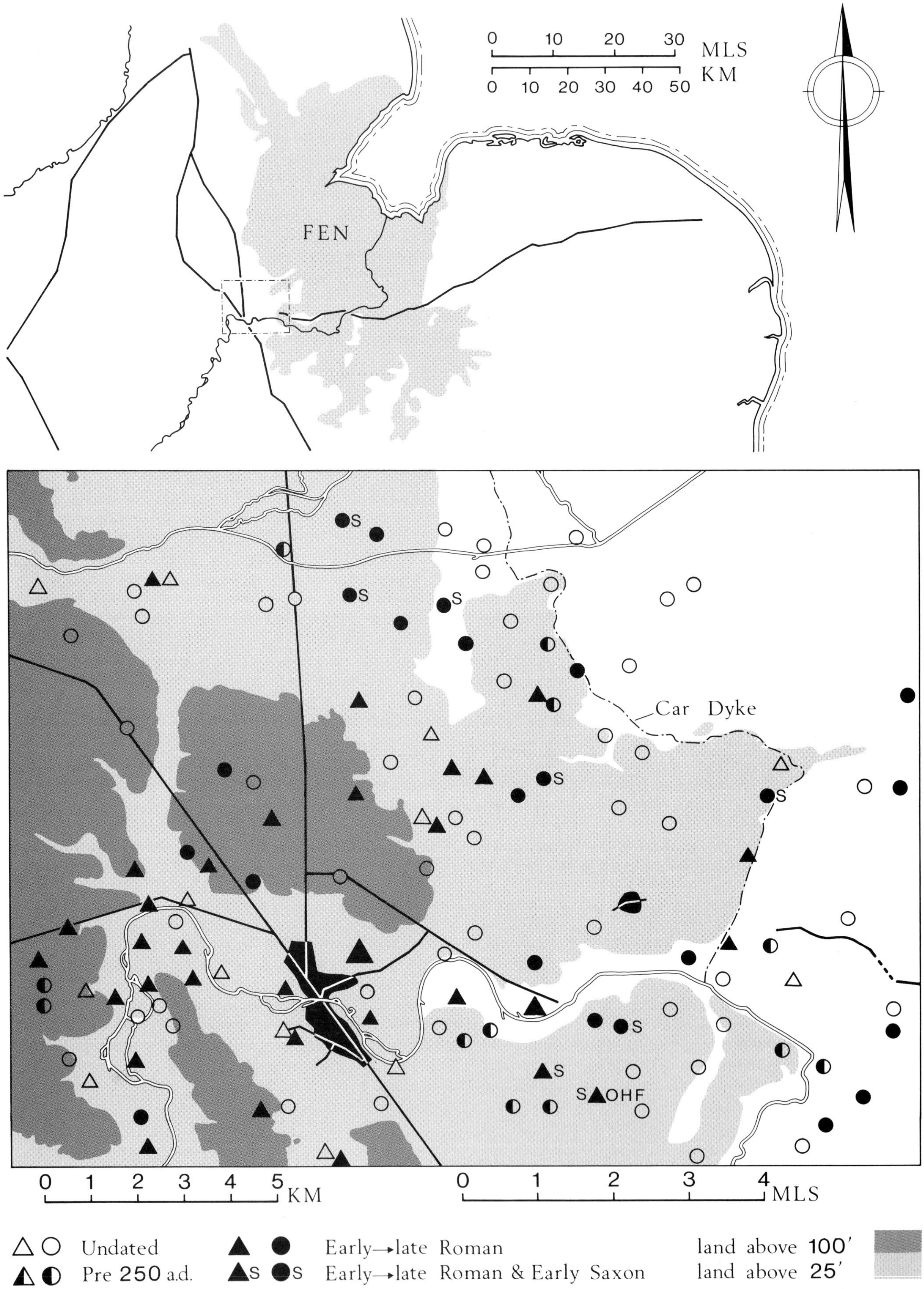

Figure 1 Location map: the Fen edge and settlement around Orton Hall Farm.

Figure 2 The site in relation to modern development.

intersections wherever they fell. Small features such as post-holes were completely excavated, but pit clusters tended to receive summary treatment once the general date range had been established. Very few features were found which could not be assigned to a period, at least on grounds of their probable association, if other information was lacking. Only about 10 per cent of the total amount of deposit was removed. The chief aim was to examine junctions and, where possible, to empty plain lengths of linear features away from furrows in the hopes of finding uncontaminated dating evidence. Obviously specialised features, such as the driers, were completely excavated and, occasionally, linear features were more extensively dug if it appeared that the information to be gained was important. This applied especially to those containing Anglo-Saxon pottery, or to those likely to contain a valuable pottery group which would prove important for dating. Some features, owing to accidents of excavation and their extent, received more than one number. Equivalents are to be found in MF3.

Work continued through two winters when one of the major disadvantages of the site became apparent: the high water-table. Along the northern edge of the stripped area, the water in summer was only *c*.0.5m below the surface of the field and heavy rain would make it impossible to work there. In winter the water-table rose by *c*.0.3m and the same area became flooded for months. Consequently, any deep hole on the site was bound to reach water, and more quickly the further north it was. Funds would not permit extensive pumping, especially as adequate means of disposing of the water would have been expensive. It was the presence of a handy spring as well as the high water-table which no doubt persuaded people to choose the spot in the first place, but it was also the reason why any change in plan was accompanied by a new set of ditches to carry off the unwelcome excess of water.

The finds from the site before the Development Corporation became the landowner were very kindly put at the disposal of the Nene Valley Research Committee by Mr John Hunting and are, with those from later work, now placed in Peterborough Museum where the site records are also to be found.

The main texts of this report were completed in 1985. Lack of time and finance prevented some parts of the report from being prepared: for instance, those on querns, slags, timbers from the base-frames of wells. It was intended to have a note on the pottery computer program and the use of the graphics tablet. However, in the time since they were used and the printing of this report, the computer has become no longer operable and neither it nor the 8-inch disks used are in manufacture. This serves as a warning to any who may imagine that the disks created for post-excavation work will form a satisfactory archive.

The site proved to have developed continuously from at least the middle of the first century AD to the early sixth century. It will be appreciated when the period plans (Pls I–VII) are consulted that this development imposes problems in presentation: it would be possible to divide the growth and demise of the site into a great number of periods or phases which, while detailing many events, would make the understanding of the process more difficult. The discussion has assumed that change was a constant feature of the history of the site and that, within each of the periods in which this history is presented, there was no stay to alterations. The intention behind the use of colour on the main plans (Pls II–VI) is to give some idea of the dating evidence recovered. It must be emphasised that the colouring is not intended to represent numbers of phases as such, those are arrived at on the basis of what are taken to be significant changes in plan and are described in the relevant parts of Chapter 1.

The beginning of each period represents, as far as can be told, some kind of major change which tended to impose a different emphasis on the use of the site. Thus, Period 1 is a development away from an unexcavated earlier, possibly Late pre-Roman Iron Age, nucleus to the north-west of the excavated area; Period 2 marks the first stage of the creation of a farmyard; Period 3 the consolidation of the west end of that, with the probable abandonment of unlocated elements outside the stripped area. Period 4 marks the greatest expansion of the building layout on the site and Period 5 deals with the gradual diminution of this, along with the arrival of Anglo-Saxon settlers at some point which need not be at the beginning of the period. The dating which has been applied is, as far as possible, purely archaeological, with no allowance made for alternative interpretations based upon some preconceived model for the history of Roman Britain which might appear at the time to be the most favoured. Such considerations are entered into in the general discussion (Chapter 10).

The work could not have been carried out without the active support of the Peterborough Development Corporation which, along with the Department of the Environment, provided the funds for the excavation. The post-excavation programme was funded by the Department of the Environment, and latterly by English Heritage. The Development Corporation, through the liaison officer appointed by it, Mr Laurie Campbell, provided the housing for the diggers and the first building in which the Committee could store its finds. Later, through a working party, initially chaired by Mr E. Schoon and then by Mr David Bath, a fully equipped headquarters building was provided from which the full-time staff of the Committee were able to work until 1988. Sincere thanks are given here to the Development Corporation for all its support and encouragement, especially at a time when its own finances were becoming constrained without the calls on them diminishing. The Committee also thanks the London Brick Company for giving permission to use one of its roads to get to the site: in the first place, the new route spared the farmer's pasture and reduced the number of gates which needed to be shut, and in the second, allowed access when new housing development would have prevented it.

Although the excavation was carried out over what seems now to have been a long time, the working force seldom exceeded ten and was often only five or six. This small number largely explains why the excavation was not completed at an earlier date. For a short period in each of the years 1973, 1974 and 1975, the site was supervised in my absence by respectively Dr Simon Esmonde Cleary, Mr Dimitri Anson and by Mr Lyle Browning who also supervised the work on the site in detail for the last nine months. Special thanks are given to the stalwarts of the excavation who were present for all or a great part of the time: Lyle Browning, Stewart Davison, Clare Kerrigan, Francis O'Neill, Calum Rollo, Stephen Speak, John Walker and Martin Wright-Lakin. Lastly, my thanks go to my wife Christine who, in those early days, conducted the correspondence, took care of pay and accounts, dealt with the recruitment of volunteers, and had her home cluttered with finds and tools.

The post-excavation work was started with the help of Sarah Jennings, Fiona Cameron, and Lindsay Rollo, and was continued by J.R. Perrin and Christine Mackreth. The plans and sections were started by Robert Boyle, but the style of these had to be altered in order to save time and were finally finished by Edward Curry. The pottery and small finds were carried through by Linda Meadows. For advice and help in the later stages of the report, I am very grateful to Dr John Peter Wild for his forbearance in dealing with potentially controversial matters; he provided invaluable advice for Chapter 9, although I alone am responsible for any errors of fact or judgement. Barbara Green was kind enough to read the section on the Anglo-Saxon pottery and, in the final presentation of the report, my thanks go to Stanley West, Jenny Glazebrook and Susanne Atkin.

General Introduction

Summary of the Site

Occupation on the site, apart from the periphery of what was a Neolithic-Bronze Age enclosure system, ran from the first to the sixth century AD. In Period 1, the stripped area lay to the south-east of an unexcavated nucleus and only contained elements of enclosures with some evidence for domestic occupation. The enclosure system developed through the second century until, towards the end of that, and marking the initiation of Period 2, a new enclosure incorporating two barns was laid out. In Period 3, one barn was used as one side of a small walled yard and a house was built along the opposite side. A new barn, bringing the number to three, was built on the south side of the Period 2 enclosure which now was provided with a new south boundary. There is evidence for an increase in the number of animals from the site, and for the conversion of one barn into a brewery. In Period 4, more buildings were added and one barn resited. There is evidence that one of the new buildings had been a mill-house and it is possible that the building next to it was used to house farm-workers in a kind of barracks. In Period 5, the Roman site became degraded with some buildings being reduced in size, although brewing on a large scale still continued. Anglo-Saxons occupied the east and west ends of the Roman main yard and gradually took over the whole plan, possibly retaining one of the barns in use all the time. The presence of a granary and a probable hall point to the site having had a relatively high status. The evidence suggests that the site was abandoned sometime in the earlier sixth century at the latest.

Summary of dating of the main periods

Period 1 Mid-first century — *c*.175
- Phases *a* and *b*: mid-first and later first century AD
- Phase *c*: late first — early second centuries
- Phase *d*: early second century
- Phase *e*: first half of second century
- Phase *f*: mid-second century plus

Period 2 *c*.175 — 225/250
- Phase *a*: *c*.175 — *c*.200
- Phase *b*: *c*.200 — *c*.225?
- Phase *c*: *c*.225? — *c*.250?

Period 3 *c*.225/250 — *c*.300/325

Period 4 *c*.300/325 — *c*.375

Period 5 *c*.375 — early sixth century

Phasing and Dating Method by J.R. Perrin

Owing to the size and complexity of the site, it was initially divided into four 'blocks', each of which had a preliminary phasing based on stratigraphy and association. The pottery was then examined and approximate dates given to the various features and layers within each 'block'. A secondary phasing was then prepared by correlating these dates with the sequence suggested by the earlier phasing. The pottery was then re-examined in its new order and such additional dating evidence as was available from items such as samian ware, mortaria and coins was included. This gave a third phasing from which most of the obvious discrepancies and errors had been eliminated. At this point the four 'blocks' were merged to give an overall site periodisation. This was rechecked and the pottery and finds which appeared not to fit were re-examined. In this way all but a relatively few layers and features were eventually assigned to one of the five main periods and the two sub-periods though, not surprisingly, some could not be slotted in, and others were phased less confidently.

The phasing process was largely dependent on pottery dating, although the site showed a clear, logical development in layout and structure which provided a good framework into which the dating could be placed. Those involved in the phasing were acutely aware of all the inherent problems of using pottery dating alone, not least those of circularity.

Note

The sample system, as well as the large number of features, over 1,000, and layers, nearly 3,000, imposes some difficulties in presenting a formal report because adequate discussion leads automatically to the listing of many features and more layers. Simple tabulation takes up too much page space in these cost-conscious days and the insertion of these details within the text itself would interrupt the reader's concentration and prove unacceptably tedious. Therefore, so that the information upon which the discussion hinges is not lost in a virtually inaccessible site archive, the relevant batches of layers and their features are presented in a numbered series of groups which are given in the microfiche attached to the report (MF1). The group numbers are set in brackets without further designation.

The microfiche also lists all layers with their feature, period and dating, where present (MF2); all features in their periods and with the layers arranged in sample order (MF3); all post-holes are listed with dimensions in order of their context groups (MF4); and all small finds with their layers, features and periods, where this can be ascertained, along with brief identifications (MF8). Those small finds selected for report are specified as well as their figure number, if illustrated, and reference number in the small find report. The pottery selected for drawing and those groups which are important for the site are indicated in the lists of the layers. If readers wish to use the microfiche extensively, they would be advised to have a print-out of the layer-listing by them, as that is the only one giving, though briefly, the dates on which much depends in the detailed argument.

The abbreviations used throughout the report are:

F: Feature

L: Layer

(39) *etc.* Context Group

It is emphasised that the last provides a primitive immediate cross-reference system between the full text and the microfiche listings.

Geology by A. Challands

The geology of the site varies, north to south. Most of the northern half of the site lies directly on Oxford Clay of the Upper Jurassic age, although random patches of thin drift deposits of Third Terrace river gravels are also present. The north-east corner of the site is covered by head which overlies both the Oxford Clay and the Third Terrace gravel

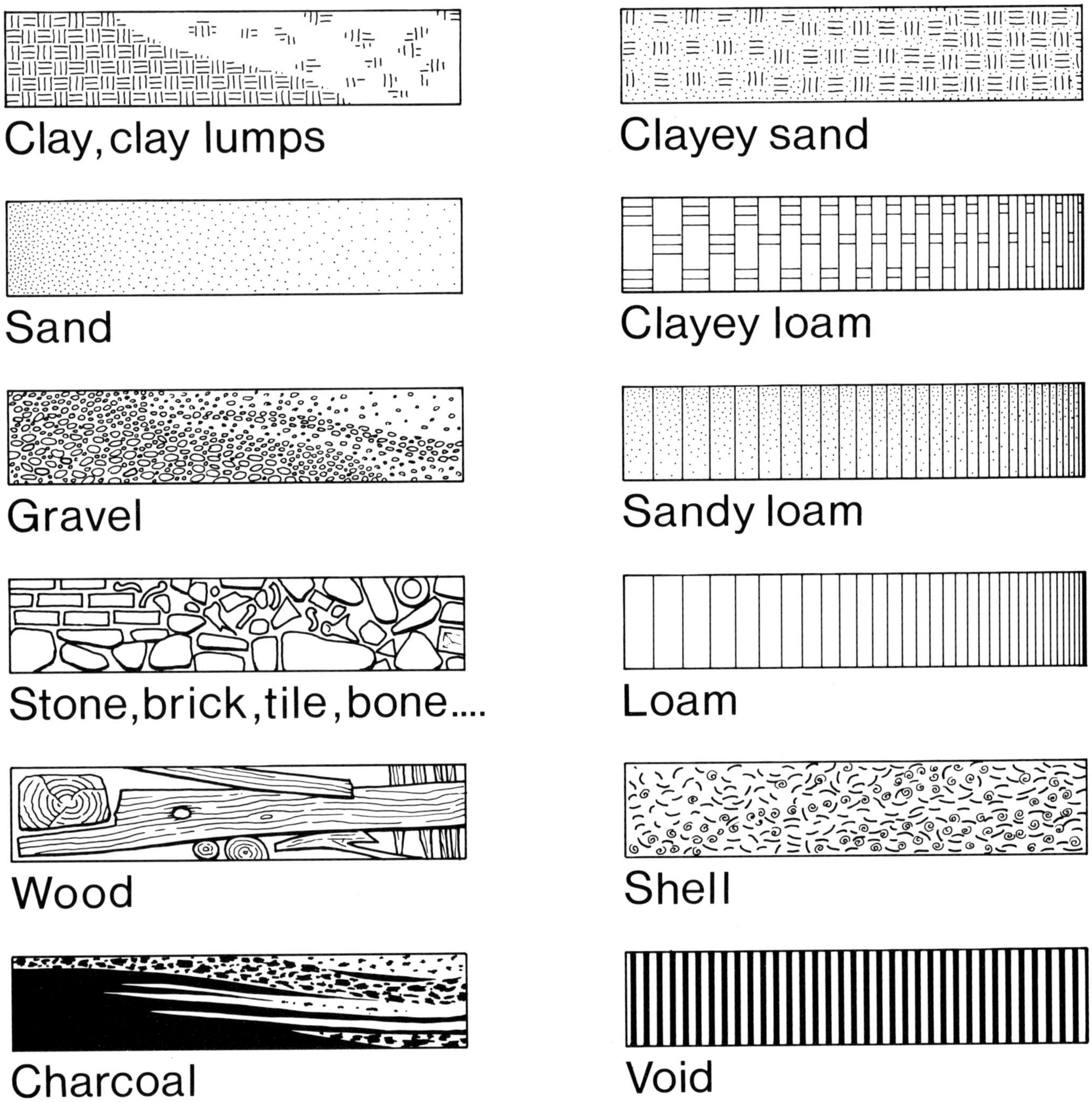

Figure 3 Section conventions used in the report

deposits. The head in this instance may have been derived from the Third Terrace gravels fluvial action. In the southern portion of the site, somewhat thicker deposits of Third Terrace gravels overlie the Oxford Clay. An exposure in the quarry face forming the southern boundary shows a depth of 2.44m of Third Terrace gravel deposits.

The gravels and underlying Oxford Clay have been subject to periglacial action which has caused frost heaving and the formation of cryoturbation structures. This has led to perched water-tables in the deposits below and around the site.

Topography

The description of the underlying geology of the site will have made it clear that the site lies in an area which has relatively heavy soil and, at the point at which it was located, was badly drained. Relief is not a noted feature of the topography of the area close to where the Nene once debouched into the Fens (Fig. 1) and, at *c.*19.5m OD, the site lay on a gentle north-facing slope on the south side of the valley whose base, *c.*2.25km directly north of the site, is about *c.*5.5m OD. If the ground had been largely cleared of woodland, a fairly good view would have been had along the valley in each direction, but the bitter north-east winds driving straight off the North Sea and across the Fens would have been a trial without cover of some form, as the volunteers who worked through two winters found.

The characteristic of the geology which almost certainly determined the location not only of this site, but also of at least one other to the west, Monument 97 (Mackreth, forthcoming), was what is in effect a held water-table. This broke out as a spring immediately to the north-east of the site until modern times when the London Brick Company pit uphill to the south disturbed the natural conditions. That the ground below the site always tended to be damp is borne out by the fact that, prior to redevelopment, the fields to the north and east of the site remained under pasture at a time when more land was being converted into arable. However, the field in which the site lay had been ploughed once during the Second World War but was considered to be generally poor ground, except for the area covered by the Roman farmstead. The extent to which the high water-table was recognisable on the site can be seen by the location of the two ponds which, having been dug, filled up naturally with ground water. It may well have been that, as the site declined and the drainage ditches were no longer fully maintained, the wetness of the ground was a contributory factor in its final abandonment.

Chapter 1. Description of the Site

I. Pre-Period 1 activity

(Pl. I)

There is little to say about any of the features which are placed here. Their stratigraphical position, where this is directly expressed, was at the bottom of all sequences and the chief characteristic, with the possible exception of the pond, F225, was that they bore no relationship to the succession of events described in this report. In only two cases was a feature succeeded by another: F7 cut F8; F225 cut F227. The dating evidence was meagre and, for the gulleys, virtually non-existent.

The earliest definable layout on the site consisted of F8 with its branches F227 and F544 (E1). F8 was narrow, *c.*0.4m– 0.5m wide, and ran about 70m on an east-to-west line from the eastern edge of the excavation and then turned sharply north to become F544 which disappeared amongst later features. About 34m from its first appearance to the east, a branch, F227 (Fig. 28, [78]), ran south. It was cut by the pond F225 (Fig. 14, [2]) and failed to reappear. Wherever sectioned, the gulley proved to be steep-sided and almost as deep from the excavation surface as it was wide. The fills were uniformly silts from which any humic content had leached away. The limited plan recovered seems to have formed the south-western part of an enclosure probably around, or associated with, the original spring. The dating from the gulleys themselves rests exclusively on flint flakes not distinctive enough to date them definitely earlier than the Iron Age. However, the general flint collection from the site (see Dr Helen Bamford's report, MF8), although not a large sample, was markedly late Neolithic–Early Bronze Age in date. There was evidence for flint-working on the site and there was also an anomalous, very high, proportion of retouched flakes. These are signs that the collection was not derived from casual loss away from an occupied site, even if the small proportion of working waste suggests that there was no intensive occupation. The greenstone axe (Fig. 60, No. 30) also fits this period and the copper-alloy razor (Fig. 60, No. 31) could also fit. Despite the lack of better evidence, it seems reasonable to associate the artefacts with the gulleys. The general lack of other features assignable to this early horizon on the site may point to a seasonal use.

The gulleys were cut by a deep feature, filled with a hardly altered clay, F7 (E2). The feature was only cursorily examined as it clearly did not belong to the Roman site and was difficult to deal with because of the high water-table. The only find was a piece of pottery with some grooves on its surface, not really large enough for clear identification but, if Bronze Age, it could show that the gulleys can be associated with the flintwork.

The largest feature belonging to this phase of the site was the pond, F225 (E3). The pottery content was sparse but unequivocal in its Iron Age character. Most of the pond had been cut away by another in Period 4, F200 (Fig. 14, [1]), but enough survived to show that it was similar to both F200 and the Period 3 pond, F500. It seemed to have been deliberately backfilled in the main with gravels. The signs were that it had been roughly 11m by 10m with a basically rectangular plan. As with the later ponds, the ground water was high and constant enough to have kept it filled even in the severest drought. Although the pond could be argued to have survived into Periods 1 and 2 to be replaced by F500 in Period 3, the lack of *any* pottery belonging to the first two periods should be enough to show that it had been backfilled before Period 2 and almost certainly before the major developments of Period 1, phase *c*.

Only two other features are assigned here: F248 and F88 (E4). The relationship of the first with the pond and the earlier gulley branch was ambiguous because of the spreading top fills of the Period 4 pond. F248 was a gulley later than F227, but not necessarily stratigraphically earlier or later than the pond except that the first-century pottery in it points to it having been later. There was no plan relationship with any other feature and no continuation was found in either direction and it is given to pre-Period 1 activity as no associations can be seen with Period 1.

F88 was a post-hole on the line of F49 belonging to Period 5, but at this point the gulley had been largely ploughed out, so destroying any direct relationship. At first sight it looked as though it ought to have been another Period 5 post-hole, but its fill contained no less than ten flints, the highest concentration in a single context on the site and at odds with any other similar feature. The flints suggest that it should also be included in the earliest defined occupation and associated with the early gulleys.

II. Period 1, mid-first century–*c.*175

(Pl. II)

Phases *a* and *b*	mid-first and later first century AD
Phase *c*	late first–early second centuries
Phase *d*	early second century
Phase *e*	first half of second century
Phase *f*	mid-second century plus

Summary

There seems to have been an earlier nucleus to the north-west of the excavated site which, by the time of the Conquest at least, had spread sufficiently far for part of its ditched enclosures to have penetrated into the corner of the excavation. There is little evidence that much use was made of the rest of the area. By the latter part of the century, at least one round house lay outside the original main enclosures, and a small ditched enclosure with at least one round end with an entrance lay nearer the spring to the north-east of the site (phases *a* and *b*).

The beginning of the main sequence of developments leading to the formation of a Romano-British farmstead consisted of a long east-to-west ditch running right across the site and attached to two sets of new enclosures at the western end. There may have been an entrance through the long ditch sited at the division between the eastern and western sets of enclosures (phase *c*). What survived of the earliest site was swept away in a redefinition of the enclosures, the eastern one being extended to the south (phase *d*). The round house at the west end was demolished and replaced to the east by another. To its south was an area in which lay two ovens and some pits. The eastern enclosure was then divided into two (phase *e*).

The final stage of Period 1 was marked by the recutting of part of the ditch-lines of the eastern and western enclosures with either modifications to the previous entrance or the provision of a new one for the first time. This was accompanied by the digging of a new ditch which seems to have divided the original corridor between the enclosures into two, possibly for a new system of stock control (phase *f*).

Phases *a* and *b*

When parts of one period are buried beneath, and are cut about by, later ones, and also relate to a main focus outside an excavation, there are problems of interpretation. The presence of an earlier focus to the north-west is shown by the pottery found in the early features in that area of the excavation as well as residual sherds in later phases. The principal identified feature belonging exclusively to phase *a* was ditch F775 (1) (Fig. 14, [3]) much of which had been cut away by later ones on the same basic alignment. These suggest that the original version, of which F775 was the only survivor, defined the north-east corner of an enclosure whose southern end was not found. These later versions, Fs797, 588, 688 and 826 (2) show that the site was under active occupation through the latter part of the first century. F1037 (Fig. 14, [4]), running away southwards, should have been part of a large enclosure, with subdivisions F814 and F1018 (3) datable to phase *b*. All of these ditches, where they could be seen under the later recuttings, were about 1.5m wide and *c*.1.1m deep (Fig. 14, [5]) at most. It may be that only the latest was deep, F775 being hardly more than 0.7m. There was no direct evidence for an entrance unless the out-turned west end of F775 marked one side, but there was no sign on the other side in the same general line.

Within the enclosure there was, apparently, a round house defined by a gulley, F875/922, (4) whose western part had been cut away by a furrow. It had also been cut into by the Period 5 nine-post structure, and by Mr Dakin's excavation of that (MF6), and overlaid by Period 3 yard deposits. The width of the gulley was *c*.0.8m at most and it was about 0.4m deep. There was an entrance facing slightly north of east and, in this respect, the structure matched the later house further east. No trace of structure was found in the gulley and it is assumed that this was for drainage. The southern side of the entrance had a slightly out-turned end. The approximate area defined by the gulley was *c*.56m^2 and, allowing for 0.5m width for any structure, the internal area of the building would have been of the order of 44m^2. No trace was found of any structural element or of any activity nearby, except for a shallow pit lying on the south-east side, F894 (5). In the north-east corner were traces of features which may have belonged to these phases (6); most were undated and are assigned to Period 1 because they either lay at the bottom of the sequence or had the same general alignment as the ditch to the east. All had been severely damaged by the plough.

The hardest area to interpret lay to the north of the main early enclosure. It is not certain if there had been another enclosure here and what would have defined its east and west sides. F588, where it ran out of the excavation, could have been a replacement of a mid-first century and earlier ditch, but there were no residual finds to support this view. The undated ditch, F767 (7), running north out of the excavation, had a definite butt end which should show that it was contemporary with a version of the early ditch. The absence of dating evidence may well be a sign that it was early and part of an enclosure, its end forming one side of an entrance, F775 marking the other. How F814 fitted in is uncertain as it could not be traced as it turned away to the north-east.

The east side is less easy to determine, later deposits allowing only a cursory examination of earlier periods and a deep furrow removing a good deal of information. The main line for this side could have been that represented by F554/801, later to become the west side of the Droveway in Period 2. However, what happened at the south end is obscure. F554 appeared to have been about the same size, but was almost certainly enlarged in Period 2, so removing the original version, if it existed, except for the ditch elements F771/809 (8) (Fig. 14, [6]). The ditch was *c*.1.2m wide and *c*.1.1m deep and ran from the corner of the first enclosure across to the F554/801 line, but its pottery suits phase *c*, not phase *a* or *b*, and, as F554 is logically a replacement of F771/809, the dating of that becomes difficult. It is discussed under phase *c* and again in phase *e*.

The presence of an enclosure north of F775 is suggested by a clutch of apparent post-holes containing mid-first century pottery and a pit, F858, the finds from which will not easily allow a date after *c*. AD 75 (14) (Fig. 4). It contained an uncharacteristically high number of small finds (MF8, Catalogue Nos 5, 7, 10, 97, 159). Two brooches are earlier than AD 75 and the rest of the material also suits this dating.

The features formed part of a collection immediately north of the end of F775. Many belonged to the second half of the first century and some were undated. If all are put together, they represent what may be defined as phase *b* activity, some belonging to a structure although this could not be defined. The boundaries north and south were reasonably clear, but furrows had removed the east and west boundaries, the series not emerging either east or west from the furrows. The minimum area covered was *c*.6.5m by 6.25m. If F684, cut into the fill of F775, belonged to the set, the best date for the bulk of the group would be phase *b*, the earlier pottery being residual. But F827 may be a warning that all should be assigned to phase *f* with some containing residual second-century material. However, whatever function was associated with them could have lasted for more than one phase.

A closer analysis (see Fig. 4) fails to reveal even part of an identifiable building. Three features having a specific common characteristic makes interpretation difficult: Fs684, 685 and 686 each contained the base of a pot which had been burnt *in situ* and then filled with clay (9). It is tempting to relate these to F687, a hole in which a large natural iron concretion was buried, and then to F854 (10) with its very large collection of bones, and to see either a ritual significance or, more mundanely, an association with cooking in general, but the burying of the iron-pan mass, F687, is less easy to explain in that context.

There was evidence for a succession of activities. One of the pits containing burnt pot bases replaced F758 and

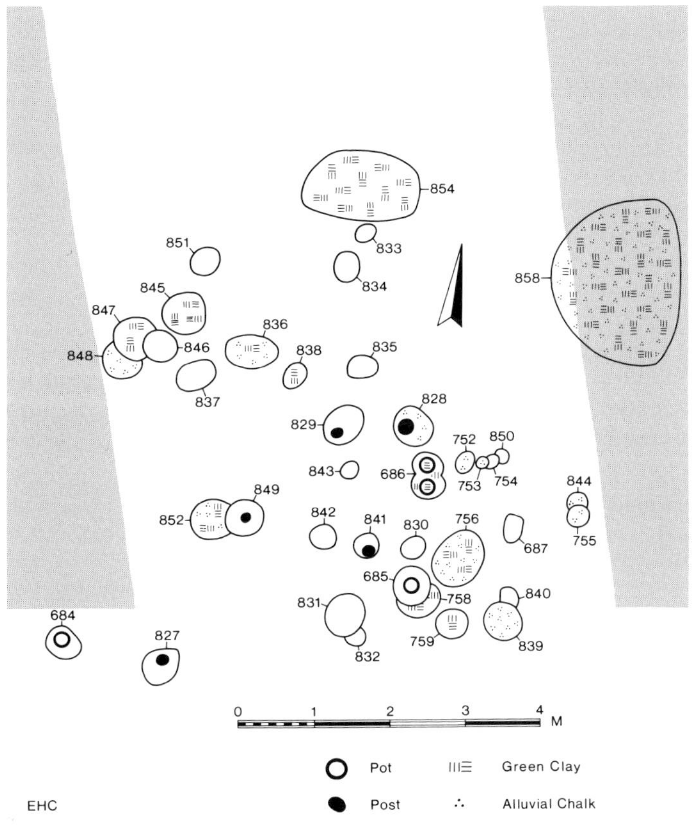

Figure 4 Period 1: specialised pits and possible post-holes in the north-west corner of the site.

this, had it been found anywhere else on the site, would have been counted as a straightforward, badly preserved, post-hole. Burnt clay occurred in F849 and F851, the first being secondary to F852 (12), but this may have been due to burnt clay being in the topsoil of the site when they were created. One possibly significant constituent of the filling of several features was green clay lumps or flecking. The dating applying to most of these features is mid-first century (* in (9)–(15)). The clay ties in one hole with a pot, F686, with two secondary features, F758 (11) and F847 (13), which, in turn, can be tied in with deposits of the second half of the first century, Fs685, 686 (9), 687 (10), 837, 845 and 848. The last of these is potentially important as it was replaced by F847 which was, in turn, replaced by F846 (13). These have the same date as pit F854 (10), which contains the extraordinary bone assemblage (see The Bones), including several articulated limbs, and F858 (14) whose date runs up to *c.* AD 75.

The presence of flecks of alluvial chalk (@ in (11)–(15)) seems to belong to a sequence in which they occurred alone and then, in some secondary features, with green clay lumps. A primitive ordering emerges suggesting that there was a sequence of events too detailed to suit a building and more fitted to a singular, long-term function which led to replacement at frequent intervals of one or two features. The unburnt sides of these holes should not be against their having had fires of some sort in them, as several features which should have been associated with fire extensively, were actually unburnt (see Chapter 3.II).

A survey of the medieval pottery industry and its markets (Moorhouse 1981) dealt in a summary fashion with the uses to which pots were put. In this, attention was drawn to recipes involving fermentation and distillation (*ibid.*, 115–18). While the parallel between what is described there and found here is not exact, there is a close enough relationship to suggest that the general principles involved may have been known in rural Britain in the first century AD. The flecks of raw green clay, as well as the pieces of burnt clay, could well have been associated with luting lids or pots together.

Some form of shelter should have been provided for whatever took place here, but its form, and which features should be assigned solely to it, is unknown. The alignment of the group was visually east-to-west, but the fragment of ditch, F855 (16), may give the best indication for this; the tongue of ditch, F646 (17), may also have been associated. If this was a kitchen area, the question arises as to where the food was eaten: presumably this would have been done in a normal domestic structure rather than in a separate building of unidentified type. This probably lay outside the stripped area to the north-west: the arrangement of ditches in the excavation seems against the western of the two identified houses being used, even if its dating would suit. The dating of both the round houses covers phases *a* and *b* and the major change in phase *c* appears to coincide with the end of most of the activity here.

Only one major feature belonging to phase *b* remains to be mentioned, F10, a ditched enclosure lying well to the east. Its dating belongs to phases *a* and *b* (18). The plan is reasonably secure, except for the west end which was cut away by a furrow as well as ditches dug in Periods 2 and 4. The east end was semi-circular with an entrance, 1.5m wide, at the mid-point. The sides were straight, parallel and 11m apart. The maximum surviving length inside was 16m. The section of the ditch was 0.75m wide by O.28m deep and had a shallow curved profile (Fig. 14, [7]). While there was no trace of any structural element in the gulley itself, for the deposits showed that it had filled more or less naturally, there was, on either side of the entrance and seated in the ends of the gulley, a clay plug which was interpreted in 1971 as having been for a post. Although the west end was cut away, a rounded end similar to that to the east should be assumed: there was enough ground left to the west of the furrow here to show that the side ditches had not run straight on. The minimum internal area, with the west end reconstructed to match the surviving one, was *c*.200m.

Had only the east end been found, it would have been interpreted as having been part of a house and, if it had belonged to a recognisable type of structure, the suggestion might have been allowed to stand. Even in conjunction with a hedge or a fence, the defining ditches do not seem to have been large enough to have acted as effective barriers against animals. What may have been post-holes in the ends of the gulley might suggest a porch for a house, but clay would be an unusual fill for a defunct post-hole on this site.

Phase *c*

It is to phase *c* that the acts which led to the uninterrupted development of the site can be assigned. The main feature of the new scheme was that the site was, as far as the excavated area was concerned, divided into two parts by a ditch running east-to-west, the area to the north remaining largely undeveloped until Period 2, possibly because of the high water- table. Starting at the east end, the sections (Fig. 14, [8], [9]) show that there were two main phases, the pottery indicating that they were fairly close in time. The original line of the ditch, F13, coming in from the east, was recut to the north on an almost completely new line, F40 (19). Both versions of the ditch were relatively slight, the second being largely a scoop 1.02m – 1.8m wide and only about 0.3m deep. Both may have been associated with a hedge set next to hurdles to begin with: no substantial bank could have resulted from the spoil. The two could be traced to the east where F13 was largely cut away by later features (Fig. 15, [10], [11]), but the variable course of these ditches means that F13 was lost completely and F40, now separate, was cut into (Fig. 15, [12]). Although no good connection exists, the ditches located at the extreme east end of the excavation (Fig. 15, [13]) can be equated with known ones.

To the west, the earlier of the two ditches, F912, swung round to the south and was lost in later disturbances. This behaviour in relation to what seem to have been phase *c* deposits (20) in ditches swinging eastwards from the line of F969 points to an offset entrance between the two and subsequent developments tend to reinforce this impression. The other side of such an entrance may have been formed by the sinuous ditch, F771/809, whose pottery (6) places it in this phase and which would otherwise make little sense. The effect would then have been of a funnel entrance with a left-hand skew. This would provide a partial answer to the problem presented by F554, the phase *a* material in that becoming residual; the ditch could then have been a phase *e* replacement of F771/809. It would also explain why there was so little phase *b* pottery present, and why the phase *c* pottery occurred above phase *f* material (21).

As for the possible entrance in the recut version of the long ditch, no definite break in F40/957 was discovered, but one is assumed on the basis of the behaviour of the earlier ditch and later developments. South of the eastern side of this area lay the traces of a single large enclosure — Fs909, 1025 and 906, with possibly F907 which cannot be tied into any other scheme, and which had large gaps east and west next to the F40 line (22). This end of the enclosure was complicated by separated ditch lengths demonstrating some elaboration: Fs900, 950, 958–60 and 920 (23). As the north end of F906 cut F920 which, in turn, did not easily go with F13, the whole of this enclosure could have been laid out after F13 had been replaced by F40, unless the constriction in the north end of F906 represented a recutting of the enclosure here to suit the new arrangement. The ditches were all of the same basic section and were generally 1.5m wide and up to 1m deep (Fig. 16, [14]). There was little sign of activity within the enclosure. One elongated pit, F930 (24), was found, but its pottery belongs more to phase *b* than *c*. As a pit would presumably have been quickly backfilled while a ditch would remain open to receive rubbish of the time of its use, it may be that the digging of this pit occurred at the beginning of the phase rather than before.

The conjectured funnel entrance is so closely matched in Period 2, and an entrance seems so logical in phases *e* and *f*, that the lack of proof for a gap in F40 need not be an embarrassment. However, the course of F40 developed a sinuous deviation from its main line and became indistinct some 27m from the line of F969, and it is here that any entrance would presumably have been sited. The approximate position of the west side is implied by the behaviour of other ditches in later phases. It was at this time that the earliest version of a ditch running from F969 towards the last known point of F40, F737/766 (25), was dug. This provides the context for the creation of F554 which had a clear butt end with the new east-west line (Fig. 16, [15]). A single post-hole, F921 (26), partly cut away by later developments in this phase, may have been related to an entrance belonging to the F40 ditch and could have been the sole relic of a gateway. It lay north of the F13 line and should not, therefore, be related to that, especially as that entrance seems to have been 20m wide at the point at which the post-hole lay.

The making of an entrance would have been accompanied by the digging of the enclosure to the south-east with entrances to east and west at the north end. That at the north-west was divided into two by means of F920: one at the north-east end, 2.7m wide, and the other, 2.5m wide, to the east with F906 forming the other side. The plan shows a narrowing of the ditch which suggests an extension, F923, dug to block the gap. Some evidence for the recutting of F906 was recovered at its north end. The arrangements down the side of F40 showed some care and the impression given overall is that the area was laid out for sorting and controlling stock of some sort. Ditch F900 formed an offset gap entrance 3.4m wide, with the end of F920 giving, at this point, a choice of four directions. F900 along with F40 formed a short track, and just round the corner at the east end was another entrance, 2m wide. The widest entrance may have been controlled by a gate, as some post-holes, Fs916, 924 and 965, can be assigned to this or the next phase; others may also have been associated but they are undated (27). The pottery from the post-holes belongs either to phase *c* or *d*, and possibly arrived when the posts were removed (28).

The post-holes fall into two basic groups. In neither did any evidence for post-pipes survive. One group lay against F900 on the south side. The two latest posts may have existed together, possibly to stiffen their side of the gate structure. They replaced Fs918 and 919 which were of the same general size and also the same basic centre-to-centre distance. The later posts could, like the earlier ones, hardly have been bigger than about 0.25m across, the narrower dimension recorded being due to both having had their upper parts cut away by the secondary set. On the north side of the entrance were certainly two more posts, with the possibility of two others so shallow that they survived to be planned but not to be numbered and excavated. F952 was only 0.08m deep and their lack of depth may have been due to their being less deeply founded, or set in the toe of a bank along the south side of F953.

None of these features cut any other and it cannot be demonstrated that they were arranged in pairs, but the possibility exists: F952 with the mark to the west, and F965 with the mark to the east. On the other hand, these pairings do not correspond with the posts on the south side, only the eastern one being approximately in line. The gap, north-to-south, between the two groups was about 2.5m. The northern set could, conceivably, have formed part of a separate gate to the north, 1.6m wide at most, but only if the entrance gap between Fs906 and 900 had been oblique. Such an arrangement would provide a context for the closing of the eastern gap with the recutting of F13 by F40 and the provision, seemingly, of an offset entrance between Fs920 and 912 further west. There was another small pit or post-hole, F924, about 3m to the west of F916. Most of this had been cut away by a wide and shallow delve, F925. F924 is dated to the early second century, F925 is undated. No clear resolution was possible, but this may have been a series of post-holes matching the western marks of the north set.

The initial enclosure may only have been a simple one, but extra ditches at the south-east corner are dated to this phase: the early version of Fs1043 and 1045 (29). Their junction with the enclosure was destroyed by the Period 3 well. In view of the phase *d* alterations, these two could belong to that phase, the requisite sherds not being present. Before the next phase began, or even as part of its first works, F950 replaced F900, doing away with the east entrance and providing another about 2.5m wide near the middle of its length: very little pottery was recovered (23).

Phase *d*

The phase *c* arrangements proved good enough to last for at least a generation before being modified to an unknown extent: much of what may have been phase *d* could well have been cleaned out in the succeeding phase. The new enclosures west of the site's centre were extended southwards with F1088 as their new limit (30), and F1043 and F1045, discussed in phase *c*, may belong to this time, but the pottery is not helpful. Unless there is a major influx of new material, new features may only contain pottery derived from earlier schemes and, without other evidence, can only be dated by the actual sherds in them. In the case of phase *d*, very little ceramic material was found and this possibly distorts the amount of alteration.

Consideration of the probable life of the western house provides an example of this kind of problem. It could have continued until replaced by another to the east in phase *e*, but it could also have passed out of use before then and the lack of dating evidence for this phase is due to diminished domestic activity. But deposits in the ditch facing the entrance and dated specifically to the early second century may show that the house had remained in use. Unfortunately, the house ditch had been largely emptied by Mr Dakin and the finds are now unidentifiable. Other deposits lay in the upper fills of the defunct phase *a* features as well as in the F969 ditch (31). If pottery is likely to have derived from nearby activity, the house along with whatever was represented by the 'post-holes' to the north (9–15) could have continued. There is, however, a hint that the western enclosures were recut, most of the evidence from the northern part being removed during later cleaning out in Period 1.

Phase *e*

What has been presented tentatively in phase *d*, becomes definite now. The phase *c* plan was enlarged with a growing emphasis on the area which was to become, from Period 3 onwards, the Main Yard of the Late Roman farmstead. Whatever may have been the case concerning the western house, it cannot have lasted for long and was replaced by another, F525–526 (32), to the east. The only hint of domestic activity near the western house was a pit largely removed by the Period 3 north wall of Barn 1: F994 (33). The eastern house was poorly preserved, being cut by a furrow and partly lost in the Period 3 pond. The internal diameter was about 6.5m and the optimum area 33m^2. The entrance was about 2.2m wide at most. The gulley or eavesdrip channel surrounding it had been cleaned out at least once as the bifurcated end of the north side of the entrance showed; no division could be seen in the silty fill, nor any trace of structure either in the gulley or within its circuit.

The eastern house was not in an enclosure and was associated with a wide and shallow gulley running away to the south with a slight return west, F1000/1001 (34). The gulley had been cleaned out more than once, obscuring an early version on a different line and with a shorter course, F1023–1024 (35). The main line ended in a set of features, F1002 which cut Fs1001, 1012 (36), and a lobe attached to F1001 which suggested post-holes, but without the confirming post-pipes.

The only hint of domestic activity near the house and in its ill- defined yard was provided by two sorts of feature. The first was a pit sequence, Fs1013–1015 (37) (Fig. 16, [16]), which yielded little information. The second consisted of two 'furnaces': Fs1081–1082 (38) (Chapter 3.II, **2** and **3**). These were so far from the house that any roofed structure over them would have been independent but evidence for this had completely disappeared. A building here may have filled the gap between F1000/1001 and F909 and may have used the apparent post-holes at the end of the former or have been timber-framed. The discussion of the site in Period 5 suggests that there may have been an Anglo-Saxon building here and all these remains, including the pits whose fill was darker than virtually all Period 1 features, could have belonged to it.

The phase is marked by the increase in the number of enclosures, all enhancing the established character of the occupation. The main western enclosure was recut (39) with the introduction of a new east-to-west ditch within it, F993 (40); the eastern set were enlarged to the south, Fs1049, '1083' and 1032 (41), and subdivided by Fs914 and 929 (42). The suggested entrance through the long phase *c* ditch, F40, is supported by the rearrangement of the controls on traffic which would have moved southwards through it. The east side seems to have remained where it had been in phase *c*, but the west end was redefined by a wing ditch running east from the F969 line, F766/675, although the actual excavated end belongs to the phase *f* layout.

F993 was the first version of a feature which appears finally to have passed out of use in Period 4 (F1006, (206)). The successive stages, beginning with Period 2 when the first barn was built, show well in section (Fig. 16, [17]). F993 ran to the east to join F969, the final version here of F1036/1037. The frequent recuts of this major ditch-line also show well in section (Fig. 16, [18]) which, while oblique, reveals the shift westwards which sharpened the angle of the corner it made with the east-to-west line.

It was probably now, if not before, that the F554 ditch coming into the site from the north was dug. The pottery is ambivalent, but would suit this phase best as the earlier material does not demonstrate a simple progression (21). The gap (Fig. 16, [15]) between the butt end of Fs554 and 766 could have been filled by a bank thrown up on the north side of F766. The undated but parallel ditch about 28m to the west, F767 (7), may also have been dug at this time. Within the site, traffic seems to have debouched into a large enclosure with an entrance immediately to the east surviving from phase *c* and with post-holes, possibly belonging to a gate, containing pottery generally of phases *d–e*, Fs916 and 924 (28). How much of the ditch F906 was still in use is uncertain, as phase *e* pottery was confined to the new linking ditch at the extreme south end. If the pottery is a good guide, there was a very wide entrance at the south end. However, there was a set of post-holes (Fig. 5) lying between F1032 and the ditch belonging to phase *f* to the west. These may have belonged to that phase, or the ditch may have had its place here: phases *e* and *f* may be telescoped if the backfills of phase *e* contained mainly residual material.

The pottery from the post-holes is not really compatible with phase *f*, but post-holes would have been sealed during a phase while a ditch would have been open. That the post-hole series had run right across the gap between Fs1032 and 969, could not be shown owing to later activities. Although it may be better to allocate the post-holes to phase *f*, they are given to phase *e* because of the pottery in them (43). One, F1055, can be dated to phase *d*, but the pottery was probably residual. Another, F1060 (44), also datable to phase *d*, lay in such a position that, taken in conjunction with the ditches next to it, a whole separate set of sub-phases could be argued.

The impression that the set of post-holes were not in being all at one time is enhanced by the detail that the survivors lay mainly at the bottom of a furrow. None had a trace of a post-pipe, the fill having a uniform matrix. However, a primitive sequence emerges when the additional elements of the fill are looked at (Fig. 5) even though only one post-hole, F1053, cut another, F1058. Both eastern ones, Fs1060 and 1059, were more closely associated with the enclosure ditch than the rest, the second being partly cut away by what might have been an additional ditch, F1032. Both were distinguished by having orange sandy patches indicative of deliberate backfill with lumps of natural. To the west lay three post-holes, Fs1055, 1057 and 1056, with a dark altered silty natural fill. F1058, cut by F1053 is linked with F1054

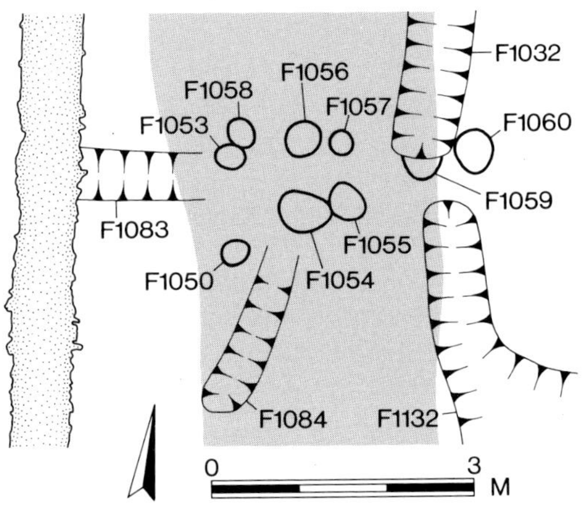

Figure 5 Period 1: probable gateway system in main route across the site.

as both contained lumps of green clay. The last two, Fs1053 and 1050, had paler fills than the rest and the latter also contained limestone pieces which may have derived from the packing round a post.

What all these features represented is equivocal and their relationship with the gulley F1084: F1054 lay in or under its north end. The excavation of its south end suggested that there might have been a post there. The gulley, as it survived at the very bottom of a furrow, was only 0.5m wide and less than 0.1m deep. It ran from F1054 for a distance of 2.8m on a slightly curving course and may have been produced by removing a series of posts once forming one side of a major gated structure opening into the enclosure, its other side being cut away in later works. How the wide gap into the enclosure had been closed can only be guessed at, perhaps F1083 belonged to an arrangement to do this. Only the smallest trace survived of this gulley; it was shallower than F1084, but as wide. It lay at a higher level, being best preserved between the furrow and the Period 4 boundary wall.

Continuity between phases *d* and *e* is shown by pits Fs1111, 1130–1131 (45) lying south-east of this area. As a group, they seem to have been in the earlier phase and to have ended in phase *e*. They also formed a representative sample of those generally found on the site. The largest, F1111, was 2.5m by 1.7m and only about 0.5m deep. The sampling of the feature suggested that there had actually been at least two pits here. In common with practically all the pits on the site, there was no identifiable rubbish content. The other pits were discrete and varied from 0.7m to 0.9m in diameter. Only F1164 had any great depth, 0.6m, and the other two, Fs1130 and 1131, were only 0.25m and 0.28m deep. Even allowing for the lost topsoil or surfacing of the site, these were hardly more than scoops.

A narrow entrance, only 1m wide, lay in the western side of the eastern enclosure. Inside, and to either side, were the remains of post-holes, Fs915, 926–928, marking the site of a gate. The greater number of replacements on the north side suggests that it had been hung on that side. The south post, F915, was 0.18m in diameter, set in a pit 0.53m in diameter and packed round with pieces of limestone rag. The post was only 0.25m deep, while the pit was 0.42m, showing that a post-pipe will often not survive in very shallow post-pits. On the north, there had been three posts, the earlier two, Fs927 and 928, being almost completely obliterated by F926. The first two were not well enough preserved for comment and no post-pipes were seen. The replacement had been removed and what had probably been the stone packing round the post had been stuffed back into the resultant hole (46).

The distance of *c*.0.7m from the centre of the post-holes to the near edge of the gulley points to what had probably happened generally to spoil dug out from ditches and gulleys on the site. No evidence was found for fences in association with enclosures. The nature of the subsoil was such that the one-time presence of a simple dump bank would hardly ever show in the section of a ditch. However, the post-holes here suggest a bank along the east side of the ditch with the gate placed at its approximate crest-line. There can be no doubt that the line of Fs914 and 929 was intended to form some kind of barrier, yet they were only 0.9m and 0.7m wide respectively. Their depths were between 0.2m and 0.3m. Even allowing for the loss of the original topsoil, the scale of both bank and ditch seems inadequate as a barrier against animals. There ought to have been something else, and the gate points to this. The general lack of evidence on the site would suit something like a quickthorn hedge or a set of hurdles planted with hedging stock in the top of a bank as the most likely form of physical barrier immediately obvious to a casual visitor at the time. Whatever form of barrier was used, it did not inhibit frequent changes in the layout of the farm whenever there was a need.

At the south end of the site was a short length of ditch, F1117 (47). Again, this was slight, being only some 0.25m deep, even if it widened out to 1m at its western end. It had a clear end to the east, but not to the west. The east end would suit a redefinition of either a phase *c* or *d* ditch, F1043 (41), and it may be that the intention was to have a short length of track, like that at the north end of the enclosure, with a similar constricted entrance at one end, this time at the east.

Three areas of loosely defined activities assigned to this phase remain. Possibly the most coherent of these consisted of a ring of post-holes, Fs1071–1079, associated with a hollow, F1070, and lay in the south-west corner of the enlarged eastern enclosure (48) (Fig. 6). None is dated, but the whole set survived long enough for two posts at least to have needed replacing. They were sealed under what seemed to have been a surviving topsoil which contained nothing later than phase *f* pottery.

The depth of the posts varied between 0.1m and 0.04m, the diameter varying between 0.4m and 0.24m and no pipes survived. Taking the outside edges of the group, the maximum dimensions were 2.2m by 2.1m. The largest gap was where the pit F1070 lay, and measured a clear 1m across. The plan of the post-holes was more pentagonal than circular. The fills fell within a narrow range of soil type, possibly the product of deliberate removal. F1070 itself was aligned south-east/north-west along its major axis, 1.4m long and 1m wide. Its depth was only 0.2m and its sides were shallow curves. The whole looked more like a wear hollow than a pit and its fill better suited this: initial wear producing the barely altered natural at the bottom, subsequent use creating the more altered and darker material of the main fill. Whatever such a structure was for, its use seems to have been sufficient to wear through the original ground surface into the natural below.

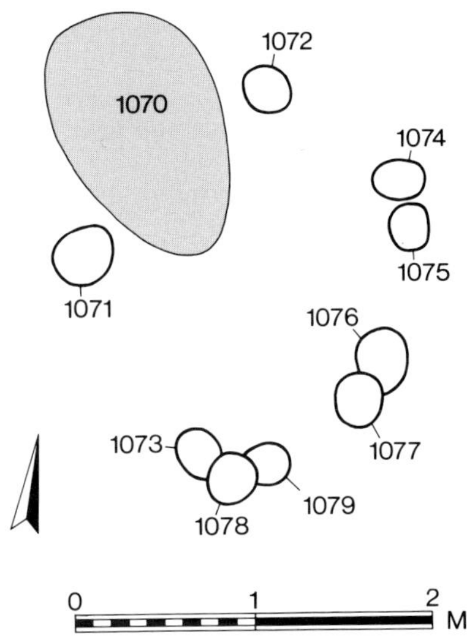

Figure 6 Period 1: structure and wear hollow in the southern part of the central enclosures.

In the corner of the next enclosure or traffic lane to the north-west were two pits so spaced that they could conceivably have formed part of a traffic control system: neither revealed any sign of structure, Fs910, 997 (49). However, the first had ash mixed with sand and burnt clay pieces in the bottom which might point to rubbish disposal. But the second contained pieces of limestone rag in the three layers filling its depth of 0.2m. Only the darker top layer, possibly a later infill of a developing sag, had any finds. If their real function was related to the use of the north-to-south route through the site, the only feature with which they could be read is F968 which was unequivocally phase *f* and not necessarily at the beginning of that.

Lastly, even further north-west, and south of the problematic post-hole area described under phase *b*, was a short length of very shallow gulley, F673, which had a northern but no recognisable southern end. The real date may not be represented by its pottery which would suit an earlier phase. To the east were three undated post-holes which are unlikely to have post-dated Period 1: Fs740–742 (50). The area had been considerably eroded and these traces may only have been the deepest of a series. Whatever their date, they should have been earlier than the final phase.

Phase *f*

As in phase *d*, there was no large-scale development, there being only recutting and redefinition: whether it has a separate identity from phase *e* has been discussed. The layout of phase *e* survived virtually intact except for one apparent omission and one important addition.

The excavation failed to produce, in plan, any evidence for a house having being present by the end of Period 1. The dating of the gulley of the eastern round house and associated features show that that had passed out of use. But that there had been a more substantial house later than that is shown by the final fills of Fs813 and 969 (L1773–4, L1818, Fig. 16, [19]; L1582–1583, Fig. 17, [20]) under the later sag infilling. These contained window glass and small lumps of a very sandy burnt daub once finished with a fine white plaster coating. The fragments were very small but unmistakable, and it was also here that the first major collection of samian occurred even though, in quantity, it was not large (51). The burnt fragments of a form 37 bowl can be dated to *c*.160–190 (Fig. 107, No. 5). The demolition of this house and the filling of these ditches provides the date for the beginning of Period 2.

The phase *f* house probably lay south of the main east-to-west ditch-line first laid out on phase *c*. How far south cannot be known, but none too far from the fragments of its cladding and it may be that F559/678 was associated with it. This was a shallow and narrow slot with a small amount of pottery of mixed date, and the phase *f* material from within the building on the north side of the Period 3 Small Yard (see below) derived from Period 1 features destroyed during the building of that (52). Sealed beneath the destruction fills in Fs813 and 969 was the oven, F901, (39)(51) (Fig. 16, [18], [19]; Chapter 3.II, **1**), which had been dug into the earlier fills of F969. Its siting suggests that it had not been in a building, but some sort of temporary roofing could have been placed over it and the oven may have lain close to the destroyed house.

An assessment of both the dating and the distribution of glass and samian reveals that there was only one close focus for both and that was in the western part of the site. This emphasis is repeated in the distribution of beakers datable to this period (Figs 112 and 113). The glass from the site contained little third or fourth-century material and is of more use than the evidence of the samian, as most of that should have been lost by the end of Period 1. Periods 2 to 5 reveal the same concentration of glass in and around the Period 3 Small Yard, but suggest that it was obviously residual and becoming dispersed. There was a secondary centre for both glass and samian in the general area of the north-east corner of the Period 2 Main Yard. These deposits could represent survival of individual pieces into the beginning of Period 2 itself. The distribution of mortaria does not show a marked concentration, but there were light emphases on the sites of the two round houses.

The dating of the pottery from F554 points to the south end of the feature having passed out of use, but the plan of Period 2 (Pl.III) strongly suggests that this may be deceptive (see p. 11). Similarly late deposits in the southern limb of the eastern enclosure marked the end of the period there, and the absence from these of any trace of building materials tends to support the proposition that the building from which they came lay further west (53).

The new and important feature was F968 (54), a ditch running north-to-south parallel with and between the ditches marking the route through the site. The north end was clear, but the south lay outside the formal area of excavation. It defined a narrow track, a minimum of 2m in width, on the east and a wider one, 6m, on the west. The latter was probably for traffic destined for an area south of the excavation, the former for beasts intended for a nearer location. In the latter context, it is possible that the post-holes (43) at the south end of the narrow lane, and given to phase *e* because of the dating evidence, belonged to a gated structure which might make better sense in this phase. The furrow and the Period 3 yard surface obscured the south end of F1032 proper.

F968 was replaced in Period 2 by a line of posts and then, in Period 3, by a stone wall and these severely damaged its remains reducing it to *c*.1.1m wide and 0.5m deep under the wall of Barn 1, but only 0.6m wide and 0.2m deep under the line of the yard wall (Fig. 17, [21]). Most of the dating evidence was probably also removed and it is the earlier sequences of Period 1 which place F968 at the end. Later activities, especially the building of the Period 3 wall, preserved some of the original topsoil which contained pottery belonging only to Period 1 (55).

Unphased features

Incidental detail lying north of the long east-to-west ditch F40 is difficult to assign to any particular phase although the pottery places them in Period 1. Samples of the linear hollow, F555 (Fig. 17, [22]), with its probable northern termination, F493 (Fig. 17, [23]), show that it could have been placed in phase *c*, except for some sherds belonging to phase *f*, but these may have been trampled in much later. The feature could not belong to a later period as it was cut through by the Period 2 enclosure ditch. Similarly, the pit, F608, 0.8m in diameter and 0.1m deep, and the slot F648, 0.4m wide by 0.14m deep with a U-shaped profile and a surviving length of 7m, are put here on the grounds that there is no later plan association which suits them and the pottery at least belongs here. Under the floor of the building added to the Period 4 rectangular building (see p. 23) were preserved some elements of topsoil which, like the features already mentioned, only contained Period 1 pottery and so are placed here (56).

III. Period 2, *c*.175–225/250

(Pl. III)

Phase *a*	*c*.175–*c*.200
Phase *b*	*c*.200–*c*.225?
Phase *c*	*c*.225?–*c*.250?

Summary

The main new feature was a large enclosure defined by a major ditch, with a proper droveway and two attached barns. As before, the colour on the plan indicates the pottery dating and not phases. It is upon this and stratigraphical relationships that the following discussion of what forms the individual phases of the period is based.

Two main difficulties affect the discussion. The first is that Period 2 is the earliest identifiable stage of what was to become the Late Roman farm. Unlike the plan of Period 1 which stands out because of the major change initiating Period 2, later modifications so overlaid the earliest arrangement that it is difficult at times to be sure what should be the correct plan form. This is especially true of the west end of the site.

The second difficulty concerns the dating of the pottery. Much of this would seem to belong to the latter part of Period 1 rather than be certainly later than *c*.175. Two conditions produce this effect and the small enclosure in the centre of the site illustrates one of these. It overlay the northern area of the Period 1 enclosures whose terminal date is given by pottery in its southern boundary ditches (42 and 53). The latest date in these groups is mid-second century plus, yet the pottery in the gulleys of the small enclosure is largely of the first half of the second century with only a little which can be assigned to the second half. Had the enclosure been completely isolated, it would have been given to Period 1. The pottery should, therefore, be regarded as being residual from Period 1. The other point is that the enclosure overlay the major east-to-west divide of the site and only really makes sense when seen in relation to the new Droveway of Period 2. If this is the case here, how much of the rest of the pottery from Period 2 is also residual?

At the other end of the period, there are still difficulties in deciding which assemblages belong to the middle of the third century and which forms are solely fourth century. Part of the problem lies in the more uniform character of pottery at the end of the second century and it is this which has led to the designation of AD 225/250 as the terminal date of Period 2.

Three phases can be roughly defined in Period 2. Firstly, the laying out of the main ditch and the creation of the small enclosure in the centre which may have had a short life ending within the second century. Secondly, there was the development, and the initial sequence, of a series of small enclosures along the northern edge of the new main one. The pottery suggests that this came to a close fairly soon after 200. Thirdly, a new barn was built and a process of simplification of the small enclosures began. As for the end of Period 2, the new work of Period 3 took place at a time when the forms and fabrics of the pottery begin to look late third or early fourth-century in character. However, as the site developed organically from the beginning of Period 2 until it was abandoned, there is a fine dividing line between what is called phase *c* here and the initial developments of Period 3. On the whole, the pottery supports the adopted view: that from the ditch around the new barn, Barn 2, marks a period of initial use falling within the time span of the period, as only rubbish from its use after it was built would get into the ditch.

The plan of the west end of the site in this period poses difficulties. Barn 1 contained very little which can be assigned to Period 3 or later as nearly all the features in it are datable to the second century in general hence the pottery may, again, have been residual. As the area was extensively used in later periods, it is unreasonable to argue that chance alone dictated that no later sherds should be found.

Next to no pottery came from the barn's post-pits to provide a secure *terminus post quem* for its building, but the layout of the site almost demands that the barn should have been put up in Period 2 in timber, and given stone walling in Period 3. Such a rebuilding would have followed the clearing of the site, thus wiping out existing internal features unless they had been cut deeply into the floor. The stone walls cut into the ditches surrounding the timber barn. The fence running north from the east end of the barn cut into the late Period 1 ditch and was overlaid by the stone wall ascribed to Period 3. The fence line helps to clarify the ditch layout in the western part of the site. In dealing with Period 1, there was a fair amount of discussion on the ditch which clearly became the west side of the new Droveway and in Period 2 there was little activity nearby producing pottery. Indeed, the further north samples were cut, the less pottery was found. The south end of the ditch reads well with the end of the post-hole row and it is surely not coincidence which caused a gate to be sited here in the Period 3 plan.

The principal feature of phase *c* was the continued expansion to the east following the trend in that direction shown in Period 1. The phase *c* developments formed a prelude for Period 3, part of whose arrangements were to be a systematising of phase *c* as well as a proper incorporation of Barn 2 into the central area of the farm.

Phase *a*

The changes in Period 1 were increasingly aimed at the creation of bigger yards pivoting about a major access route dividing the enclosures into two groups. These seem to have been an elaborate system of animal management. In essence, all this was swept away when a single large enclosure was made. The south side was not found and the western one presents problems of interpretation, but its line was perpetuated in one form or another in Periods 3, 4 and at least part of 5.

The minimum area of the enclosure was some 7,400m^2 and it had a major entrance, the Droveway, in the north-west corner. This was slightly funnel-shaped, 7m wide at the north edge of the stripped area and about 9m where the side ditches began to swing away at the south end. Lying diagonally across the Droveway was a line of post-holes running north-west from the east ditch to about half-way across (Fig. 7). In the western part and to the north were other posts and features which did not form any obvious pattern (57). There was very little dating evidence and, apart from the main line, there is no guarantee that the others should be added to, or even read with, the line itself. A lack of a sequence of replacement argues against a long life and the plan does not lead to an easy reconstruction of a gated structure for animal control. However, that there had been one seems likely in view of the care which had apparently been taken in the Period 1 entrance areas and the lack of evidence for a similar system south of the Droveway, although the phase *a* arrangements (see below) there might imply hurdles.

Only one of the holes had a post-pipe: F700. Most of the holes were less than 0.1m deep and any trace of posts would have been lost in erosion of the site. However, F693 looked as if the post had been dug out, the hole being left to silt up naturally, and this may have applied to most of the posts. The primary line was from F693 to F698 (Fig. 7). Posts Fs709 and 710 which continued it both had a characteristic in common with those lying further north: the addition to their fill of burnt clay, charcoal flecking, sometimes dense, and pieces of limestone, all of which might argue for a later date, but equally for replacement. Only F712 offered any evidence for direct replacement:

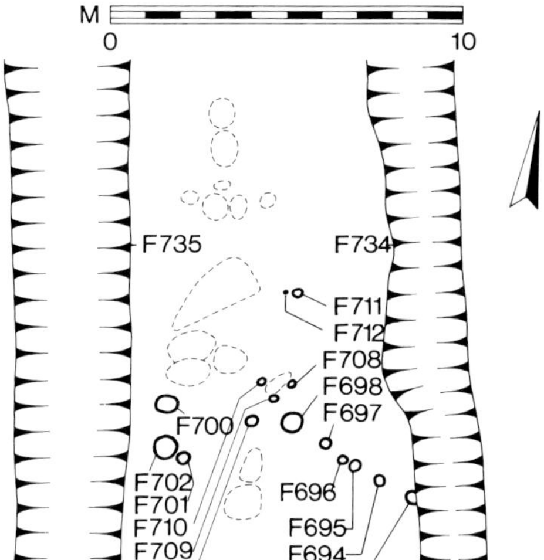

Figure 7 Period 2: Droveway, Period 3 features in broken outline.

two hollows in the bottom separated by a ridge, but being only 0.11m deep, the section was uninformative.

If the line represented a fence, traffic would have been forced to the west side and so replacement at this end would be expected. The isolated position and deeply founded nature of F700 also argues for special emphasis on this side.

The main enclosure ditch was, on average, *c*.2.5m wide and up to 0.75m deep. There was no evidence to suggest on which side a bank may have been. On the east side, the fill running south from the well, F218, was mainly redeposited natural, possibly from a bank, with little sign of silting beneath (58) (Fig. 14, [1], Fig. 17, [24], Fig. 33, [108]). The same ditch was almost certainly found in machine-cut sections by the hedge beyond the south boundary of the excavation and seemed to be turning west. A quirk in the plan (Pl.III) of the main east ditch suggests that there had been a butt end on the south side of an entrance and a hint further north of another would suggest an opening *c*.3m wide. Any entrance here had been cut across by a new length of ditch. Perhaps significantly in Period 3, an entrance was provided opposite this point in the new boundary ditch. As will be seen, the ditch north of well F218 had been heavily modified (see Fig. 20, [35]).

In the new Main Yard only one major independent feature was formed. This was a small enclosure measuring *c*.13m by 16.5m inside (59). Its gulleys were very narrow and shallow, being on average only 0.5m wide and *c*.0.25m deep (Fig. 18, [25]), too slight to have formed a boundary in themselves and with no trace of a posted structure in their fill. There was an entrance in the south side, slightly off-centre to the east and just over 2m wide.

The dating from the gulley fills was basically residual, but pottery from the north-east corner is firmly of Period 2. Nevertheless, it could be argued that the enclosure itself really belongs to Period 1, the later pottery occurring in the last surviving hollow of an earthwork being eliminated. This would mean that the Period 1 long east-to-west ditches had gone and that this enclosure had been superimposed. However, a view of Plate II yields no good relationship with other parts of the plan, whereas Plate III shows that it is well placed in relation to the Droveway, the west boundary of the new yard and to the curious enclosures attached to the north boundary.

This point is illustrated by F964, the ditch running south-east from the remnant of the main Period 1 east-to-west ditches, which had no independent function other than to deflect traffic to the west. If, however, it is read with the enclosure, the two together would have formed a funnel entrance or exit *c*.11.5m wide from a concourse area at the mouth of the Droveway. If this reading is correct, ditch F964 should be assigned to phase *a*, its generalised second-century dating suiting this stage of development (60).

F964 apparently cut a post-hole, F981, which bears little relationship with the Period 1 plan, but may have served a purpose in Period 2. It was on the edge of the gulley and erosion of the side of that would have allowed the post to stand until the gulley had been backfilled after Period 2 pottery had accumulated in the vicinity. No post-pipe was found and the contents looked as though the pit had been filled with displaced limestone packing after the post had been removed (61).

That the appended enclosures of phase *b* had predecessors in phase *a* is suggested by the pottery in ditch F422 (62). This would fit Period 1, phase *c*, but was almost certainly residual. The placing of the feature in Period 2 follows from its plan relationships. The north end stopped just short of the new main boundary ditch. Even if this was fortuitous, the south end of the ditch veered eastwards to run into the line of the phase *b* ditch F324. The relationship between the two was lost in a furrow, but the basic similarity of one to the other is evident.

The complex development of these small enclosures is ensured by a few ditches and gulleys which were largely cut away by later features and only partly seen in the sampling of the Period 4 gravel surfacing: F436, phase *a* dating; F444/445, no date; F459, no date; F557/558, basically phase *a* dating (63). These should be relics of the earliest arrangement here and F444 at least should probably be read with the small enclosure.

The structure of Barn 1 (Chapter 2) was, it is argued, entirely of timber in Period 2, the stone walls of Period 3 cutting into the silted fill of the boundary ditches which only ran along the west and north sides. The date of the finds from the features within and from the lower part of the post-hole rows points more to phase *a* than to either *b* or *c* as no third-century material was recovered. This was generally absent in the area of the barn, but occurred in the developing fills of the ditches along the west and north sides (64). The small quantity of pottery from the structural elements is largely indeterminate second-century (65): if the building had been put in an area not in general domestic use, there would presumably have been a lack of material to become residual. Once the structural elements had been created, they would not have been open to receive pottery belonging to the use of the building.

However, this condition should not have applied to the features within. The dating of these, which are assigned here because the barn is the only known building to provide a roof for them, either predated the barn or belonged to its earliest use. Not less than five were hearths or ovens (Chapter 3.II, **4–13**) and it is difficult to see how these could have operated satisfactorily in the open air — the only one which would seem to have been outside any building was F901 (39)(51) of Period 1 (Chapter 3.II, **1**). None needs to have been used in conjunction with a pottery-using activity and therefore none needs to have received sherds which truly represent its period of use. However, the character of the pottery suits that occurring elsewhere in phases *a* and *b* (66). On this basis, it is hard not to see the barn as having been here from the beginning, and therefore the basic plan of the west end of the site in this period also being present.

The layout of the features in the barn (Fig. 35) should reflect the management of the interior. As no floor level survived, the picture must be a partial one. With the exception of F1019 (Chapter 3.II, **6**) in the west end of the south aisle, no feature lay in the western half, and only F1008 (*ibid*., **4**) seems to bear a direct relationship to the barn as a whole. It lay slightly west and north of the mid-point of the middle bay. Its size alone marks it out from its fellows and its siting suggests that it was intended to have had a general effect throughout the barn. The rest of the features lay at the east end and favoured the nave although most lay between the easternmost post in the south row and the approximate position of the east wall. Whatever the purpose of the barn later, the concentration of these features in this period is against the stalling of animals. A central hearth may suggest living accommodation in the barn, but F1008 could conceivably date to Period 5 (see p. 89).

The continuity from Period 1 into Period 2 is shown by the post-row (MF4) which replaced the Period 1 ditch F968 and which ran from the north-east corner of the barn for a distance of 19m (67). Although damaged by Period 3 stone walls and a furrow, the row was relatively well preserved. The southernmost post was partly sealed under the corner of the stone barn walls and should, therefore, have been placed in relation to an earlier bounding wall for the barn. Despite the damage caused by later activities, the generally good condition of the post-holes is a testimony to how well founded the posts had been: the deepest post-pit, F988, was 0.55m. The four settings retaining their stone packing ran along the east edge of the now filled ditch with F988 at the north end. There may have been two stages of development in the post-row, the first running from the corner of the barn down to the end of the gulley, the second taking the line to the centre of the later gateway.

When the yard wall was put up, the posts had been sawn off or dug out and, in most cases, where the wall line ran directly or partly over a post setting, that had been solidly packed with stone. None of the post-settings showed any repair or replacement during the life of the fence. Only the pairing of F963 with F962, and F985 with F988 suggests any need for maintenance.

The gateway of Period 3 was pre-figured in Period 2. The south side was represented by the last post and its northern edge was formed by the eastern return of the last version of F969, as that was not finally filled until the end of Period 2 (68). What is revealed is a passage, *c*.8m at the north end and 6.5m at the south, running from the entrance down to the two eastern bays of the barn. The comparative narrowness of the later gate makes the width in Period 2 seem generous. But the real width would presumably have been fixed by a bank thrown up from the ditches. These were of sufficient size for the spoil from them to have been a problem unless formed into a bank. The end of F554 and the edge of F675 were so close that there could have been no bank between them, but one on the south side of F675 seems reasonable and would have reduced the width of a track to 4 or 5m. The corner in the ditch-line moved gradually north-westwards (Fig. 16, [18]) and in Period 2 was almost square. The track suggests the location of at least one door into the barn. F969 in this period had a butt end just short of the barn's north boundary ditch. Later arrangements so damaged the west end of F991 that no sign was left of any entry into the barn.

The remaining elements of phase *a* consist of incidental details at the west end of the site. North of the ditch F813 was a slight north-to-south ditch, F774. It was 0.3–0.4m deep, 0.8m wide at the south end and only 0.4m at the last point it was seen to the north and seems to have disappeared in phase *b*. In the same general area and running west from the Droveway was a small undated gulley, F736. It is assigned to Period 2 because of a lack of activity here once the Period 3 plan was laid out, the area north of the new boundary ditch being largely abandoned until Period 5, and there being no grounds for placing F736 as late as that.

The dating of the rest of the ditches is reasonably consistent, each containing indeterminate second-century pottery along with material belonging to the earlier third century at the latest (68). The indeterminate material in the tops of Fs969 and 675 contrasts with the easily datable Period 1 assemblages (39)(51). The west ditch of the Droveway seems to have been kept clean with the result that there has always been difficulty in assigning it to its proper period. The changes marking the beginning of Period 3 led to the filling in of the southern end and there is little pottery which can be properly given to Period 2. However, some may have accumulated in the central part of the exposed length (69). The ditch itself was wider than that to the east: the earliest version was about 2.35m wide and not more than 0.6m deep beneath the surface of the stripped site. However, this could represent a Period 3 version. Recuts were few (Fig. 18, [26]) and it looks as though they silted up slowly, but heavy recutting in later periods would be undetectable because of the lack of relevant pottery: this became available in Period 5 by which time the west side of the Droveway had passed out of use.

A deep sump, F815 (Fig. 18, [27], [28]), had been sunk where F548, a Period 3 ditch, overlay the west side of the Droveway. The feature was remarkably straight-sided and square, *c*.1.35m by 1.35m, and seemed to have silted up naturally. Its shape showed no distortion caused by use and, as the water-table was penetrated during its excavation, it may have been a well. The easiest explanation is that a timber steyning was removed when the feature was abandoned. When the hole was dug is not known, but the pottery from this part of the ditch layout points to a date late in Period 2 and the Period 3 ditch, F548, ran across its site: it had gone out of use by the end of the period (69).

Phase *b*

Apart from F964, the rest of the ditches at the west end survived throughout Period 2. The features which belong to phase *b* marked the beginning of the main development of the small set of enclosures attached to the north boundary ditch. We have seen that some elements could have belonged to a set in phase *a* (62), (63), but the plan was not recoverable. In phase *b* the plan can be divided into two parts. F324, the eastern ditch, was possibly a replacement for F422 and ran south from the main boundary to return east to define an enclosure in the north-east corner of the main enclosure (70). The east end of F324 was lost when the Period 4 pond, F200, was made (Fig. 19, [29]). The ditch itself was recut and it survived into Period 3: the only odd feature was a post-hole, F472, sealed under the deliberate fills of both F324 and the north boundary ditch. The F324 ditch seems to have been recut later, the junction with F189 (Fig. 19, [30]) was explored to a limited extent and the fills here almost certainly represent the deliberate backfilling of the main ditch at the end of Period 3. The dating recovered was not helpful (L925, (70)), but the relationship of the ditch with the layout to the west makes it fairly certain that not all its fill should be given to Period 3.

The set of enclosures attached to the main north boundary ditch developed through both Periods 2 and 3 and into Period 4 which makes the establishment of the earlier stages a little difficult. Consequently, only those parts which, according to the pottery dating, passed out of use in Period 2 are discussed here and the fragmentary evidence only yields sense if it is assumed that there had been a coherent scheme.

A narrow enclosure aligned east-to-west, and represented by F440, with its entrance at the west end, persisted for a while. The initial enclosure may have been completely separated from the northern boundary ditch, but it had probably been joined from the beginning to F189 by a linking ditch, F446, recut in phase *c*. It may also have replaced a western doubling of F324, represented only by F436 and could have been later than the small enclosure away to the south-west. It was much smaller than the latter: if F440 belonged to an earlier version of F437, it would have been *c*.2m by *c*.14m. The pottery in F440 was generalised second century (71).

F440 was replaced by F439 and this would have widened the enclosure to *c*.4m. Unfortunately, the early version of F439 dealt with here had no dating, being largely cut away by a Period 3 ditch on almost the same alignment. The west end of the new ditch returned north leaving a wide gap between its end and the main boundary ditch. The earlier F440 may have behaved the same way. The next stage seems to have been to replace F439 by F556 in order to reduce the whole of this system to a series of enclosures having a basically common south edge: Fs511, 556, 324.

However, as the dating on Plate III shows, there are ambiguities and it may not be so much the last phase here as one of the first phases of the next period. Even so, considering the relatively plentiful dating available in Periods 3 and 4, and the developments in those periods, it is better to accept that an incomplete history of recutting was recovered and that the pottery dating of F556 represents the deliberate backfill at the end of Period 2. For instance, the pottery of Fs437 and 446 (72) has a marked phase *b*–phase *c* bias which does not seem to suit F556. However, Plate VII shows that the damage caused by furrows was extensive and no physical connection between F437 and F556 survived.

The dating of gulley F511 to the west suggests that it belongs solely to these later arrangements (73) and its layout supports the proposition that the whole of the series may have been more uniform. F511 cannot belong to Period 3 as it was completely superseded by the new arrangements of that period, being cut by ditch F481 (Fig. 19, [31]).

Taking all the northern enclosures into account, three or four small areas can be seen. The largest is in the north-east corner of the main enclosure and would have had a maximum area of *c*.250m^2, but it may not have been a single one as there is a short section of ditch damaged by later features running south from F189 which might have once divided it: F506 (74) (Fig. 19, [32]) whose final version cut the initial fills of F189. The course of this minor ditch was lost in a Period 5 feature but would have been interrupted by a group of pits, Fs236–240, 244 and 246. One of these contained pottery assignable to phase *c*, while the rest, where there was any material, had indeterminate second-century pottery (75). The features north of the main enclosure ditch which may have formed a track (see below) also suggest that there may have been a causeway across the main ditch. The latter had been heavily recut in this area, the only sector where this had taken place (58) (Figs 19–20, [33], [37–38]).

The next enclosure to the west, between Fs324 and 446, had an entrance east of F447 *c*.4.25m wide, with F436 as its original possible south side, and an area of *c*.70m^2. The last enclosure lay between Fs446 and 511 with the final south side being F556 and it is this enclosure which lay at the entrance of the elongated one once bounded to the south by F429 and later by F440. But it can be argued that the F511 boundary was formed so that something like the same area could be placed against the main northern boundary.

Further west was a short length of ditch, F635 (76), which may have belonged to this overall scheme. Other features lying north of the main boundary ditch, F189, also suffering from a lack of definition, belong at the earliest to phase *b* and will be covered here as little sense can be drawn from them in phase *c*. They included three post-holes, Fs474–476, almost entirely removed by the construction and fitting out of the large Period 4 rectangular building and the others lay east and west of that.

The east set of features consisted of a series of curved narrow and shallow gulleys, Fs228, 229 and 230 (Fig. 19, [33]), which look as though they defined a short and irregular track (77), and which may have replaced an earlier one of which only the slightest traces, including possible post-pits, could be seen (78). The northernmost of the gulleys appeared to be the latest. Under the north-east corner of the Period 4 building, were three features which may have formed part of a timber structure, but were too irregular and badly preserved for this to be certain: Fs521–523 (78).

Possibly better traces of a building were found west of the Period 4 structure. F450/589 was a slot only *c*.0.48m wide and *c*.0.1m deep in the north-to-south part, and *c*.0.42m wide by *c*.0.09m deep in the east-to-west (80). West and in line with F589 was a projection, 0.38m long, its darker fill suggesting the presence of a post-hole. A trace east of the furrow also suggested that F589 had been *c*.5m long. Further west was a gulley, essentially parallel with the slot, which returned west at its south end: F451/597, with a possible narrow entrance (81).

These features north of the boundary ditch may provide a context for the small pits there, also divided into east and west groups; some were small and shallow enough to be survivors from a structure rather than anything else. In the east group was a bed of clay, F321, interpreted as a hearth (Chapter 3.II, **14**), which was cut by a post set in a post-pit, F320; the only pottery in these placed them in Period 2 (82).

Phase *c*

The chief feature of this phase was Barn 2 (see Chapter 2). Its site was marked by two rows of major posts set in individual pits. At their north end lay a drier, a vat base and some other features east of the drier whose pottery shows that they belong to Period 3. The relationship of the post-rows with the ditches bounding their site to the north and east yields a plan of a barn with an additional bay at each end, aisles half the nave width (but see Chapter 2, pp 66–8) and its main entrance in the south wall. This was the only barn which did not ultimately have stone in its walls or its posts set on pads.

The dating of the structure itself came from undoubted repairs and the initial date of the first use of Barn 2 is provided by the pottery from the ditch, F65, on the east and north whose contents date from the later second century to some point in the middle of the third (83). This, read with the pottery from the later deposits in the main boundary ditch (58), indicates construction towards the end of Period 2. The small size of F65, 0.9m wide by 0.45m deep (Fig. 19, [34]), not only emphasises the quantity of pottery found but also makes the fairly close dating more secure (see Chapter 6).

F218 appears to have been a well from which the steyning, presumably of wood as there was little stone in the fill (Fig. 20, [35]), had been removed. It lay immediately west of the south end of the hypothesised barn wall. The feature was roughly square measuring *c*.2m along each side and was at least 1.8m deep beneath the surface of the stripped site. It penetrated well into the water-table and would have needed a prepared wooden steyning to be inserted quickly as the character of the natural was such that the archaeological section of the well collapsed almost as soon as the water-table was reached. The feature blocked F189 whose fills were banked up against the north well structure. That this had also happened on the south was less clear. The initial arrangement seems to have been a clay dam across the ditch (see Pl.III and Fig. 20, [35] for its tail), the fill behind this probably coming from the digging of F218. The clay of the dam sealed the base of the ditch which was stained green, suggesting that the ditch had received quantities of phosphate-rich liquid.

The pottery in its backfill only ran up to the middle of the third century (84). This might point to the feature having gone out of use at the end of Period 2, but a properly maintained well would have been cleared of waste and secured from rubbish disposal. Therefore, the actual dating evidence is likely to have been residual and the well may well have lasted throughout Period 3, especially when the probable use of the barn in that period is considered (see Chapter 9, pp 230–1).

If Barn 2 had a water supply, it would seem likely that Barn 1 also had one. Even though no candidate is datable to Period 2, F1094 south of Barn 1 deserves consideration. The pottery in it (125) belongs to the earlier part of Period 3, hence its assignment to Period 3. However, that material marks its disuse and it may have started out as a well in Period 2 and then been severely modified later. It is shown dotted on Plate III and in full on Plate IV.

The main ditch south of F218 was allowed to go out of use, but the stretch north of F218 was maintained throughout Periods 2 and 3. The sections (Figs 20–21, [36]–[40]) show this clearly, the dirty fills occurring only in this section down the west side of Barn 2 and just round the corner. They did not run under the large rectangular building of the Period 4 north range.

Barn 2 lay outside the main Period 2 enclosure in an area largely devoid of anything which can be dated to this period. The ditches north and east were for surface water drainage, the eastern arm extending to the estimated position of the barn's south-eastern corner. South of the barn and about 6.5m from the probable south wall line was an east-to-west ditch, F125. It almost certainly joined F189, but the junction was cut away by the Period 5 Sunken-featured Building. To the east the ditch became shallower and disappeared, but it seems to have set a limit to the area associated with the barn and suggests that that was only approachable from the east. The dating evidence from the ditch supports that from F65: the barn was in use towards the end of Period 2 (85).

All that remains to be dealt with are miscellaneous features, pits, odd lengths of gulley or ditch, occurring sporadically or in groups, mainly across the main enclosure. They all, either individually or as groups, add very little to the period plan. They are placed here because of their ceramic content and, in a few instances, their stratigraphical relationship with later features.

Beginning at the east end of the site, a single pit, F196, predates Barn 4 of Period 4. It cuts ditch F65 around Barn 2 and so could belong to Period 3, but its pottery is not more closely datable than as being of generally the same date as F65 (86). Next to the inner edge of the main enclosure was a group of pits, F261 (Fig. 33, [109]) and Fs337–8, with little pottery (87) and none assignable to the third century or later in an area where later sherds were plentiful.

Thereafter, only isolated features remain and are dated purely on their content, or are sag infills of Period 1 features (88). Lying out in the Main Yard between the east boundary and the small enclosure lay three pits, Fs402, 403 and 433, which could be better described as shallow scoops (Fig. 21, [44]). On excavation, F433 turned out to be a group of pits (Fig. 21, [41], [42]). The poor quality of the information yielded by the sampling showed that detailed work was not justified and, in this, the group was typical of most on the site. There were at least three pits here, the sections showing their general character. F510 up against the main enclosure ditch near the mouth of the Droveway was found inadvertently when the floors of the Period 4 building there were sectioned (Fig. 21, [43]). F601 further east was only partially seen and was also sealed beneath Period 4 deposits, as well as what was the final Period 3 fill in the main boundary ditch (Fig. 46, C). The dating of all was firmly Period 2 in character (88).

South-west of the last two pits was a large, very shallow one, F955 (89), cutting the phase *a* small enclosure. On its eastern side and running north was a narrow and shallow gulley, F947, 0.4m wide by 0.2m deep. This, were it not for the phase *c* pottery in it (90), could have been a remodelling of the phase *a* enclosure: had it been, more should have survived. South of the pit F955, an original topsoil deposit, F480, was preserved which contained only Period 2 pottery (91).

At the east end of Barn 1, the first of two pits, F1022, cut into a Period 1 ditch and was sealed by yard deposits of which the earliest belong to Period 3 (92). The other pit, F1134, to the south, again cut a Period 1 ditch, this time of the latest phase, and was itself cut by a Period 5 ditch (93).

In Period 1 it was reasonably easy to identify the centres of domestic occupation or their general location. In Period 2, there was no real evidence either in the form of structure or in the distribution of particular vessels of either glass or clay. The best location for part of the period may have been in Barn 1 where the cluster of small features may have been a product of domestic use, but Barn 2 contained no evidence. The slots and pits north of the main enclosure could point to domestic occupation here, especially if there had been a causeway near the north-east corner, but no plan can be reconstructed from the damaged features there. However, the weak focus for both samian and glass noted in the discussion of these in Period 1 (see above) generally in the north-east corner of the site might provide some further support for this area. The distribution of Period 2 mortaria also indicates activity in this area. The scattered distribution of samian in Periods 3 to 5 deposits in the same area suggests that whatever domestic focus might have been there had become so degraded after Period 2 that any better hope of identification was lost. However, living accommodation here might provide a reason for what seems to have been a very out of the way siting for Barn 2. Such a development would then allow Barn 1 to have served in phase *a*, for the shift to the north-east corner to have taken place in phase *b*, and would have been the spring for building Barn 2 nearby in phase *c*.

IV. Period 3, *c*.225/250–*c*.300/325

Pl. IV

Summary

As the site had become largely static, there are no identifiable sub-phases in the same way there were for the previous periods. Period 3 covers that part of the third century in which pottery dating poses problems. These are reviewed, as well as the difficulties of the overlap between the later third century and the early fourth, in the pottery report. Caution was exercised when the material was looked at, and some dates may be considered to be conservative. Those deposits, only datable by material which seems to run from the later third to well into the fourth century were carefully scrutinised and reduced in number, particular attention being paid to stratigraphical relationships and, where these failed, to the character and overall date of complete sequences. A distinct change in the wares taken to belong to the middle of the fourth century associated with the alterations which took place in Period 4 offers encouragement: Barn 2 passed out of use at the end of Period 3, the latest possible date for its use being somewhere in the early fourth century. Material sealed by the Period 4 rectangular building taken to have been put up at the same time as Barn 4 could run up to the middling years of the fourth century and, as the end-date could be set at 350, the final date could be expressed as 325/350. However, this would make Period 4 unreasonably short.

The main change in this period is the increase in the number of buildings with the use, apparently for the first time, of stone for all or part of their walling. With the creation of a walled area between Barn 1 and the new building to its north, the site developed two major functional areas. The old main enclosure is now called the Main Yard and the one between Barn 1 and the building to its north, the Small Yard. The change marks a consolidation of previous use as well as an increase in the facilities on the site.

Barn 3 was built in the second half of the third century, the dating coming from pottery in its structural elements and the features in and around it. There is no direct dating for Barn 1 and the building to its north, and the suggestion that both go together rests upon the overall character of the plan at the west end of the site. The date derives from the ditches which seem to have been dug to suit them.

The small enclosures in the Main Yard were replaced by a simpler system, new and larger ones being provided east of the Main Yard. Some may have been added to the south and west sides, but not apparently to the north. The new scheme had a new east ditch with Barn 2 opening directly into the Main Yard which now had a new south boundary. A pond was provided for the first time.

In other words, a recognisable farmyard was created. The north-to-south traffic route in Periods 1 and 2 was consolidated with the addition of metalling along the western side of the Main Yard. Much of this was in stone, but there were traces elsewhere of a gravel surface. There is nothing in the plan itself to suggest that the fundamental economy of the site had changed, only that its use was intensified.

The eastern side of the Main Yard

South of Barn 2, the whole of the boundary ditch was moved to the east, Fs202, 410, throwing access to the barn open to the Main Yard and showing that the door into the barn was most probably at the south end of the west aisle. The new ditch varied in width from 1.2m to more than 1.6m and had an irregular depth running from 0.45m deep to 0.75m in random hollows (Figs 21–22, [45–47]). The eastern edge was cut away consistently by Period 5 developments so divorcing it from the new enclosures to the east. As none of these was completely stripped, their sizes cannot be assessed. Recutting and modification in later times limited the amount of dating evidence, but the earliest material lies generally in the third century (94). The end-date of the pottery in the Period 2 east boundary (58) sets the earliest date for Period 3, and the difference in dating between that and the new one shows that the two could not have co-existed.

The south ditch of the new south-east eastern enclosure, F31, was at least 2.5m wide at the edge of the excavation and 0.85m deep (Fig. 22, [48]). If the width was important, the depth was less so as, further west, it had never been more than 0.6m beneath the stripped surface of the site (Fig. 22, [49]). There was, as earlier, no convincing evidence for a bank on one side rather than another. Irregularities in width and depth were general and can be seen in recuts of the north ditch where it turned north at the east end (Fig. 15, [13]).

Whether or not there had been an entrance in the east ditch of Period 2 (see p. 10), one was certainly provided in the new east ditch opening into the south-west corner of the central enclosure. The gap was *c*.4m wide and had a proper gate structure (Fig. 8). The posts belonging to this, Fs378–384 and Fs391–398, seem to have been replaced on several occasions and have all been assigned here, as only one post produced any pottery (95). An arbitrary

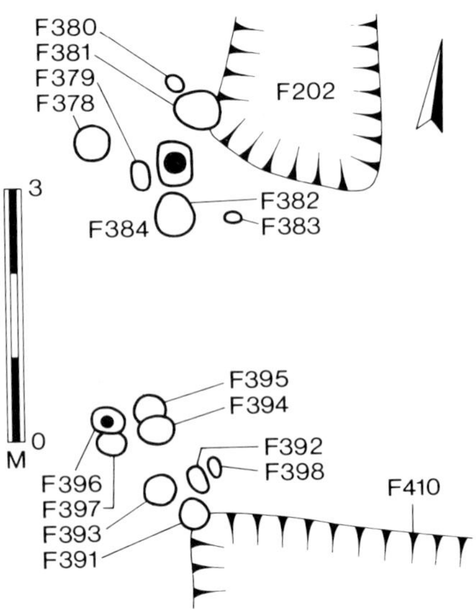

Figure 8 Period 3: gateway into the eastern central enclosure.

division between Period 3 and Period 4, let alone Period 5, is not advisable.

Both groups of post-holes lay at the bottom of a furrow, hence not all may have been preserved, the depths recorded varying between 0.06 and 0.22m. As at least 0.25m had been lost in the furrow, and allowing for the original topsoil, it should be assumed that the recorded posts had been deeply founded. The only evidence for periodisation occurred in the southern group: F394 cut F395; F396 cut F397. Out of the fifteen, evidence for posts survived in three, and only in one case, F396, was it clear that limestone pieces had been used for packing. However, one hole in the northern group, F379, had held a post which had not been a plain upright. It had inclined to the west and may have been a brace of some kind. If so, the lack of a post-setting to suit this may show that several post-holes were eradicated in later ploughing.

The spread of these two groups on either side of the entrance and out to the west may point not to simple paired posts but to an arrangement with short fences across the ditch ends designed to prevent an animal from turning and falling. If this was so, it would imply that any bank or hedge lay east of the ditches. F28 (Fig. 23, [50]) (96), as a major feature, came to a clear end about 3m from F202 possibly marking the inner edge of a bank down the east side of the enclosure. The small gulley running across the gap between the two larger ditches, 0.4m wide, was counted as being part of F28 during excavation and produced a single sherd of basically undatable pottery (97). The gulley may have had no lasting value, having been dug as a marker during the laying out of the system. Normally, such a marker would have been dug away during the making of the finished feature, but in this instance a trace could have survived, if it had been sealed under the upcast from F202: its fill consisted of redeposited natural with a lot of gravel and lumps of unbroken natural.

A bank on the east side of F202 would have made the management of beasts inside easier. The undated pit lying in the gateway, F399 (98), may have been for a post used in conjunction with hurdles to ease handling animals. A hint of a bank on the north side of F31 (Fig. 22, [48]) should be discounted as one there would have compromised the entrance. An area of ill-defined gravel (99) in the entrance area may have been metalling. The fill of the Period 2 boundary ditch was consolidated so much that the resultant hollow filled up with soil containing pottery generally of the fourth century (100) and may be evidence for intensive use.

Barn 2 carried on in use to the end of the period, its replacement being taken to define the beginning of Period 4. The new ditch system south of the barn seems to have excluded its original drainage ditches which continued to silt up. By the end of the period they had lost all effectiveness (101) (Fig. 19, [34], Fig. 23, [51]). However, a new ditch, F11, running away to the east, was introduced. It was more vertically sided and flat bottomed than most other ditches, being 0.9m wide and 0.40–0.45m deep and may have been recut, possibly in Period 4 (Fig. 23, [52]). The picture was confused by furrows and at least one Anglo-Saxon pit. Its dating will hardly allow it to be later than Period 3 (102) and no evidence for a way through its line was found. Not all of it lay within the excavation and, if read with the north return of F77, would yield an enclosure some 17m by 21.5m with an area of 365.5m^2.

The structure of Barn 2 was probably refurbished now as the very small amount of pottery found in the fills of two post-pits (103) should have derived from the use of the barn: it is dated *c*.250–*c*.300. Only features sunk into the ground in the barn were found and all lay at the north end. They consisted of a drier, F156 (104) (Fig. 48, Chapter 3.I, **1**), cut through by a Period 4 ditch; a set of what are interpreted as hearths or features once associated with conventional ovens, Fs258–260 (Chapter 3.II, **15–17**), without dating evidence (105); and what looks at first sight to have been an oven, F105 (106) (Chapter 3.II, **18**), but which probably supported a vat.

The whole formed a tightly knit group all associated with fire and betraying partial sequences except for F258 which had no stratigraphical links. The other three were sited around the eastern perimeter of the drier. F260 was tucked into the angle between the main part of the drier and its stoke-hole; F259 partly overlay the wall of the stoke-hole (Fig. 53); while the hollow in front of F105 was carefully angled to pass round the corner of the body of the drier. The only other peculiarity in the barn was the care taken over the southernmost post in the east row and, in this respect, there was a marked similarity with the precisely similar post in Barn 4 (Chapter 2, p. 58, p. 63).

The earliest date for the drier, provided by the material from its construction, is in the second half of the third century. Following the discussion of the dating of 'developed' driers in Chapter 3, it looks as though this is one of the earliest known. Its introduction probably marks a change of use. Only the final fills of the vat support, F105, had any dating and this falls in the latter part of Period 3. The close siting of all these features could argue for a close association in date and, as none can be given to Period 2, the change may have been quite marked. The pottery from the Period 2 boundary ditches was generally domestic, while that of Period 3 was mainly culinary. This again suggests a change in function in this area and one which is reflected in the Period 4 layout.

The north side of the Yard and the Droveway

The original ditch of the Main Yard had no clear end by Barn 2 and the evidence is that it had been regularly cleaned out where it turned away west from the side wall of the barn. The lack of clarity further south may well be due to the continued survival of the well, F218. Although some silting in the ditch took place (Fig. 20, [36–39]), the small amount of pottery recovered suggests that there was little rubbish around. The bulk of the deposit in the ditch was due to deliberate backfilling in preparation for the buildings put up in Period 4 (107). Again there was little sign of there having been a bank except in one section (Fig. 23, [53]) which might indicate one on the south side.

The faint traces of possible buildings north of the boundary ditch noted in the Period 2 discussion (p. 12) were repeated in this period. The slot, F504, was accompanied by pits, Fs304–305 and 527–528, and the trench F471 (108). Pits Fs303–305 cut through a Period 2 feature, F228, and were, in turn, sealed by the last versions of F189 (Fig. 19, [33]). None of these formed any coherent plan and the extensive remodelling of Period 4 and the degradation in Period 5 did a lot of damage. The three post-holes in this area assigned to Period 2 (82) do not easily suit the slots near which they lie.

F471 was a steep-sided ditch or trench 5m long (Fig. 23, [54]) containing a fair amount of limestone rag, but no evidence for any structure. The bulk of the fill was very dark and silty, but without a sign that it had accumulated gradually. F504 was 0.2m wide and 0.15m deep with a basically rectangular section towards its south end. It looked very much like a beam-slot. Its eastern edge was regularly formed, the western less so and more difficult to define. North of F471, it narrowed to 0.15m and became less rectangular in section. It was cut by F471. The north end was complicated by F528 which looked like another beam-slot, but, although definable as having been 0.3m wide and 0.05m deep, it had less definite sides. One end lay just west of F504, the other was lost in a furrow. If these had belonged to a building, its plan was not recoverable.

South of the main ditch and immediately east of the mouth to the Droveway lay a badly damaged spread of metalling or topsoil (109) firmed with limestone, preserved where it had sunk into a fill in the Period 2 main boundary ditch (58). Not only does its presence go against there having been a bank on the south side of the ditch, but its character also suggests fairly intensive traffic nearby.

In the Droveway new arrangements had come into being (Fig. 9). These can be divided into two types. Firstly, a set of shallow scoops which may have been for posts set either singly or in groups. None was effectively deeper than 0.15m, and all had undifferentiated fills mainly compacted unlike those of Period 2. All were less than 1m across and tended to form a north-to-south line. As only three contained any dating evidence (110), there is little guarantee that they should all be put together, but has been done because they had the same character. They have been separated from those assigned to Period 2 because of their differing plan. Their dating, however, suits either Period 2 or 3. The layout of the features suggests the presence of a reserved area along the east side of the Droveway.

Secondly, there was a set of shallow pits with outliers (111). Two, Fs706 and 707, were not ordinary rubbish pits and therefore the rest, Fs730, 731 and 703–705, may not have been either. F706 was cut by F705, had its sides lined with clay, the bottom being of natural clay. Its plan was triangular, being 2.5m long and 1.9m across the base. It lay across any possible entrance between the Period 2

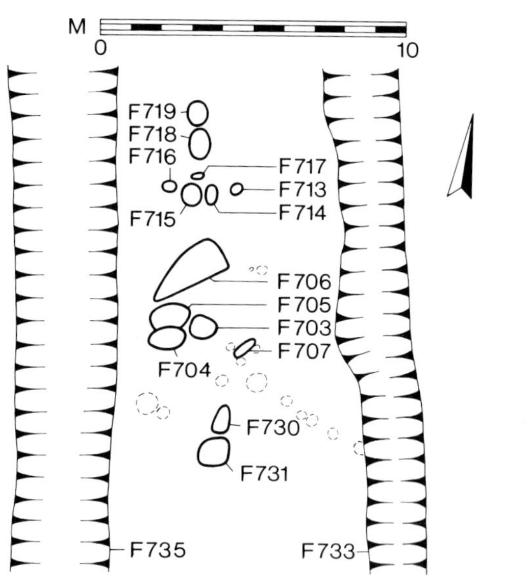

Figure 9 Period 3: features in the Droveway, Period 2 features in broken outline.

post-holes and the other features given to Period 3. Its fill was undistinguished, but contained limestone rag: the use of this in quantity dated only from the beginning of Period 3. F707 was much smaller. Although it was next to some of the Period 2 posts, its fill and burnt character suggested that it was most unlikely to have been in use when they were. F706 might have been some form of water tank, but F707 should have been a hearth (Chapter 3.II, **19**).

In the mouth of the Droveway were ditches associated with post-holes. The dating of these leads to some doubt about their correct sequence. The signs were, however, that the overall layout in the Droveway began with a set of post-holes and ended with the ditches or gulleys and posts at its entrance. The pottery indicates the order as the ditch junctions with the west Droveway ditch were lost in the large hollow which developed there in later periods and which was itself cut by later ditches (Fig. 18, [27]).

Ditch F810 may have come first, as it contained material belonging to the second half of the third century and this also occurred in a single pit to the south, F643 (112). The latest form of F810 showed that a narrow gap, about 1.75m wide, had been left at its east end. As the east end of F810 was merely where the bottom ran up to the stripped surface of the site, it may once have run through to the east Droveway ditch. As uninterrupted access from the Droveway into the Main Yard is implied, the first gap may have been on the west and this would suit the bias suggested by the post-hole layout in Period 2.

No posts associated with F810 were found. However, there were posts set in the ends of the gulleys to the south. The original line of these was L-shaped, springing from the western side of the Droveway and returning east to leave a gap of about 6m between it and the east side of the Droveway. The original gap may have been narrower, but later developments prevented confirmation of this. The undated F659 post-hole (113) may have been associated with this entrance. The feature lay under a furrow, thus removing at least 0.2m–0.25m of its depth, at the approximate mid-point of the suggested entrance and was 0.7m north-south, 0.4m east-to-west and only 0.15m deep. Its shape, along with the limestone rag in its fill, suggested that it was more likely to have been for two posts than to have been a pit. Returning to the narrow gulleys, a wide and deep furrow made the development difficult to understand. The final form, F771, provided only a short wing running due south from the junction of F801 and F548. The dating of the pits immediately to the east suggests that it may have lasted longer than either the final system used within the Droveway or F810 (114).

As for the rest of the L-shaped line, the earliest version may have been F680 which, largely cut away by F652, was only 0.1m deep. F652 was more substantial although only about 0.45m wide. Apart from a dirty redeposited natural in the bottom, the bulk of the fill was made up of deeply set limestone fragments which, taken in conjunction with the post-hole sequence Fs653–654, may have been infilling a slot from which timber-work had been removed. F654 was the earlier of the two post-holes at the east end and was set in F652. The post was replaced by F653 which had also been removed, the resultant hole being stuffed with limestone. The impression given is that something substantial was removed when the east end was shifted some 3m to the west. However, F653 may have been contemporary with a second version of F659 to provide two equi-spaced posts in the widened entrance.

The new version, F661, was V-shaped in profile and about 0.6m deep, like the original feature. It contained less limestone and the backfill was redeposited natural. There were two post-holes at the south end, F681 and F683. These survived as imprints in the base of the final fill which, because of its dating, has been assigned to Period 5. No packing survived and it is only the similarity in siting with the post-holes to the east which places them in this period. Their fill would mark their disuse as belonging to Period 5 (see p. 37 and (295)). The fill of F641 to the east was similar, but dated to Period 3 and it may have been sag produced by consolidation of F652. The final version, F771, was very different in character being a ditch, on average 1.5m wide and 0.7m deep. Again a furrow had destroyed the relationships with the features just discussed, but its dating was clearly later than the Period 3 element in those. However, as it had been an open feature, this is to be expected. The possibility exists that what has been described here as a reasonably coherent, if largely not understood, sequence runs through from Period 3 to Period 5, and that the yard wall here belonging to Period 4 actually had a major entrance all trace of which was ploughed away. The presence of coin C37, dated 330–340 (see Coins), in F652 shows something of the ambivalence in the sequence and the latest arrangement is in the description of Period 5.

Three pits, Fs642, 644 and 682 (115) lay in the area between these features and F810 to the north. Their sizes and filling did not suit their having been post-holes. However, their chief characteristic was their extreme shallowness, their depths varying from 0.13m to 0.2m. This was not due to deeper medieval ploughing, for all three were sealed by a ridge.

The enclosures in the Main Yard

The emphasis of arrangements in and at the mouth of the Droveway was on deflecting traffic to the east towards the ditched enclosures next to the north boundary. The pond, F500 (116), shows that beasts were let into the main area. Whether animals were allowed free range is open to question as there was also a new well, F254, out in the open. The optimum size of the pond seems to have been about 17m by 14m and it was hardly more than 0.75m deep (Fig. 24, [56]). The pond was deep enough to penetrate both the winter and summer water-table levels, hence a water-supply was ensured.

Turning to the small enclosures to the north, the impression is that the north-eastern one established in Period 2, F324, continued in use, the ditches finally filling up in Period 3. The junction of F324 with F189 showed a homogeneous material running from one into the other (Fig. 19, [30]), almost certainly deliberate backfilling in preparation for Period 4: part of F324 had been recut and only became filled in Period 4 (Fig. 24, [57]). The evidence suggests that F324 continued to define an area in the north-east corner of the Main Yard.

The digging of pits in this area, noted in Period 2 (see above), continued (117). There may have been many more pits than are shown on Plate IV: F359 was entirely sealed under the Period 4 pond and the excavation trench which revealed the pit also produced a post-hole (Fig. 23, [55]). Along the eastern edge of F324 was a series of post-holes. Dating evidence was either entirely absent or ambiguous: they could all have belonged to Period 2 (118). The row ran north-to-south from the main boundary ditch, F189, to a point in line with F437, the northern boundary of the enclosures to the west, which itself passed out of use during the early part of Period 3. The posts may have been a part replacement of F324 (Fig. 24, [59]) and so may have dated to the later part of Period 3 or even into Period 4 (see p. 23), in which case any association with what lay to the west may be mistaken.

The post-holes fall into two classes: those with evidence for posts, and those without. Of the former, the stump of F423 had rotted in position, that in F424 had been rocked out. F427 lay in the same line as these two, otherwise the others did not conform easily with a single fence line. How the rest of the line, had there been one in Period 3, behaved is open to conjecture as any extension to the north would have been cut away in Period 4.

The enclosures to the west were both simplified and enlarged. The last vestiges of the earlier site were finally filled up (119) while the latest east boundary was recut to suit a larger, single enclosure detached from the boundary ditch. The south side was F438 and the west side ran further north than the east finishing *c*.2.5m from the boundary ditch (120): this was either an entrance, or the toe of a bank. The east side returned west for about 3.5m and, if the plan is to be read with the post-hole row, or F324, further east, an 'L'-shaped entrance area some 4m–6m wide emerges. The relatively simple scheme offered here seems to have been the result of development as a short length of east-to-west gulley, F578, was cut across by later Period 3 activity. The dating in the gulley pointed to an early stage in Period 3 (121). F578 may have been laid out in conjunction with a version of F437, but furrow damage and Period 4 alterations destroyed the overall layout. However, F578 may have formed part of a trackway drawing traffic directly from the Droveway.

The next stage was to expand the enclosure both south and west, the new south side cutting part of the developing fill in the pond, F500 (Fig. 24, [56]). The newly laid out plan remained largely static, except for the east side. It seems that this had always been shorter than the west end, leaving a gap of 13m–16m between it and the main boundary. The dating from the enclosure (122) shows that the new plan was more-or-less maintained until new arrangements were needed in Period 4 (Fig. 24, [60–63], Fig. 18, [25], Fig. 25, [64–67]). As with the north-eastern one, pits were dug in the enclosure (Fig. 25, [68]). The pits themselves contained little to indicate what they were for; any rubbish content must have been organic (123). They do not mark the disuse of the enclosure as they were occasionally sealed by the changing ditch layout (Fig. 24, [63]).

The Small Yard

Although the changes in and around the Main Yard were important, the replacement of the untidy Period 2 scheme on the western side with a completely new plan would have been the most impressive. Barn 1 was given stone walls. North of the barn, almost parallel and with ends coterminous with that, was a new long building lying *c*.37m away. Between the two was a walled yard about 850m^2 in area. This is the Small Yard mentioned in the summary above. The east wall was best preserved and was apparently inbuilt with the new building, but the junction with the barn was lost in a deep furrow. There was also a west wall represented by a robber trench running away from the new building, but otherwise lost in a furrow.

In the centre of the east side was a gateway some 2.3m wide between the foundations: it may have been wider above ground. The gate was presumably double-leaved, each leaf hung from a large pier 1.7m east-to-west by between 2.2m and 2.5m north-to-south. The gate was probably set at the east edge of the opening allowing each leaf to be folded back against the pier without projecting beyond. The size of the piers and the wall suggests either a timber lintel or an arch over the opening, the piers being big enough to take the thrust of the latter. There was probably an opposing gate in the west wall matching the layout in the yard at Whittlebury (Pretty 1852, 107–14). F898 and F899 in Period 5 (309) may have marked the approach to the gate.

General wall foundations were shallow, except when taken down to the bottom of earlier ditches (Fig. 17, [20]) in the north building. Even so, this was not a consistent feature, but as the major Period 1 phase *a* ditches were so treated at the internal wall junction, the superstructure was probably of stone to the eaves.

The function of the new north building was probably to provide living quarters (see p. 70) and it is referred to from now on as the House. Nothing can be said about its quality as only a few footings along with robber trenches remained. Later periods and ploughing had effectively planed away all floors or internal arrangements other than the single cross-wall which was built as substantially as the main walls which everywhere were more deeply founded than those of Barn 1, itself better built than Barns 2 and 4.

No direct dating is available for all this new building: the yard wall produced one deposit showing that it was later than the early third century and the rest were less useful (124). There is actually no evidence that the stone walls of Barn 1 were of the same date as the rest, but it is hard not to see the new western layout as part of a single scheme.

The best indication of the date by which the barn at least had been altered is given by a feature excavated by Mr Dakin (MF6) (Fig. 35) and destroyed in the process: a brick-lined drain in the south wall by the third pier from the east. A gulley ran through it from some unknown point inside to a large sump, F1094 (Fig. 26, [69]), which may have begun as a well belonging to Period 2 (see p. 13). Half the sump remained to be excavated and, although the bottom four layers were sterile, the dating recovered was consistent. It had been filled by the end of the third century (125) and the drain should have passed out of use by then. As it had been integral with the stone wall, that was in place well before the end of the century.

Any Period 2 well provided for Barn 1 was replaced by another to the north in the Small Yard where, in the south-east corner, there was a sequence of wells. The original construction deposits were badly damaged by each succeeding remodelling leaving, after the demise of the site, a large cluster of deep holes filled mostly by destruction products. These were masked by yard deposits and the mud generated in Periods 4 and 5. The wells were only sampled and the full story of their development was not completely recovered, partly due to the collapse of one of the main sections. While one may have been sunk through the defunct F969 (126), the evidence was far from conclusive and the first well on this side may have been F1016 (127). This was built of stone and was, as F254 seems to have been, of the normal Nene Valley pattern in being square and based on a timber frame, which was all that remained after most of the lining had been removed, probably for reuse in the new Period 4 well, F1052. The dating of F1016 places it in the second half of the third century.

The foundations of the west wall of the barn cut into the fill of the ditches assumed to have been coeval with the Period 2 plan. Here again the date is third century (Fig. 16, [17]: F1017). The new west ditch was becoming full during the second half of the third century (128). This ditch seems to have been a replacement of another slightly to the east, ending short of the south-west corner and passing out of use in the second half of the third century. East of the sump was a shallow trench, F1121, and a pit, F1113 (127) both of which ceased to be used at the same time and provided, along with F1094, one of the best groups of pottery from the site (see p. 143). A furnace was salvaged just outside the formal excavation and this passed out of use at the same time: F1092 (128) (see Chapter 3.II, **20**, p. 82). There was no sign of any structure which might have contained it. The disuse of these features mark a change of use in the barn at the end of Period 3 and the change of function at the beginning of the period may be related to the building of a new barn, Barn 3.

West of the Small Yard were some new enclosures which may have been laid out in Period 3. Only the ends of their ditches lay within the area of formal excavation. One ran up to the north- west corner of Barn 1 and cut the new drainage ditch down the west side (131) (Fig. 16, [17]: F1006); it was undatable beyond that, but as the barn's ditch seems not to have run north beyond it, it may have been a renewal of a ditch first laid out now.

The other new ditch, F777, was dug to end at the north-west corner of the House and its dating (132) at latest is third century. It was soon replaced by another, F548 (Fig. 14, [5]), which ran past the House to join the Droveway and so finally doing away with F554. That had been the end of the west Droveway ditch and may have been retained for a while, the House and Small Yard being laid out immediately to its west. However, the upcast from F548 was used to block and part-fill at least F554–F772 (Fig. 18, [27]), and F548 was itself of the same date as F801 whose line it continued. Dating from F548 begins within Period 3 and runs on, although there was not much pottery: rubbish from the House seems generally to have been disposed of elsewhere (133). As F548 could only have been laid out after the House and the early ditch F777 had come into being, a third-century date for the House is appropriate.

The metalling both down the west side of the Main Yard and in the Small Yard also helped to provide a date for the plan in this area. The metalling was badly damaged by furrows and modern ploughing, the latter removing much of it from the ridges; it was often only preserved where it had sunk into the consolidating fill of earlier features (Fig. 16, [14]). It was not appreciated during excavation what these stone-laden layers really were and this led to a false impression of the dating of the earlier periods on the site (Mackreth 1978, 210–14). In the Main Yard the surfacing was of strews of stone and was best preserved where it had run up to the Small Yard. There were two separate spreads, the lower datable to the third and early fourth centuries. As the material found could only have accumulated after the stone had been laid, the dating must represent the time during which it was in use.

It was from the mire beneath patches of the second metalling that the sherds used for dating came. The date is reasonably uniform and fits with that of F548, belonging essentially to the third century and running into the early fourth. The second metalling belongs to Period 4, but there were hints here and there that some repairs had taken place before then and there were isolated signs of activity not necessarily compatible with a widespread hardstanding: F1047 (Chapter 3.II, **21**) (134). Elsewhere, a clear distinction could not be made (Fig. 17, [20] east end, L1624, and Fig. 16, [15], L1624, west end, L1564) and sometimes the metalling was detectable only as a general layer of scattered limestone fragments (Fig. 16, [18–19]).

Barn 3

The barn (Chapter 2) was built north of the mid-point of, and parallel with, a new ditch along the south side of the Main Yard which, although the proof was lacking, was probably contemporary with the new east side ditch. Little survived inside the barn of either floor levels or features. It was demolished and cut about by ditches in Period 5 and then later by furrows before the headland which sealed it developed. The dating of the barn places it unequivocally in the period *c*.250–300 (135). The new south ditch, F1089 (Fig. 26, [70]), yielded pottery from its lower fills pointing to a date somewhere in the second half of the third century (136).

Only the slightest traces of features inside the barn were left (Fig. 41) and it is not always clear to which period, 3 or 4, or even 5, they should be assigned. For convenience, they are dealt with here. Firstly, there was a gulley, F1139/1142 (137), which could have belonged to the barn, if that had been of timber alone initially (Fig. 26, [71]), and could have been a drain even though it had not been steyned and there had been no specialised deposits in it. Its course was neither regular nor of even section and it may be that there had been two ditches draining west and east respectively, or that F1042 was a recut, as it seemed to have an end in the third bay from the east. The stone version of the barn could be assigned to Period 4, a time of extensive building in stone, but is basically considered here as it had a room added after the yard wall was run up to it and the latter is definitely of Period 4.

It is only possible to detect one major door, but others certainly existed in Period 4. The evidence for a large entrance lay at the east end of the nave and consisted of a spread of stone (138) just over 3m wide along the face of the wall; it continued for a distance of 3.5m into the interior where it had a rounded end. As there was no evidence for a general floor level and the stone spread was clearly observable despite having no proper defined edge, it is argued that this was a firming up of a general earth floor inside a double door to provide a firm standing for something as substantial as a loaded cart.

The only area within the barn which may have had a floor, a stone spread rather than pitching or flagging, was the west end (139). It may once have extended further east but, if so, had been eroded in Period 5: there was sufficient stone in the top of F1142 in the centre of the building for it to have derived from a floor. It was best preserved along the south wall with no satisfactory connection with the only major internal feature, a 'furnace' base, F1135 (Chapter 3.II, **22**). There was a less substantial one in the middle of the barn, F1163 (140) (Chapter 3.II, **23**).

F1135 sealed a post-hole, F1136 (141). There were others (142), but they made no particular sense either in relation to the building or to each other. Some were sealed under the stone spread at the east end which may mean that any door here was secondary. Along the east part of the south wall was what looked like a base of a bench *c*.0.5m wide, F1170, with a stone edging; it ran from the corner to near the line of the north aisle posts. There was a better built one, F1160, at the west end of the north aisle (143). These may seem too substantial for seating and could have been the bases for work benches. The upper layer contained some general fourth-century pottery, but the lower can be dated to the second half of the third century.

At some time in the life of the barn, a wall, F1169, was built across the nave between two of the main piers, separating the two eastern bays from the rest of the barn. The wall was undated (142) and was almost certainly secondary and may have continued across the aisles: it was preserved only where it had been cut into the fill of F1142. The wall presumably represented a division of functions in the barn. No reason for the change in the position of the second pier from the west in the south row suggests itself. It was replaced slightly to the north-east of the original position, F1162 (145).

The Period 3 plan shows that, apart from the Droveway coming into the Main Yard from the north, there was another main line of access from the south. The actual entrance lay almost certainly towards the south-eastern corner of the Main Yard and outside the excavation, if not actually in the corner itself. This is indicated by gravel metalling along part of the east margin of the Main Yard and the position of Barn 2 in the corner of the redesigned yard with its entrance facing south. Barn 4 of Period 4, the replacement of Barn 2, was sited in the same corner and was also approached from the south suggesting a well established route, and this would suit an eastern entrance into Barn 3.

The new barn was accompanied by a new well, F254 (146) lying about 17m north of the barn. The structure of the well had been completely robbed out (Fig. 26, [72], drawn before the collapse of the section) except for the square wooden base-frame for the stonework. The date of the destruction of the well is hard to assess as the only pottery in the backfill was almost certainly residual. It was, at the earliest, near the end of the third century. Its elimination would suit the major re-organisation of the site at the beginning of Period 4. The construction date is fixed in the second half of the third century.

Two features lying south of the F254 well, and largely cut away by medieval furrows, seem to have been associated with post- holes. The impression given by these features is that there may have been a building here perhaps supported on posts and sill beams. They formed two approximately parallel lines, on average 5.5m apart internally, both ending to the north on the same east-to-west line. The length of the east side, F1080/1046, was 7.75m to the centre of the post here, and the west side, Fs1068–1069, was 5.4m long. F1080 was more or less uniformly 0.6m wide and had a clear rounded end. The post-hole, F1046, lay north of the north end and cut the construction pit of the well. There was no evidence for a post and what was left of the trench or gulley also bore no sign of a structure in it. F1069 was the same depth and width as F1080 and, had the furrows not removed a great part of their volume, each would probably have been seen to have had a minimum depth of 0.35m. In the case of F1069, its post-hole, F1068, was set in the south end of the gulley or trench. The north end of F1069 was not clearly defined, the bottom of the feature rising until it merged with the top surface of the stripped site. The material from them was meagre and consisted of sherds belonging generally to the first half of the third century (147). However, as F1046 cut the construction of F254, that at least must be Period 3 or later. Putting them all together, they could have formed something like a well house. This was the only time during the known life of the site that there was a completely isolated well and it might be that some special provision was called for.

Between Barns 1 and 3 were three pits of which F1109 was more a large delve, not unlike F955 in Period 2, and contained pottery of the first half of the fourth century; it could have been dug in Period 4. The second, F1108, lay slightly north and its pottery had a more third-century bias. The third pit, F1123, was cut by the yard wall of Period 4 (Fig. 17, [21]) and contained pottery of the late third and early fourth centuries and, as such, provides some of the evidence for the terminal date for Period 3 and the beginning of Period 4 (148).

The new south ditch disappeared out of the excavation in both directions. What happened to the east is unknown, but it may have been F1101 at one stage and had, therefore, a butt end at the south-west corner of Barn 1 (149). The ditch, F1099/1100, west of the barn, ran on to the south and seemed to be inclined to the east when it left the area of formal excavation, and may have been a secondary version of the main south ditch — it contained later pottery than F1101 (150). There was in fact a succession of ditches (Fig. 27, [73]) and this matches the recuts found in the main south ditch, hence, despite the *lacuna* between the two areas, it is probable that we are dealing with the same ditches in both areas. A trial-hole outside the formal excavation and between the two parts of the system revealed part of a series of recut ditches containing only generalised third-century pottery (136). The position is unclear: a sample of a Period 5, F1114, ditch further to the south-east located an earlier ditch with early third-century material whose line did not conform with the Period 3 ditches (151).

The only sign that the area south of the new southern boundary had been divided into enclosures is provided by the single butt end found just west of Barn 3: F1154. This cut the developing fill of F1089 (Fig. 26, [70]) and its dating matched the others of the period, running as it did from somewhere in the second half of the third century into the early fourth (152).

The House would have been the probable domestic focus on the site for this period, if it could be certain that it had been one. If mortaria were used mainly for food preparation, the major concentrations were along the north side of the Main Yard and opposite the faint traces of a timber building north of the boundary ditch. Another lay east of the House and west of the Droveway, and there was a smaller one between the two barns on the south side of the site. Three possible *foci* are present and all three in their ways represent uncharacteristic buildings, or in the case of Barn 3, a barn whose internal arrangements were unlike those of the other three. The main focus was to be represented in broad terms in Period 4 by the large rectangular building in the north range whose construction may have destroyed almost beyond recognition an important element in the Period 2 and Period 3 plans. The second was associated with the House, even if it lay outside the Small Yard.

Finally, during the period, various earlier ditches acquired pottery in the hollows created by the consolidation of their fills (153), and vestiges of topsoil receiving pottery during Period 3 were as rare as they were in all other periods (154).

V. Period 4, 300/325–*c*.375
(Pl. V)

Summary
The date for the beginning of the period comes from the deposits marking the end of Period 3. The end date has been set at *c*.375 because of the bias of the pottery occurring in the later deposits in the period (The Pottery, pp 160–4, pp 174–7).

The principal changes in plan were the creation of the north range of buildings and the building of a boundary wall round the western part of the Main Yard, but not all were put up at the same time. The second north-east barn, Barn 4, and the great rectangular building to its west formed a consistent element of a north range even though they were not joined: both had parallel axes, the rear walls of each appeared to be in line and the front walls were sufficiently so for the proposition to stand.

A building west of the rectangular building seems to have been secondary as the south-west quoin of the latter was built with no allowance for a continuation to the west, though the turning of a quoin may only mark constructional convenience. There had been a building south of these two, but only one wall survived. No connection survived between the north yard wall and the building attached to the large rectangular one, but the west end, if projected, overlapped the end of the Period 3 House. The same applies to the south yard wall in relation to Barn 1. What happened at the junction of the new wall with Barn 3 is unknown. The only other structural innovations were the addition of a room to Barn 3 and the provision, possibly, of pentices there.

The Period 3 pond was backfilled and replaced by another in the north-east corner of the Main Yard. The minor enclosures of Period 3 in the Main Yard may have completely passed out of use leaving only one ditch running away east from the old Droveway which also ceased to be well defined.

Relatively few layers are allocated to this period, partly because many relevant deposits occurred at or near the surface of the site. They were either very badly damaged by the plough or could not be disassociated from Period 5 activities. The latter point is borne out by the abiding impression that a lot of the material in the much greater number of layers assigned to Period 5 came from earlier features which had become degraded.

The east side of the yard and Barn 4
Alterations dating to Period 5 show that the ditched enclosures must have survived through Period 4 although next to no pottery can be allocated to them (155) (Fig. 22, [47]). The ditches became largely filled with silt and there was no focus of activity nearby to generate rubbish which could be dumped there. The continuation of the east side of the Main Yard to the new barn is a sign that the older

eastern enclosures were still in use and the section of that (F164, Fig. 27, [74–75]) suggests that the older ditches had become much shallower. The gate (Fig. 8) into the central enclosure was not eliminated until Period 5 and it is likely that some of the post-holes, discussed in Period 3 (see above) and largely undatable, should be assigned to Period 4 (95).

F102 (156), south of the south ditch of the main enclosure, was more of a narrow slot than a gulley. It averaged *c.*0.35m wide by 0.3m deep at the east end but became shallower to the west, where it was cut by the Period 5 grave, and finally petered out after showing a tendency to turn south. To the east, it disappeared into F31 which seemed to have silted up so sealing the filled slot. Its function is not known: but close to where it ran into F31 one post-hole was seen and this could have been all that was left of an upstanding barrier around the south enclosure.

The replacement of Barn 2 by Barn 4 (Chapter 3) shifted the nearest centre of activity to the north and the eastern boundary was run across the abandoned site of Barn 2 up to the new building. There was more pottery in the new ditch, F164, than in the rest of the boundary (157). That it was not all residual, deriving from the demolished barn, is suggested by the shift forward in the dating, the lower material belonging to the middle of the fourth century.

The state of the main posts of Barn 2 may have been a reason for its replacement — this was the only barn whose posts had not, at a later time, been put on pads — but the trigger was probably the creation of the large rectangular building. Direct continuity of function between the two barns can be demonstrated. The drier in the old barn was replaced by a double one in the new. The old well was replaced by a new one, F172 (158) (Fig. 27, [76]), square in plan with stone walls based on a timber frame. The construction deposits contained pottery dating only to before *c.*200. This could be a sufficient sign that there was little pottery use in Barn 2 in Period 3 thus enhancing the value of the sherds found in the ditch across its site (157). The one feature which was not seemingly re-established inside the new barn was the vat base, F105. A further sign of continuity is shown by the inexplicable changes in the position of the southernmost post of the east row, F181, F197, which is a precise reflection of what happened to the equivalent post in Barn 2.

Of the barn's walls (Fig. 43), only enough survived to show that they had had stone footings: F187, part of the east wall, and F188, a fragment of the west (Fig. 20, [39]). Outside to the east lay F166 (159), a shallow feature without precise edges or bottom surface, full of gravel in a silt derived from the natural subsoil. The feature was parallel with wall of the barn, its centre *c.*1m from the face of that. All the indications are in favour of it having been an eavesdrip channel. If so, the barn had been provided with very generous eaves.

The dating of the actual structure is weak. The post-holes and later pads contained hardly any finds. That the post-pits did not is hardly surprising as the barn was basically on new ground. What there was ran from the late third century into the fourth (160). Only one pad, F183, produced later evidence and that was of the first half of the fourth century and, like most of these cases, was only sealed by topsoil. The main dating for the new barn, apart from the final use of the old, came from the last fills of the old Main Yard boundary ditch, F189 *etc.* (Pl. IV) (161) (Fig. 20, [36–37], [39], Fig. 21, [40], Fig. 28, [78]) which had been sealed under the new building. The pottery belongs to the late third/early fourth century.

When first built, the posts of the barn were set in pits and carefully packed round with limestone rag. The site was very wet and this led to a major reconstruction in which the posts were put on pads. Some of the posts may have been sawn off suggesting that the lower parts were rotten. Capillary action in the posts would have made the lower parts above ground wet and therefore they would have had a shorter life than would be expected for posts set in much drier ground. Normally it would be tempting to equate a major alteration in a building with a new period; thus the change in the structure would, here, be given to Period 5. However, as that seems to have been a time when maintenance was at a low level, it is better to equate the change in periods with the time when the barn lost its west aisle and two northern bays (see Period 5 below, and Fig. 43). Therefore, the change in structure is taken to be an intermediate stage in Period 4: F190 shows that the change occurred before the barn was cut down in size (see Chapter 2, p. 63, Fig. 44).

As for the internal features, the character of the driers is discussed in Chapter 3 (I, **2** and **3**). Next to no evidence for date was recovered from their structure (162). Apart from those, there was a series of hollows and a slot near the south-easternmost post. These argue for heavy activity producing a marked disturbance, not only of the floor, F173 (163) (Fig. 28, [77]), but possibly also of the wall itself, F195 (164). In a line between the post and the one to the north were two features which might have been the remains of an informal partition, Fs198/199 (165). F196, allocated to Period 2 (86), may have been added during Period 4, the pottery coming from the Period 2 ditches it had cut. Thereafter, there was only the fragment of what may have been topsoil sealed beneath any floor in the building. The slight dating evidence (166) pointed to a late third/early fourth-century date.

Outside the south-east corner of the barn were two post-holes, Fs193 and 257 (167) which may have been in use with that. F193 lay at the extreme south end of the east face and in line with the end of the ditch F164 which was noticeably shallower than the rest of the ditch. It is possible that F193 was the sole survivor of a post-row cut away by the ditch. As for F257, no immediate function appears: it was by itself and, as it lay outside the Main Yard, would seem to have been disassociated from any activity there. The axis of the post-pit was about 10 degrees north of east and it is conceivable that a post here could have supported a pivoted arm for lifting water from the well, F172, in other words, a type of shaduf. The advantage of this would have been that, in a well no more than 0.7m across, and possibly with a constricted opening at the top, a bucket could have been dropped in and raised in a curved trajectory through the centre of the opening.

The wet site of the barn demanded ditches around it to carry off unwanted water and these needed constant recutting. The wetness of the site showed well as, from the initial excavation of the drains at the north end in 1971 until the closure of the site in 1974, the flues never dried out in the summer and were flooded in the winter: the lowest water-table would have been somewhere about 0.6m and the highest about 0.3m beneath any floor in the barn.

The new ditches were Fs165/128 on the east, F108 on the north and F148 on the west. The eastern (Fig. 28, [79]) and northern ones were cleaned out many times, and were sometimes continued to the north. Of these, F138, seems to have been the only one to have retained any material dating to this period. The north ditch, F108 (Fig. 28, [80]), ran west and ended at the north-east angle of the rectangular building so providing a boundary to the site here. A high water-table, coupled with the fact that only in the barn were there features sunk deeply into the ground, may be the reasons why a ditch was dug down the west side of the barn for this must have restricted the use of the narrow yard here, especially as it was sited so far from the barn's wall.

The small amount of pottery in the ditches was reasonably consistent with the dating of both the beginning and the end of the period: all, except for F108 where it formed the north boundary west of the barn, passed out of use with the changes placed in Period 5. The earliest pottery is third century and may have been present already when the ditches were first dug. The main fills had pottery of the first half of the fourth century and the latest pottery, from the west side ditch, ran into the middle years of the century (168). One small area of topsoil survived and this had fourth-century pottery (169).

The Rectangular Building and the yard to its east

The building (Chapter 2, Fig. 45) was large, some 24.5m by 20.7m. The walls yielded little pottery, possibly contaminated, dating generally to the fourth century (170). Most of the walling had been removed by Period 5 and later activities (Fig. 28, [81]). However, some features in these damaged areas only have meaning if they had been part of the construction of the building itself: F279, stuffed full of stone, lay where the projected east wall crossed the prehistoric long ditch. Of all these, only F508 yielded any dating and that was to the second half of the third century (171). The remaining structural element was the entrance, remains of whose posts securing the bottom of the frame survived (172). The main dating evidence must remain the final filling of the ditch (107) which lay under the floors inside: up to the middle of the fourth century, matching the fill of the ditch to the east. Although the floors themselves are allocated to Period 5 because of the pottery associated with them (Period 5, p. 37), the sequence should have started in Period 4.

The internal features are hard to understand because of the extensive damage they suffered after Period 4. Along the east side of the building were two sets of north-to-south gulleys, Fs277 and 243. The latter had an arm, F242, running west to a point roughly level with the east side of the south door. The earlier of the two had pottery of the first half of the fourth century, the second of only general fourth century (173). Two pits which might have had posts in them, Fs524 and 517, had been largely cut away by the two gulleys (Fig. 28, [82]) and contained no pottery (174). The earlier of the two gulleys not only cut into the top of the backfilled boundary ditch under the building (Fig. 24, [58]), but also fitted neatly into the corner of the building itself: the gulleys belonged exclusively to the building and it is likely that the two pits or post-holes did as well.

When first recorded, F242 was sealed beneath the stone flooring inside the building with no relationship with either of the gravel floors. A section further west suggested that it had cut a gravel floor, but which could not be decided because of the erosion of the site. The sequence began with the pair of pits. Then came the gulley F277 which not only changed course to run across the pits, but also so neatly bisected them that it is hard not to see more than just a coincidental relationship between them. Finally came Fs243 and 242. What these arrangements mean in functional terms is difficult to see.

The function of the rectangular building is unclear, and not made easier by the difficulty in reconstructing its above-ground form. Whether it had more than one purpose is equally hard to see. In terms of a single function, it would have served as living quarters best as the floors inside seem better suited for this than for stock. Some support for this may be offered by the distribution of mortaria in this period which is unequivocally along the frontage of the building (Fig. 116).

The Small Yard contained nothing which could be put into this period. An interpretation of the use of the yard arises from the probable use of tank F426 (see below). The ditches on the east and north sides have been dealt with above. The boundary on the west was the building itself, but the south side was different: the evidence points to there having been something like a fence whose remains were very shallow. These lay east of F426 and were cut away by a furrow to the east which also removed any relationship with the ditch on the west side of Barn 4. The only sherd found (175) is basically late third and early fourth century. The simplest explanation for these features is that they belonged to a fence, but there were inconsistencies which might suggest otherwise.

Starting at the east end, F307 was a slot with a squared west end, 1.15m long, 0.3m wide and only 0.11m deep which had slight traces of possible post-holes associated with it, none of which could be resolved. To the west lay F308, another slot, with a squared east end, 3.25m long, 0.3m wide and 0.12m deep at most. Set at the east end were post-holes Fs322 and 309. The first survived essentially as a print at the bottom of the slot, there being no clearly detectable difference in the fill of each. F309, on the other hand, cut the fill of the slot. At the west end lay the post-hole F323 which, like F322, seemed tp have been an integral part of the slot which may have run further on to include F325, possibly the last trace of a post-hole. The last feature, F326, badly damaged by a furrow which prevented it from being fully defined, was a slot, *c.*0.3m wide and only 0.03m deep, running south from the extreme west end of the main line.

There appears to have been a basic symmetry in the layout: a long slot with holes set at each end and minor holes or slots to east and west. Such care seems at odds with something as mundane as a fence and more like part of a wall for a timber building. The central part would, in this case, have been the basis for a door. Had it been earlier than Period 4, it would have been the north wall of a building ending at the ditch of the Main Yard. If it had belonged to Period 5, its fill would have been almost black in colour like the rest of Period 5 deposits in this area and not the pale soil which actually occurred. A Period 4 building here could not have run south because of the new pond, but it could have run north as the old ditch had now been filled. The absence of any trace of an east or west wall could be explained away as being due to furrow damage, but the lack of a north wall is a disadvantage: Fs325 and 326 survived to be just identifiable despite the furrow. Therefore, the safest course appears to be to accept it as a fence line.

The north-east corner of the Main Yard
The largest feature on the site, F200 (Fig. 14, [1]), was a replacement of the Period 3 pond, F500. The description of Periods 1–3 will have made it clear that there had been a gradual development in the layout and character of the Main Yard from a close-knit collection of enclosures to a looser arrangement of a few small ones in a much larger enclosure. These catered for specialised functions and were apparently fed from the Droveway.

In Period 4, the abandonment of anything resembling animal pens suggests that either the latter had been closed or drastically reduced in importance. The shift in the site of the pond may have had something to do with F500 having become unsatisfactory, but possibly more because there was a need to have a largely unencumbered Main Yard: the F254 well disappeared at the same time. Plate V shows the final shape of the pond after erosion of its sides: the original size is unknown. However the Main Yard was used, the access to Barn 4 was respected, the strip between the pond and the east boundary conforming well with the suggested Period 3 track. The removal of the internal enclosures raises the question of whether the function they had served had been transferred elsewhere. If that should be expected to be still attached to the site, then it may have been moved to enclosures along the north side where there had obviously been extensive replanning. Therefore, the importance of the Droveway may not have diminished, it now serving an area much further north.

Returning to the pond, three pits, Fs264, 265 and 412, (176), partially overlaid by the latest fills of the pond, were probably early in the period. Their relationship with the pond is explained once it is realised that the edges of that were progressively degraded during use. The spoil from excavating the pond probably filled the north boundary ditch, the character of the backfilling there being in favour of this. The pottery in the fill of the pond, apart from a small element of obviously residual material, showed an initial date for the accumulation in the late third/early fourth centuries. Thereafter, it acquired material slowly and higher deposits yielding a date consistent with the end of the period still left the pond in a just usable state, the topmost deposits were partly natural and deliberate infilling some time in Period 5 (Fig. 23, [55]). The relevant Period 4 deposits ran to the end of the first half of the fourth century, with a hint that the actual date was later (177). The dating is confirmed by the latest deposits in the Period 3 pond, these arriving there when the hole had been more or less levelled up (178).

West of the pond lay F434 (179), a short length of ditch dug along the east side of F435, a boundary of one of the Period 3 small enclosures (Fig. 29, [83]), but turning east at the north end and with a butt end at the south. The pottery from it was abundant, forming one of the best groups on the site. The terminal date was *c.*350 or later (The Pottery, p. 163) and showed unequivocally that it was not related to the Period 3 arrangements. The only late deposits nearby were those in the south leg of F324 (180), the ditch whose origins lay in Period 2.

F434 appears as an isolated feature in the plan of this period. At base, it may have formed some form of traffic control at the entrance to the new building. The return east at its north end could not have been long as the feature failed to reappear on the other side of the furrow running through the door. Between the building and the ditch, a gap about 3m wide seems to have been left purposely, and it may have been meant to deflect traffic coming from the south-east. The posts and post-pits (118) cutting F324 (Fig. 24, [59]) assigned to Period 3 may have a better place here forming, along with F324 whose pottery runs late enough, part of a barrier on the east side of this corridor. Certainly the latest fills in F324 were more like those in this period than the last in being much darker (Fig. 24, [57]). Perhaps the real function of F434 lay with the area to its west as it lay at the east end of a possible building of which all that survived was F483 and in this case would have helped to terminate a passage between that and the buildings to the north.

The last feature to be covered here is F426. This was very badly damaged by a furrow with only part of the west wall surviving, its clay floor set below the stripped surface of the site defining the original layout. The clay ran up the inner face of the fragment of wall and the full reconstruction would place a stone wall all the way round, the internal faces being lined with the same clay as that used in the floor. This would form a tank or cistern whose height would have been conditioned by how it was to be used as much as how it was to be filled. The siting of the feature between the south-east corner of the rectangular building and the possible fence to the east shows that it formed an integral part of the Period 4 plan. The pottery in its construction (181) is late third/early fourth-century with fourth-century material in its fill.

It may seem odd to propose a water tank here when a well lay only 18m away, and anyone standing 4 to 5m to the south would have been in the pond. The feature should not have served either the Main Yard or the barn, was awkwardly placed for the rectangular building, and only the yard to the north had direct access and no obvious water supply other than stagnant water in the north ditch. F426 may have been ideal for long-necked animals, horses and the like: some of the odd features and post-holes, given to Period 2 because of the finds in them, may have belonged to some lean-to shack or hovel used to house animals.

The north-west corner of the Main Yard, and the area north
Beginning with the stone building (Fig. 47) (Chapter 2, p. 72–4; Chapter 9, p. 226–7) added to the rectangular structure, this lay between the south-west corner of that and the ditch to the west: F581. The only walling to survive was part of the west wall and the return east at its south end (182). The latter showed that the south wall had been in line with that of the rectangular building. The whole had been of stone throughout as the west wall showed: the old boundary ditch below had been completely dug out and the stone foundations carefully stepped up its sides. The position of the north wall is inferred from the end of the ditch F609 and the treatment of a series of pits which should have been in the northern part of the building. These were either stone-filled or had stone in their upper parts. The pottery in them would have placed them as isolated features in Period 1. In Period 4, they have a context as the quantity of stone is uncharacteristic of Period 1 and of Period 2, outside the major posts in the two barns. There were other pits in the neighbourhood of

these, but without stone, suggesting that the rest had been dug out and replaced with packed stone.

There is a possibility that the stone structure replaced a timber one. Only the west end of the south wall of the stone building was found, the rest of the line being taken up by F512 (183) whose edges were straighter in line than those of an ordinary gulley. Its width was 0.45–0.6m and 0.15 deep at most (Fig. 17, [22]). The form of the feature was more regular than that of any of the gulleys on the site. The fill was also uncharacteristic, being largely compounded of gravel with some limestone rag set in a matrix of silt. The east end was rounded and once had had a post set in it, the post having been removed, leaving a gap of 0.44m between the post and the wall of the rectangular building. The west end of F512 disappeared into a furrow, but if it had continued to the west, it would have been cut away by the stone wall there. A hole, assumed on excavation to be part of F512, cut down under the furrow and was probably a post in its line. If so, the minimum length of the feature would have been, centre to centre of the posts, 13.2m. The bottom of the feature stepped down 1m from the east post, coinciding with a widening in the north edge of the slot which contained a mass of stone. This may have been the backfilling of the hole resulting from the removal of a post; the stone matched the fill of the eastern post.

F512 could not have been the bottom of a poorly dug foundation trench for a stone wall but bears some resemblance to the linear features on the south side of the narrow yard to the east. If those had been a fence, then the probability is that F512 had been one as well. Floors under the stone one (L1015, Fig. 21, [43]) in the stone building did not belong to a timber predecessor as they sealed the stone (L950). The floors in general were only preserved where they had sunk into the consolidation hollow of the earlier boundary ditch. The first floors were of gravel and stone chips and were laid directly on the scraped surface of the natural as had been the case in the rectangular building whose gravel floors also had an admixture of stone chips. They ran through to the latter showing that there had been no intervening wall. The dating associated with them, the first half of the third century (184), was earlier than the final stone floors of the stone building: late third–early fourth century. The presence of essentially only one sherd per pot, and few of those, suggests that all were residual.

Inside the stone building, and put in before a floor, were three stone pads all of the same character and size: being *c.*0.85m by 0.65m (see p. 72). The pottery from these as well as what is taken to be stone firming up of old pits (see above, Pl. V, Fig. 47, and (185)) along the northern side of the building, was earlier than *c.*200. The stone floor, however, produced sherds of the late third–early fourth century at the west end and general fourth and middling fourth at the east. As the pottery was in and under the floor, none should represent use of the floor (185). The earliest possible date for the building is 325.

The remaining features within the building were post-holes, two at the east end and three set between the western pair of pads (186). The sparse dating suits that in F512. The fills of all were remarkably similar as well as their shapes which were basically rectangular. None had any sign of a post-pipe.

Outside and north of the added building, ditches were dug to form a small enclosure using the standing buildings as the other sides. The west ditch, F609/655, has already been used to define the northern limit of the added stone building. Whether a gap had been left between that and the building is unclear. The stone-filled pits suggest not, and a gap of *c.*1m next to the estimated quoin position of the large rectangular building to the east makes the matter opaque, but on the whole the gap is likely to have been the result of the degradation of the site in Period 5 and later. The enclosure formed by these ditches was probably entered from either or both of the buildings forming the south and east sides. It did not survive for long, the pottery from its ditches apparently being largely residual: running up to the early fourth century. A pit, F596, dug through a branch running out of the excavation, F598, contained fourth-century material. Deliberate backfill at the eastern extreme contained early fourth-century pottery (187) and this was cut through by a Period 4 redefinition of a pit, F587 (Fig. 29, [84]).

The plan of the period clearly shows that the Droveway, as far as it lay within the excavation, had been drastically altered. The yard wall, F757/780, assumed to have largely closed the gap between the added stone building on the east and the House to the west, is only known because of the part which had been deeply founded where it had crossed the old Period 1 ditch adapted to form the west side of the Droveway (Fig. 18, [26]). Despite the dating of the pottery found in it, up to the middle of the third century (188), it could hardly have been earlier than the added building to the east.

An entrance through the wall on the line of the old Droveway is suggested by the remodelling of the east ditch of that and the provision of a ditch, F542/482, inside the Main Yard. The east Droveway ditch, F733/581, was dug on a new line east of the old. It curved round at the south end to leave a gap of about 2m between it and the end of the added stone building. The ditch yielded sparse dating evidence, including residual material. However, its south end contained mid fourth-century sherds which probably derived from a recut which removed the junction with a small length of east-to-west ditch to the north (F720) containing definite pottery of the first half of the fourth-century (189) and a coin of 341–348.

The recut almost certainly crossed an entrance to the enclosure to the east which was marked by a large collection of post-holes as well as metalling which had sunk into the soft fill of the earlier boundary of the Droveway (Fig. 10). The posts provided little dating: late third–early fourth century (190) which means that the date of F720 (189) almost certainly marks the use at least of the gateway suggested by the features which occur only where the metalled strip crossed the earlier ditch. The strip was nearly 5m wide and stretched for 6m along the defunct ditch. A modern land drain ran diagonally through the collection and some may have been lost in the recut of the ditch. The surviving post-holes were poorly preserved, none showing any evidence for a post. Fs729, 726 and 723 were a minimum of 1m across and can hardly, in themselves, be described as post-holes, but the collection could mark a succession of gates each about 4m wide. Not all need have belonged to Period 4: some could also have marked a gate in Period 5.

The west ditch of the Droveway was filled in and the ditch along the north side of the House was recut (Fig. 14, [5], Fig. 18, [27]). The latter had already been extended eastwards to narrow the original Droveway entrance to

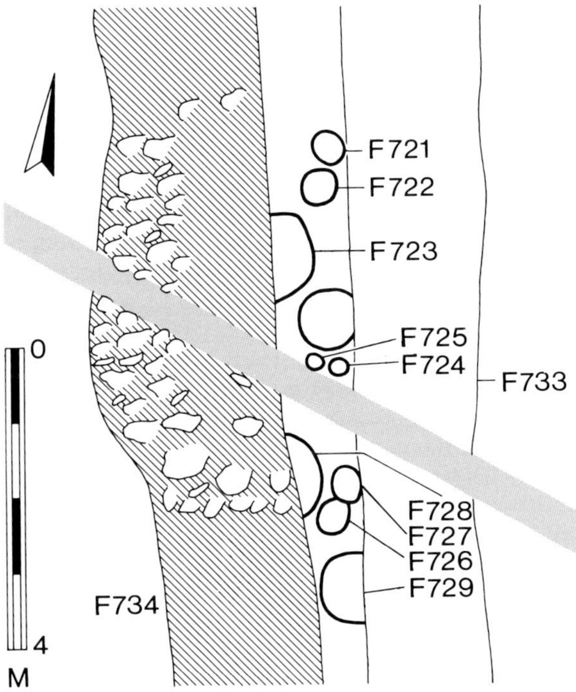

Figure 10 Period 4: gateway into the eastern central enclosure.

c.4.3m wide. The dating is unequivocally fourth century (191).

Inside the Main Yard, the internal enclosure system was disrupted by the erection of the building south of that added to the north range. All that survived was part of a wall, F483, no more than 5m long. The pottery from it is general fourth-century (192). Associated with the wall and west of it was a thick stone bed heavily cut about by the plough and not surviving beyond F482. Once thought to have been a floor, its character did not suit what little survived of any floor save for that in the added building to the north. The layout of the ditches as well as possible post-settings will not allow the wall to have been an eastern wall. It should therefore have been a west one and, as has been suggested, a building here could have run through to F434. If so, none of the small enclosures here would have survived, and their dating would suit this.

The probable post-hole, F502, just west of the south end of F483, contained material running just into the fourth century (193). The fills betrayed no sign of a post, but its depth and size do not suit an ordinary pit. Had it belonged to Period 4, it might have been related to a gate in the small enclosures here, but there was no other evidence for such features.

A new ditch, F482/542 (Fig. 24, [60], Fig. 25, [65], [67]), ran from the south-west corner of the added stone building to the south end of wall F483. Taken in conjunction with a building represented by F483, the purpose of the ditch may have been to deflect traffic from the re-entrant angle formed between that and the one to the north. The dating yielded by the ditch, and the accompanying features running across the line of the yard wall, covers the first half the fourth century and, if there had been no entrance here, that wall is unlikely to have been built before the middle of the century (194).

The ditch was not a continuous feature: towards the north entrance it broke down into a series of short inter-cutting lengths on slightly differing lines (Fig. 29, [85]) with a distinct suggestion of its having ended in at least one post, F583. The signs point to an entrance between it and the end of F581 leading to the gable wall of the added building. The features at the end of F542 were badly degraded. Enough, however, was found to isolate F579, F583 and F584. Of these, the first, an elongated hollow, was the most complex. Its final fills represented a destruction of what may have stood here. No direct evidence was found for any posts, but the fills represented the result of inter-cutting holes which finally formed an irregular hollow with a large amount of limestone in its upper parts. The other two features only partially matched this, being simpler in form and content and much shallower, not exceeding 0.25m in depth.

The large amount of limestone and the thorough grubbing out of whatever had stood here was almost certainly due to the construction of the yard wall: it would have needed firm foundations. This was demonstrated where the wall crossed the early west Droveway ditch. That had been dug out almost to the bottom and the foundations for the wall laid up the sides. The wall line actually ran over F584 leaving the other two features outside the Main Yard. There should still have been an entry into the Main Yard and the soft areas of these features were probably dressed with stone to improve the surface of the approach.

The remaining features to be described were fragmentary, did not form a reasonable plan and are placed here because their dating will not let them be later. Neither did they bear any good relationship with what had gone before, or what was to come. Parts of the yard surfacing, however, belong to this period and may have been laid when the yard wall was built. That may also have marked the disuse of F482/542 (195), possibly before the end of the period for, not only were pits cut through the ditch, but its fill was consolidated by a layer of stone (Fig. 24, [60]). The two pits concerned, Fs470 and 516 (Fig. 25, [66]), appear to have contained pottery deriving from the ditch itself (196). Four pits lay generally south of all these features and had no relationship with anything (197). Lastly were two odd features, Fs632 and 647, whose interpretation is far from clear, and Fs603 and 477, both undated post-holes (198).

F632, south-west of F542 ditch, was part of a curving gulley whose full course was not found. The other, F647, lay further west and was incompletely preserved. It consisted of a clay-lined feature set in the ground about 1.5m wide inside (Fig. 29, [86]), with a rounded southern end and of unknown length. It may have butted the yard wall of this period: the west side ran straight to a point on the projected south face of the wall suggesting that it was later than that. The pottery in the feature pointed to a date later than the building of the wall (198). A simple interpretation would be that it had been a water-tank of some form. Its resemblance to F706 of Period 3, which lay well out in the Droveway (Fig. 9), suggests that F647 was a direct replacement.

Barn 1, walled yard and the House

We do not know how much was actually recorded when Barn 1 was first excavated (MF6) and the House at the other end of the walled yard had certainly been very badly damaged both by robbing, the siting of the large Anglo-Saxon building, by post-medieval ploughing, grubbing out stone walls and finally by Mr Dakin's excavation. However, most of the damage seems to have taken place in Period 5. As for the Small Yard, the Period 3 hard surfacing could have been in sufficient repair for rubbish to have been degraded on its surface. The area may have been kept clean or disturbances in Period 5 destroyed much of what might have been given to Period 4.

Pottery from two of the post-pads in Barn 1 belongs, at worst, to the beginning of Period 4. The presence of the sherds suggests that the barn received repairs at this time, but their extent is unknown (199).

In the Small Yard, well F1016 was replaced by another, F1052, closer to the barn. The pottery belonged essentially to the first half of the fourth century with an early enough bias to suit construction at the beginning of the period. The new well was square (Fig. 29, [87]) and the fact that it remained partially intact suggests that it may not have been directly replaced by another (200). It was covered by a dirt deposit belonging to the churned-up surface of the Small Yard which was cut by a Period 5 ditch, F991 (Fig. 29, [87]).

A ditch, F1033, east of the well, ran away to the north-east and pottery in it points to Period 4 (201). The east yard wall did not survive here to be excavated, depriving us of knowledge as to how the two were related. As the yard wall had not been removed in this period, F1033 should perhaps be assigned to Period 5. However, the wall may have been locally breached by a door allowing access from the Main Yard to the well or a small opening made in the wall bottom to let the gulley act as a drain.

Further north on the east side of the Small Yard was a linear feature, F573. It was 7.8m long, varying from 0.4m in width at the north end to 0.7m at the south end and had a V-shaped section (Fig. 14, [6]). No trace of structure survived in it. However, it could have been for a sill-beam (202) being well defined and straight, but as it was far from parallel with the east yard wall, this seems unlikely. Thereafter, there was a single fourth-century post-hole, F883, and various small spreads belonging to relics of the yard surfacing and containing Period 4 material. One deposit was a sag-infill of F813, a Period 1 ditch (203).

In the House, there was only a general scatter of late third–early fourth-century pottery in a disturbance, F799, in the centre of the east end possibly representing Period 5 use as it must have been below any floor. There was no definable feature and the pottery probably lay in a degraded area. However, treating the sherds as being better than just residual, a date at the beginning of the period is appropriate (204).

South, west and north of the western buildings and yard, a scatter of pottery and a few features told of some activity. To the south was a gulley, F1098, and two post-holes cutting that, Fs1105–1106. The pottery is only general in character with a slight leaning towards the end of the period (205). The gulley cut through the backfilled sump of Period 3, F1094 (Fig. 26, [69]). The small amount of pottery contrasted with that belonging to Period 3 here and matched the generally low incidence of Period 4 pottery over the whole of the western side of the site. Excavation south of Barn 1 was mainly salvage but, even so, the small amount of identifiable Period 4 detail here would suit its isolated position once the south yard wall had been built.

West of the Small Yard only two deposits yielded material from this period. One was in a recut of F1006 through the soil build-up against the wall of the barn in the ditch running away from the north-west corner of the barn (Fig. 16, [17]). The other deposit seems to have been an infilling of a hollow in a long disused feature (206).

The best evidence for some care and maintenance and use of the House came from the area to the north. It consisted of the cleaning out of the boundary ditch there, F548, right through to the end in the old Droveway area (207) (Fig. 14, [5]). By the House, the ditch contained only generalised fourth-century pottery, but further east, the pottery became later and at the ditch end, it was mid-fourth century. This implies that the ditch was most easily reached from the east through an opening in the north yard wall and this has already been argued for.

Barn 3 and the south side of the Main Yard

The construction of the south yard boundary wall was similar in character to the north wall: generally shallowly founded except where soft early features were encountered (Fig. 17, [21]). The dating material from the walls was not properly sealed and was mainly general fourth century, but with a specific first half of the fourth century content from the barn's east wall (208). A room had been added to the west end of the barn (Fig. 41), but its relationship with the yard wall was destroyed by a furrow. However, if the line of the room's west wall is projected south, it meets the last trace recovered of the yard wall, suggesting that the yard wall came first. No dating evidence came from the walls of the room and there were no deposits inside (209): Period 5 activities here were extensive and would have reduced the room down to its foundations.

Two lines of posts running parallel, firstly with the west end of the north wall of the barn and, secondly, with the south yard wall were found. The first series overlapped only the two western bays in the barn and ran westwards at a distance of about 3m from the barn to just beyond the line of the west wall of the added room. A furrow here removed any trace of a continuation. The provision of an extra post-hole, F1172, at the east end should show that there had been a closing wall here, even if the pentice itself had been open to the north (210). Unlike the second series, this one revealed a history of repair and the surviving evidence showed that the posts were round. At least one, F1120, had been back-packed with limestone after the post had been removed and in two other cases, F1150 and F1159, the remains were definitely of rectangular pads made of coursed stone. F1159 had been converted from an ordinary post-hole with limestone packing. The distance between the two posts, 2.6m, was the widest in this row and it may have marked an entrance into the timber structure, if that had not been an open pentice, and perhaps into the barn beyond. The next post position east of this element in the plan was the corner of the building, and that had been replaced twice, evidence surviving for the final removal of the post.

The second series ran in a line about 2.5m away from the yard wall. While both series may actually have belonged to Period 5, the lack of dating evidence and their plan relationships have led to their being placed in the last fully Roman period (211). However, the possibility that the second series at least may belong to Period 5 is discussed below (see p. 40). There were five or six posts and four of these had had squared posts firmly packed with limestone pieces. The maximum length of the row was 13.7m, but if F1096 is discounted as too poorly preserved, the length was 11.6m. The two eastern bays averaged 3.4m in width.

The furrow which destroyed much of the west wall of the added room also removed any evidence that the two lines of posts had been connected by another series along the face of the added room.

The barn itself had no deposit which could be dated to this period. There were also few in the southern part of the Main Yard including a stone-filled pit, F1122, against the west yard wall dated to the first half of the fourth century (Fig. 17, [21]). Out in the yard was a small pit, F1027, datable generally to the fourth century, and an infill of a sag hollow over part of a Period 2 ditch containing material belonging to the beginning of the period (212).

That a fair amount of structural work can only be Period 4 or later and be accompanied by so little general activity is disconcerting. Although the site was obviously badly damaged by medieval and later ploughing, general deposits allocated to Period 5 because of their Anglo-Saxon pottery content also had a lot of residual material, and it is more than probable that Period 5 use churned up the top deposits on the site so intermixing Period 4 pottery with that of Period 5.

VI. Period 5, *c*.375–early sixth century

(Pl. VI)

Introduction

Dating evidence for fine division of the site into a coherent sequence covering well over a hundred years is minimal. There are three basic reasons for this. Firstly, the top deposits of the site were badly plough-damaged from at least the time that ridge and furrow was laid out across it. Secondly, the relative quantities of Anglo-Saxon datable material is very small when compared with Late Roman, and in itself does not show a development. Thirdly, the sampling technique used on the site was least effective when it came to the kind of problem created by the first two conditions.

Therefore, no summary is given at the beginning of the description of the site in Period 5, and tentative conclusions only are offered after all the site has been looked at (p. 40).

The features were not only generally shallower than those of previous periods, but often had relationships with the Period 4 plan which were hard to define. How true this may have been in reality is difficult to estimate as the sampling system used almost certainly prevented some features from being proven to be Anglo-Saxon because the single sherd which would have been needed failed to be in the excavated part. The lack of marked development in fourth-century Roman pottery on the site adds to the difficulties. Much rests in this period on the cutting of Period 4 features by later ones, Anglo-Saxon pottery occurring in others and specific forms of Roman pot seeming to have belonged exclusively to the features which can only be placed here. The chief one of these is the thick-walled shell-tempered bowl (The Pottery, p. 174).

The only tangible evidence for a direct interaction between Roman and Anglo-Saxon is a sherd from a mortarium in an Anglo-Saxon fabric (see Anglo-Saxon Pottery, No. 15). There were Roman goods in use during undoubted Anglo-Saxon occupation, but no final evidence that these had not been gleaned from elsewhere. Otherwise, the argument for close association of the two cultures rests on a possible chronological overlap between latest Roman material and earliest Anglo-Saxon. Dating can only be subjective and receives only limited support from probable close juxtapositions of Roman and Anglo-Saxon features on the ground. No feature sustains the view that the Anglo-Saxons used a Roman site in full working order, but the accumulation of detail in the following description leads inexorably towards that view.

There is no incontrovertible evidence that there had been a distinct break in occupation: too many of the proven Anglo-Saxon features only make sense if Roman structures were still standing. The impression is that the Romanised site had become debilitated before Anglo-Saxon pottery arrived, but it was not possible to show that this must have happened before Anglo-Saxon occupation of all or part had become established. The old Main Yard seems to have been left almost as bare in Anglo-Saxon times as it had become in Period 4, the few Anglo-Saxon features in it probably representing late colonisation. The crowding of the margins of the excavation by Period 5 deposits, many with Anglo-Saxon pottery, still left the central area light of emphasis even if the features there were not late. Although the area stripped provided an almost complete outline plan of the Roman site, the areas used by Anglo-Saxons clearly ran beyond this and there can only be a partial image of the site as used by them.

The plan (Pl. VI) is, again, presented with the dating evidence distinguished on it. In the four preceding periods, the dating arrived at was the product of a consideration of the slightly erratic picture presented by the individual samples. The system worked tolerably well in those cases and there was a direct progression from one period to another, even from the beginning of a period to its end. However, here, while the sequence of development which seems to belong to a purely Roman presence can be treated as before, those elements which seem to be wholly or partly Anglo-Saxon have had to be given in a different form.

The ratio of Anglo-Saxon pottery to Roman is approximately 1:25, which means that a small sample from a genuinely Anglo-Saxon feature may not have furnished the proof that it was so. No Roman pottery was being made, and almost certainly not being used by the end of occupation of the site. The progressive degradation through Period 5 meant that some features tended to fill firstly with residual Roman material reduced to small sherds, usually abraded, and then, every now and then, the presence of Anglo-Saxon sherds, also abraded. The study of the implications of residual pottery in site collections is in its infancy and its precise significance here is obscure. Nevertheless, some attempt has been made to give an impression of the degree of Roman residuality, but this is highly subjective and open to gross error: F81 was almost completely excavated, most of the Roman material found must have been residual yet included two parts of a complete Roman pot and few of the Anglo-Saxon sherds could be described as not being abraded. The condition of the pottery would suit the ditch post-dating the Anglo-Saxon period completely, but the associations in plan go against this.

On Plate VI, the Roman content of the features containing Anglo-Saxon pottery has been presented in three gradings: Anglo-Saxon and Roman, in which the Late Roman content is only general fourth-century material; Anglo-Saxon and Late Roman; Anglo-Saxon with Roman residual material. Only in the last case can it be safely assumed that the feature is definitely post-Roman, even if features in the other two categories can be similarly assigned. The assumption has been made that 'Anglo-Saxon and Roman' should be earlier than 'Anglo-Saxon and Late Roman'. The reader is asked not to accept unthinkingly what is shown, but the presentation can only be altered by reworking the material in the light of greater knowledge, not only of the pottery, but also in the handling of residual quantities.

The last point to be made here concerns undated features not having any stratigraphical associations: to put all of these on a separate plan would have severely impeded argument where reasonable certainty existed as to the correct period. This is made clear by looking at the post-holes making up the only relatively complete Anglo-Saxon house on the site. The integrity of the plan, as well as its form, declare what it was, but only two sherds of Anglo-Saxon pottery were present. It could be argued that these arrived after the building had passed out of use and so are not useful for dating. Such a fine discrimination would not only demolish a reasonable case but throw the rest of the site into confusion.

The eastern side, centre and south

(Figs 11 and 12)

It is salutary, in the light of what is to follow, to consider that this part of the excavated site is the most coherent of all in plan, sequence and interpretation of function. Discussion of previous periods has shown that there had been next to no incident in these enclosures which could be dated earlier than Period 5. Most of the undated features shown are post-holes. As the evidence points to fairly intensive Anglo-Saxon activity, it is perhaps not surprising that there should be such an abundance. But the question of the start of the sequence is very much a matter of speculation: did the Anglo-Saxons move into a yard which was already partly domestic in use, or did they move there because it was 'clean' and therefore apparently Roman features are really Anglo-Saxon ones containing only residual Roman material? If so, how are the purely Roman features elsewhere on the site to be treated?

The distribution of mortaria highlights the difficulties (The Pottery, p. 182, Fig. 117). There was a marked drift right down the east side of the site where there had been very little before. Practically all the sherds came from Anglo-Saxon features, especially the latest ones and that these should not be largely residual is shown by the concentration in F1065 in the south-west corner of the Main Yard. Period 4 here produced very few mortaria sherds and the dating of those in the ditch is distinctly in advance of Period 3.

To deal with the enclosure ditches themselves first, the main one was recut at least once in the Anglo-Saxon period and the last version of the Roman system on the south side had been greatly reduced in size by the time it had begun to acquire Anglo-Saxon pottery (213) (Fig. 22, [48–49]). At least two main periods were present. In the first, the original entrance into the central enclosure was maintained. In the second, the ditch system was carried across the old entry and a new one provided further north. Only one major ditch was present down the west side and there may have been some constraint which forced the cleaning out to be confined to the initial line (Figs 21–22, [45–46]). However, subsequent developments close to the Anglo-Saxon House seem to have obliterated the top of the ditches: L346 (Fig. 22, [46]) is what is left of a general layer over the top of the recut ditch (L385) and the slightly shallower hollow over the earlier ditch to the west, F202. The sherds were both Anglo-Saxon and Roman, all small and abraded, arguing for abandonment of the eastern enclosure, but not of the site as F318 and F319 (Fig. 11, for location, see Fig. 12) cut the deposit. Along the north side were traces of separate phases (Fig. 15, [11–12]); three could be detected (Fig. 14 [9], Fig. 15, [10], Fig. 23, [51]).

That not all features should be regarded as discrete from others is shown by F76 (Fig. 29, [88]). Dug in the side of F77, the return north of the central enclosure's north side, the pit was allowed to remain open. The bottom deposit in it spread out along the bottom of the ditch showing clearly the absolute contemporaneity of the two (Fig. 30, [89]) and both filled up *pari passu* (214). The pit was plainly not just for rubbish and many pits must have been dug for the material forming the subsoil and then backfilled with spoil and any waste immediately available. Most of the pits on the site seem not to have had a rubbish content, but it is hard to demonstrate for what other purpose they were dug. F76, on the other hand, was quarried into the side of the ditch where the gravel subsoil could be seen to be of reasonable quality. Once the gravel had been taken out, there would be little inclination to backfill it just in case more material was needed. The further course of F77 may be represented in Figure 15 [13] on the east side: the central lower fill was reminiscent of F77 towards and at F76. This part of the ditch system was abandoned in favour of a new east limit to the main eastern enclosure F78 (Fig. 15, [12]).

The new gateway (Fig. 11, for location see Fig. 12) lay in the centre of the west side of the enclosure. The post-holes contained no dating evidence (215), but that they belonged to Period 5 is shown by F319 at least cutting the fill in the new ditch. All the post-holes had the same

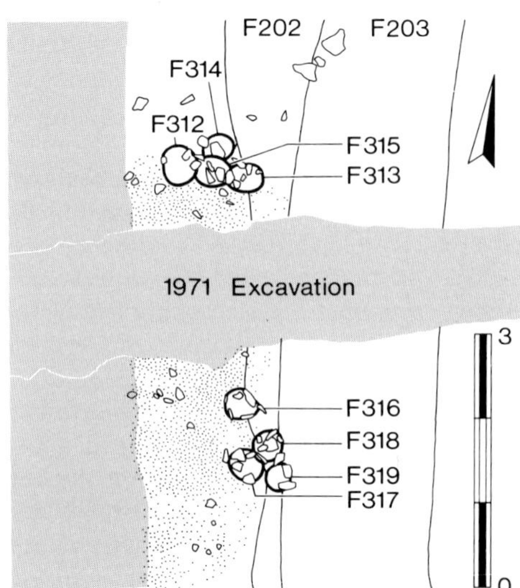

Figure 11 Period 5: Anglo-Saxon gateway into the main eastern enclosure.

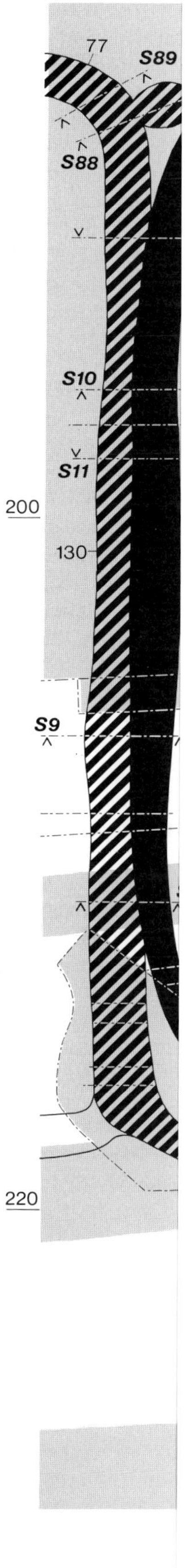

Figure 12 Period

character which differed from most of those in this and earlier periods: each was more carefully dug and lined with upright pieces of limestone rag. The gate had a minimum width of *c.*2.75m, and the plan of the groups suggests a succession of gate positions.

In the south group, F316 had the size of its post preserved; the posts of the others having all been removed. F316 was by itself, the others were clustered, but there was only evidence for one relationship: F318 cut F317. In the other group, only F312 retained evidence for the post itself, F313 having a trace of a post-print. Both F312 and F313 cut F315 which, in turn, cut F314. The sequence of possible pairs does not repay discussion and a standard gate width is not unequivocally present.

The latest version of the ditch was marked by the dark soil which filled it. This ran down the west side of the central enclosure (Figs 21–23, [46], [45], [51]) and returned a short way to the east at the north end. It ended at the edge of the 1971 excavation, 'Area A', where a machine had been used to strip everything down to the gravel base (Figs 14–15, [10], [9], [11]). On the south side, the dark soil hardly ran east of the eastern end of the Anglo-Saxon House (Fig. 22, [48], [49]). The critical junction with the east side of the southern enclosure was damaged by a furrow which had cut away all the dark deposit (Fig. 22, [47]). Either the dark soil did not run down the west side of that enclosure and its ditch was not recut, or the dark character was derived from use in the central enclosure — the discoloured soil finding its way into the ditch from the general ground surface. If so, there would not have been an internal bank and the activities producing the effect were confined to the western side of the enclosure.

The interior of the enclosure presents a confused picture (Fig. 12). A domestic presence is shown by the only reasonably complete Anglo-Saxon house plan (Fig. 55) on the site and by numerous other post-holes which, at best, form relatively straight lines with the occasional suggestion of a corner or junction. At worst, they are dispersed and form no hint of a coherent plan. Most of the post-holes were scattered over the western side of the enclosure, there being very few along the eastern side of the excavation. Amongst these part plans were several shallow ditches and gulleys whose sequence and dates are not easy to disentangle.

The earliest of the ditches seems to have been a redefinition of F28 which wandered away to the north at the west end: F129 (Fig. 23, [50]). There was little pottery and most belonged generally to the fourth century, but part of the fill near the centre of the exposure of the feature contained some from the second half (216). As the originalfeature has been assigned to Period 3 on the basis of its pottery, it was probably still visible as a shallow earthwork and reused to divide the enclosure, leaving an entrance at the west end, unless there had indeed been an internal bank. F28 should possibly be moved to Period 4 to close the chronological gap, but the change in course suggests a recut following a fairly long interval of non-maintenance. The deflection north may have been related to Anglo-Saxon use here (see below).

Fs16 and 17 were narrow gulleys which were very shallow and contained pottery only datable to before 300–350. The first was, at most, 0.38m by 0.12m, but the second was only 0.2m wide and about 0.05m deep. F34 was a shallow hollow which cut F17. The fill of these was a fine textured silt and it is this which served to associate all three. Along the east side of F17 was the trace of another slot, F33, which, at best, was only 0.2m wide and 0.07m deep. It was too badly damaged to tell what its overall plan may have been or what its relationship with F17 was (Fig. 30, [90]). However, it could show that division along this general line needed to be maintained (217). If these linear features had been beam-slots, the kind of structure they represented is obscure. The basic alignment was at right angles to ditch F129, the north end was lost in the area stripped in 1971, while, at the southern, the relationship with F28 was ambiguous. There was next to nothing representing activities to the east, but there was to the west. The three features together do not form a single plan as F17 cut F16 and was in turn cut by F34, and the best interpretation may be to see a barrier of some kind here.

Although a furrow removed the junction, the next major element of the plan should be later than F28 and possibly the recutting of that: the ditches, Fs18 and 35. Both averaged *c.*0.5m wide and hardly more than 0.25m deep. The former may have joined the east-to-west ditch AF45 (AF is a feature in the 1971 Area A excavation, not renumbered later), but a furrow and the 1971 stripping removed the evidence. The pottery in the ditches was general fourth-century (218), but as nearly all the Anglo-Saxon activity recovered lay west of the main line, they read better here than in earlier periods. The two ditches did not meet, having clear terminals 1.25m apart. This should have been an entrance, but the sections of the gulleys were too shallow to see what kind of barrier they represented. No traces of a gate were found and it is doubtful if later ploughing could have removed them unless they lay in the furrow on the east side. If AF45 was associated with these ditches, the question of function arises again: it was a shallow scoop up to 1.1m wide and less than 0.25m deep (Fig. 15, [12]). F18 and F35 could, conceivably, have been the foundation trench for something more substantial in timber, but its form is hard to assess: the fill was undifferentiated throughout.

East of the F18–F35 line four features occurred: two undated post-holes, a hollow and a pit dating to after AD 350 with a coin of 330–335. The two post-holes were dissimilar: F63 had vertical sides and a flat bottom while F64 had a dished profile. F63 could have been essentially only a post-print. The hollow, F37, was apparently tucked into the angle between Fs28 and 18. It had an irregular shape and contained only general fourth-century pottery (219).

Pit AF48 (Fig. 30, [91]) was, unlike others, not assumed to have been dug for rubbish. It was as near perfectly circular as it is possible to make a pit, 0.94m in diameter with nearly vertical sides down to a depth of 0.93m, and it could have been lined although no trace was recovered. As it was not deep enough to have penetrated the water-table, the pit was not a well, but may have served for storage. The central fill had a fair amount of large bone and was clearly rubbish disposal. The small amount of pottery from it was of interest as there was no known Roman occupation in this area from which it could have been easily derived, unless it lay outside the excavation to the east. None was seen, however, when the new roadway was continued eastwards.

Turning to the rest of the central enclosure, if F28/129 was only recut west of F18–F35, it could be read with F49 (220) which stopped at that line. F49 was not one of the latest features as it had been cut by post-hole F89. The two ditches together fanned out from the central part of F18–F35 and may have been intended, together, to drive traffic towards the possible entrance. However, F28/129 was both wider and deeper than the others and this is against it being related in function to them. F49 was remarkably similar to Fs18 and 35 and, although apparently cut by F35, the nature of the soils was not in favour of there being a significant difference in date. F49 should not have been by itself and it should perhaps, if seen in relation to the area to the south and south-east, be read against F408 (225) or the Anglo-Saxon House itself. But F408 was more like a drain and F150 may suit the case better. The plan shows that there may have been another Anglo-Saxon house immediately west of F35, and F49 may have been a screen for that and, in conjunction with F150, have formed part of an offset entrance to the southern area here.

F54 was another ditch containing only general fourth-century material (221). It ran obliquely to the south boundary ditch and clearly into the latest versions of that, although not necessarily the last one (Fig. 22, [49]). If it had been for drainage in connection with either the certain or possible Anglo-Saxon House, it would provide a little evidence for the position of the south-east corner of the former. The ditch was deliberately backfilled over the remains of a 'spit-roast' beast. The skeleton was articulated but without its head, tail and most of its limbs and had obviously not been disturbed by scavengers.

In this area lay F36, a gulley whose content was later than the others considered so far (222) and as it was cut by at least one late post-hole, cannot belong to the latest phase. It was not obviously a drain, but is not easy to make into a fence as there is nothing to put with it.

Near the west side were two ditches, Fs143 and 344 (224). The earlier, F344, was cut by F143 and two pit groups. It had a slightly sinuous course from north to south very close to the west boundary ditch, possibly running into it at the north end, but a pit prevented certainty. The other end was only 0.75m from one of the posts of the Anglo-Saxon House and just over 2.0m from the east ditch. The ditch was deeper than most of the gulleys to the east and its plan would better suit a drain. The only dating evidence available was generalised fourth-century pottery.

The later east-to-west gulley, F143, deepened markedly towards the east boundary ditch. F143 did not belong to the latest phase as there was none of the dark soil marking that in the main ditch (Fig. 30, [92]). The earlier version of F143, F150, was about 0.4m wide and 0.05m deep, matching F49 and so forming a neater entrance. The west end of F49 widened out in such a way that there may have been a post-setting, possibly matched by the post- hole F147 (223) which lay just to the west of the end of F150. The dating of F143 shows a shift forward to the second half of the fourth century and matches the dating of the pits to the north and so brings F150 forward to suit the date of F49.

Only one house plan was restorable and assignable on good evidence to the Anglo-Saxon period (224) (Fig. 55). It lay in the south-west corner of the enclosure and was aligned east-to-west. Its dating rests entirely on two Anglo-Saxon sherds, the small Late Roman content being nondescript. The discussion of the building (Chapter 4) proposes central opposed doors in the south and north walls. If early in the Anglo-Saxon phase, access into the main part of the yard would presumably have been past the south door as, at about 0.6m away, its west wall lay too close to the west boundary ditch for easy passage. If later, the approach would have been through the new gate to the north.

F408 may have been a drain in use with the House. No trace of a lining was found, and it may have been boarded over to carry traffic passing along the south side of the House. Any boarding would have run from the house for about 3.5m after which the width of the gulley increased markedly. Its fill was of general uniform pale silts containing only a little fourth-century pottery. There may have been an earlier undated (225) version in the form of F357 joining F31 further west. Its south end had been cut away by the same furrow that ran through the House. Both features predated the final recut of the southern boundary ditch.

The rest of the structural elements in the enclosure were either fragments of post-rows or individual post-holes with little relationship either to each other or to more definite lines. As there is little chance of being able to reconstruct house plans, attention is drawn to groupings which may have been parts of houses, but some of the lines could have represented fencing although none resembles either of the two patterns discussed below (see p. 90).

The best combination of lines of posts for a house lay between the House and F18–F35 (226). The two post-rows could have defined the north-west corner of a structure 7m north-to-south by 2.75m east-to-west. Each line had only four post-holes and only the shorter had the close spacing found in the House. The spacing in the other line was more or less regular and may have been part of a wall with shallow posts keyed at intervals by deeper ones of which only some of the latter survived. Two inserted between the survivors would produce a spacing very much like that in the House. None showed any sign of a post-pipe. The post-holes in the shorter row were clearly deeper than those in the other line and of a more definite uniform, oval shape than the others. The deepest, 0.43m, was the joint post of the two rows. The three posts in the north-to-south line were between 0.06m and 0.12m deep. If F54 had belonged to whatever structure stood here, the limit of the north-to-south length was of the order of 7m at most and F90 could well have been a corner post. The southernmost post was the only one which showed any sign of a repair or replacement.

Two other lines lay in the same area. The first was 3.9m long and ran north-to-south (227), the southernmost post being *c*.3.3m away from the nearest other one and it is only because it was on the same line which places it here: the gap could have been filled by others which had not survived, the surviving intermediate posts were only 0.01 to 0.02m deep. All four were about 0.2m across. The other line (228) was 2.15m long and ran slightly north of east from F373. There were four posts, but the east end was really two posts so closely set that one might have been a replacement for the other. The sizes of the posts varied and gave little sign that they were to be associated. F94 in the middle was comparable with the holes belonging to the House both in size and in profile. Fs48 and 50 were only 0.01 and 0.02m deep. In the case of the first row, there was a single post-hole, F372, which lay at an angle close enough to 90 degrees and at the right sort of spacing for it to have formed part of a south wall running away east (229). Its near vertical sides and flat bottom matched the end posts in the line to which it may have belonged as well as resembling most of the posts in the House itself.

Close to the south-east corner of the house, nearly parallel with the south wall and only 0.5m away from the corner, was another line of posts 3.5m long (230). Five were given formal numbers, but a dark mark east of the end of the line was planned which conformed in plan and general depth to the rest and fitted in with the spacing of those. The mark was so shallow that it disappeared when the weathered site was finally cleaned for excavation. There is a good chance that the row had belonged to another house. The line was very nearly parallel with the south boundary ditch and a building having the same width as the House could have been fitted in without any post cutting the fills of the latest version. No evidence was found for an east end: F90 (226) was too far away to help fill out the plan.

Towards the west end of this row and immediately south of it was a clutch of other post-holes all close set but which could have belonged to other lines, none being long enough to be useful (231). In any case, the closeness of these lines to the House itself would seem to indicate that any other house here should have belonged to a different phase.

The only other real possibility for a building lay in the north-west corner of the enclosure where two lines formed a 'T'. Their alignment and their layout in relation to the main boundary ditch points to any building here having run away to the north-east (232). The fills of these post-holes had a slightly greater content of burnt products in them, including burnt gravel in F250. Pit F137 (Chapter 3.II, **24**) with its high slag content may provide a context for a building here as the unusually large quantity should have come from somewhere close by: slag was usually found as single small pieces. The pit may have been part of, or associated with, an Anglo-Saxon smithy nearby as there was a sherd of Anglo-Saxon pottery in it (233). The possible building stood about 1.3m away to the north-east and seems to have been the only building in this area and it seems logical that a potential fire hazard should be isolated from other buildings and down wind from those. The arrangement of the ditches may mark a separate enclave with an entrance at the west end of the F129 ditch. The east side was possibly formed by F18 or the slots Fs16 and 17.

All that remain are isolated post-holes whose dating is exiguous (234). In the microfiche, the posts are arranged where possible in groupings lying near the lines already mentioned, with all the unrelated ones being gathered at the end. None betrayed a character markedly different from those already discussed and their quality ranged from faint, F117, to good with a post-print, F92. Two post-holes showed a succession seldom found in this period: F144 was cut by F142 (Fig. 30, [93]).

There were few pits. A cluster lay just inside the late gate, two were by themselves and a third, F223, containing only general fourth-century pottery (235), lay just south of the Anglo-Saxon House. Only F137, dealt with above, in the second group is an exception to the view that all were probably for rubbish. The main group forming the cluster (Fig. 30, [94–96]) contained pottery later than 350 and a Late Roman pewter flagon (MF8, Catalogue No. 108). The two latest pits here, Fs346 and 347, had definite Roman residual material in them and it is likely that all the Roman pottery in the pits was residual (236). Pit F222, by itself, contained middling fourth-century pottery (237).

The other features, Fs42, 341 and 361 (238), were not obviously post-holes, being on the whole larger, yet they were smaller than the obvious pits.

The southern enclosure

Like the central enclosure, only part of it lay in the excavation. Only the very end of the west ditch was exposed and, as there was some difficulty in establishing its proper profile, it may not have been fully excavated. The fills were largely an undistinguished pale silt, a contrast with the latest version of F31 itself (Fig. 22, [47]). It had almost certainly been recut within Period 5, as the penultimate version (F410) contained sherds dating into the second half of the fourth century. The slighter F409 succeeded this, cutting across the corner and containing Anglo-Saxon pottery. The Anglo-Saxon stage here predated the recutting of F31 (239). Where the southern limit of the enclosure lay before Period 5 is not known. In Period 5, there was a main southern ditch with more than one stage in its history: F80, later replaced by F81, both with Anglo-Saxon pottery. Three ditches ran north from this boundary of which only one can be proved to have been Anglo-Saxon, but, as the others are meaningless without one or other of the south boundaries, they must also have been Anglo-Saxon. Two main stages can be detected and these relate to what may have been the two main Anglo-Saxon stages of the central enclosure.

First came ditch F80 running across the south end of the excavation. It had been recut (Fig. 31, [97–98]). The Roman content was general or mid-fourth century in date, but Anglo- Saxon pottery was present in the earlier version (240). Running away slightly east of north was a shallower gulley, F30 (241) which was cut, not by the latest version of the south ditch of the central enclosure (L44, Fig. 22, [48]), but by a major earlier version (L113) which, in the sample here, was basically undated. It is hard to see how F30 could have cut such a straight course if F31 had been in use at the same time: the central enclosure was probably open on this side running through to F80.

Further west lay another ditch, F55, whose course was nearer to being north-to-south. Its relationship with F80 was in some doubt as its depth did not exceed 0.1m: the fills of F80 were subsequent to those of F55, but it is possible that this was the product of differential silting in contemporary features due to the unequal depth of each. At the north end of F55 was a gap of about 2m between it and the recut final version of F31. The ditch contained Anglo-Saxon pottery (242).

Ditch F81 seems to have been the second stage (Fig. 31, [99–100]). All of it lay in the excavation and ran along the south side of F80. The two were not connected and no other feature cut, or was cut by, F81. The contents of the ditch suggests it was filled when Roman pottery and a good deal of the Anglo-Saxon sherds had been largely reduced to abraded small sherds (243). This may suggest that the ditch was one of the latest features on the site, even belonging to a time when the site had passed from memory. But a complete Roman vessel had been broken into two, with one piece being found near the bottom and the other near the top. Neither part is heavily abraded on the fractures and the pot was worn when it was discarded.

The relationship of F101 (244), a ditch running north from F80 at least with F81 is assured: it would have joined

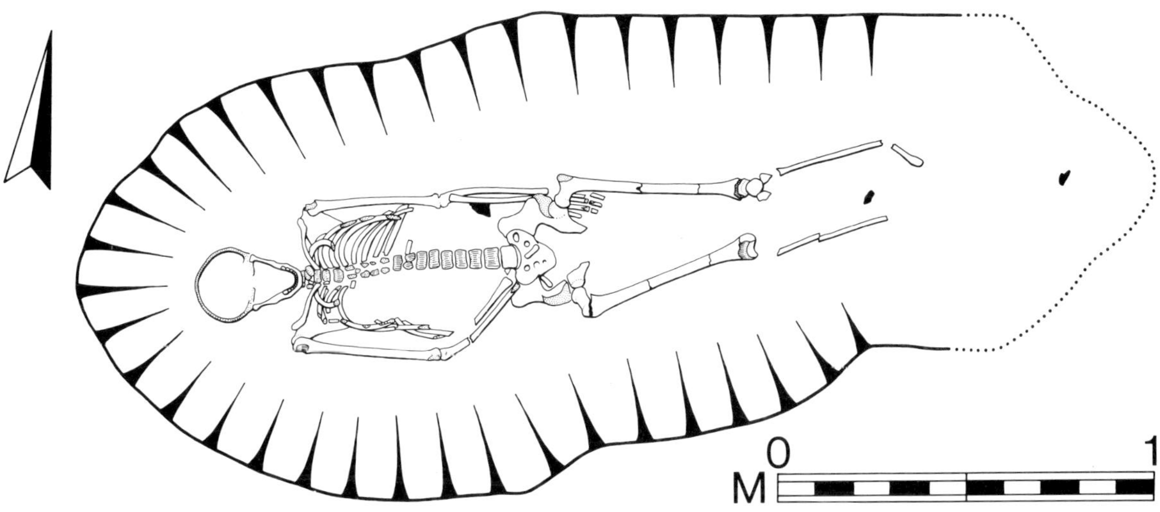

Figure 13 Period 5: adult female inhumation.

the west end of F81, had it been continued. The pottery in F101 was sparse and general fourth-century. F101 cut the first version of F80 and, as that had Anglo-Saxon sherds in it, F101 can safely be deemed Anglo-Saxon.

These five ditches can be reduced to a two period scheme, Fs80–30–55 and Fs81–101, and the cutting of what is assumed to be the earlier of the two by the second main Anglo-Saxon phase of the enclosure to the north suggests a degree of cohesion in site planning absent from most of the rest of the site. But F55 stopped short of F31 and was generally in line with F36 which may suggest that it was later than F30 and belonged to the final recutting of F31.

F101 also did not join F31 and it is possible that each only ran up to or into the toe of a bank along the south edge of F31 thrown up when the last version of F31 was cut. That F101 was later than F55 can be deduced from the hearth F83 which cut F55 (Fig. 31, [101]): the hearth would then have been associated with the post-holes to its east and the area of activity defined to the west by F101 and to the south by F81.

The presence of human adult bone as well as the remains of at least two infant burials in F55 (see MF11) does not seem to suit a purely domestic presence south of F31. Only one formal burial was found in the excavations: F79 (245), an inhumation of an adult female laid out east-to-west with the head at the latter end (Fig. 13). The pottery in what was left of the grave fill was general fourth-century. The alignment may have been dictated by the ditch next to it and, had the final version of the ditch had a bank on the south side, the burial would have been under that. In terms of what can be deduced of the arrangement of enclosures, the grave was bounded by Fs31, 55 and 409 or 410, but the absence of other adult burials makes it difficult to see this as having been a cemetery area. The two infant burials were at the north end of F55 (242), along with an adult femur. This would make them Anglo-Saxon although the single bone could have been displaced from an earlier context.

It seems logical to put all traces of burials together and assign them to a single phase in Period 5. Whatever the date, it does not suit domestic activity in the southern enclosure which perhaps should also be seen as a single phase and, as hearth F83 cut the ditch with the infant burials, placed after the disuse of F55. This would suit F81 with F101 as having been the new boundaries.

The evidence for domestic occupation falls into two parts: post-holes and a hearth at the eastern edge of the excavation; and post-holes further west with the hearth cutting F55. The only features in the whole area to be directly dated to Anglo-Saxon times were the post-holes Fs72 and 73, one replacing the other (246). The post-holes only overlapped by 0.03m and it was impossible to tell which had come first, if either had. F72 contained the pottery and some burnt stone, the other had a slightly higher incidence of charcoal. They were 0.3m from the eastern baulk and the only feature which can be associated with them is the hearth F32 (Chapter 3.II, **25**), some 3.25m to the north. Although it contained only Roman shell-tempered pottery, there was enough of it to show that it was being used when the hearth was remodelled (247). The lack of any other demonstrable Roman domestic occupation before or in Period 5, coupled with the detail that the bulk of the features were Anglo-Saxon, suggests that the hearth and its pottery were used by the alien culture. The simplest answer, even if it contains an awkward component, is to associate 'Roman' with 'Anglo-Saxon' and see the whole as a single sequence of usage confined to Period 5 and dominated by the Anglo-Saxons.

The suggestion made above that all the domestic activity south of F31 is very late in the site's history does not suit such a theory. However, these three features may have been much earlier than the rest as they lay just east of F30 which passed out of use before the end of Period 5. Most of the post-holes lay at the west end of this enclosure along with the hearth, F83 (Chapter 3.II, **26**), and the only dating material from the set was Roman pottery (248). The only dating by direct association was for the hearth. This was a hollow (Fig. 31, [101]) cutting into F55 which contained Anglo-Saxon pottery. The hearth, or rake-hollow in front of an oven (Chapter 3.II, p.__), was therefore Anglo-Saxon in date. As for traces of buildings, there was a group of post-holes of which four formed a line 4.9m long (250) and three were perhaps too shallow for their plan shapes to have much value: Fs87, 99 and 100 varied between 0.03 and 0.08m in depth. The fourth, F86, was larger in plan and deeper. The remaining four post-holes here were more or less random in disposition (250). Two, Fs84 and 85, were large and shallow, the third, F97, was more like those belonging to the House while the last, F98, was similar to F97 but was only 0.08m deep. Nothing useful can be said about two post-holes south of Fs80, 104 and 106 (251), or about the short length of gulley, F82, which contained fourth-century pottery (252). It was 0.4m wide and less than 0.2m deep, its south end was clear, but the other end was lost in a furrow.

To the west, and against the west edge of the next enclosure was an undated elongated and shallow pit, F376, 2.75m long, 0.55m wide and only 0.1m deep. The upper fill contained some burnt limestone rag and the lower a fair proportion of charcoal, ranging from large pieces down to a comminuted element in patches, as well as burnt clay. The matrix material was generally a slightly altered natural. The pit is put here because there is no other period which would suit; there was late activity nearby shown by a post-hole, F411, cutting the filling of the latest ditch to the west (253).

Barn 4, and the north-east corner

Once the identifiable remains of one or two enclosures have been dealt with, the rest of the Roman site presents an amorphous appearance, an impression reinforced by the indistinct history of the latest buildings. The self-contained appearance of the north-east corner of the Main Yard in Period 4 disappeared. The boundary south of Barn 4, between the old Main Yard and the area to the east had completely silted up except for a small hollow, F157, which was itself to disappear in Period 5 along with the last real phase, also containing residual Anglo-Saxon pottery, of the old boundary ditch (254). The loss of the boundary was replaced by a great deal of activity east of the barn.

The principal change may have been economic: the needs which the Period 4 farm had served had diminished or altered, the barn was reduced by an aisle and two bays (see below), and the pond, F200, continued to silt up and could only have survived through the period as a noisome patch of weed-infested mud (Fig. 14, [1], Fig. 19, [29], Fig. 23, [55], Fig. 31, [102]). The impression is of a progressive down-grading of the Roman site, and when it ceased to be either Roman or to have a recognisable Roman content could not be discerned.

There was no direct structural evidence that Barn 4 had been radically refurbished at or near the beginning of the period; there were no datable floor deposits. There was, however, a new system of drainage ditches cutting through part of the site of the full barn and the double driers (Fig. 43) and the layout of these respects the four southern bays without the west aisle. The simplest interpretation is that the barn had been reduced in size. Had it passed out of use all at one time, the differential survival of the walls would be hard to explain. Only one main length remained to be excavated: the central part of the east side, F187. The only other fragment was part of the west wall, F188 (Fig. 20, [39]) where it had sunk into the soft fills of the old boundary ditch. It is possible that water problems led to the making of a new drier before Period 5, but it is easier to read both the act of reduction and the building of a new drier as part of the same scheme, especially if the function of the north-east barn from Period 3 into early Period 5 has been correctly identified (see Chapter 9).

The west aisle was cut through by F162 (Fig. 20, [39]) and the new north end defined by a new ditch, the earliest version of which was F211 (Fig. 31, [102]). The Period 4 ditch, F164, may have been recut as well (Fig. 27, [74–75]), but the dating is not good enough for this to be certain. The dating of the ditch system (255) was markedly different from the equivalent system in Period 4 and elements continued well into Period 5 (compare Pl. V with Pl. VI). The Roman pottery at least was of mid to late fourth-century in date and contained Trent Valley material, a symptom of the changes which can only really be associated with Period 5.

Inside, the new drier had two phases, represented by Fs159 and 160 (Chapter 3.I, **4–5**). The later version of the drier had been emptied in 1971 with the loss of dating evidence, but the original stoke-hole contained pottery of the middle of the fourth century (256). A partition, F256, had apparently belonged to the first version and ran from the north-east corner of the drier to the east wall (257) defining a reserved area, *c.*3m by 1m, in the north-east corner of the reduced structure. The slot was straight, 0.4m wide and 0.15m deep, with an almost rectangular section. The feature was definite enough where it survived for its absence west of the drier to be certain. The dating from it was general fourth-century. The second phase of the drier was accompanied by a vat base, F161 (Chapter 3.II, **26**), which sat partly on the fill of the old stoke-hole and whose construction cut the slot. The pottery from the vat base is Late Roman (258).

The well, F172, apparently continued in use in Period 5. The final stage was its partial demolition and its filling up with an almost black mud into which two large stones were thrown (Fig. 27, [76]). They may have come from the conjectured Mill-house (Chapter 2, pp 72–4). Like the other wells on this site, there was little dating evidence in it and it seems to have been kept clean and protected until finally demolished: a complete pot which had been smashed was mixed in with the mud (259).

The ditch system along the west and north sides of the reduced barn was complicated by elements which seemed to mark a track *c.*2.5m wide leading east from the area immediately north of the barn. Much of the eastern part of the track was cut away in the 1971 machine-stripping of the site. Two main periods can be seen. In the first, the ditches ran right through the stripped site and out to the east, Fs20 (Fig. 31, [103]) and 3. There was only general fourth-century pottery in the northern of the two, but the southern set contained material dating after 350 (260). However, when the southern ditch was recut, F126, its dating proved to be earlier, matching that of the northern ditch (261).

The later history of the ditches in this general area is obscure. The ditch along the new north side of the barn ran into the south track ditch, but only after it had changed its alignment, and it may have been the same ditch as the southern ditch to the east. The end of the recut version of the south ditch, F5, deepened to the east and ended in what can fairly be described as a sump which received Anglo-Saxon pottery (262). There had been a history of recutting (Fig. 31, [103]) and it is unlikely that the shallower west end would have been defunct before the whole was abandoned. It seems best to relate to this stage a poorly dated ditch, AF5 (263), cutting through the north side of the track, which ran away north out of the excavation. Its end was about 3m from the end of F5, possibly forming the south-east entrance into an enclosure. The plan is not improbable and two undated post-holes, AF11 and AF14 (264), could have been related to some form of closure system.

The new west ditch of the barn continued beyond the new north ditch to join the south side of the track and turning west as F149 to run into F108 which ended at the original north-east corner of the rectangular building. The west ditch was recut through fills of the south track ditch and contained pottery of the middle and late fourth century (255). The ditch forming the north side of the small yard between the barn and the rectangular building was recut later (Fig. 31, [104]) showing the boundary was still needed. The latest ditch fills contained Anglo-Saxon pottery (265): the rectangular building was not reduced at the same time as the barn.

While the broad line of development can be seen, some incident is hard to fit in. Of two minor ditches, one, F6, was cut by the latest version of the south track ditch, the rest was lost in a furrow. The other ditch, F212, is undated, and was a short length running from the north-east corner of the reduced barn to the ditch F171 (266) which was later recut across the line (Fig. 31, [102]). Roughly parallel with F171 and to the north was another, F192 (267), which ran through the F168 drier. Both ends were lost in 1971 trenches. There was, however a faint trace beyond the eastern one, possibly indicating a junction with F126. The many ditches and their recuttings in this area show that the water problem needed constant attention.

There was a considerable amount of incident in the area east of the barn and north of the central eastern enclosure, bearing in mind the change caused by the deeper stripping of this area in 1971. The traces consisted of one gulley, complete and of individual plan; some pits, nearly all isolated; what may have been a hearth or two; and several widely spaced post-holes.

Most of the gulley F12 had been lost in the 1971 stripping, the dating evidence coming from the undamaged west end (268) which was between 0.5m and 0.6m wide and 0.36m deep. The pottery was consistent in being after 350, and as the sherds were large, must have come from activity nearby. The gulley was east-to-west, returning north down the slope at each end and was presumably for drainage, though for what is difficult to see unless it had been dug round the uphill end of a rectangular building. The length of wall which could have been accommodated would have been about 6.75m. Two post-holes, AF21 and AF18 (269), lay about 10m and 13m from the ditch, and could have been on a line with a west wall, but the distance and their isolation is against this. On the other side was a short length of east-to-west gulley, AF24 (270), one end lost in a furrow.

The remaining post-holes (271) in the area also yielded no plan sense most probably because of the 1971 stripping: most of the Period 5 post-holes elsewhere were very shallow which should be a guarantee that others less deep were lost when the site was stripped in 1971. AF40 and AF41 were respectively 0.3m and 0.43m in diameter, but only 0.04m and 0.025m deep. Of the others, F24 cut through a Period 1 ditch. Both Fs21 and 22 contained posts whose pointed bases had pressed through the bottom of the pits, F21 having had an extra stake. F214 lay north of the reduced barn and was sealed under part of the drainage system there. F26 cut into F5 which already contained Anglo-Saxon pottery (262).

Away in the south-east corner of the enclosure was a short length of a linear feature, AF39, which might provide a context for the post-holes, AF40, which lay directly in line with it. AF39 was described as having round ends. It was 0.48m wide at the top and 0.2m deep. It had a flat bottom 0.19m wide and the sides were at a constant slope. In the bottom was a layer of dirty natural under one of brown loam (272). Its profile could suit a beam-slot, even if a further 0.2m were to be added to it to compensate for the 1971 stripping.

The only pits which were directly phaseable to Period 5 were AF2 and AF3. These cut through the north ditch of the track and contained a relatively small amount of pottery dating to after 350. They were deep, the high water-table preventing their bases from being reached. They had not been deliberately backfilled and, while their sides were too shallow for them to have been ordinary wells, they would have been permanently full of water. Their fills suited slow infilling with interleaved slippages of natural from the sides and they could have been water-holes: deep steyned holes were not needed in this area. The only pottery was Late Roman (273), which was sparse and came from the topmost fills and all could have been residual. There need only have been one hole at a time with one major clearing out and, as such, these could post-date the filling of the formal well, F172.

The remaining pits were scattered: one, AF31, was definitely Anglo-Saxon and the section (Fig. 23, [52]), together with the finds and the fill show that there had been another at the junction of two earlier ditches, 'F11', although this was not identified at the time of the excavation. The last pit, AF22 (274) contained no dating evidence at all.

Along the northern margin of the excavation was a set of features whose periodisation in part is in doubt. The character of two was different from all others in this part of the site. F2 was a narrow gulley or slot, 0.45m wide and 0.25m deep, somewhat reminiscent of F232 further west which had Anglo-Saxon pottery in it (302). The fill was so similar to that of F1 that the two may have been linked, but the 1971 excavation removed the evidence. Slot F1 was 0.45m wide, and 0.15m deep and undated. In its bottom were four equally spaced circular depressions, possibly the impressions of posts, although no sign of any post-pipes were visible in the fill above. Immediately east of F1 lay AF6, an undated pit. The final two definite features consisted of a post-hole, F43, replacing a delve, F44, of indefinite shape and only 0.08m deep. Neither was dated (275). Finally, there were two patches which may have been hearths (276). AF20 survived only as a few pieces of burnt clay forming a patch, while AF37 remained as a triangular area of dark soil.

The Main Yard

Most of the features here assignable to Period 5 were pits. There was also a limited amount of evidence in the form of structures, mainly along the eastern margin, that there had been some colonisation, probably later in the Anglo-Saxon period. Beyond these, the relevant deposits were largely those accumulating in hollows of features either long dead or passing through their final disuse. The amount of Anglo-Saxon pottery was actually small, more coming from the areas around it. In this, the distribution is very reminiscent of the Roman site from the end of Period 2.

Chief of these was the pond, F200. Dug at the beginning of Period 4, there was no evidence that it was in a usable state in Period 5. Much depends on when the boundary south of Barn 4 was swept away. Once it had gone, the pond was probably largely filled: the type of occupation in the north-east corner seems to have been incompatible with the management of animals. The dating of the final deposits of the pond moves inexorably from general fourth-century pottery, through middling to late and then, finally, to a time when Anglo-Saxon pottery was in common usage (277).

The disuse of the pond is emphasised by four pits. The first, F345, lay where the bottom of a furrow crossed the southern margin of the pond, its deposits being correspondingly shallow. The content of the pit was residual belonging to the middle of the fourth century (278), later than the creation of the pond and the pit's fills derived from that. The second and third pits, Fs241 and 302, cut through the pond (Fig. 23, [55]), and contained little pottery predating Period 4.

The last pit, F252, cut the northern edge of the pond. It had steep sides (Fig. 32, [105–106]) and, at *c*.4.5m across, was very large. Although waterlogged, there were very few organic remains, and these were mainly very small pieces of wicker lying randomly in the lower part and could have come from a lining. The east-to-west section (Fig. 32, [106]) shows, in the lower part on each side, evidence for a backfill behind a lining, the backfill on the west having slumped. The character of the pit suits a well, built in a technique at variance with that of proven Roman ones on this site. The dating evidence was meagre, was entirely Roman and of the first half of the fourth century (279). The pit was not only next to the probably Period 4 water tank F426, but also cut that. That F252 succeeded the Roman well attached to Barn 4 is uncertain, but it is logical to see in the change of location and technique of construction a straight replacement of F172 by another constructed in an essentially non-Roman manner.

The generally low incidence of Anglo-Saxon pottery leads to such difficulties. There were only two deposits on the whole site which could be said to have had it in abundance. One lay at the south margin of the site (F1114, (329)) (see Anglo-Saxon Pottery, Group 1, Nos 1–20) and the other, with a wrist-clasp (Catalogue No. 246) in the sag hollow of the long-disused well north of Barn 3, F254 (280) (Group 2, Nos 21–40). The very mixed and small size of the Roman sherds in each suggests that these had been rubbish lying around. It follows that, if a pit was dug where there was little Roman rubbish, and proportionately less Anglo-Saxon, there would be little or no chance of dating the pit correctly.

Two pits almost fall into this category, but, by virtue of one or two sherds, belong to Period 5 and to an Anglo-Saxon phase. Both Fs263 (Fig. 32, [107]) and 505 were big enough, before excavation, to have been Sunken-featured Buildings (Rahtz 1976, 70–3), but neither was sufficiently regular nor, despite a search, had the slightest trace of any structure either in or around it. The small amount of Roman pottery in F263 was general fourth century in date, but in F505 it was entirely of Period 1 (281).

These pits and the pottery in the hollow of the demolished well have been mentioned first because the question arises of where the focus of activity lay to which each or all may have belonged. The view taken here is that pits were not normally far from the centres which generated the rubbish thrown into them and in previous periods it has been reasonably clear where *foci* of activity were to be found, but this is not true of Period 5. As both the pits were dug in an Anglo-Saxon phase, it seems inappropriate to look at Roman centres. The use of an informal hollow, F254, for rubbish does not suggest that it had come far, but there was, ostensibly, only one inhabitable structure in the Main Yard.

On the east edge of the yard lay F224 (Fig. 55). This is interpreted as a Sunken-featured Building (see Chapter 4) as it was the bottom surface of the pit which had been trodden and the three major deposits making up the fill produced a small amount of Anglo-Saxon pottery (282). The bottom main fill derived from the natural subsoil, without it having been weathered in (Fig. 33, [108–109]). The middle fill, with its green staining, points to the hollow having become a dump for unwanted waste, possibly as a periodic store for manure. The final fill looked more like ordinary topsoil with a high amount of gravel. Around the feature were the remains of a structure (283). A pit, F226, just west of the hole contained Roman and Anglo-Saxon pottery. The condition of the Roman pottery was such that it may not have been redeposited residual and in this it was at variance with the condition of the Roman sherds in F224: the pit was probably earlier than F224 (284).

The isolated siting of the Sunken-featured Building may be more apparent than real. To the south was a basically undated row of post-holes, with some outliers (285). The line was 8m long, the northern part consisting of a fairly regularly spaced series forming a line some 4.5m long. It is possible to detect a rhythm in the arrangement. For instance, the first, middle and last differed from the others, but not by much: the two intermediates at the north end were circular compared with those which bracket them. These were set at an angle to the row and tended to be rectangular. The southern intermediate pair were both oval with their long axes being north-to-south. The southern-most post was D-shaped. These differences may not be allowable, but the outliers were all circular, none exceeding 0.09m in depth, and as a group they differed from the row. The interpretation of these posts is difficult: there was no sign of any line at right angles to the three points which may have formed ends of a building.

Some slight traces of activities using Anglo-Saxon pottery and others with only general fourth or late-fourth century Roman pottery occurred near both F224 and the post-row. Most were infilling of the consolidated hollows of long disused features (286), but there was a small group of pits which may have belonged to Period 5, though they contained no datable material, Fs266–271 (see Fig. 12) (287), except for F266 which had some fourth-century pottery.

Away towards F254 was what was arguably a furnace, F999 (288) (Chapter 3.II, **28**), in the sense that it was linear, burnt and sunk into the ground. The upper fill had pottery only of the second half of the second century, the lower had an Anglo-Saxon sherd as well as good fourth-century pottery. It is likely to have had a roof over it, and that it might have been associated with other features. As for a building, there is only one structural element dated to this period, F1038 (289). Its size and shape compared fairly well with other Anglo-Saxon post-holes and its preservation may have been because it had been dug into the soft fill of a Period 1 ditch. Its apparent solitary state may be misleading for other structural elements, themselves not well dated, may have been associated.

In the description of Period 1, the likelihood that there had been a building south of the eastern round house was mentioned. Its site then would have been partly defined by Period 1 ditches and some post-holes (290) which need not have been so early (see p. 5). There were not enough post-holes to properly define a building, their layout not suiting a regular shape and their character being variable. With the exception of F1044, a rectangular shape can just be described. F1038 would have defined the south-west corner, the eastern end would have been marked by the trio of posts Fs1003–1005, and F1026 could have been part of a north wall, the whole measuring roughly 14m by 7.55m. The furnace, F999, lying in the conjectured north-east corner, seems hardly enough to justify the proposition. But, if the pottery found is no true guide to the date of features, furnaces Fs1081 and 1082 (Pl. II, Fig. 53; Chapter 3.II, **2, 3**), previously assigned to Period 1 (38), may have been as late as F999: they were parallel with the main axis of the hypothetical building, and lay at an equal distance from the side walls. The Anglo-Saxon pottery in F254 was immediately outside and the lack of more in these features may have been due to their structure actually being above-ground: there was only one Anglo-Saxon sherd in the lower fill of F999.

One possibly significant detail is that the access route down the east side of the Main Yard put forward in the description of Period 3 seems to have been repeated in Period 5 as 'metalling' occurred there. This is clearly seen in the upper part of the Period 2 ditch, F204 (Fig. 17, [24]); to the east of that over F410 (Fig. 22, [47]); and in the central eastern enclosure (Fig. 30, [92]). The gravel mentioned as having been in the top fill of F224 (Fig. 33, [108]) may also belong to it, in which case it was late in the period. Traffic would account for the wide wear hollow, filled with a more gravelly soil than that below, over the abandoned east ditches of the Main Yard (Fig. 22, [46]). The hollow was confined to the area next to the Anglo-Saxon House.

The central part of the north range

This is the area west of Barn 4 through to the west side of the old Droveway. The development of a series of gulleys over the northern part of the rectangular building has been used to determine the extent of any surviving part of that building well into this period.

The disengagement of the topsoil from the latest deposits over the building was not aided by the three furrows cutting through it, and the later activities over the northern part also made the site of this particular building difficult to deal with. The structure was greatly reduced and produced no dating evidence, all that there was came from the floor levels inside.

The lower of the two floors was of gravel (Fig. 23, [54]) and the upper of small pieces of limestone rag laid flat (Fig. 19, [30], Fig. 24, [58]). The pottery dating begins in the mid-fourth century (291). The stone floor produced general fourth-century pottery as well as some sherds of Anglo-Saxon. The deposits listed (292) were only in the southern part of the building (shown stippled on Pl. VI) because the deposits in the denuded northern part did not have the same dating value (293). The dating of the mortaria as well as its distribution covering the inside of the building should show that it had survived into Period 5. The difference in preservation between the south and north parts suggests that the latter had been abandoned, leaving a strip about 10m–11m wide still roofed. No trace was found of a new north wall: it may have been timber-framed, or it may have been of stone, but not deeply founded.

Apart from the limit of reasonably well preserved walls, the approximate position of a north wall can be estimated from the layout of the ditches cutting the northern part of the building. That the whole series is later than the rectangular building is shown in Fig. 28, [81] where the latest version of the enclosure crossed the line of the wall. F541 was replaced by F543 (Fig. 34, [110]), both running away south-west from the enclosure. F543 just faded away and it looked as though each had been used to separate the ground south of the enclosure into eastern and western areas as well as being for drainage. In that case, it is legitimate to ask what was being drained: to take water away from standing buildings? However, their value as boundaries is more obvious. If they had separated two parcels of land, what formed the rest of the division? Other ditches would have survived, but standing walls defining three sides of a smaller roofed structure makes sense.

West of the rectangular building, no direct evidence for survival of the added building and the yard wall was found. However, there is F660, a strongly founded post about 0.25m square, seated on a stone sole-plate 0.5m down in a pit 1.1m in diameter, and packed round with tightly wedged pieces of limestone rag (Fig. 34, [111]). In the packing was a piece of Anglo-Saxon pottery (294). The post lay on the line of the north boundary wall in the area of the opening suggested in the description of Period 4. If the post had belonged to a building, other posts set in a similar way should have been found: there was none. Therefore, it may have associated with the opening through a boundary which could hardly have been other than the wall itself. The old north boundary wall should, therefore, be assumed to have been standing. If so, the south wall of the building to the east should have continued the line up to the rectangular building.

We return to the problem of the gulleys discussed under Period 3 (see p. 16–7). These on the whole make better sense there (114) but the pottery from the top fills of the east-to-west part, F641, is definitely fourth century (295). F641 would appear to carry F661 with it especially as the post-prints, Fs681 and 683, lay at the junction of that with F641 at the base of F652 whose dating is definitely Period 5, like that of F661 itself. Yet both gulley and prints were the immediate successors of the same sort of features to the east, and the date of the fill of the gulley between, F641, fits between the elements put into Period 3 and F661 here.

In the area of the old Droveway, the ditch along the north side of the Roman House, F810/548, was finally filled, the pottery running into the late fourth century (296). There was evidence for a gate (Fig. 10) from the old Droveway into the area east of that behind the Period 4 north range. A wide band, about 5m long, of stone had been tipped into the soft upper fill of the earlier ditch and there were post-holes between that and the Period 4 ditch. These have been discussed (see p. 24). None of the post-holes was properly dated (190) and some at least should be assigned to Period 5. The stone 'hardcore' in the earlier ditch became muddy itself and pottery dating after 350 came from the accumulating dirt (297). Clear deposits of this date were not common and the presence of one here strongly suggests that the gate was still in use and, therefore, the whole of the boundary in which it was set.

In the area north of the rectangular building, three stages could be seen in development of the features there. The first two consisted of the laying out and modification of an enclosure with a branch to the south. The first version, F532, had a squared south-east corner with an entrance, but swept away irregularly to the west once the branch, F541 (Fig. 34, [112]), had been reached. The south-east corner was complicated by a pit, F533, before the corner was cut off. Even before the enclosure was laid out, one pit, F534, had been dug. It was undated and backfilled with clay and stone (Fig. 34, [113]). The scale of the gulleys is shown in Figure 34, [114]: F531 was hardly more than 0.2m deep and the earlier one only 0.12m. The date of the first ditch system is only general fourth-century, but the recutting contained middle and late fourth-century pottery, possibly only arriving there when the next stage was put in (298).

Inside the enclosure was a probable rubbish pit: F518 (299). The dating evidence from it was again general fourth-century, but the evidence came only from the topmost layer. Even so, it suited the first stage of the enclosure.

In the second stage, the south-east corner was recut, F536, on a line cutting off the original corner (Fig. 29, [84]) and a new branch to the south-west, F543, provided. The pottery was definitely middle to late fourth-century and came from rubbish already on the site, much having been found in the earlier version (300).

The third stage marks the disuse of the enclosures as both versions were cut by a round pit with nearly vertical sides, F537. It was 1.9m in diameter and at least 0.8m deep. There was little pottery, but one Anglo-Saxon sherd (301) lay in the top (303). The high water-table prevented full excavation, but it may have been a wicker-lined well: few other types of pits penetrating the water-table would have needed such steep sides. To the east and aligned on the centre of the pit was a slot, F232, 0.2m to 0.25m wide and only 0.05m deep with a rounded base. The only sherd from it was Anglo-Saxon (302). The straight line of the feature, coupled with its slight section, suggests a beam-slot rather than anything else, but it lay too close to the edge of the formal excavation for any associations to be found. Having run across two ridges, it failed to appear on the third one east of the pit.

Within the area of the original rectangular building were two shallow features (303). F529 was 1.69m north-to-south and 1.25m across. It consisted of a bed of stones set in a clay reminiscent of the base of the probable water tank, F426 (see p. 23 (181)). Little sign of any upstanding walls survived although there were stones laid to an edge along the north side and there were the faintest indications of a wall on the west. The stones were sealed in clay and the whole seems best interpreted as the base of a feature designed to be water-tight — there were no traces of burning. The whole was sealed by the strew of stone from the last remains of the rectangular building ((293), L1056) and there was no sign of there having been an associated floor.

The other feature, F539, was at least 3m by 1.8m with an essentially east-to-west axis, and was less well-defined than F529. Its fill, however, was of the same character. Unlike Fs426 and 529, whose top surfaces were set below floor level, the top here was part level and part higher where the general ground level fell away into a hollow. The bedding may have been to support some kind of level superstructure. F539 was not sealed and the sherds found in its make-up, all small and residual in character, argue for a fourth-century date. Both lay north of the suggested new wall and F539 may actually have been an outside feature and so possibly a replacement of the Period 4 tank, F426.

The Small Yard and the House

The Roman House had been dismantled before the end of Period 5 (Fig. 14, [5]) with much of its foundations having been robbed. This is demonstrated by a large timber building lying obliquely across its site. Only part of the north wall of the House in the central and eastern sectors was found which may suggest that the yard had been extended across the site of the House. The dating evidence from its robber trenches should, therefore, belong to the time when it was demolished: the material recovered was general fourth-century, with an Anglo-Saxon element (304). This is the only dating available for the new timber building which seems to have respected the line of the north boundary wall of the old Main Yard. Its relationship with the east wall of the Small Yard (305) is also of interest: the footings of that survived right up to the south wall of the timber building which could suggest that the two had existed together.

The full form of the timber building is not clear (see Chapter 4). Furrows, along with Mr Dakin's excavation trenches, did considerable damage. The posts seem to have been more substantial than those used for the Anglo-Saxon House and were, in general, set in post-pits and packed round with earth, the replacements sometimes having limestone rag set round them. These replacements point to the structure having had a longer life than the Anglo-Saxon House to the east. There were sufficient pairs of replacements (306) to suggest a complete overhaul and enough preserved post-pipes for it to be reasonably certain that the building was either abandoned or the posts sawn off at ground level during a general demolition. Inside was a single post-hole, F645. Another, F800, against the north wall line may have been a repair of some sort (307). F751 out to the east may represent an annex to this building.

F751 was a shallow but steep-sided gulley or slot, varying from 0.5m to 0.75m in width, returning west at the south end. The corner was well rounded which does not suit a beam-slot, but there was no evidence for posts except for what may have been some replacement or repair in the form of a single post-hole, F747. Its fill was identical with that of the slot and it must have been deeply founded as it lay at the bottom of a furrow. The dating from it was generalised fourth-century (308).

The presence of a furrow removed any evidence to show what had happened to the west wall of the Small Yard, but it had been taken down before the site was abandoned as its line was crossed by a series of gulleys running from the south-west to the west end of the timber building. The parallel course suggested by the fragments looks like a track skirting Barn 1 to run obliquely across the old Small Yard, the west side turning north at the end to cut the robber trench of the Roman House. This part, F822, was basically undated, but could not be separated from the rest, F676. The west side, F872, contained only general fourth-century pottery, but in the southern part of F676 were several Anglo-Saxon sherds.

Running west from the possible track was a set of gulleys. Two of these have already been mentioned as being, possibly, the sides of a track leading to a gate in the west wall (see Period 3, p.___). The definite end of the southern one, F898, against what would have been the wall face is suggestive. As for F899, it disappeared into a furrow and it either finished against F676 or more probably in line with its partner. None of these features were well preserved and it was not possible to detect separate phases. Most belonged to the fourth century, only the west side of the possible track having Anglo-Saxon pottery. Of those to the west, F677 can be dated to after 350 (309). F649, which appears to lie in the line of the east side of the possible track, had nothing to do with that as it had a burnt rim and may have been a hearth (Chapter 3.II, **29**). It was undated (310).

Within the old Small Yard there was direct evidence for a multiple posted structure and evidence for a small rectangular building (see Chapter 4), F865. The latter consisted of a platform, largely of stone, approximately aligned with the east side of the probable track and with a width matching that of the Anglo-Saxon House at the east end of the site. Only one post-hole was found: F864 at the north-west corner. The bottom and middle layers of the platform produced middling fourth-century pottery, the top one a little Anglo-Saxon pottery. The post-hole was undated (311).

The other building had nine post-holes, arranged in three rows of three (Fig. 57), and an additional one to the south which was the only one to produce Anglo-Saxon pottery: the others had been excavated by Mr Dakin (MF6) and the finds can no longer be identified (312).

Other post-holes in this area assigned to this period had little dating; a few contained sherds earlier than Period 5 (313). The holes were unlike those belonging either to the large timber building to the north or to the nine-post structure, lacking any limestone packing, and only one had any evidence for the post, F908 (Fig. 34, [115]) which may have had two phases. The relationship of these posts with the nine-post structure seems assured by their plan and they are taken to have been part of a fence of Anglo-Saxon type (see Chapter 4). A furrow down the west side would have removed any posts there and the east side was damaged by another. The siting of the probable fence and building was so precisely placed in the south-west corner of the Small Yard that the chief Roman features were most probably still present.

Very close to the centre of the old Small Yard was a major post, F896, carefully set in a pit, F897. It had no associations and could only be dated to the fourth century (314): it could belong to either Periods 4 or 5. The pit was very large, the post itself was 0.45m in diameter. It was either akin to a flag-pole, or a very stout post to which horses could be tethered, or both.

The disruption of the Roman system, implied by the replacement of the stone House and the probable track coming from the west, is more or less demonstrated by some partly explored features to the north of the formal excavation. It was only the presence of Anglo-Saxon pottery which caused them to be examined. One was a ditch, F855, with some Anglo-Saxon pottery in its upper fills. It has been assigned already to Period 1, but had at least remained as a topographical feature beyond the end of the purely Roman period on this site. The other was obviously a platform of the same kind as F865 in the Small Yard, F871 (315) (see Chapter 4). Anglo-Saxon occupation can be shown to have run right through to the west end of the Roman site: exploratory trenches further west, however, failed to find anything which could be plotted.

As for the yard surfacing, there was no major access of new material after Period 3. There was some which suggested patching in Period 4, but none for Period 5. However, the final deposits were of consolidated mud (Fig. 16, [15], [19]) containing mixed fourth-century and Anglo-Saxon sherds (316) which show that the surfaces were usable and used. These deposits only survived down the east side (Fig. 17, [20]): by the end of occupation, changes along the western side had destroyed all traces.

The reason for the siting of the nine-post structure and the large timber building at the west end of the site is not immediately obvious. Both wooden structures give an emphasis here which could be said to repeat the Roman one in which the House was balanced by Barn 1. The reason may be that the security offered by the Roman Small Yard was still available in the early part of the Anglo-Saxon period.

Barn 1, Barn 3, and the area between the two

The principal question is to what extent either or both buildings survived into the purely Anglo-Saxon period and much depends on the ditch systems which can only be allocated to Period 5. To a large measure, these only make sense if other plan elements were still present and therefore probably in use. The ditches show that Barn 3 had gone by the end of the period, but what happened to Barn 1 is much more difficult to assess.

The ending of the surviving footings of the east wall of the Small Yard precisely at the wall line of the timber building should only have occurred if the stone wall had been retained: the rest of the stone walling under the timber building had been completely robbed, as has been mentioned, after Anglo-Saxon pottery had arrived. To rob foundations without taking down a standing wall immediately next to them is, by itself, peculiar. If the north end of the wall was kept, so should the south end have been and this could only make sense if the barn was also kept. Only the ditch F991 and the alignment, and position of the nine-post structure with its probable fence, provide any evidence in this area that this might have been the case.

F991 ran along the north side of the barn. Initially, it belonged to Period 2 alone, but was recreated in Period 5. The undoubted Period 5 phase was confused by Mr Dakin's excavations. Some changes in this area had been introduced in Period 4 when a drainage gulley ran through the east wall. The layout of F991, and F1035 to the east, was different, the line running through the yard wall to turn south in a weak curve. The abandonment of the wells (Fig. 29, [87]) in the Small Yard seems to have happened by the beginning of Period 5, and presumably removed a need for special drainage. The new ditch defined a discrete area outside the east end of the barn in conjunction with the first versions of F1065; the length along the north side of the barn was probably for drainage, conceivably from the barn roof.

The fill of the whole length of the ditch is instructive: the pottery in the east end of F1035 was only generalised fourth-century, but by the time the line of the yard wall had been reached, it was definitely middling to late fourth-century (317). Inside the yard, the dating became essentially Anglo-Saxon with post-350 undertones. Anglo-Saxon pottery was in both the final fill of F991 and under a bed of stone which, if it had come from anywhere, should have been derived from the wall of the barn itself (318) (Fig. 34, [116]). However, as the ditch ran further towards the line of the west wall of the yard, the dating began to revert to being purely Roman. The barn was therefore probably still standing and the greatest amount of Anglo-Saxon pottery was between the east wall and the nearer fence line of the nine-post structure. It seems unlikely that rubbish would have come from that, or have been carried from the building away to the north, and the lack of Anglo-Saxon pottery around the building, F865, may indicate that little pot was used there. In such a case, the best source would have been the barn itself. However, precisely when F991 was dug in Period 5 cannot be determined.

The principal feature in the south-west corner of the old Main Yard was the ditch, F1065, dug after the arrival of Anglo-Saxon pottery on the site. The earliest ends were F1132 to the west and F1133 to the east, the former having an intermediate state: F1061 (Fig. 34, [117]). Both the versions at the west end had Roman pottery dating after 350, but the earlier end to the east contained Anglo-Saxon (319).

The initial east end, F1133, ran up to the north wall of the room added to Barn 3, the last version, F1065 itself, turned south to cut the wall. The first west end, F1132, returned north for a distance of 3m and this was replaced by F1061. Both ended 3m from the Main Yard wall, the last version, F1065, leaving a gap of 6.3m. The first east end obviously respected the room tacked on to the barn which should, therefore, have still been standing. The various west ends bear little relationship with anything, unless the Roman walls were still standing as well. The first west end formed a short funnel entrance with the yard wall. The main length of both early and late ditches was parallel with the south boundary wall and so should have separated this area from the ground to the north and east, but only effectively if the other sides were defined by Roman walls. The last version, F1065, contained remarkably little Anglo-Saxon pottery, and had a low Roman content, largely general fourth-century, but not clearly residual enough to be given the latest classification on Plate VI. However, there was a fair number of small finds (MF8, Catalogue Nos 26, 53, 76, 99, 123, 128, 129, 147, 164, 211, 225, and a coin of Arcadius, C51). It is because some or all of the Anglo-Saxon pottery was residual, therefore, that the ditch may have been one of the latest anywhere on the site.

Within the enclosure, only one feature, F1103, can be given to Period 5. It was a sunken hearth (Chapter 3.II, **30**) containing very little, but had at its bottom some Anglo-Saxon pottery (321). It should not have been out in the open and the only obvious roof would have been that of the pentice structure already assigned to Period 4, but possibly of Period 5 (see p. 27).

One result of this discussion is to see that Barn 3 survived into Period 5 and, as the early east end of F1065 had had Anglo-Saxon pottery in it (319), into the Anglo-Saxon phases. Further support for this view comes from a gulley, F1149, along the south, or uphill, wall of the barn some 11.25m long and almost exactly equidistant from each end of the barn. The ditch should have taken water from the barn's roof. Amongst the sparse generalised fourth-century Roman pottery in it, and specifically from the lower fill, was an Anglo-Saxon sherd (322): it could only have got there when the feature was fully open: the barn should have been standing when such pottery first arrived on the site. But it had been taken down before the end of the period as the east and south walls of the barn were breached by ditches and the west room was cut by the final form of F1065.

The ditch cutting the south end of the east wall was shallow, F1146. The pottery from it was general fourth-century with a little Anglo-Saxon (323). At the west end of the south wall were two ditches, one, F1128, cutting the other, F1129 (Fig. 34, [118]). Both ditches had pottery dating after 350 in them (324). Before these had been dug, the final hollow of the Period 3 boundary ditches had filled up. The dating is not specific, just general fourth-century with a hint of late material (325). The full course of the earlier of the two ditches, F1129, is not known, but the latest layer in F1089 to the west may have belonged to it: it contained Anglo-Saxon and mixed Roman pottery mainly in small sherds (326). The purpose of F1129 is obscure unless it should be read with F1065. In that case, its north end forms one side of an offset entrance, about 5m wide, with the last east end of F1065. Therefore, the south boundary wall would have been demolished, but there was no sign of what formed the rest of the enclosure to the west. It is hard to fit F1128 in and not enough was seen of F1146 at the west end to make a comment.

The last two features to be described were both gulleys or ditches and belonged to activity south of the main site. The first, F1153 (Fig. 26, [70]), came to an end just before meeting F1129. The dating of both F1153 and that of the adjacent part of F1129 was general fourth-century (327).

The other ditch, F1114, was parallel with F1153 and a continuation of its course would have passed the south-west corner of the walled yard. Apart from the undoubted Anglo-Saxon content of the top layers (Anglo-Saxon Pottery, Group 1, Nos 1–20), and including ten *tibiotarsi* of a species of redshank (see The Bird Bones), the rest of the feature proved difficult to date and there may have been a recut of a long defunct feature surviving as a hollow (328). The mixed character and quantity of the finds, including a buckle tongue with a zoomorphic end, an iron reaping hook and the Barred Zoomorphic comb (Catalogue Nos 27, 142, 251), could point to a major Early Anglo-Saxon area south of the stripped site, outside both the field and development area of the New Town. In this area is a brick-earth quarry with a conveyor belt and an access road along the north side: there is no chance of further work there. However, the isolated siting of the other major group of Anglo-Saxon pottery in the Main Yard, should show that to automatically assume that an important area of the Anglo-Saxon site has been lost could be a mistake.

Summary

The description of the site will have shown that the presentation of a simple phased development would be unwise. The problem is compounded by a lack of good evidence for just how much of, and in what state, the Period 4 layout survived. The difficult nature of the dating, based on poorly understood developments in Late Roman pottery as well as an Anglo-Saxon assemblage largely incapable of being divided into chronological stages, coupled with residual quantities of both, further confuse the issue. There were, however, traces of sequences of events which can be used to arrive at a primitive scheme for the period. A broad view depends on two details: firstly, how much of the Roman site was intact early in Period 5; secondly, how the detectable successions can be related to yield a credible organic development covering the 100 to 150 years the period seems to have lasted.

The Roman layout on the east side of the site seems to have survived in a sufficiently good state for new functions to have been placed there without undue disruption, but it was there that a proper Roman element was hardest to detect. Most of the features were open and could only have acquired material deliberately thrown into them or derived adventitiously from rubbish-producing activity near by.

Barn 4 was reduced, a new drier and vat-base being provided, and the ditch system was replanned and altered more than once. The area between that and the next building to the west was maintained, its north ditch being tied to those round the barn. The large rectangular building almost certainly survived in a truncated form. The developments along the rest of the north side of the site imply that the yard wall, and the south wall at least of the building at its east end, also carried on into Period 5. On the west side, the Anglo-Saxon remains make little or no sense unless the Barn 1, the Small Yard, and the House at the north end were at least present at the beginning of the period. And the same goes for the yard wall and Barn 3.

Three phases are suggested which beg the question of the precise relationship of the two cultures involved:

- ***a*** Roman, possibly with the first introduction of Anglo-Saxons;
- ***b*** An apparent purely Anglo-Saxon use of the site still basically in Roman form;
- ***c*** Anglo-Saxon use of a site showing signs of radical change.

The only fixed points in the chronology are the limits of Period 5 itself: *c*.375 into the early sixth century. There is no dating for the intermediate stages. The choice of *c*.375 as the beginning of Period 5 has been discussed (pp 163–4, 174) and the assessment of the end date depends exclusively on a few Anglo-Saxon small finds (Catalogue Nos 243–250) whose dating is not close and only generally supported by the characteristics of the Anglo-Saxon pottery (see below). The discussion of Periods 3 and 4 showed that the Main Yard was cleared of major incident until, at the beginning of Period 5, the disuse of the pond meant that it had become empty and remained so for phase *a*. There are signs that some colonisation in the yard took place in either phase *b* or *c*, even if the large building in the centre is dismissed.

As with any large site with an unbroken history, there is bound to be an element of ambiguity in what constitutes the precise difference in plan between one period or phase and another. The phases are only general propositions, and may not have lasted the same time in all parts of the site or even be represented in some areas. But it can at least be assumed that all three phases were present, even if disentangling one from another is impossible at times.

Starting again with the east side, but looking at it as a whole, the suggested sequence is that in phase *a* occurred all those features containing only Roman pottery, but which were demonstrably not later than any with Anglo-Saxon. Also included are those which may mark first Anglo-Saxon use of the site. The plan, therefore, has the central enclosure, but with no good south boundary: activities could have spread south perhaps as far as F81. This would allow the identifiable Anglo-Saxon House, phase *b* or *c* itself, to have been roughly contemporary with the isolated features against the east baulk further south.

The original Roman southern enclosure could have been divided into two, the south-western portion being reserved for special uses, one of which seems to have been for the disposal of human bodies. The latter would be in phase *b* in the south and contemporary with the recutting of the ditch of the original central enclosure. Phase *c* in the south would mark the introduction of domestic activity represented by post-holes and the demonstrable Anglo-Saxon hearth. This would go with the elimination of the subdivisions further north apparently containing only Roman material, suggesting that they did not last to the end of phase *a*, except for F129. Other phase *c* elements would include the introduction of the gate in the middle. Domestic activity in the central enclosure would then be in phases *b* and *c*, when the Sunken-featured Building was created, though it had passed out of use before the end. During phase *c*, generally, the integrity of the Main Yard was breaking down, its east side beginning to disappear.

Along the north edge of the site, Barn 4 was certainly reduced in phase *a* and a new drier built; the well next to the barn continued; the rectangular building was cut down in size; and the yard wall and part at least of the building added to the west side of that would still have been standing. There were three minor phases behind the rectangular building which, at first sight, divide evenly between the three major phases, but the activity split more probably into two: the enclosures followed by the possible Anglo-Saxon well and the slot. If this is the more likely course, the well would suit phase *c*. The enclosures would belong entirely to phase *b* which would mean that the reduction of the rectangular building had happened by the end of phase *a*. The north ditch of the area to its east would have continued to the end of phase *a* and possibly beyond and could have been replaced in phase *b* or *c* by the slot, F232.

How the barn fitted in is less clear. Its existence in phase a seems assured. Both the barn and its well, F172, would have passed out of use before phase *c*, the latter being replaced by new water points in the Main Yard, and on the extreme north-east of the site. This arrangement would allow one to have served the Sunken-featured Building and perhaps the area east of Barn 4. This would suit phase *c* with the loss of the ditch separating the area from the Main Yard.

Although an extensive area was stripped north of the Roman House, little can be said about what happened there as only salvage work took place and the platform, possibly for a house, was only investigated because of the Anglo-Saxon pottery lying there. If there had been an effective barrier to the north, formed firstly by the Roman House and then by its replacement, the platform would have probably belonged to phase *c*. The general lack of Anglo-Saxon pottery in nearby ditches would suit this as all of these could have been silted up by the time that Anglo-Saxon pottery was being discarded here.

It is in the Small Yard and the buildings to the north and south that the crux of the interpretation of the site lies.

If the nine-post structure and its probable fence were deliberately placed in relation to the Roman site, then enough of that survived to be useful and the posted building would belong to phase *a* or *b*. The way in which Roman walling seems to have survived in close conjunction with the timber building overlying the Roman House again suggests the continued presence of the Small Yard. The ditch along the north side of the Barn 1 could have been a boundary to fill the gap created by its demolition, but the sequence of deposits in it, the layout and the succession of ditches east of the barn are hard to explain unless that was still standing. Both barn and stone house could belong to phase *a*, but the two major timber structures probably suit phase *b*, in which case, the barn at least may have continued in use. Both the major Anglo-Saxon buildings had limestone rag pieces packing their posts and, even if this is a weak link, it could be a mark of contemporaneity with the demolition of the bulk of the Roman House.

The track cutting diagonally across the west boundary of the Small Yard seems to have respected the end of the Barn 1 which may, therefore, still have been standing, but the track would belong at best to phase *c*. The building of another house in the Small Yard may mark phase *c* activity as it seems to have its end aligned on the track. The occurrence of the greater number of the Anglo-Saxon metalwork finds in the western part of the site would go with a marked phase *c* presence here as the material belongs unequivocally to the latter part of the Anglo-Saxon occupation on the site, especially as most of this material is minor and probably the result of casual loss during everyday work: the focus of activity should have been somewhere nearby.

The southern part of the site can be reduced to a simple scheme: Barn 3 standing through at least phase *a*, as its south ditch began to receive Anglo-Saxon pottery before it was demolished. Also in phase *a*, part of the Main Yard was cut off from the rest by ditches which respected Roman buildings. In phase *c* the south barn had gone and the ditches extended to cut into it to redefine the enclave, although Roman walling may well have still been standing at its west end.

The proper character of the old Main Yard was hard to define. Certainly pits and the Sunken-featured Building, and possibly a large timber building in the centre, belong to it. Such activity here could have been the source of the Anglo-Saxon pottery found in the hollow of the long-destroyed Period 3 well. Otherwise the quantity of Roman sherds here was negligible and residual, a condition which would suit phase *c* admirably. Even so, to propose such a colonisation might be a sign that the Main Yard was still useful enough to have provided a limit.

Overall, there are many permutations, and it will always be a matter of opinion as to whether the evidence is strong enough to bear the weight of so much conjecture. The site was not easy to periodise and in Period 5 was impossible to treat in a normal fashion. It is a salutary thought that, had it not been for the Anglo-Saxon presence, little indeed could have been made of the final stages of the Roman site. Had there been only an increasingly debilitated use of Roman pottery, coupled with a growing residual factor, the temptation would have been to close the site down between *c*.375 and 400, the coin of Arcadius along with two of the House of Theodosius (The Coins, C51–C53) being the only sign of a date as late as the end of the fourth century. It may be doubted if any amount of sophistry could have carried the life of the Roman site into the fifth century without the aid of the 'foreign' material which, while it shows that Orton Hall Farm cannot be easily dismissed as just another Roman farmstead, imposes its own obligations in interpretation.

VII. Post-Period 5 activity
(Pl. VII)

No detectable trace was found of any use of the site between the end of Period 5 and the creation of ridge and furrow. This was preserved in pasture until the Second World War, after which it reverted to pasture until the early 1960s when a bounty was paid for every new acre brought into cultivation, and ploughing continued until the land was taken for development.

Not only was there no dating evidence for the laying out of the ridge and furrow, but there was so little medieval pottery on the site that, if manuring was carried out, the source of the manure was not related to ordinary household activities. This low incidence of evidence for the close connection between domestic sites and the hoarding of manure, human or otherwise, persisted well into the nineteenth century when there was an increase in the amount of pottery (P1). However, the quantity was still small and this should be related to the fact that the site had become permanent pasture, a condition possibly dating back to the Middle Ages.

There was no good evidence for how the furrows were originally marked out although Fs385 and 386 (P2) at the very bottom of furrows, and traces of similar features elsewhere but not numbered, were either the very bottom of deliberately dug gulleys or only the result of extra deep ploughing now and then. No specific major act was really needed as the use of ephemeral markers for the first ploughing were all that were needed: once ploughed, the furrows would have been evident on the surface of the new fields ready for the next year.

Comments made throughout the descriptions of the five periods show that the furrows had caused considerable damage and the degree of this can be quantified as a minimum of 42 per cent of the area of the stripped site being occupied by furrows. Allowing that damage would have been continued through the depth of the final deposits of the site along the edge of each furrow, it cannot have been less than 50 per cent. Although the ridges, because of their increasing elevation, came, in time, to protect what lay beneath, the initial ploughings would have taken some toll and the field would only have been workable if there had been progressive clearance of stones from the site, a process which began again when it was brought under the plough in modern times.

The furrows, save one which could not be traced through to the south edge of the site, ran right across the excavations. There was no sign that in plan they had a regular sinuous reversed S curve, although there was a slight tendency in one or two. A headland had developed along the northern edge of the hedge forming both the boundary to the field and the development area of the Development Corporation. It is possible that there had been a further sub-division of the field locally, either to cut down on the length of the furlongs or to exclude the wet ground down the slope from the headland. F29 may have been inserted into what could have been a particularly wide ridge at the east end of the site.

Despite the impression of regularity which surviving ridge and furrow presents from both the ground and the air, there was a surprising amount of variation to be seen at Orton Hall Farm. The furrows varied from 1.8m to 4m, taking the average of their individual widths, and, on the same basis, the ridges varied from 2.4m to 7.4m. It is possible to detect on the plan two sizes of ridge: those less than 5m, averaging 3.59m, and those greater than that, averaging 6.26m. If the spacing from centre to centre is looked at, the same kind of answer is produced and there may have been two standards used in laying the lands out: those at the west end, and perhaps the east end as well, with the group between being deliberately narrower.

The next major event was the division of the site by what can be described as close boundaries: F479 (P3) and the fence row (P4). It is not certain when this took place and it should not be assumed that the date of the Enclosure Award, 1728, or shortly after, provides the obvious answer, but it is possible that the solitary post-medieval pit, F651 (P5), belonged to a time when the arable had become pasture: it seems unlikely that a pit would have been dug in a ploughed field. Only one complete division of the suggested new layout was present. It was 83m wide and coincided with the width of the set of narrower ridges already mentioned.

The fence row contained little dating evidence, but enough to show that part of it must post-date the site. The row ran through the site parallel with the furrows on each side right up to the edge of the headland on the south side. The character of the holes and the spacings between them do not form a consistent series. The entries in (P4) show which holes belong to the basic row. They had a spacing varying from 4.9m to 5.25m, the commonest being 5m. None of these had a trace of a post or of packing and it is in this series that the small amount of dating evidence was found. Some of the holes — AF32, AF36 and F45 — were deep enough for a trace of a post to have survived, had it been allowed to rot in position, but the absence of one in them suggests that the whole series had been dug out to throw two closes together before a hedge could form. Two of the other post-holes still retained evidence for posts, both packed round with stone: Fs38 and 39. These, along with Fs62, 58–9, 69 and 71 may belong to Period 5, but they make better sense here. Fs38 and 39 may have been designed to suit a gate, but the sizes of the timbers in them, 0.09m by 0.08m and 0.1m by 0.08m, do not seem large enough.

The last definite phase of land improvement was part of a system of land drains (P6). F19 ran down a furrow parallel with the fence row while F760 ran down another furrow to the west and only in the northern part of that. Neither was linked on the site to an extensive series of drains, but this could have happened to the north down the slope below the spring line. It is a comment on how wet the ground there was that the southern part of the site was thought dry enough not to need draining: even in 1973 when the brick pits to the south must have reduced the supply of ground water, there was still enough to have kept full one Iron Age and two Roman ponds.

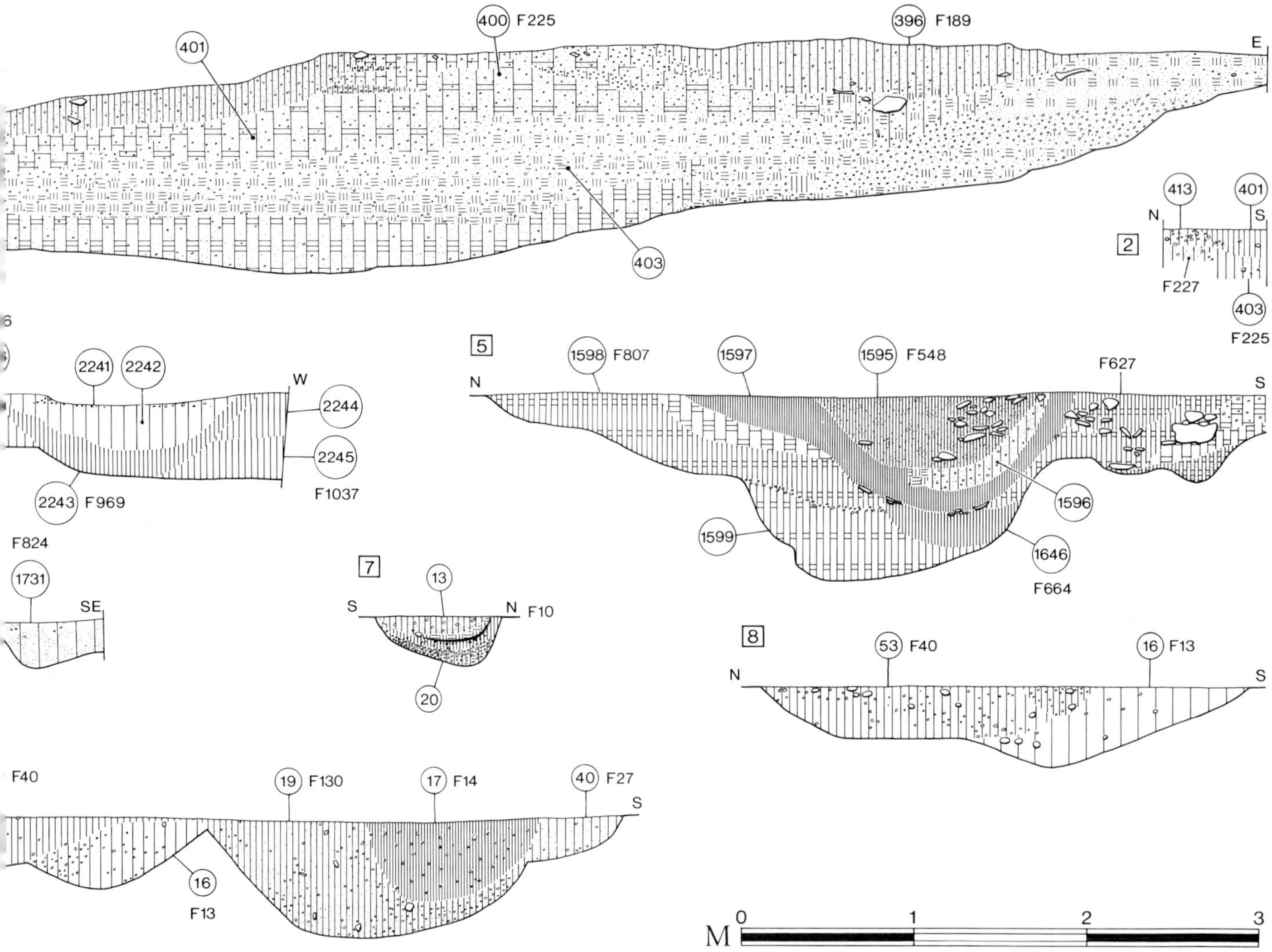

Sections [1]–[9].

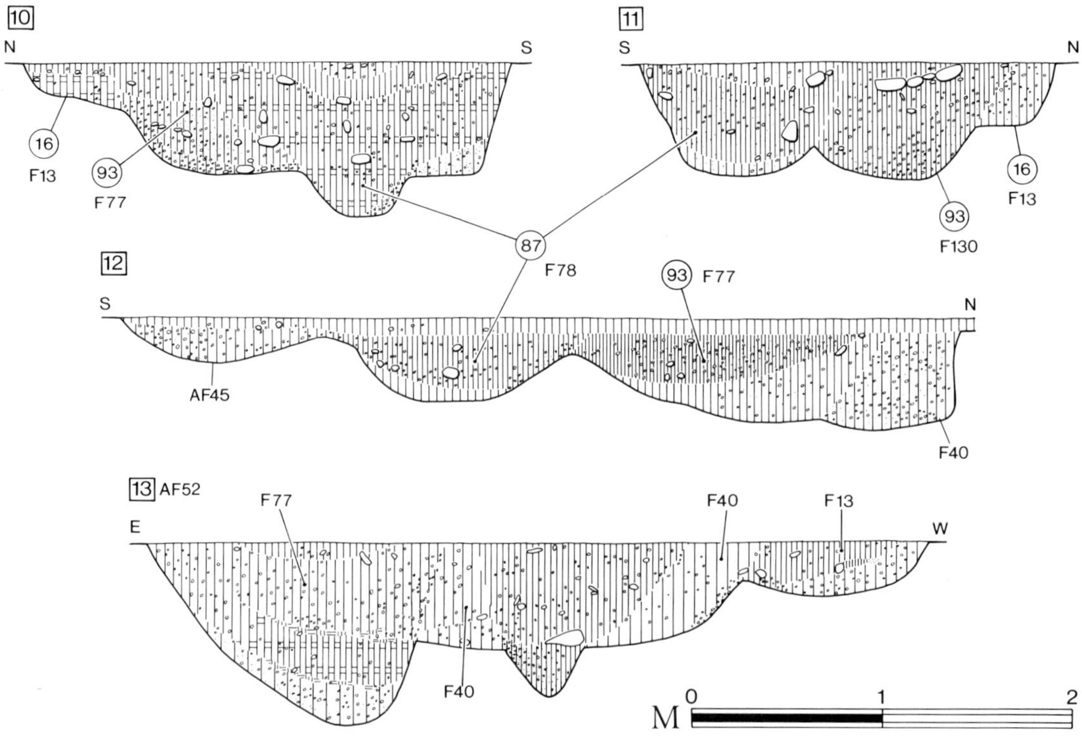

Figure 15 Sections [10]–[13].

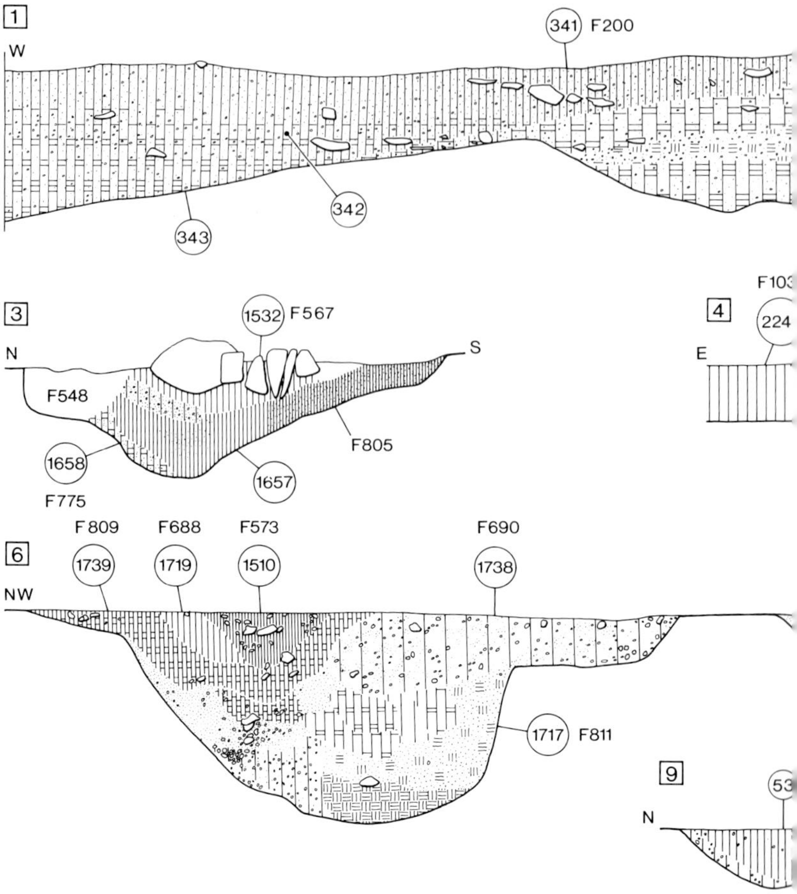

1
W
341 F200
342
343
3
1532 F567
N
S
F548
F805
1658
1657
F775
4
F103
224
E
F809
F688
F573
F690
6
1739
1719
1510
1738
NW
1717 F811
9
N
53

Figure 14

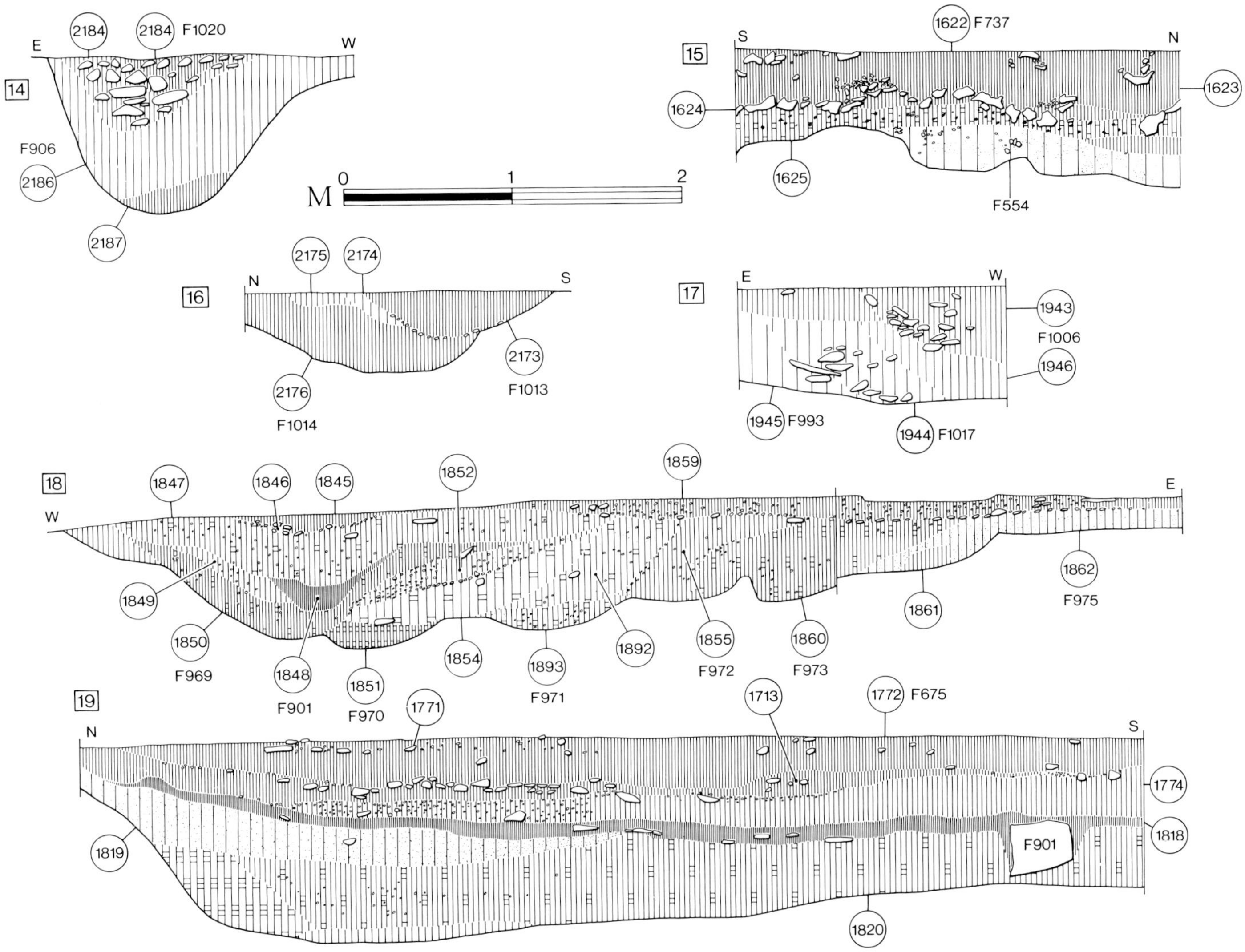

Figure 16 Sections [14]–[19].

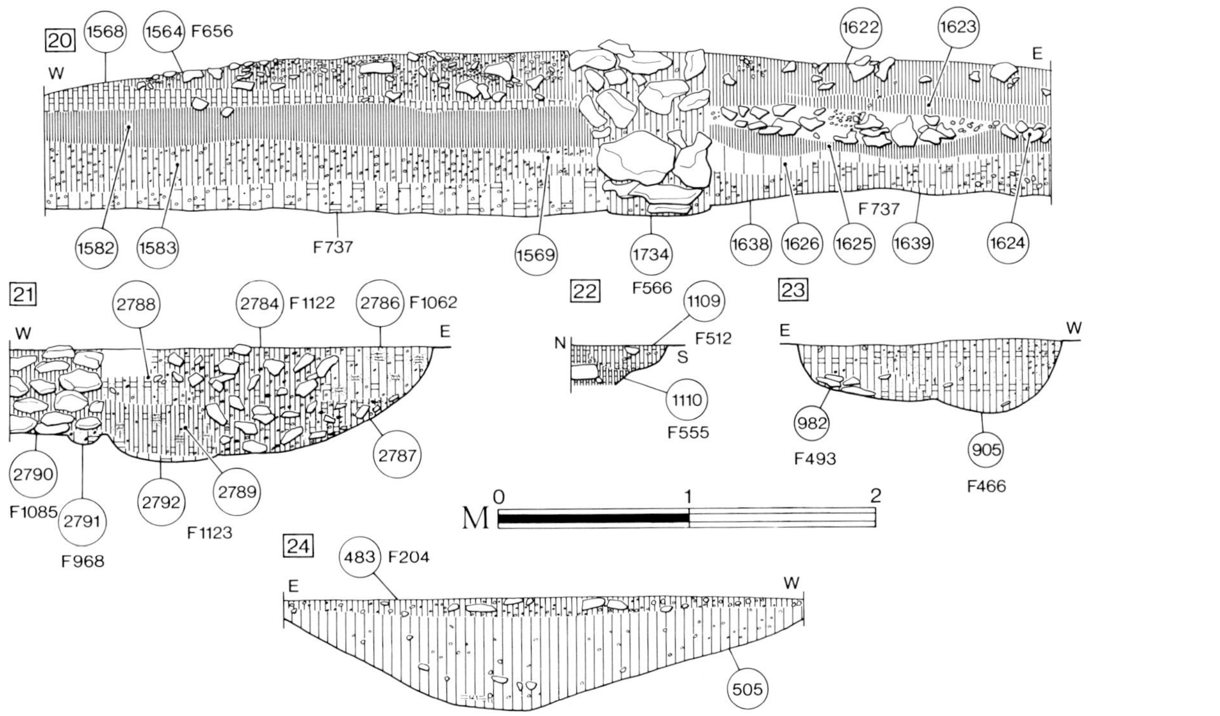

Figure 17 Sections [20]–[24].

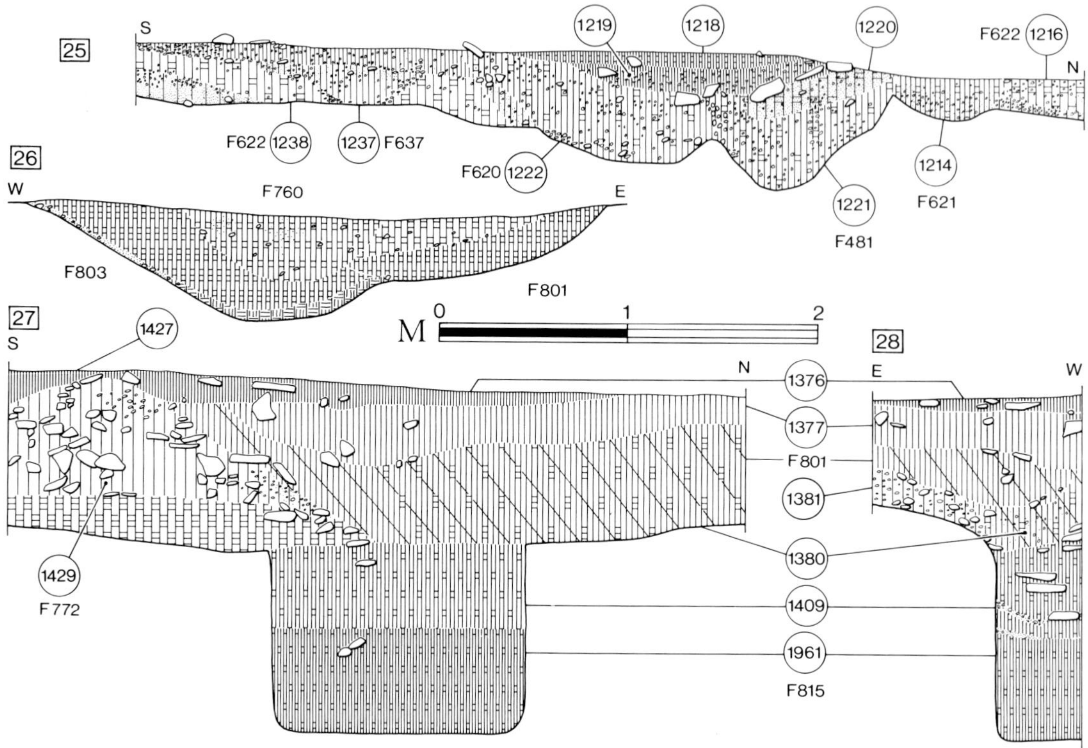

Figure 18 Sections [25]–[28].

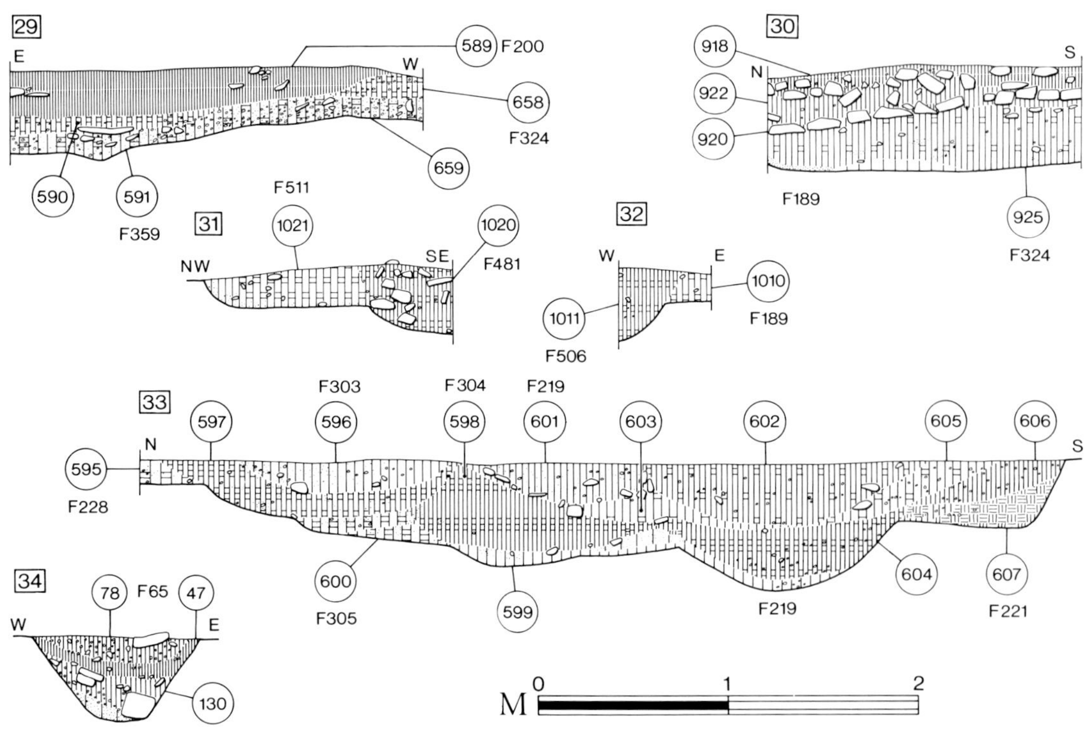

Figure 19 Sections [29]–[34].

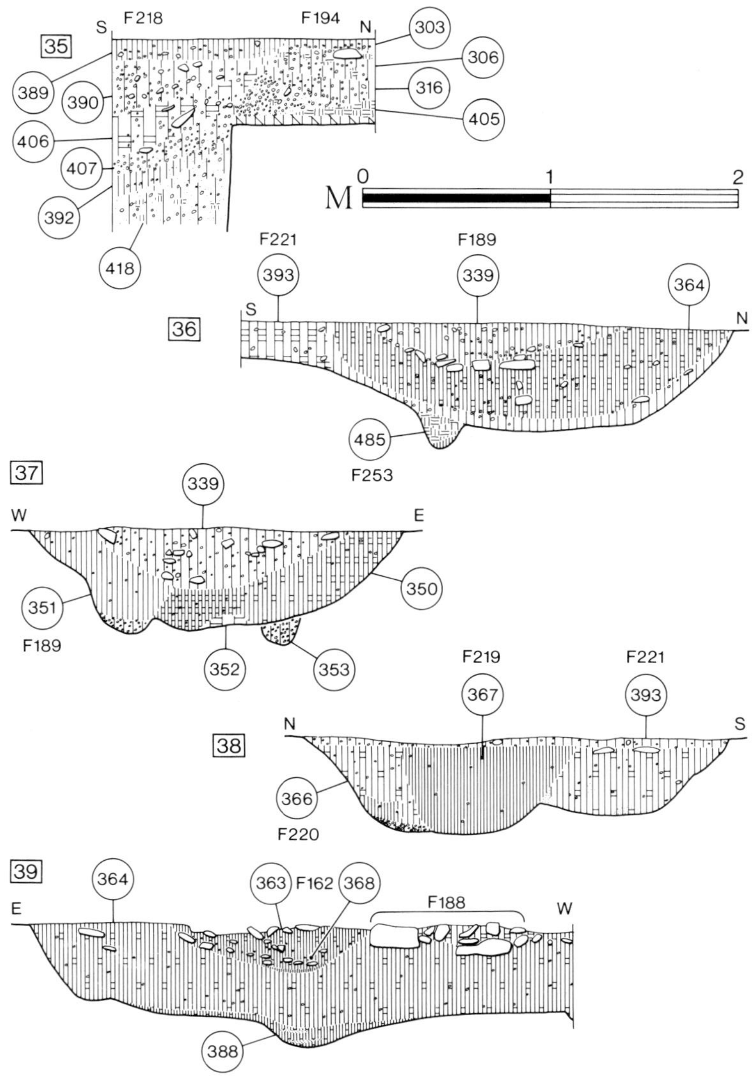

Figure 20 Sections [35]–[39].

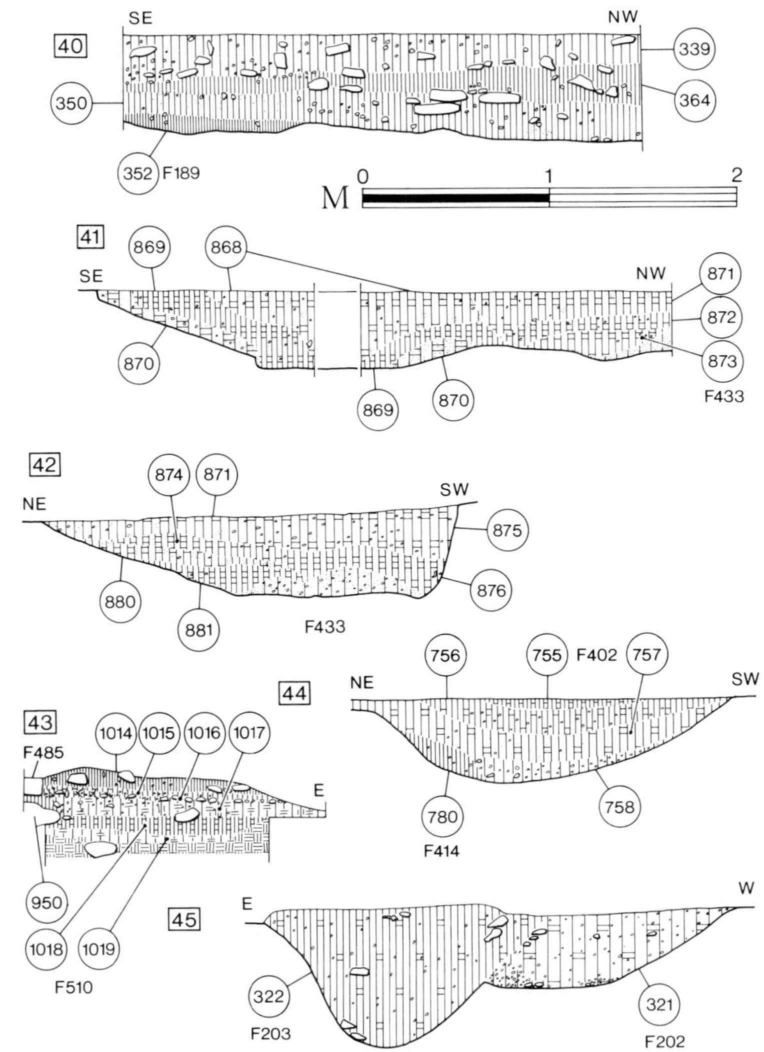

Figure 21 Sections [40]–[45].

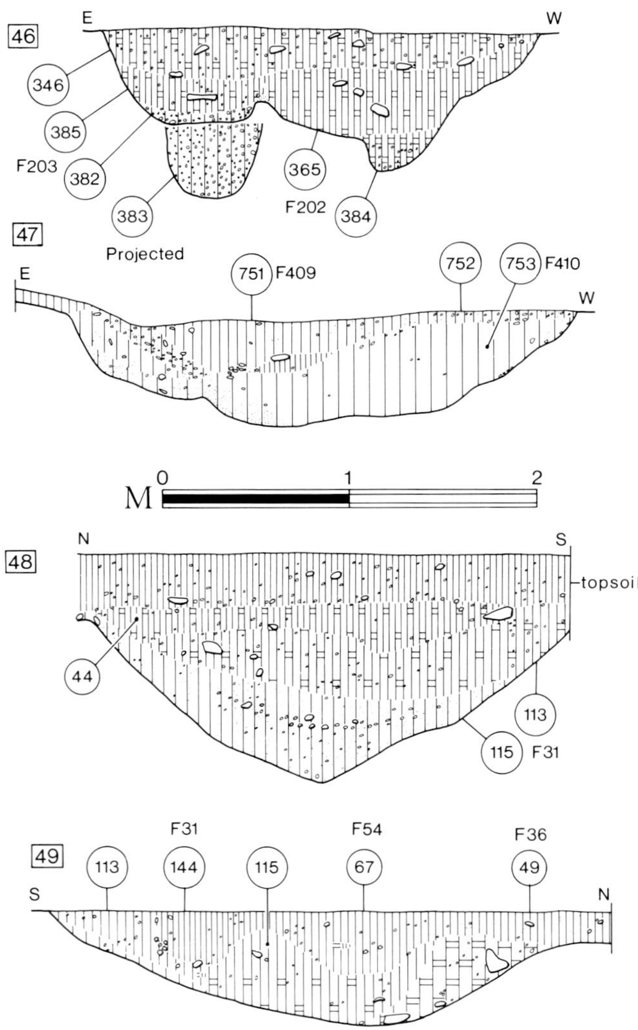

Figure 22 Sections [46]–[49].

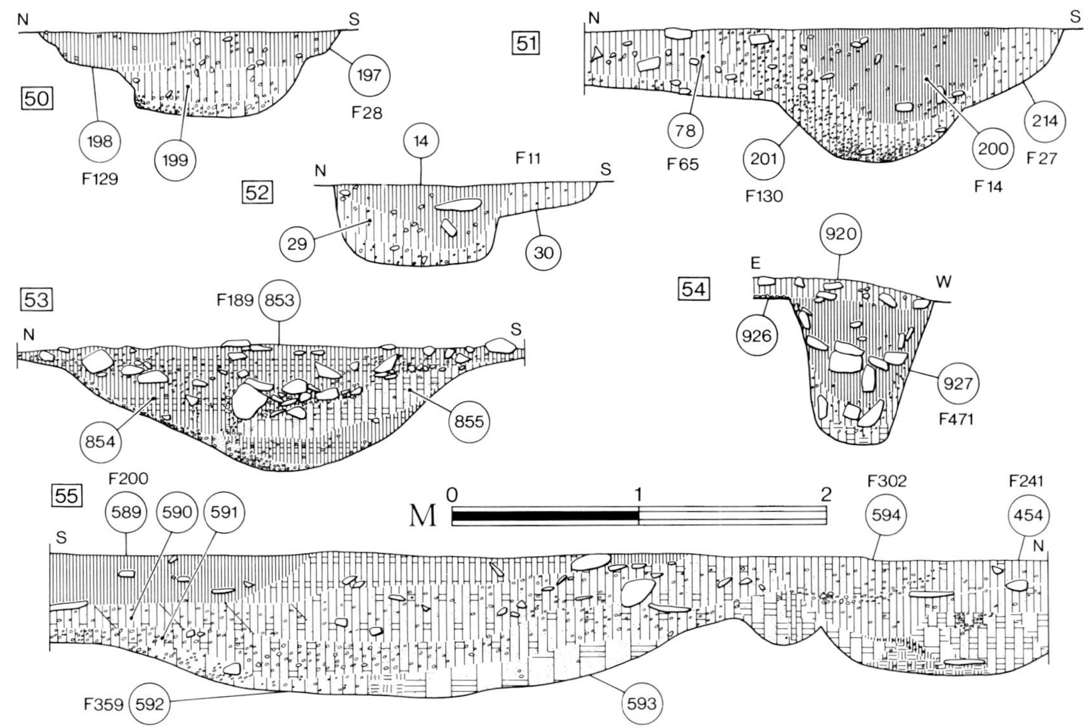

Figure 23 Sections [50]–[55].

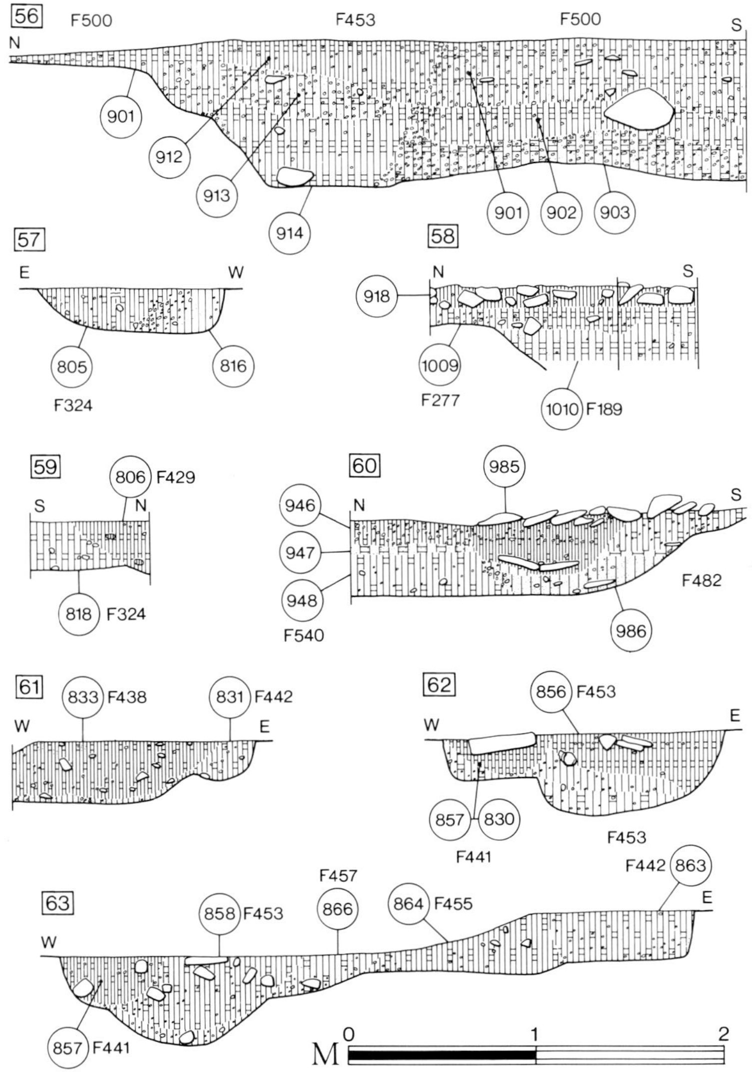

Figure 24 Sections [56]–[63].

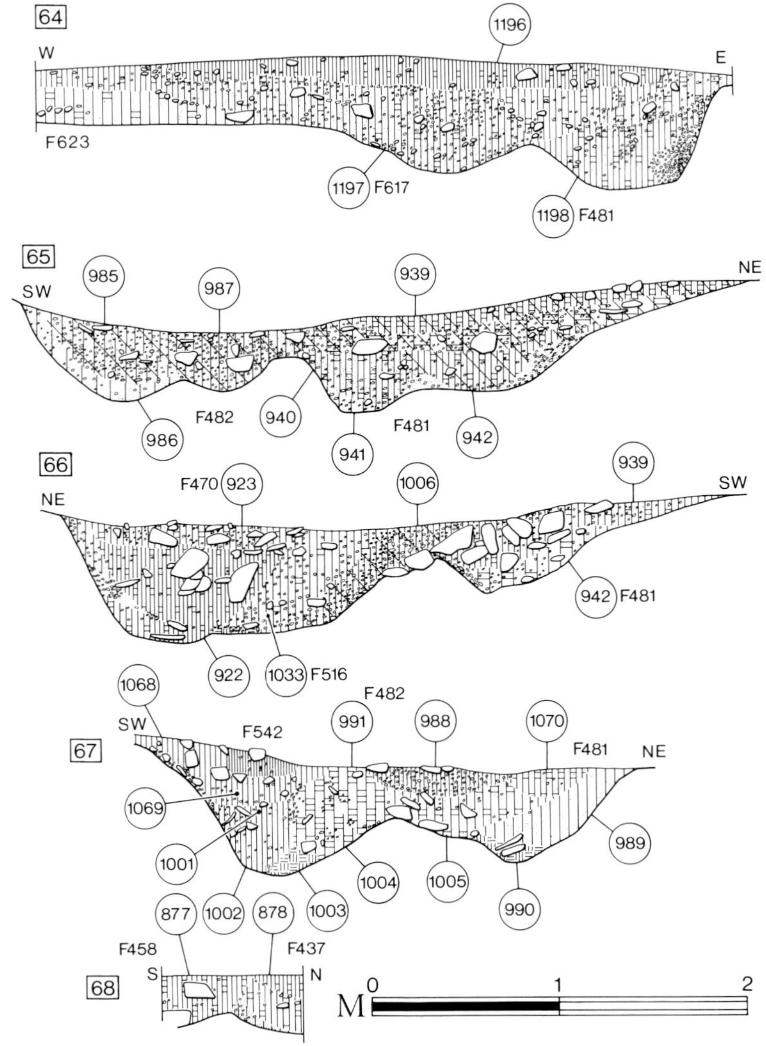

Figure 25 Sections [64]–[68].

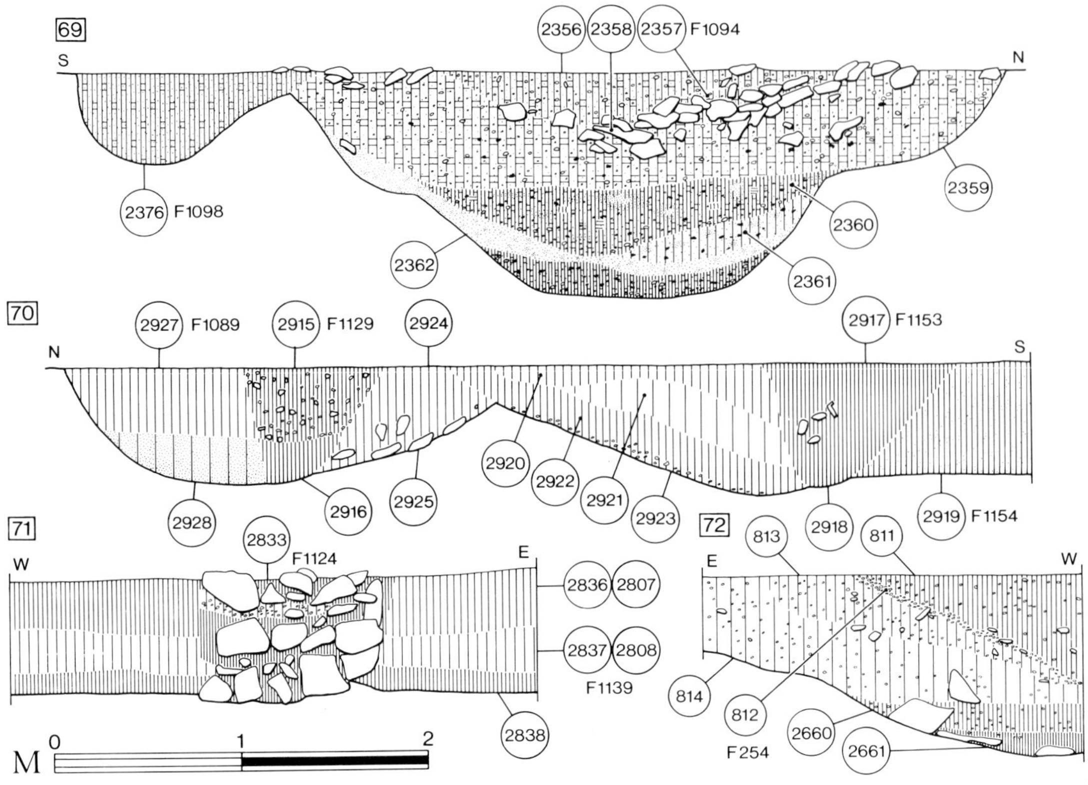

Figure 26 Sections [69]–[72].

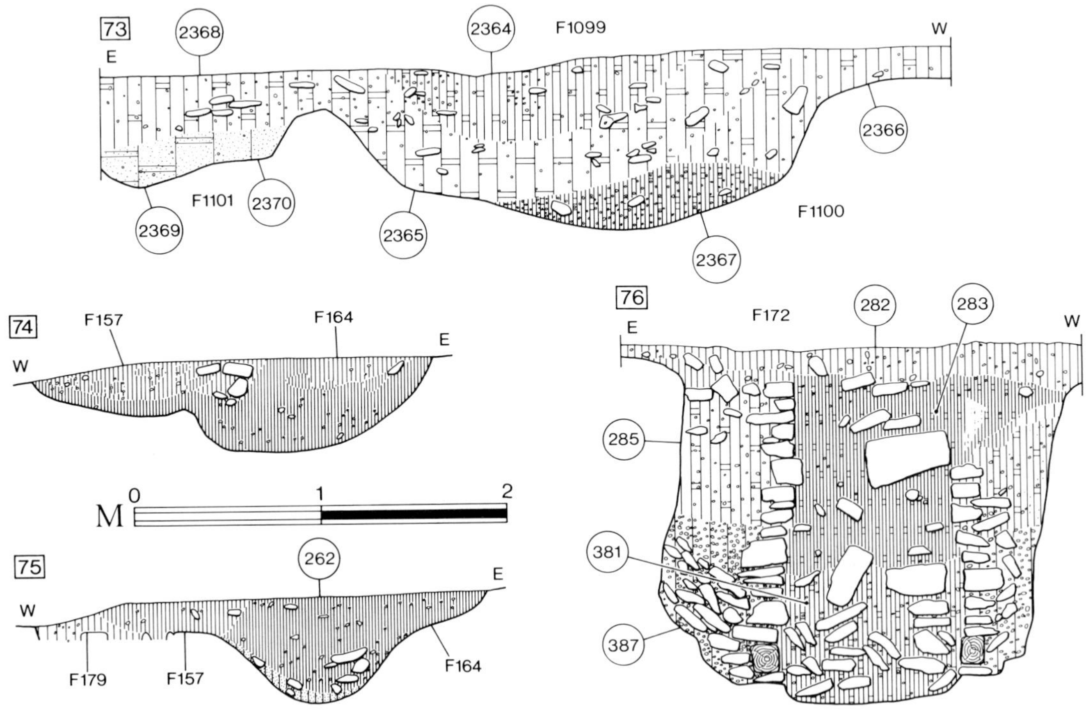

Figure 27 Sections [73]–[76].

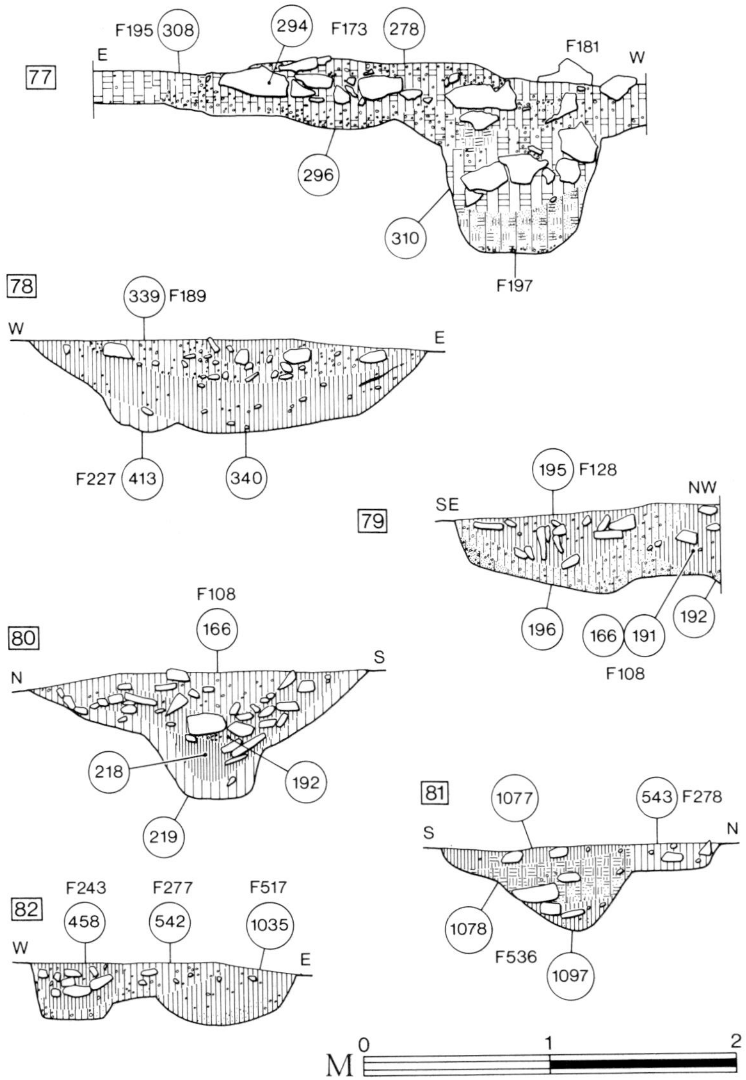

Figure 28 Sections [77]–[82].

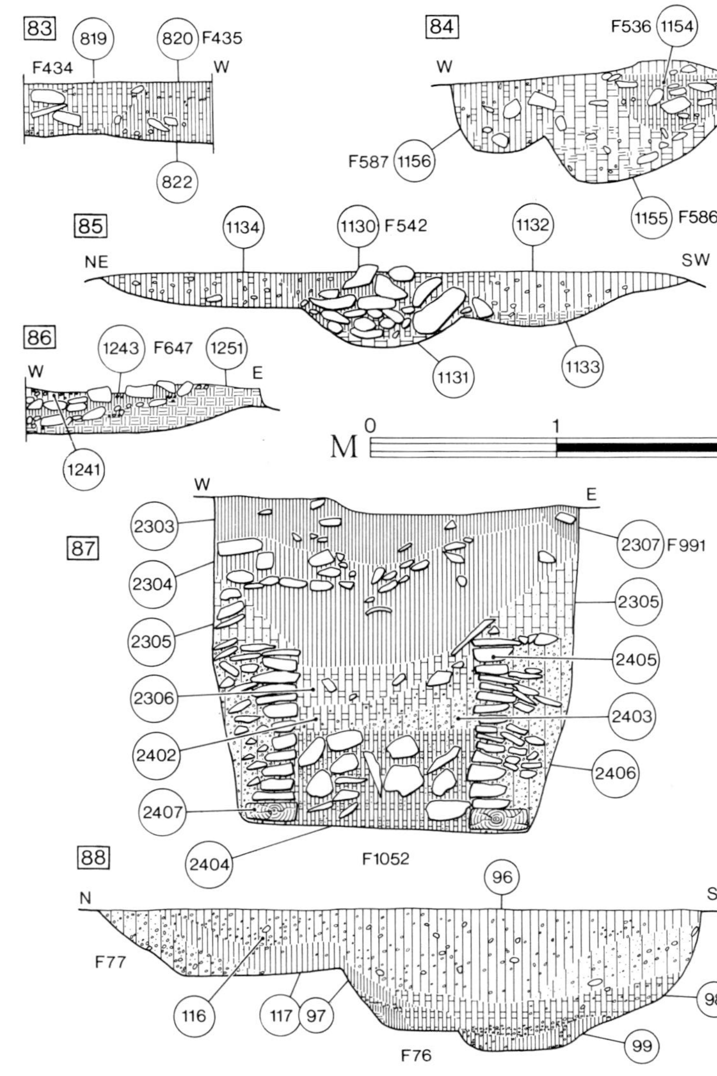

Figure 29 Sections [83]–[88].

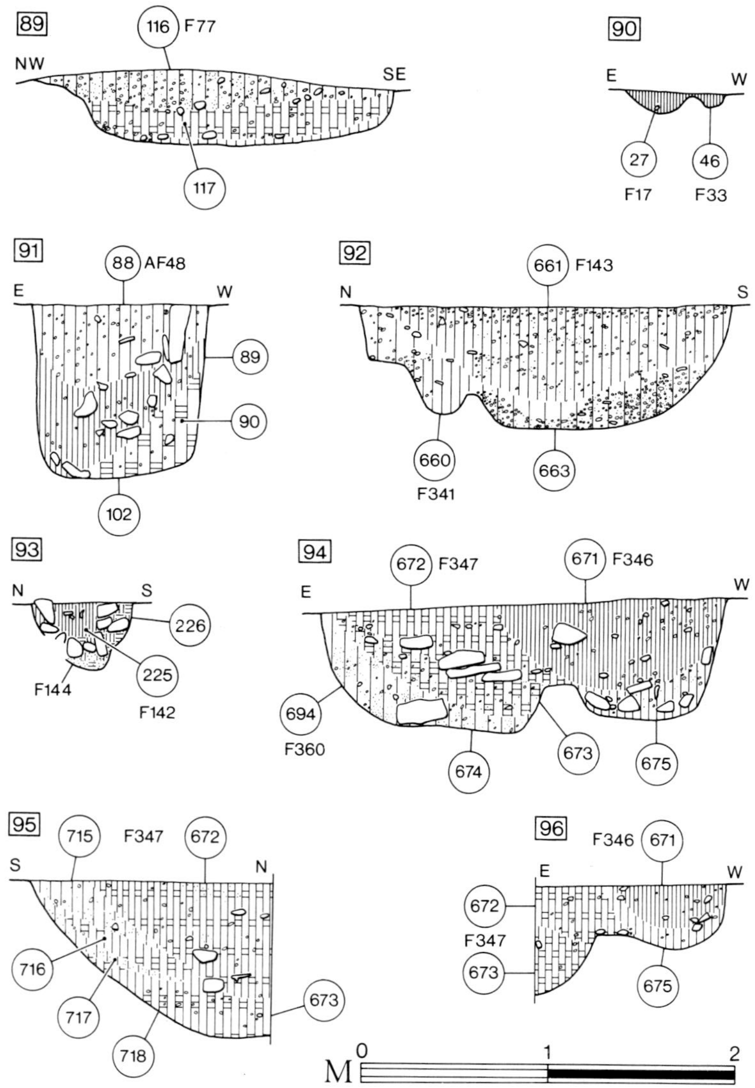

Figure 30 Sections [89]–[96].

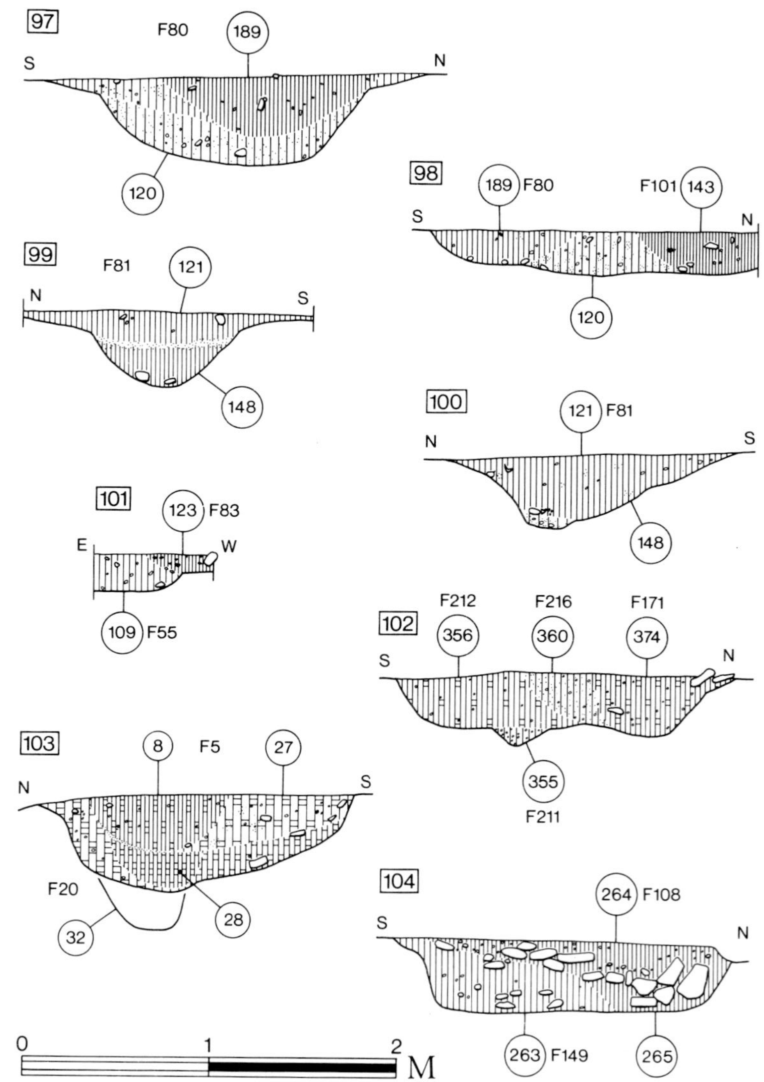

Figure 31 Sections [97]–[104].

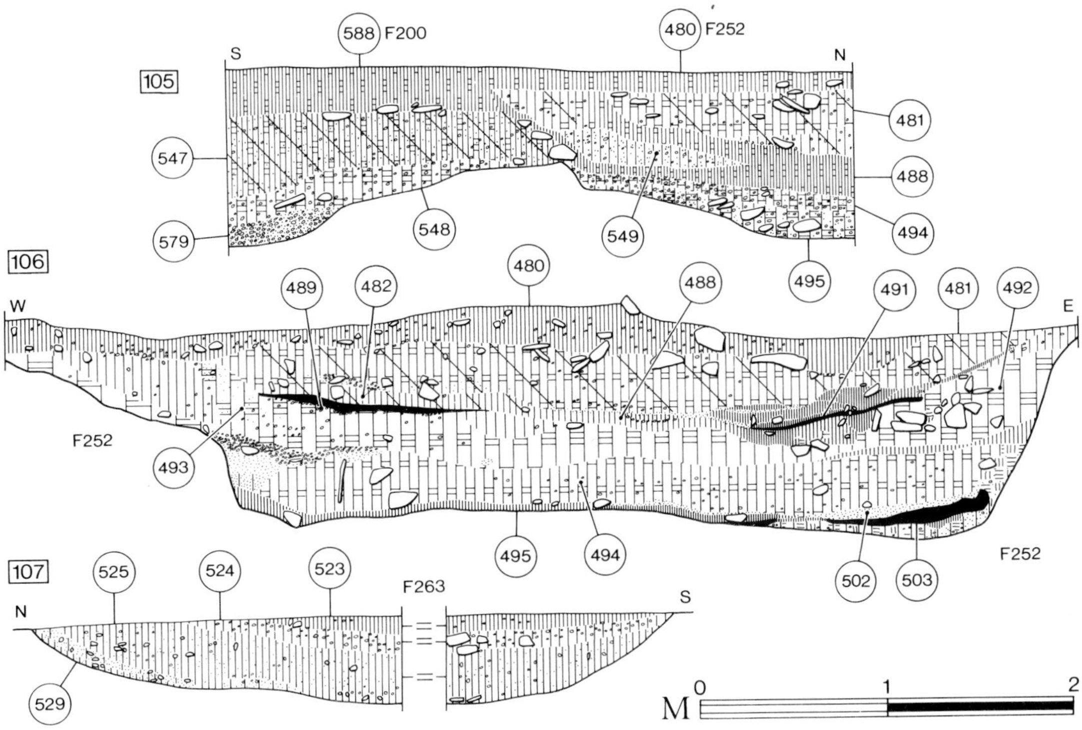

Figure 32 Sections [105]–[107].

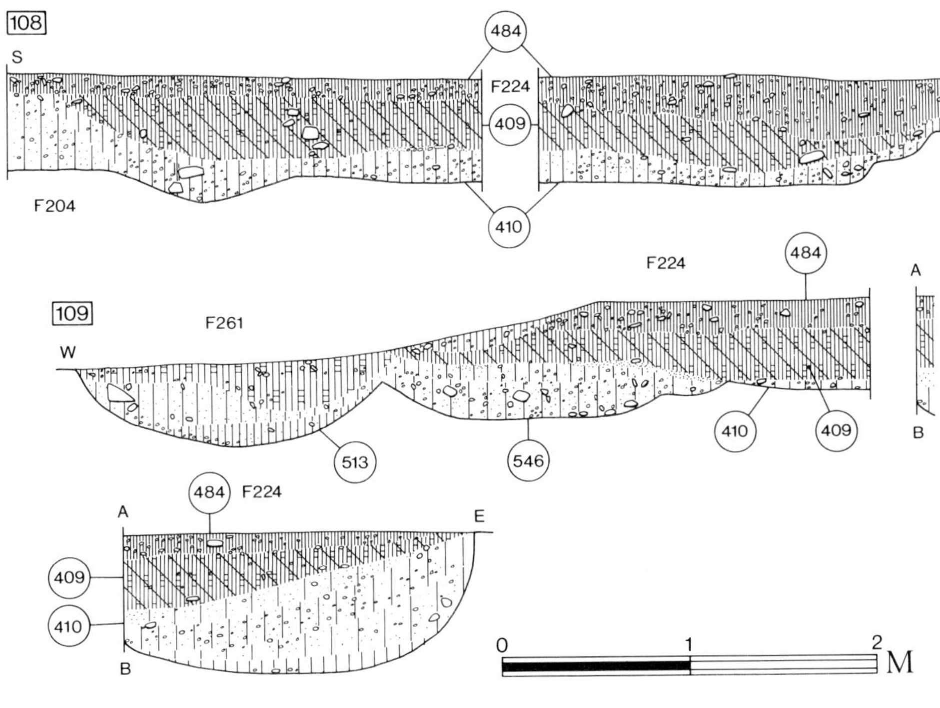

Figure 33 Sections [108]–[109].

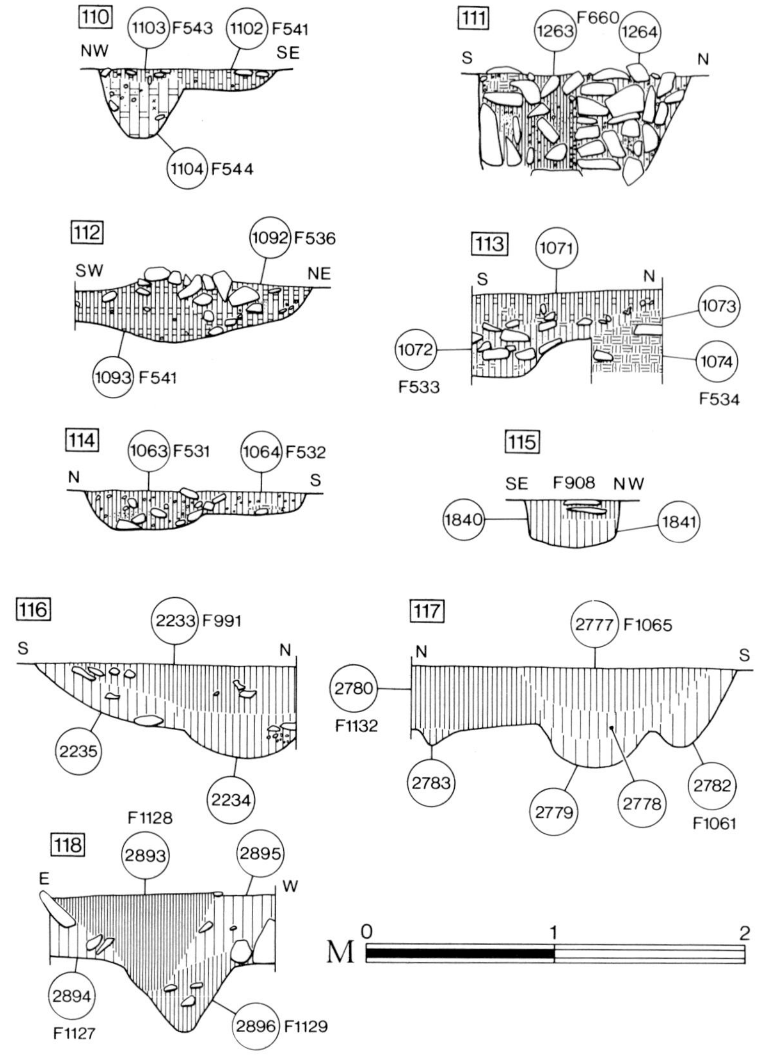

Figure 34 Sections [110]–[118].

Chapter 2. The Roman Buildings

I. The Barns

Barn 1
(Figs 35–37)

Periods 2–5
Largely stripped by Mr Dakin, only some of the interior remained undisturbed and, as photographs taken in the 1960s show, much of the structure had been removed to improve the field. Further damage occurred in 1971 when the main structural elements were exposed using a JCB which, along with the site being left open until 1973, caused further damage to the post pads. The best preserved parts of the walls had been at the west end, the north and south walls only surviving where they had been sealed under ridges. By 1973 virtually all the south wall had disappeared and the west and north walls had been reduced. Most of the post-holes and bases inside had been disturbed in the 1960s and only photographs had survived. No floor levels survived and there was no direct evidence for points of entry.

What was left of the barn walls showed that they were all basically of the same width: 0.75m. None of the Period 1 ditches seems to have been fully dug out to provide deep foundations although, at the west end, these were fairly deep along the margins of Period 2 ditches. The care taken in building the drain in the south wall was not to provide good edges for that as the rest of the drain remained unlined. It suggests that the walls above were to be of stone to the eaves. The sites of possible doors have been discussed in the individual period descriptions.

Although no direct evidence survived, the dating of the internal features which places them in Period 2 suggests that the barn had been first built then in timber. The substitution of stone walls will only suit the developments at the west end of the site in Period 3, and the integration of the barn into a new layout may account for what looks like a conversion to a different function.

Unlike the other three barns, this had six posts on each side. As in Barns 3 and 4, there was evidence that the earth-fast posts had been resited on stone pads. The nave width averaged 7.55m, the aisles were narrow in comparison (see below) and the bays averaged *c*.3.4m.

The post-rows
(Figs 36 and 37)

The north row, from west to east

F935 The post-pit was about 1.15m in diameter and 0.64m deep. The profile was bowl-shaped and the post itself, *c*.0.3m across, placed in the centre without any obvious stone sole-plate unless the one shown in section was intended to serve as such. The post had been packed round with close-set limestone fragments, set in a dirty redeposited natural. Half had been removed by Mr Dakin. No evidence survived of a post-pad, possibly due to the furrow running over it. The base of the post may have rotted, the rest being removed. The section, however, could show that an earlier post had been dug out, so removing nearly all the original packing, to make way for its replacement: there was noticeably more stone and clay around the upper fill of the post-pipe than round the darker lower fill.

F936 The post-pit was about 1.05m in diameter and 0.58m deep. The pit was flat-bottomed and had near vertical sides. The post had been set slightly off-centre and was about 0.25m in diameter at its base. The packing round the post was similar to the bulk of that of F935. There was, again, no evidence for a post-pad succeeding the post which, from the profile of the post-pipe, may have been rocked out, the hole being backfilled with soil from the surface of the site.

F937 The diameter of the post-pit was *c*.1.1m. The pit was 0.6m deep with an almost flat bottom and vertical sides. The post had evidently been removed, but seems to have been placed centrally. The packing was a dirty redeposited natural containing very little stone. The lack of the latter, no doubt, made it easier to remove the post and the hole created was backfilled with dirty soil and stone, the latter increasing in density towards the top to form part of a poorly preserved post-pad. The stone in the packing seems to have been in the upper part only and the very wide pipe, *c*.48m across, could have resulted from the replacement of one post by another, set slightly to one side, which was in turn removed.

F938 A trench, dug by Mr Dakin, had cut into the periphery of the post-pit destroying its diameter at the excavated surface. It could not have been less than 1.15m. The pit was 0.92m deep with a flat bottom and steep sides. The original post had been set in the centre and was 0.3m in diameter. The lower fill, 0.45m deep, of the post-pit was of dirty redeposited natural without any stone. The rest of the packing was a mixture of stone set in clay, but with a large amount of dirty loam and gravel occurring sporadically. The change in the packing suggests that there were two periods with the replacement post being no more than 0.25m across which is emphasised by the upper post-pipe having leaned towards the east. The depth of a second post-pit would have been no more than just over 0.5m.

F939 The post-pit was about 0.95 in diameter, 0.8m deep and had a flat bottom and near vertical sides. The post was 0.2m across and had been pressed 0.03m deep into the bottom of the pit. The lower packing consisted of beds of limestone separated by a dirty redeposited natural with occasional lumps of clay; the upper part was a solid bed of clay with little stone. The two parts of the packing recalls that of F938, but there was no good sign that there had been more than one post or that this had been removed. The surface of the finally backfilled post-pipe had been firmed up with a bed of stone and there may have been a new post, about 0.29m across, set slightly east of its predecessor.

F940 The post-pit was 1.07m in diameter and 0.89m deep. The pit was flat-bottomed and its sides varied from being vertical to being very steep-sided. There were two periods, the first post-pipe was 0.25m in diameter, the second was 0.28m. The first post was centrally placed and packed round with dirty loamy soil. Most of this was removed to set the second post at a higher level firmly secured with stone packed in a loam with lumps of clay. The second post was removed and the resultant hole backfilled with dirty soil with the occasional lump of clay under a bed of stone.

The south row, west to east

F946 The post-pit was 0.17m in diameter and 0.91m deep at most. The sides were near vertical with rounded bottom edges running into the flat bottom. The post had been placed eccentrically, and was 0.19m across at the bottom. The post was straight-sided for 0.22m where it ran through a packing made up of redeposited natural. Above this, the pipe increased in width to 0.22m and the packing was uniformly of a cleaner redeposited natural above a bed of stones, both possibly derived from putting in the stone walls. At the top of the post-pit was a third type of packing consisting of stone with a slightly clayey loam matrix set round a straight-sided pipe. It is just possible that this represents a third period.

F945 Only the top was revealed and this was similar to F946.

F944 No details, it had been damaged in a machine-cut trench in 1971.

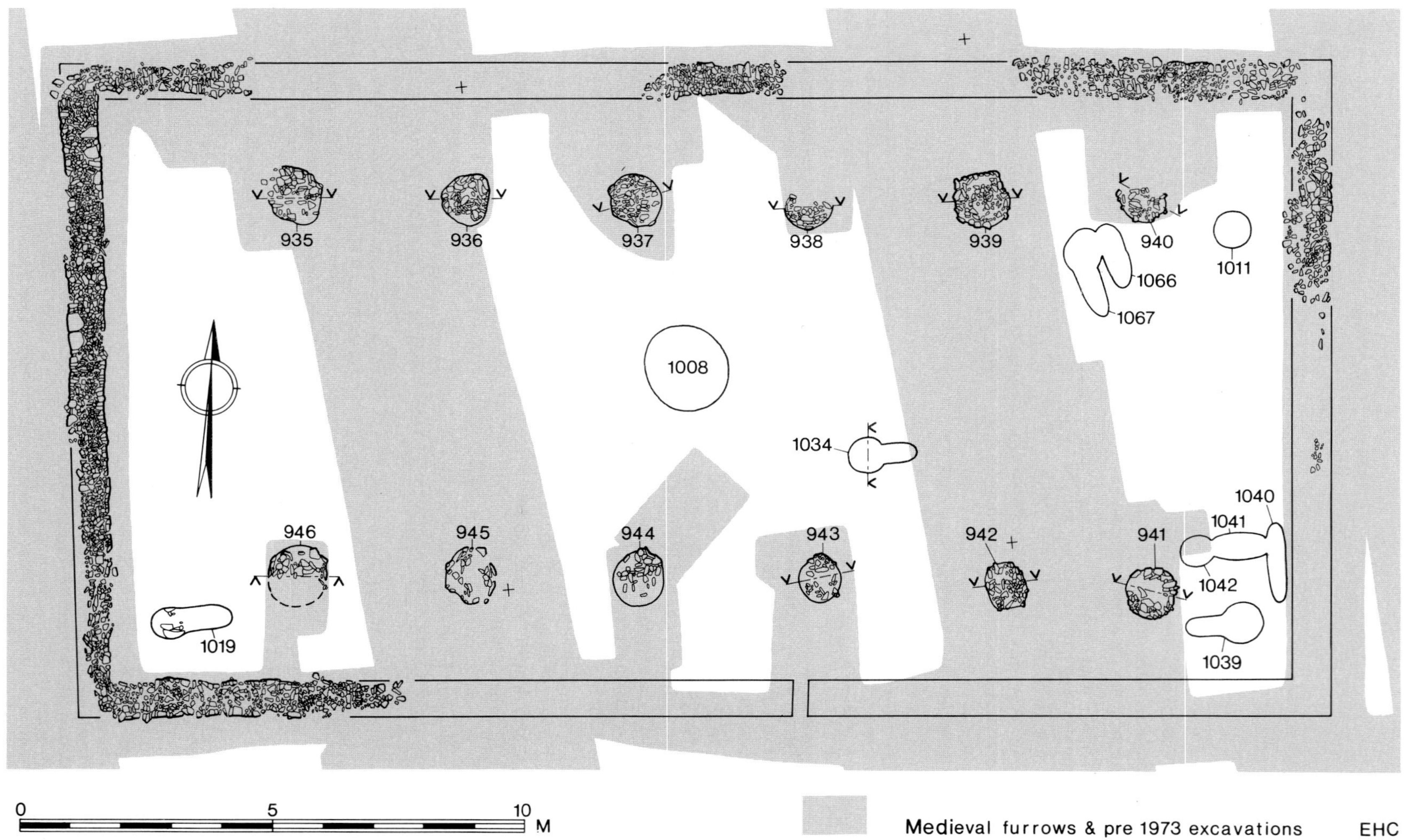

Figure 35 Plan of Barn 1. Scale 1:125.

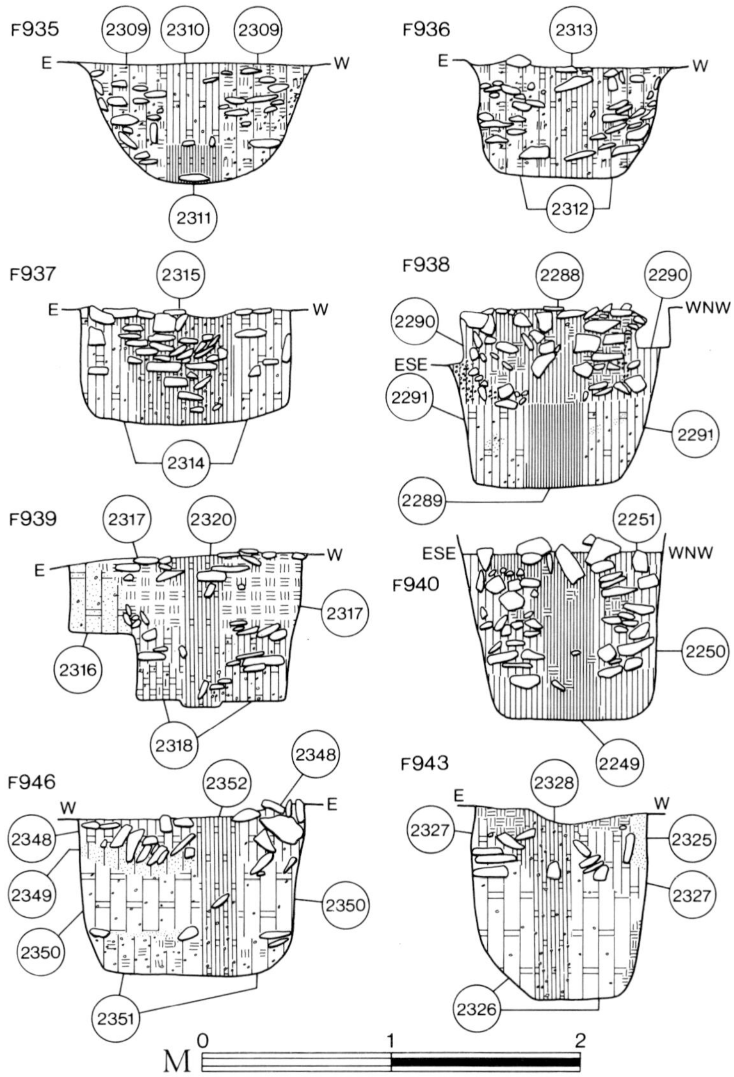

Figure 36 Sections of pier bases in Barn 1.

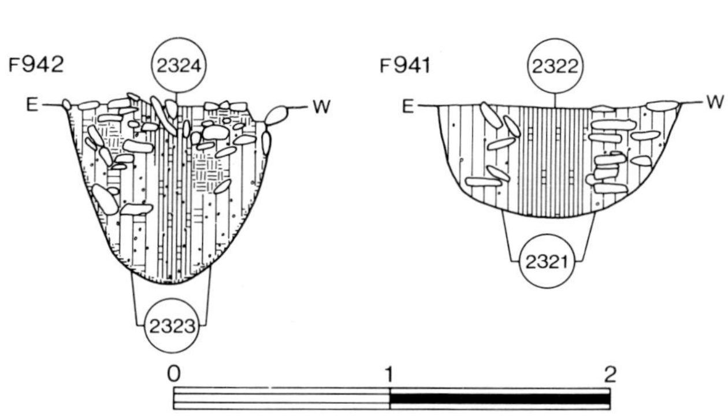

Figure 37 Sections of pier bases in Barn 1.

F943 The post-pit was 0.92m in diameter, 1.04m deep with a flat bottom and generally steeply sloping sides. The post was 0.2m in diameter. The main packing was of redeposited natural. The upper 0.4m of the fill formed a second period. The post-pipe was wider and packed round with, firstly, a bed of stone in a dirty loam and, secondly, a topping largely of clay. Mr Dakin's photograph of this post setting showed that there had been a full stone pedestal present when first exposed.

F942 The post-pit was 0.9m in diameter and *c*.0.85m deep. The profile was a half ellipse. The post was 0.18m in diameter. The lower packing was of dirty redeposited natural with a few pieces of limestone; the upper part had large blocks of clay and much more stone set in the same kind of material and seems to represent a disturbance, *c*.0.45m deep. As the pipe shows signs of displacement to the east, there was almost certainly a second period. The top of the post-pipe had been part stuffed with pieces of limestone and formed part of the thin bedding of stone which was all that survived of the post-pad sealing the post-pit found by Mr Dakin (see MF6).

F941 The post-pit was 1.09m in diameter and 0.52m deep. The post was 0.29m in diameter, and was irregularly packed round with stone set in a matrix of redeposited natural. There was no evidence for a post-pad or for removal.

The variations in the depth of the post-pits and in the size of the posts suggests that uniformity for its own sake was not important. The evidence for the number of periods present is uneven running from only one in F941 to a possible four, including post-pads, in Fs940 and 946. The shallower the original post-pit, the fewer the signs of replacement. On the other hand, as F938 for instance shows, new post-pits were also shallow, therefore, initially shallow pits could be, and probably were, completely dug out for a major renovation. And, following F938 again, the presence of stone and clay in secondary periods may mark just such an act.

Putting these indications into the context of the history of the barn, the original post-pits were mainly filled with redeposited natural and this would suit both a timber barn and Period 2. The suggested major repair, using quantities of stone, could mark the building works of Period 3. Fs940 and 946, at least, point to another period of repair before the introduction of posts set on pads. Simple division between the remaining periods would make the first Period 4, leaving the second for Period 5. However, such a primitive approach may be out of place: haphazard maintenance may have tended towards the use of the same suite of materials which could be misinterpreted as a single refurbishment. Period 3 work can be proposed with some confidence, and the post-pads need not be Period 5 and the chances are that they were present at the beginning of Period 5.

The only record of the tile-lined drain in the south wall was a single photograph which showed that it and the gulley running through it had been completely emptied. The drain had a good deal of the west wall in position, but only a fragment on the east. The base of the drain was made up of complete tiles, but only broken ones seem to have been chosen for the walls. These survived to a height of *c.*0.22m and the drain itself was no more than *c.*0.3m wide, the floor tiles just running under the edges of the walls. A cover just above the surviving remains would have been under the lowest possible floor level in the barn. The photograph showed convincingly that the stone bed in the upper part of the filling of the sump, F1094, to the north had completely sealed the exit. As the fill of the sump belongs to Period 3, the drain had passed out of use before the end of that.

The features found in the barn are described in Chapter 3. Their layout in the barn is discussed in the description of Period 2.

Barn 2

(Figs 38–40)

Periods 2–3

The structural evidence of the barn was confined to two rows of four post-holes set in pits backfilled mainly with rubble and earth. That these constituted the central part of a barn is based on the layout of the ditches: Fs65 and 189 on the east and west; F125 to the south and the return of F65 on the north. That the barn had been aisled is shown by the detail that the addition of aisles half as wide as the nave and an extra bay at each end not only fit within the enclosure, but also contain the features thought to have been housed in it.

The only incomplete part of the ditch circuit was the south end of the east side, suggesting that access was from the east with the main entrance in the south wall. The new Period 3 boundary swept away to the east along the south side of the barn and, as the F189 boundary down the west side and round the north edge of the Main Yard was kept, only the end of the west aisle was free for a door.

The lack of any sign of external walls is taken to mean that they had been framed in timber and set on, or shallowly in, the ground. Deliberate demolition could account for the absence of any stone, but the very low stone content in F65 suggests strongly that none had been used in the building.

The post-rows

(Figs 39 and 40)

The west row, south to north

F176 The pit was roughly circular, 0.6m in diameter and 0.5m deep. It had a flat bottom, neat corners and vertical sides. The post had been placed in the middle, was 0.21m across and packed round with limestone pieces set in redeposited natural. The limestone was confined to the upper part of the post-pipe and the lower fill could belong to an earlier post which had been removed. The upper pipe filled later with a very dark loam derived from the surface of the site: there may well have been two periods.

F175 The pit was roughly circular with a diameter of *c.*0.95m, at least *c.*0.5m deep and had an irregular profile. The post had been placed centrally, was about 0.21m across and had sunk by as much as 0.1m through the bottom of the pit which had been backfilled with redeposited natural with some limestone packing. The section seems to show two periods: the surviving post-pipe did not run to the bottom of the print in the base of the pit; the east side shows the backfilling of an extraction pit, the new post being fitted against the original packing, with soil derived from the surface of the site keeping it in position.

F179 The roughly circular pit had once been at least 1.05m in diameter and at least 0.57m deep. Its profile was bowl-shaped. The central post was *c.*0.25m across and had been packed round with close-set pieces of limestone rag in a dirty redeposited natural which included lumps of unaltered material. There was no clear sign of a second period unless the stone and gravel in the upper fill represent one. On the other hand, the fills of the pit may have been entirely of a second period, as the initial stone packing of the other posts in the row is missing here: the first post-pit may have been completely dug out, a new post and its packing taking up all the space.

F178 The post-pit was 0.92 in diameter and 0.7m at its deepest. It was vertical-sided on the east, but slope-sided on the west, and had a flat bottom. The post had been close packed round with limestone in a matrix of redeposited natural. It was only 0.18m across and had sunk some 0.38m through the bottom of the pit. The base of the post had rotted in position, some traces of almost completely rotten wood surviving at the bottom. There was probably a second period: the fill of the post-pipe spread out to the west, with a looser stone packing in a dirty loam on the east and the well-laid packing on the east is absent on the west. The latter may mark an extraction pit, a replacement post being set over the remains of the first.

There seems to have been a series of repairs. At least three of the posts had been cut off level with the ground leaving the stump to rot in position. The size of the posts varied and the thinnest had sunk the deepest.

The east row, south to north

F107 This had had a complex history, the post being moved at least twice to a different position. Its first position seems to have been the northernmost and so broadly matching its opposite number, F176. The post had been dug out removing all trace of its actual site. The pit had been roughly circular, 1.04m in diameter with nearly vertical sides and a flat bottom, and was 0.58m at its deepest. The next stage enlarged the pit, deepening it to 0.8m, and put the post about 0.5m south of its former site. The post had been approximately 0.25m across, but its size was hard to establish as it seems to have been deliberately removed when the third setting was created. This was *c.*0.12m further south, the post possibly being the same size as before. The second post had been set 0.73m deep while the third was set only 0.4m into the ground. It seems clear that all three post settings were packed round with stone set in redeposited natural. The latest post was almost certainly removed for reuse and only one post need have been involved in the whole sequence.

F180 The roughly circular post-pit was *c.*1.15m in diameter and about 0.6m deep. The post had been dug out from the south side but had been placed centrally. The packing of limestone set in a redeposited natural was largely intact on the north. There was a post-print 0.04m deep and 0.27m across. The edge of the print was sealed by the packing above which, therefore, should belong to a second period, the character of the main fill on the south suiting the demolition of the barn.

F177 The pit was *c.*0.7m in diameter and 0.68m deep at most. If the profile of the pit is any guide, the post had been set towards the nearly vertical south side, there being what seems to have been a print there possibly distorted when the post was removed. The backfilling was probably the remains of a limestone packing set in redeposited natural. There was no evidence of a second period.

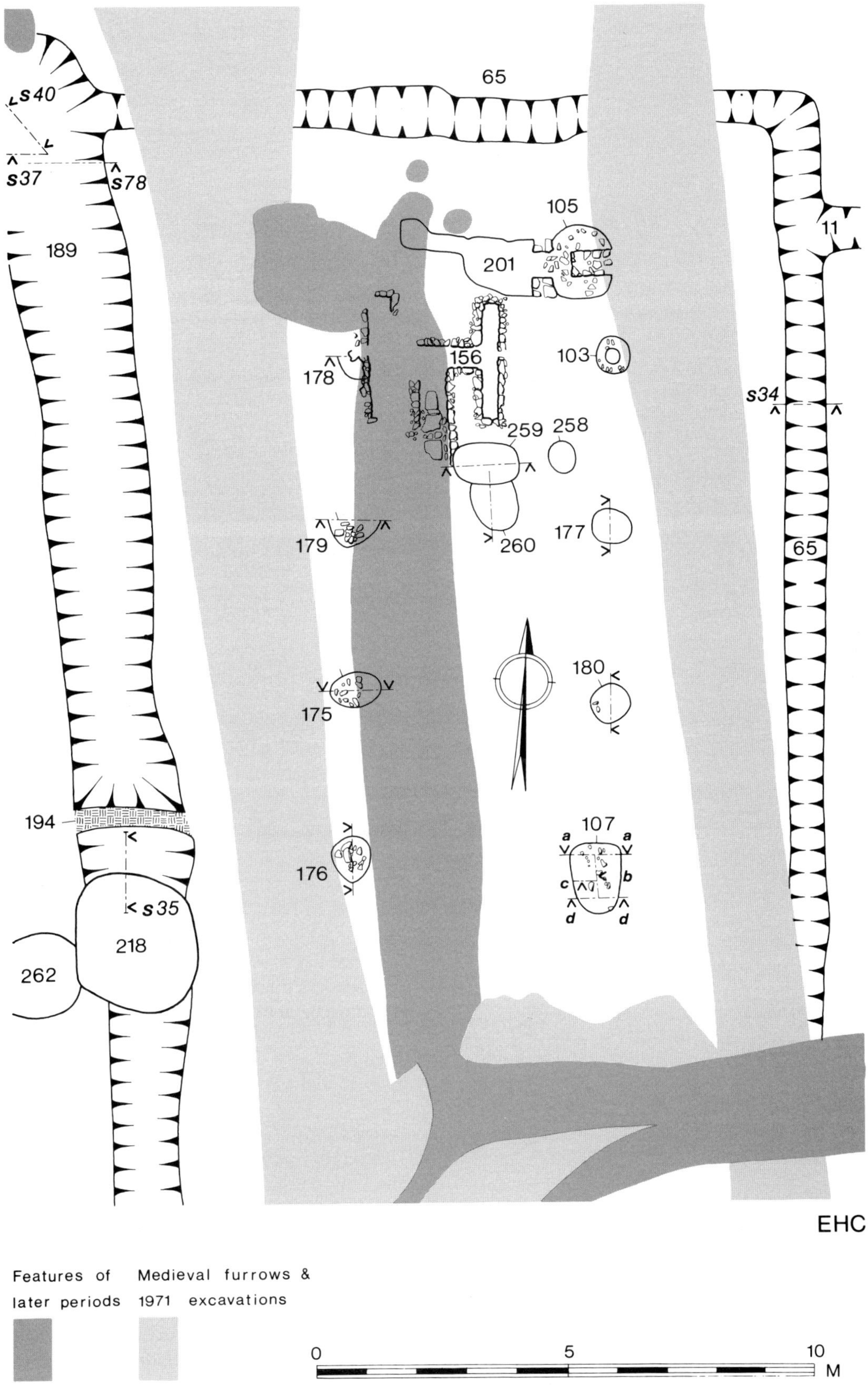

Figure 38 Plan of Barn 2, the first north-east barn. Scale 1:125.

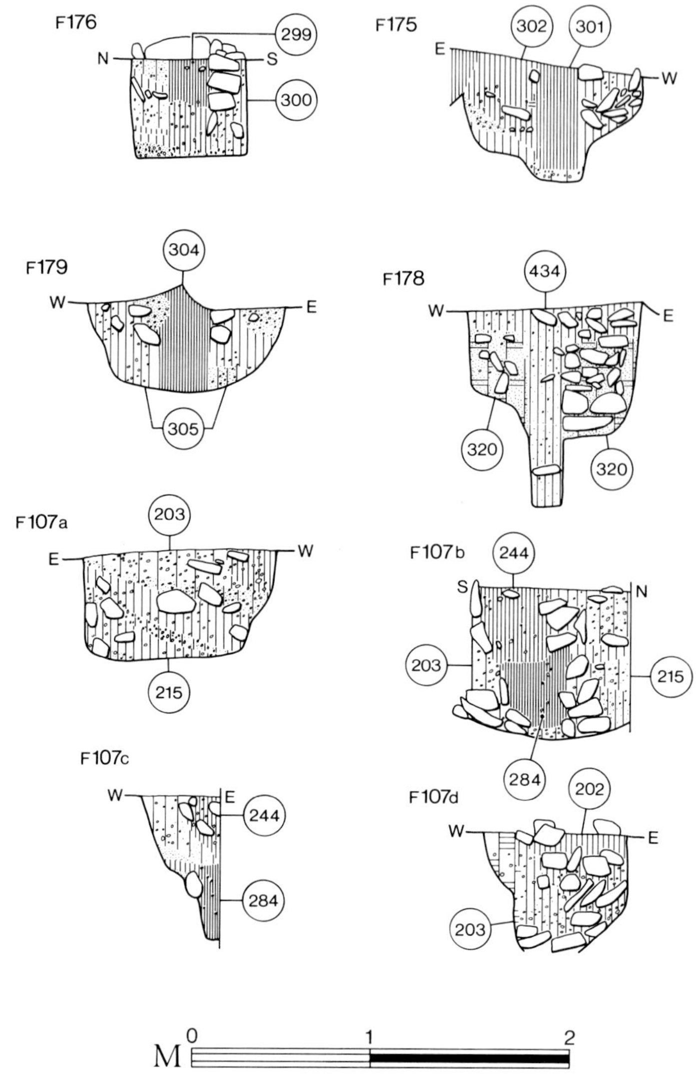

Figure 39 Sections of pier bases in Barn 2.

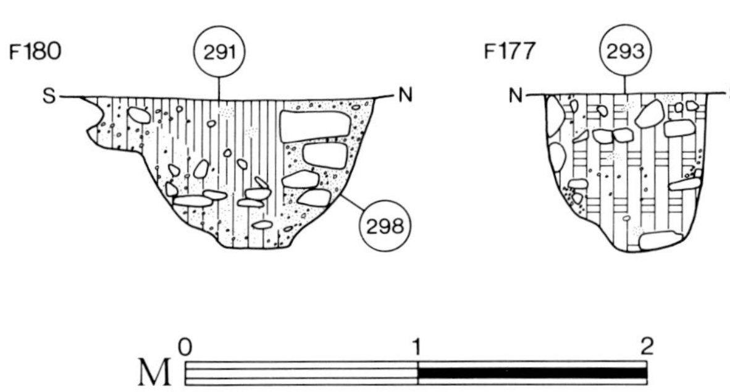

Figure 40 Sections of pier bases in Barn 2.

F103 The pit was roughly 0.9m in diameter and 0.7m deep. The original post had been dug out and the bottom of the pit filled with carefully bedded redeposited natural to seat a centrally placed post. The post was either rectangular, 0.25m by 0.19m, or had one face rounded, in which case the second measurement would have been 0.23m. The new post, set at 45 degrees to the line of the row, had been carefully packed round with limestone pieces in the usual matrix of redeposited natural.

The features inside the barn are discussed in Chapter 3. The stoke-hole or raking hollow of the vat base bent to the north as if to respect the drier. A gulley, F201, which appeared to run west from the hollow may have been for drainage. As there was no problem with ground water, it may have been to cope with spillage from a vat.

Barn 3
(Figs 41 and 42)

Periods 3–5

This was the only barn defined by remains of all four walls. The only definition which the site of the barn seems to have had was the ditch along the south side. Apart from a room added to the west end, the only other structural elements associated with the barn were lines of post-holes along its north side which may represent a pentice structure put up after the building of the west room.

In the barn were two rows of five posts and the remains of a thin wall, 0.5m wide, built across the nave at least: it only survived where the gulley, F1142, had been dug out to receive it. Three courses survived the lowest of which was made up of stones pitched across the width of the trench. The wall lay between the second pair of posts from the east and may once have divided the barn into two functional areas. The other feature in the nave was a spread of stones evidently running away from the east wall. These are thought to have been a firming-up of the floor here for

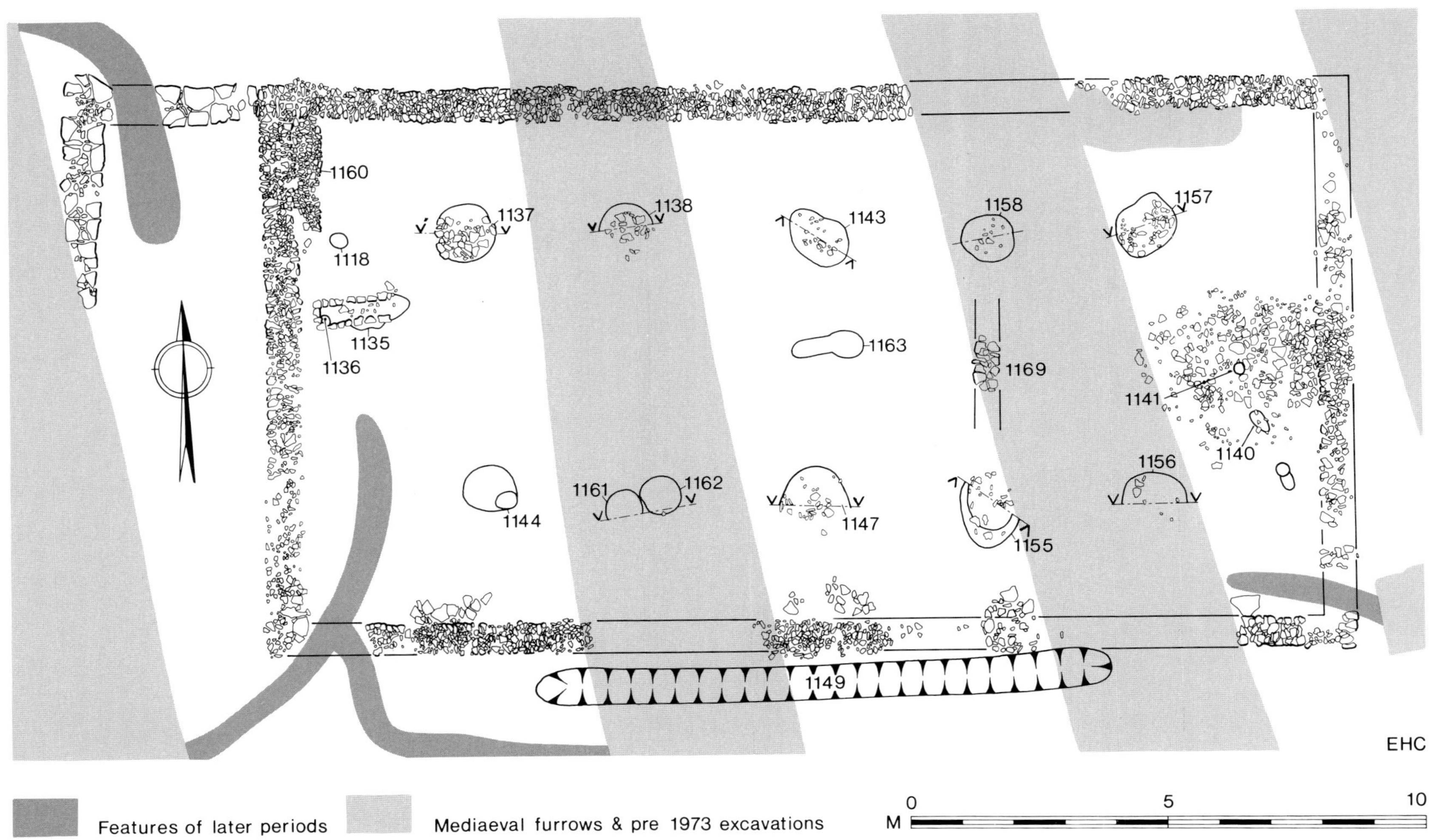

Figure 41 Plan of Barn 3, the South Barn. Scale 1:125.

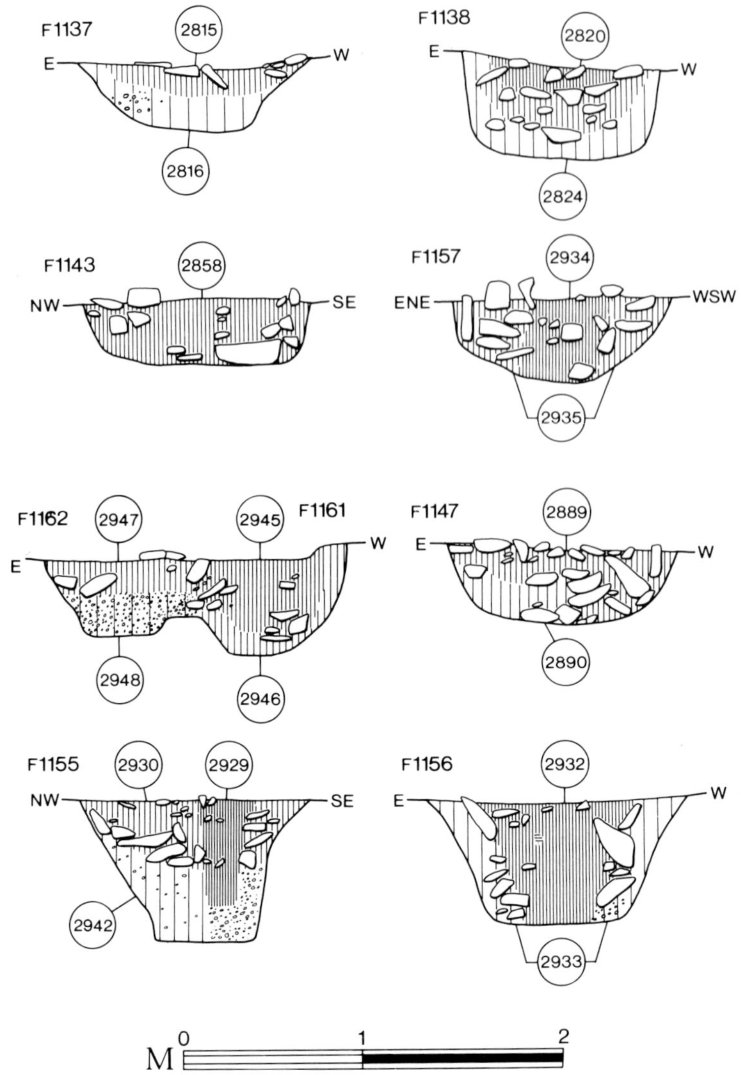

Figure 42 Sections of pier bases in Barn 3.

heavy traffic coming through a large door in the centre of the east wall. No proper floor level survived.

The west room was at least 2.9m wide inside, the room widening to the south. The line of the west wall, almost certainly conditioned by the already existing yard wall, survived for a length of 4m from the estimated position of the external quoin. Both this and the north wall were *c*.0.75m wide. No floors survived.

The post-rows

The north row, west to east

F1137 Roughly circular, the post-pit was *c*.1.24m across and only *c*.0.35m deep with shallow sloped sides. There was no sign of a post and the deposits may have resulted from the removal of the post. In the top were the remains of a bed of stones laid flat around the outside and nearly vertical in the middle.

F1138 The post-pit was about 1.05m in diameter and *c*.0.55m deep. The bottom was slightly dished and the sides very steep. There was no sign of a post. The fill was redeposited natural at the bottom under an apparently deliberate alternation of stone beds and layers of dirty, sandy-clayey loam mixed with stone. Any post here had been dug out. The stony fill and the trace of a bed of stone across the top was almost certainly due to a conversion to a post-pad.

F1143 The post-pit was roughly circular, averaging 1.2m across, and only 0.35m deep. The bottom was nearly flat and the sides were steeply sloped. The fill consists of random limestone in a matrix of dark, grey-brown loam derived from the sandy, silty-clayey natural of this part of the site. One large flat piece of limestone seemed to have been deliberately placed on the bottom of the pit possibly to seat the post which had undoubtedly been removed.

F1158 The post-pit was roughly oval with its main axis just over 1m long and only 0.22m deep. The filling contained a layer of limestone pieces in a matrix of dirty sandy-clayey loam mixed with gravel. There was no sign of a post.

F1157 The post-pit was *c*.1.15m in diameter and not more than 0.45m deep. The profile was irregular, but flat where the post had stood. The post-pipe had a marked cant slightly south of west as though the post had been pulled out from that direction. The post was almost certainly not less than 0.35m across.

The south row, west to east

F1144 The post-pit was about 1.1m in diameter and only 0.37m deep. It had a flat bottom and shallowly sloped sides. A post-print about 0.28m in diameter and 0.13m deep in the south-east quadrant. The post had been dug out. This was the only post setting which seemed to have been sealed by a use deposit (L2859).

F1161 This post was replaced by F1162, the post-pit being about 0.75m in diameter and 0.6m deep. The flat base of the pit was 0.3m in diameter, the same size as the remains of the post-pipe, and the sides curved up from its edges. The post had been dug out, pieces of the limestone packing lying randomly in the backfill.

F1162 The replacement of F1161, the post-pit was *c*.0.85m in diameter and 0.4m deep; its flat base, about 0.36m across, was possibly the seating of the post, the sides sloping irregularly up from there. The post had been removed, and the hole filled first with its gravelly packing and pieces of packing stone, and then with earth derived from the use of the barn. There was enough limestone to suggest that the final post had been based on the fill.

F1147 The post-pit was about 1.2m in diameter and 0.4m deep. The profile was bowl-shaped. The north half of the pit had a treble layer of limestone pieces placed vertically round it and set in a dark soil over a redeposited natural. This had formed part of the original packing for a post in the west part. The post had been removed and the resultant hole stuffed with stone and dark earth with signs of a bed of stone across the top.

F1155 The post-pit, *c*.1.2m across and 0.75 deep, was flat- bottomed and had near vertical sides, especially on the east where there was evidence for two post settings. The earlier lay against the side and may have been as much as 0.25m across. The lower packing was of redeposited natural. The second post was less deeply founded and set on and partly in a bed of gravel; the rest of the packing was of limestone and dirty loam. The increase in width of the second pipe, from 0.16m to O.28m may mean that this post had been rocked out of the ground.

F1156 The post-pit was nearly 1.4m in diameter and 0.67m deep. The bottom was flat and 0.75m in diameter, the slides sloping more or less evenly up from this. The fill had the remains of a heavy limestone packing for a post, the stone being set in a redeposited natural, round a wide and splayed central pipe. The base measurement, 0.35m, of this was probably the size of the post, the splay above being the result of its removal. The fill of the pipe was much darker and more organic than that in F1143, and contained lumps of green clay.

The sizing of both the post-pits and the posts in them varied as well as their depths. The details of the erratic survival of post-pipes, of replacements and of the poor record of post-pads, may suggest intermittent repairs rather than major periods of reconstruction.

Barn 4

(Figs 43 and 44)

Periods 4–5

The partial clearance in 1971 removed some evidence as it was not appreciated that the driers lay in a barn. The site was left open for two years and the growth of weeds and the development of a new topsoil cover led to a loss of a lot of stone from the pads for the reset posts.

As was the case with Barn 2, ditches were provided along all sides except the south, indicating that the main entrance was there. The ditch dug across the site of Barn 2 ended in the logical place against a wall one full bay south of the southernmost pair of posts: without a wall here, there would hardly have been an efficient boundary between the Main Yard and the area to the east. And F108 on the north also suits a similar bay extension. The digging of the new well, F172, shows that the entrance was probably at the end of the nave. This is hardly surprising, but only Barn 3 shows any evidence for an entrance in this position.

F187, the main stretch of surviving wall, was *c*.5.18m long, enough to show that there had been a sleeper wall. It was built out of pitched rough rag cornbrash and its width was between 0.6m and 0.65m. A fragment of wall, F188, on the west side was 0.65m wide and scarcely 1.4m long. It survived because it had sunk into the soft infilling of the old Periods 2 and 3 boundary ditch (Fig. 20, [39]). Both walls were so shallow that neither cut into the subsoil of the site. The bounding ditches (Fig.43) did not run particularly close to the walls themselves and may have implications (see below), not only for the reconstruction of the roof, but also for the plan of the first north-east barn, Barn 2.

The post-rows

The west row, south to north

The pairs of feature numbers in both rows below represent, firstly, the number given to the stone pad, and, secondly, the post-pit found beneath.

F182/ F205 The post-pit was about 0.8m in diameter and deeper than 0.92m — at that point water prevented further excavation. The hole had narrowed to 0.31m across. The profile may show a pit no more than 0.6m deep, the rest being due to the post sinking into soft ground. The taper of the lower part of the hole may have been caused by rocking the post to get it out. The homogeneous fill beneath L288, F182, had so little stone that any such packing had also been removed. The pad, hardly more than 0.13m deep and roughly rectangular, 0.92m by 1m, was of stone set in earth.

F184/ F207 The post-pit was 0.85m in diameter and 0.75m deep at most. The bottom was flat and the sides were splayed. There was no sign of a post, it having been dug out; the fill was a dark grey, gravelly loam with a high proportion of limestone probably from the original packing. The stone pad, F184, had straight edges on the west and south sides and measured about 0.85m across and was no more than 0.15m deep.

F186/ F209 The post-pit was 0.82m in diameter and 0.58m deep. The profile was similar to that of F184 and the section showed some evidence for the post. This was set in the middle on a stone. Had the post suited that, it would have been *c*.0.28m across. The post, once packed round with a dirty gravel, had been dug out from the west, the hole being filled with a dirty, clayey loam mixed with gravel. A circular stone setting, F186, no more than 0.13m deep and 0.73m in diameter, sealed all.

F191 A depression found at the bottom of F149 in the right position for the last post in this row. If this was the base, it had been no deeper than F209.

The east row, south to north

F181/ F197/ F198 The first of these was F197. All that survived was the dug-out post-pit about 1m in diameter and just over 1m deep. This had been backfilled with a mixture of clay, earth, gravel and stones. Its centre was *c*.0.55m north of that of F181. There was no sign of a post or any intact packing. A piece of near-rotten wood was found near the bottom. The feature was cut by both F198 to the north and by F181 to the south. F181, matching F182 in the west row, was a bowl-shaped pit 1.05m in diameter and 0.58m deep. It was filled with stone and earth, the stone being arranged as a lining with others filling the centre. Any post had been completely removed and all the evidence lost: the final effect was a firmly based stone pad for a post. F198 was another pit about as deep as F181, but not much more than 0.75 across. It was filled with stone on which sat a dressed quoin of oolitic limestone, one of the very few pieces found on the site.

F183/ F206 The post-pit was at least 1.2m in diameter and had been cut by the stoke-hole of the first version of the drier in the reduced barn. The profile of the main part of the pit was a half ellipse and 0.95m deep at most. In the bottom was a post-print about 0.25m in diameter and 0.17m deep. The bulk of the filling of the pit was limestone in redeposited natural and placed after the post was dug out, the original packing being partly intact on the north side. The rest of the fill was basically dirty soil with a little limestone. The stone post-base was set into this and occupied a pit 0.82m in diameter and 0.53m deep. The stones were small, probably deriving from the destroyed part. The top of the pad was basically rectangular, 1.2m by 1.1m, and straight on the west and south sides.

F185/ F208 The post-pit was 1.2m in diameter and about 0.66m deep. The profile was irregular with a basically flat surface at the bottom. The post-pipe was 0.26m in diameter and packed round with the same kind of material used in F206. The pipe inclined about 20 degrees to the north, possibly as a result of its removal. The pad, about 1.2m by 1m, was set in a shallow pit, no more than 0.25m deep, filled with limestone bedded in and on a dirty redeposited natural.

F190 There was no trace of a stone pad and the feature was cut on the north side by a ditch, F192. The post-pit, about 0.8m across and 0.75m deep, had nearly vertical sides and a dished bottom. The post-pipe in the lower fill seems, unusually, to have had a pointed end. The post was up to 0.25m in diameter and had been packed round with redeposited natural. The post had been part dug out before being withdrawn, the pipe being filled with a grey clayey silt, and the rest of the hole was stuffed with large pieces of limestone set in the same material.

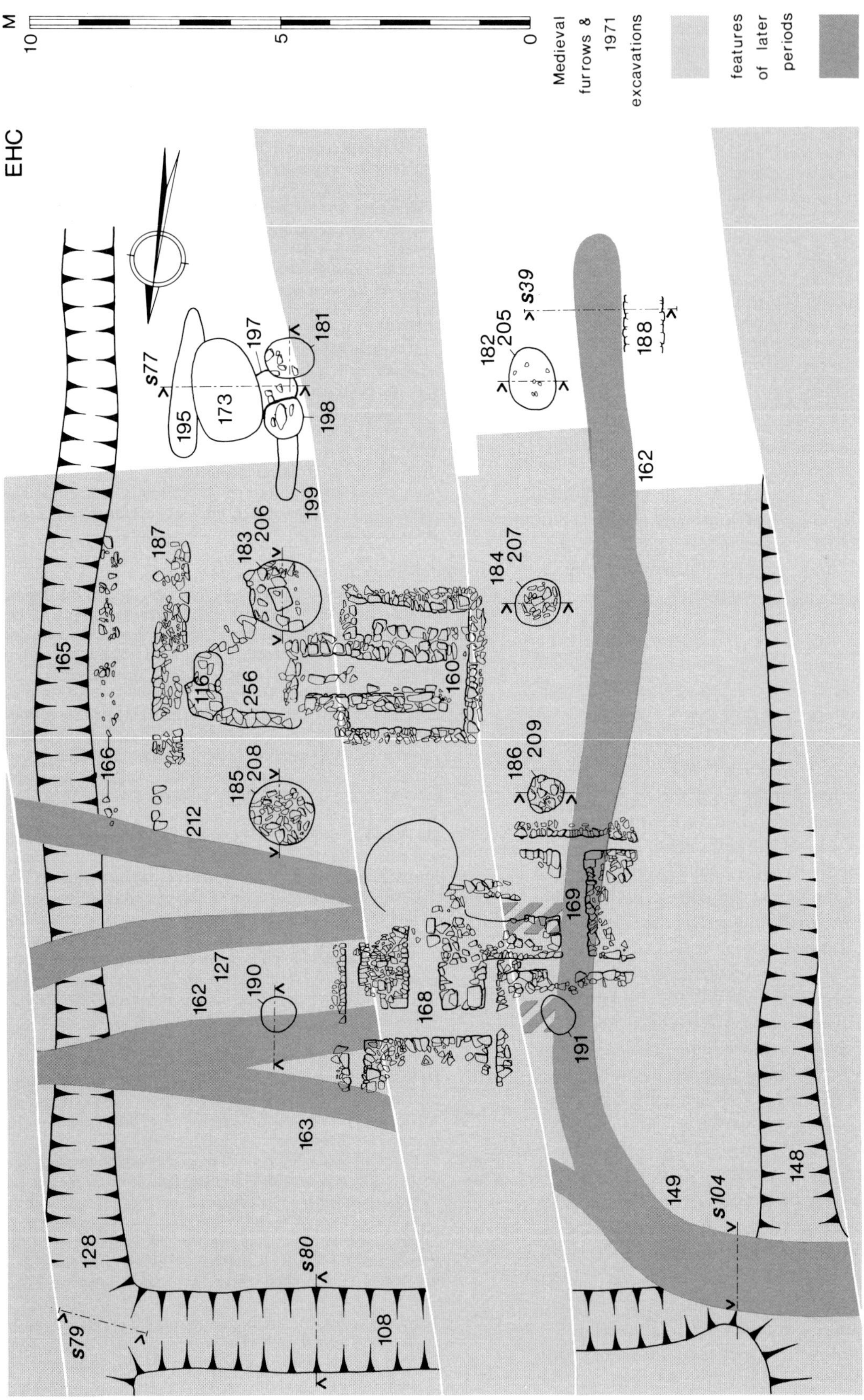

Figure 43 Plan of Barn 4, the second north-east barn. Scale 1:125.

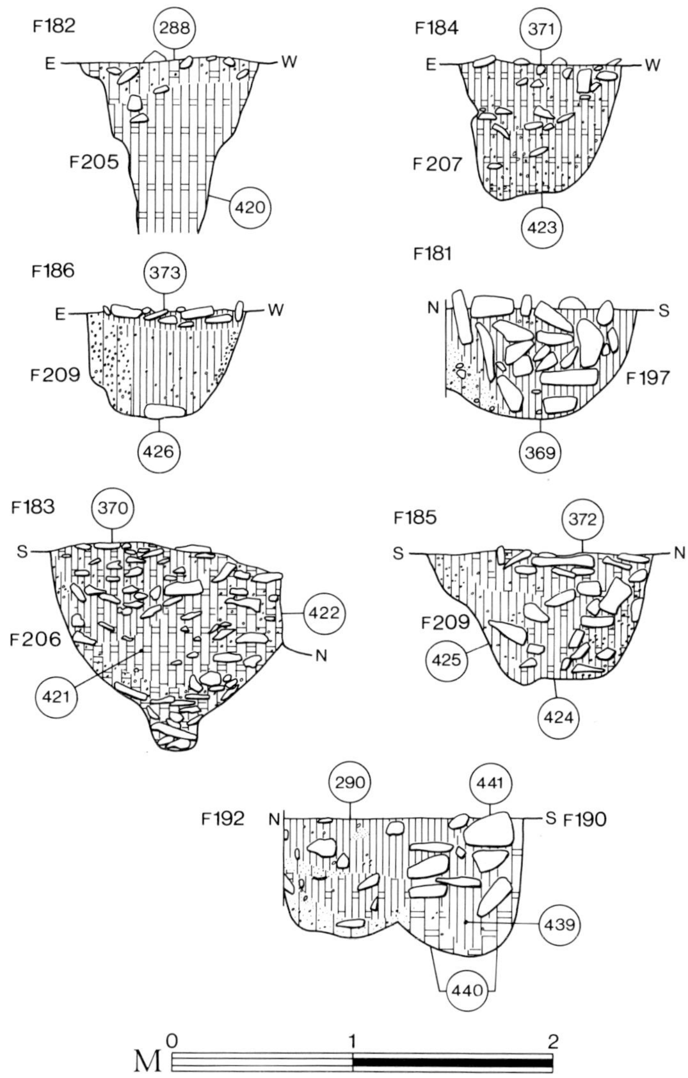

Figure 44 Sections of pier bases in Barn 4.

The destruction of the earlier post-pits was the product of removing the posts and then the forming of post-pads. The change was probably due to the absence of a firm base for the posts and the very wet ground conditions. The difference between the two post-rows in Barn 2 was, to some measure, repeated here: the eastern pads were larger than the three surviving ones of the west row. The southernmost one was larger than the other two, but was still smaller than the more or less complete eastern ones.

The driers and the probable vat base are dealt with in Chapter 3. The remaining internal features are given below.

F256 was a slot cut to the east by the vat base, F161, and to the west by the rebuilt drier, F160. The slot may have belonged to the earlier drier arrangement. It was 0.4m wide and only 0.15m deep, and could once have run to the west wall of the barn as the base of a partition, although the apparent stake-holes in its fill refused to resolve themselves during excavation. The final fill of the feature merged with the construction of the modified drier and the vat base sitting over its line shows that it had passed out of use.

The remaining three features lay in the south-east part of the barn and could, therefore, have belonged to either Period 4 or Period 5, or both. F173 cut Fs195 and 197, the latter also cut by F181, one of the versions of the south-east post of the barn.

F173 (Fig.28, [77]). A shallow depression about 2.5m long and about 1.3m maximum width, its greatest depth was 0.36m. Nothing found in it pointed to a function and its fill was mainly stone, derived possibly from the demolition of the barn wall.

F195 (Fig.28, [77]). A slot, perhaps 2.3m long and hardly wider than 0.3m, was largely cut away by F173. Its line seems to have been under the west edge of the east wall of the barn and may have been a trace of a slight foundation trench for the wall itself.

F199 Another slot running 1.35m north from F198, which may have cut it, was about 0.3m wide and only 0.2m deep. Its line was directly between the two posts of the second bay of the barn and may have formed part of a partition.

Analysis of the plan and structure of the barns

The remains of so many barns found during a single excavation campaign and all recorded in the same manner surely deserves some discussion, especially as there is no need to adapt the discussion to take disparate recording systems into account. The aim here is to see if there are any general rules in their planning and whether any deductions can be made about the methods used to build them as well as about their finished appearance. The structural evidence has already been described. The next stage is to look at the plan of each barn before comparing all four.

One point is made here to avoid repetition. The stone pads in three of the barns were big enough to allow some freedom in placing posts on them. Therefore the structure could have been better squared in plan than the posts beneath indicate. Even so, there is sufficient discrepancy for squareness in setting out not to be a criterion and the view adopted is that the structure of a barn with post-pads was an adaptation of the one seated in proper post-pits.

Barn 1

The best reconstruction of the overall outside dimensions gives 25.2m by 13m, a primitive ratio of width to length of 1:1.94. There was no thickening of the end wall to suggest a stone wall run right up through a gable.

The bay widths derive from the centres of actual posts or their estimated position where this was not directly located. The minimum width, the first bay from the east in the south row, was 2.9m. The maximum was 3.75m, in the next bay to the west. The best fit for the remaining eight bays is 3.4m. The second post from the east in the south row may have been wrongly set in the first place, but the variety in spacing suggests that precision was not a prime consideration. The width of the nave, taking all the pairs of posts, averaged 7.55m, the width varying between 7.35m and 7.7m.

Estimating the aisle width is not necessarily easy as it was presumably the structural elements which dictated the measurements or ratios used. It is assumed that basically simple proportions will have applied, such as 1:2 or 1:3. Therefore, to measure from a post centre to the nearest face of the wall may be a mistake. The wall thicknesses varied between 0.8m and 0.65m, the average being 0.725m. This fits well with most of the surviving walling and is used in the following discussion.

Both aisles are reasonably uniform and measure 2.0m to the inner face of the wall, 2.36m to the centre and 2.72m to the outer. Using these three possible widths, the simple ratios of aisle to nave are in turn 1:3.78, 1:3.19 and 1:2.77. But the question of the walls being half-timbered, or carried in stone to the top arises, and were the dimensions derived from a simple truss over the nave and aisles? In the latter case, the position of the wall-plate may have been all important. Whether the walls were of stone or timber is immaterial: half-timbering would presumably have a wall-plate at its top and one could be laid anywhere within the width of the wall. A simple ratio system probably determined where the centre of the wall-plate was to be. It is assumed that the maximum likely width of a wall-plate would be no more than the *c*.0.3m of the thickest post in the barn. As the centre-of-wall ratio has already been arrived at, only the inner and outer positions on the wall, giving aisle widths of 2.15m or 2.57m, need to be considered. These yield ratios of 1:3.5 or 1:2.94. A straight 1:2 ratio does not work, but one of 1:3 does, giving an 'aisle' width of 2.52m, the wall-plate would then lie in the outer half of the wall. This type of discussion applies to the other barns and is not repeated in detail for those.

Turning to the bay width, 3.4m has been determined as being the best fit, and using this to test the hypothesis that the position of the end walls was a direct expression of the general bay width, we find that at the east end of the barn, this dimension takes us to the centre of the wall, but at the west end it falls 0.7m short of the same line.

The sophistication of a building, to some extent, whose structure was essentially wooden can be assessed by the type and size of the timbers used. This can only be based on the post-pipes and prints found. However the wood was finished above ground, it is noticeable that no rectangular posts were found in Barn 1 and that the size of the timbers varied from 0.18m to 0.3m. When looked at more closely, these sizes show that three of the eight were close to 0.2m, three more to 0.3m and the other two were approximately O.25m.

Barn 2

Although no walls were present, the basic description has shown that aisles half the width of the nave could be fitted inside the bounding ditches. The fit of the replacement barn, Barn 4, in its own enclosure shows that such a ratio is not necessarily the right one to choose.

As the position of one post on the east side could not be recovered and the southernmost one in the same row had a complex history, determining the bay width and related matters cannot be precise. Even so, there was clearly no uniform bay width. Not only did the overall length of each post-row differ by 0.44m, but there was also no correlation between the widths of opposed bays. Only the distance between the pairs of posts across the nave was reasonably constant at 5.2m, except for the southernmost pair. Taking the average of all the bays, 3.36m, and using this to establish the original length of the barn, we arrive at 16.8m. From the nave width of 5.2m and the ratio of 1:2 for aisle to nave, the width of the barn would have been 10.4m giving an overall ratio of 1:1.62, possibly reflecting an original 2:3. And this would suit the bounding ditches. If the apparent 1:2 ratio of the overall measurements of Barn 1 is used to arrive at an aisle width, it would have been 1.6m here.

The posts themselves were mixed in size and section. Two at least were rectangular, but the rest of those located appear to have been rounded. Their sizing was more consistent than that in Barn 1. The range was between 0.18m and 0.25m across, three being of the latter size and only one of the former. The overall range in size shows that, as in Barn 1, there was little care taken to ensure that all the timbers were uniform.

Barn 3

The walls here were the best preserved of any barn and were close to 0.7m thick all round. They were reasonably well founded, matching Barn 1 in this, but not laid out to form right-angled corners. While the length, overall, was close to 21.4m, the width was greater by 0.2m at the west end than the 11.0m for virtually the rest. The plain ratio of width to length was 1:1.95, surely close enough for an intended ratio of 1:2.

The bay widths are difficult to determine as all but two had been dug out. Therefore, the assumption is that, apart from these, the posts had been set centrally in their pits, the common pattern in the other barns. In the north row, the bay widths varied between 3m and 4m. Taking the two unaltered bays in the south row, the range was still 3m to 4m. This irregularity also appears in the difference between the end posts and the end walls. In the south row the distance from the post centre to the face of the west wall was 3.1m and, at the other end, 4.15m. In the north row, both of these measurements were close enough to 3.4m for it to be obvious that it was the floor area which was more important than the regular spacing of the roof supports: providing these served their purpose, it mattered little within broad limits precisely where they were sited.

The only constants seem to have been the distance across the nave between pairs of posts and in the width of the aisles. Four of the five nave widths were so close to 5.2m that this must have been intentional and compares with the width of the nave in Barn 2. Even in the remaining case, the second post from the west end in the south row, the secondary post position conformed with this dimension, the primary giving a width of 5.6m. The distance between all but three out of the ten posts and the side walls, or the projection of their lines, was close enough to 2.2m to form a constant. The remaining three were about 2.0m. In other words, it was the alignment of each pair of posts across the width of the building which was allowed to wander and this may be significant when it comes to understanding the above-ground structure and method of erection.

As for the ratio of aisle to nave, a simple ratio of 1:2 will suit the *structural* needs if not the actual walling itself, there being a variation of only 0.24m if the inner face of the wall is taken. This could have been the line taken by a wall-plate.

Barn 4

Only parts of the side walls were left and the north ditch alone provides a check on the calculated overall length. The walls of the barn were not deeply founded and it would only have been the quantity of stone in the surrounding ditches which would have pointed to the presence of stone walls, had the barn had the same history as its predecessor.

The surviving course of the east and west wall fragments was *c*.0.7m wide. The east wall was not straight, but what bearing this has on the reconstruction of the superstructure is uncertain. The bay widths, ignoring the anomalous position of the south-west post, varied between 3.8m and 4m. The next set of dimensions, between pairs of posts across the nave, failed to conform with the tendency of those in the other barns as it increased from either 5m or 5.2m in the south to 5.7m in the north. It follows that strict calculation of proportions is not easy.

To deal with the overall length first, the average width of the bays was 4.23m and the addition of this to each end shows that there was a gap between a wall and the north ditch reasonably consistent with that between the side walls and ditches. At the south end, a wall would be just short of the construction pit of the new well put in to serve the barn. The overall dimensions arrived at suggest that the ratio of width to length would be very close to 1:2, but the layout of the posts inside is poor enough to show that, despite any intention, the achievement could not have produced exact figures.

Turning to the ratio of aisle width to that of the nave and using the average for the nave width and an assessed average for the aisle, there are three measurements running from 1.7m to 1.9m up to the face of the wall which yield 1:2.97. If the outer face is used, the result is 1:2.14. A simple ratio of 2:5 yields an aisle 2.16m wide which, it will be seen, would place the supposed wall-plate on the inner edge of the wall.

The direct relationship in sequence, siting and function of Barn 2 with Barn 4 might imply that the first had been taken down and re-erected. But this is beyond demonstration, using the posts themselves, as there were not enough dimensions recoverable in the later barn for comparison. Only three were coincident in siting and, although not precisely matching, the general sizing was close enough to offer some support, but not enough for the point to be asserted. It is suggestive that each barn had had the same number of posts and that, had the later barn been better laid out, the nave width between pairs of posts might have been more consistent with that of the first one. However, both barns can be compared in two other respects.

The first is directly observable: neither had walls as deeply founded as Barns 1 and 3. Barn 2 most probably had timber-framed walls and Barn 4 may also have had them, raised on sleeper walls. The wall thicknesses may seem rather wide for this, but 0.7m is not out of the question, especially considering the fairly erratic course of the east wall: all the stonework needed to supply was a general seating for timbers before they were installed. The poor laying out of the barn may have been due to the reuse of existing large structural elements which were being cobbled together. That the actual structure of Barn 2 also left something to be desired can be seen in the difference in the repairs to each post row.

The second point concerns the fit of each barn in its enclosure: the width of each was much the same, 12.5m serving to define the width at the south end. In both cases an aisle–nave ratio of 1:2 can be fitted in, but Barn 4 clearly did not have this. However, if the assumed ratio for Barn 4 is applied to Barn 2, the same sort of gap between the side ditches and walls appears. F166 (see p. 21) may well have been an eavesdrip gulley which was only preserved where the east side ditch swung away from the barn. In this case, the 'throw' would have been about 1.5m from the estimated centre of the wall-plate and this may have been related to the function of each barn. There is, after all, nothing inherently unlikely in the thought that, despite the vagaries of the siting of the uprights, the width of the second barn was dictated by the reuse of roof trusses.

Two details tell against the idea that one framework had been taken down and put up on a fresh site. Firstly, the very different bay widths and, secondly, the lack of consistency in the width of the nave. Three of the nave pairs in Barn 2 were 5.2m while only one in Barn 4 might have had the same measurement. Another objection might be that, if the posts from the first barn had been reused, and several had certainly been sawn off, the 'new' barn must have had less height than its predecessor. It is factors like these which advise against pressing for re-instatement though, to some extent, much actually depends on how the barns were built in the first place and this is discussed below.

Construction and reconstruction

(Tables 1 and 2)

Table 1 sets out the basic ratios which seem best applied to the barns, as well as the average bay and nave widths which can be derived from the site itself. Assessing the width of an aisle is not easy for the reasons given in the analysis of Barn 1. However, the aisle–nave ratios produce dimensions for the aisles which fit three of the structures found. As for Barn 2, because only the post-rows survived, three choices are given based on 'ideal' aisle–nave ratios. Only one of these produces a satisfactorily simple figure for overall width to length: 1:94 (1:2). The end walls of Barn 4 did not survive and it is the fit of the plan in the enclosure which produces the 1:2 given.

Apart from the probable 1:2 ratio of the overall plan, no other figure, either ratio or measurement, applies to all

	Nave	Bay	Aisle	W:L	A:N	Bay No.	Area
Barn 1	7.55m	3.37m	2.52m	1:2	1:3	7	328m^2
Barn 2	5.2m	3.36m	2.08m	1:79	2:5	5	157m^2
			1.73m	1:94	1:3		146m^2
			2.6m	1:6	1:2		175m^2
Barn 3	5.2m	3.25m	2.6m	1:2	1:2	6	235m^2
Barn 4	5.4m	4.23m	2.16m	1:2	2:5	5	240m^2

For Barn 2, 1:1.94 probably represents 1:2.

Table 1 Spacings, ratios and areas of the four barns.

Place	Nave	Bay	Aisle	W:L	A:N	Bay No.	Area
Denton	5.94m	3.22m	3.05 or 1.92m	1:2.97	1:1.94 or 1:3.1	8 or 10	*c*.285m^2?
Rapsley	5.94m	2.88m	2.13m	1:2.18	1:2.79	8	234m^2
Stroud	7.01m	3.5m	3.35 or 4.08m	1:2.74	1:2 or 3:5	12	663m^2
Wakerley	6.5m	2.28m	2.4m	1:1.7	1:2.73	9	224m^2

Table 2 Spacings, ratios and areas of some other barns.

the barns. Barn 1 stands out not only for its larger area and number of bays, but also for the greater width of nave, about half as wide again as any of the others. It might be argued that the plan of a barn was a product of the application of set ratios, even perhaps a primitive module for the width, and that the length and consequently the number of bays had been adjusted to produce the overall shape. Thus basically consistent bay widths yield 7, 6 and 5 bays for three barns. The greater bay width in Barn 4, which otherwise would have needed at least 6 bays, may be a strong argument for the reuse of the main members of Barn 2, the lack of fresh timber and the need for a larger floor area dictating the spacing.

The fairly consistent bay width of *c*.3.3m coupled with the ratio of 1:2 for the plan suggest the use of empirical rules in laying out these structures; it is the lack of precision in setting them out which points to their having been, in a manner of speaking, vernacular. This conclusion is reinforced by variation in timber size and section for the main posts and, on occasion, the setting of a squared timber at an angle to the post-row line. However, the absence of a consistent nave:aisle ratio may show that there was no fully comprehensive proportional system. In other words, only those producing the overall plan, the width of the nave, probably another for the width of a bay, were in general use, the aisle being adjusted to suit. It is Barn 4 which displays the most vagaries and, as has been argued, there may have been a good reason for this.

Unfortunately, it is not yet possible to test these hypotheses for, although several barns have been completely stripped in modern times, not enough are published to form a useful body of material. Table 2 is very limited precisely because of the lack of satisfactory parallels, but it does show that whatever rules derive from the barns at Orton Hall Farm, they do not apply universally. Just as vernacular architecture of the Middle Ages and later is regional, so it should be expected that even if ideal systems can be detected, they were most likely purely local in occurrence. The identification of specific medieval and later house types has been the result of analysis of hundreds of standing buildings, the great size of the 'sample' allowing regional and social patterns to be seen. Without upstanding remains and very few fully excavated and published Romano-British barns, there is as yet little chance of detecting significant groupings or of establishing sets of rules.

When it comes to the putting up of barns, the relative uniformity of the nave width between pairs of posts may be a sign that each nave pair, along with part or all of a roof truss, was made up on the ground and then the whole set into position. J.T. Smith, in dealing with the first aisled building at Denton, noted that the lack of alignment in the post-rows, and the uneven bay spacing, contrasted with the more regular spacing between pairs of opposed posts (Smith 1964, 78) and concluded that prefabrication on any medieval model was not appropriate.

He suggested that the basic structure was of two uprights supporting a tie beam across the nave. This would have had to have been long enough to allow the longitudinal members running along each line of posts to be straight although made up of individual lengths (Smith 1964, 79). It is possible that a plate could have run across more than one bay to provide better stability, but proper longitudinal bracing could only be provided by either horizontal or vertical diagonal members at each end. Perhaps more likely, the ends of the longitudinal series were fixed into the end walls of the building: the kind of carpentry implied by the odd sizing and shape of the posts at Orton Hall Farm suggests the absence of upright braces in each bay.

It is a commonplace of Romano-British archaeology that half-timbering was used even on substantial footings but this is more often asserted than demonstrated. Any conclusions about the superstructure of a building should be based on the examination of the depth and thickness of foundations. This point is relevant here as three of the barns had remains of stone footings, but only Barns 1 and 3 could be described as having had reasonably substantial ones. The evidence is definitely against stone having been used for Barn 2, and the slight character of the fragments of walls in Barn 4 may well mean that the basic walling above was of timber. In a district which yields abundant and good building stone, it should be assumed that foundations capable of supporting stone walls up through a ground and first floor in later times would have done so in the Roman period. Therefore, it is assumed that both Barns 1 and 3 were walled in stone. The end walls of these, where they survived, were not thickened as though there had been a gable built in stone. But there is no particular need to carry a solid wall up through the height of a pitched roof. All that may have been necessary was sufficient walling to tie in the plates along the aisles.

While the erection of a complete frame across the nave can be suggested for Orton Hall Farm, the post-holes recovered at Rapsley, Surrey, could point to a different method. Here the posts were locked into their post-pits by means of nails running from the timber into the cement grouting between the stone packing. All the posts were squared, but this does not automatically mean better carpentry as they had been set at all angles (Hanworth 1968, 23–4, fig.3). The posts could have been put up without a tie, this being hoisted into position later. The strange course of the north wall may mean that there had not been a single line of plates along the aisle. However, there are problems in phasing the site: the ditches assigned to Periods 2 and 3 around the barn may have belonged to an earlier, all timber, barn before the Period 4 walling was put in, and the awkwardness in fitting in the yard wall assigned to Period 3 may only arise because too simple a periodisation of the site has been presented. A higher order of wood-working may be indicated by structures like the Spoonley Wood barn whose posts were set in socketed stones, but the regularity of the plan may be more a product of tidy draughtsmanship than of reality (Middleton 1890, 655, pl. XVII).

Plates along each aisle, probably tied into the barn's end walls, are implied by the typical plan and it seems unlikely that there had been no structural connection with the side walls. The barn at Wakerley had remains of timber walling along the aisles. The plan (Jackson and Ambrose 1978, fig. 15) shows a basic structure reminiscent of the Orton Hall Farm barns as there was no attempt to have a regular bay width and the distance between pairs of posts across the nave were reasonably constant. The side walls were imperfectly preserved and consisted of a gulley or slot widened occasionally to form post-holes. Lines run through the pairs of posts defining the nave show that there is no satisfactory correlation between these and the wall posts suggesting that the wall structure was not laid out to conform with the main posts. But there should have been a wall-plate and the base members of the roof could be anchored to that. In other words, any framing would appear to have been horizontal and at the top of the walls and main posts. If so, the difference in bay width between Barns 2 and 4 would not be against the re-erection of Barn 2 on a new site.

A model can be proposed in which the walling and posts supported a major horizontal frame, possibly in two planes, running across both nave and aisles and along the main posts. This could have been braced in the corners and have formed a base for the roof whose trusses could have been independent of the tie across the nave. However, while the simplicity of the structure can be deduced, the actual physical appearance of a typical barn is another matter. A basilican cross-section is most unlikely and should be replaced with a single roof over nave and aisles.

Without considering the ends, there are really only two choices about how the roof was fitted. Firstly, the whole could have run across the nave and aisles which were all finished at the same level. Secondly, the side walls could have been lower. The second choice should mean that there would be a horizontal member running from the wall-plate across the aisle into the main post. The Wakerley evidence is against this and the crude carpentry and the simple erection procedure implied by the physical remains would make the frequent removal of main posts more difficult and there was obviously little to prevent fairly widespread alterations.

The actual height of a barn was probably governed by the roofing material and the pitch necessary for it. At Orton Hall Farm the complete absence of tile or stone roofers should mean the use of thatch and an optimum pitch may be set at 45 degrees. In short, a roofing system rising from a single plane covering the whole plan is preferred. This would certainly suit the single-aisled barn at the Bancroft villa (Frere 1985, 290–3, fig. 22).[1] There was no hint there of major timbers in the wall on the unaisled side of the nave, and the poor setting out of the post-pits again points to a vernacular tradition.

When it comes to how the ends of a barn may have been finished, difficulties arise. One does not normally associate large windows with barns and a consideration of the probable location of the major light sources may be useful. The features in all four barns were grouped at one end in three and at both ends in one. In two of the barns there was an emphasis on the nave. The location of furnaces, driers or whatever may have been conditioned as much by the amount of natural light available as by the use of the rest of the barn. However, the two factors may have worked together and point to the chief sources of light having been at each end of the nave.

Bays tend to run with a relatively even width to the ends of a barn, with few exceptions (Pollard 1974, 80–2, figs 3, 10; Norris and Burstow 1950, fig. 16), therefore, the roof over the aisles was not merely turned to run across each end. This is emphasised by the poor correlation between the widths of aisles and bays. Occasionally, the aisles themselves seem to have been returned across the end probably indicating a gablet giving light to the interior (Price and Price 1884; Wilson, D.R. 1968, 206, fig. 21; Goodburn 1976, 366–7, fig. 24).

If end gables are abandoned, the alternative scheme would be to provide a hipped roof. But a fully hipped roof would interfere with a regular truss system spaced roughly or exactly to conform with the main posts below as the inspection of the plans of virtually every aisled building will show. However, the ends could have been hipped as far as the first main truss leaving a light and smoke hole in the vertical face. In most cases this would still mean that the first truss at each end would not be over a pair of posts. Presumably there would be some uniformity in truss spacing according to the weight of the roofing material. Therefore, Barn 4's roof should have had more trusses than that of Barn 2. A hint that the ends of Barn 3 were partially hipped may be in the drainage ditch along the south side. It should have taken water shed from the barn roof. As the ditch did not run the full length of the barn, it may have been designed to cope only with the full height roof along the centre of the building.

There is little guidance for the height of outside walling. Barn 3 was the only one with ancillary structures: the room at the west end and the pentice along the north side. The room presumably also had a pentice roof and, if walls of a minimum of 2m are allowed and a pitch of 45 degrees was used, the barn should have had an outside height of not less than *c*.5.5m. The same height would accommodate the timber pentice. If there had been another pentice along the face of the west room, the height would have had to be increased to suit, perhaps to nearly 9m, and this is the best reason for discounting a linking pentice. A height of 5.5m for the above ground part of the main posts might be thought unnecessarily tall for a low-grade building. However, if Barn 2 really was moved to a new site, its posts must have been shortened and yet have been long enough to give a proper height to the people working inside.

It is obvious that, if maintained, barns were capable of surviving a long time. Barn 1 lasted for well over two hundred years. Only Barn 2 seems to have had earth-fast posts throughout its life, Periods 2 and 3. In the other barns there was intermittent but good evidence that posts were finally placed on pads. With so much care taken to keep the main uprights firm most of the time, it could be imagined that this was important for the stability of the structure and not just during the construction stages. But the fact that it is a commonplace to find stone bases replacing post-pits speaks volumes for the rigidity and strength of the framing, despite the generally low quality of the joinery which seems to have prevailed.

II. The House

(Pls IV–VI)

Periods 3–5

This lay at the north end of the Small Yard and was the only one overtly suitable for domestic occupation. If the building had an upper floor, the lower may not have been purely domestic. No feature in or about the building pointed directly to its function, the plan form alone suggests that it belonged to a general class sometimes called 'cottage' houses. The size of the building here, about 200m^2, however, makes the term 'cottage' inappropriate.

The remains were cut by three furrows. No floor levels survived, the whole area, with the exception of the lowest part of the north wall, having been planed off by ploughing. The only stonework visible at the surface of the site was 11m long in the central part of the north wall of the main room; footings survived at a lower level to the west. Elsewhere, where tested, the walls had been robbed out. The uniform character of the robber trenches suggests the walling had been removed on one occasion. The presence of Anglo-Saxon pottery in the south wall's robber trench shows that the building was not taken down until after that ware had arrived on the site. The House was almost certainly immediately replaced by an Anglo-Saxon timber building (Chapter 4, p. 87).

Sections across the Period 1 ditches, where these passed under the junction of the north wall and the internal wall, showed that the foundations of both had been treated with as much care as those of the rectangular building and its adjunct where these crossed the Period 2 main boundary ditch. Therefore, the building had almost certainly had stone walls to the eaves and there is no reason why they should not have run up through a first floor.

The measurements of the House cannot be exact, as so much of the plan was represented by robber trenches. It was *c*.24.6m long and *c*.8.25m wide, giving an overall area of *c*.203m^2. At the west end lay a room *c*.3.6m wide internally. The surviving north wall was *c*.0.8m wide conforming with the general width of the robber trench.

Arguments have been advanced that the proportions of Roman buildings may have provided a model for Anglo-Saxon buildings (James *et al.* 1984). If the few instances cited (*ibid.*, fig. 12, 1–9) demonstrate anything, it is that the use of the square was an ideal rarely achieved as the width of internal walls fail to fit uniformly into the scheme. Closer analysis of a greater body of information may produce general rules, but the discussion of the plans of the four barns (see above) shows that simple determinations will not do. The application of the 'rule' to the present building is difficult as the position of wall faces is not precisely known. However, the internal width, laid out from the east end, will allow three and a half squares, but the internal wall does not fit in with the division.

The thickness of the walls would easily have supported an upper floor which means that the simple appearance of the ground floor may be false as there may have been timber partitions, but the stone-walled room at the west end may go against this. Timber partitioning on an upper floor would be suitable and the west room could have been a stair-well. However, stairs could have been run up the external face of the ground floor.

III. The Rectangular Building

(Figs 45 and 46 A–B)

Periods 4–5

Measuring some 24.5m by 20.7m maximum, there was no sign of any roof support inside. The presence of floors inside, especially the gravelly first ones, show that it had been covered in some form. Any roof could have been supported by posts based on stone blocks set at ground level which, once they had been removed, would leave no evidence. With an area of 507m^2, it was by far the largest building on the site although there is no guarantee that all of it had been roofed.

The building was poorly preserved. Damage done in Period 5 was compounded by the four furrows cutting through it, removing much of the line of both the east and west walls. The degradation in Period 5 would, however, have left little of either. The site had been prepared by backfilling the north boundary ditch of the Main Yard and stripping any topsoil from the interior. The wall foundation across the ditch had been carried down to the bottom and stepped up the sides. The wall was 0.9m wide and was quite capable of being carried up through a ground floor, and higher, in stone.

The only surviving feature in the walls was the large south door. It was inbuilt with the wall and was obviously intended to be strong, as the posts were firmly fixed in the ground. A furrow ran through the door cutting away most of it, but the section (Fig. 46) of the east side shows the relationship of the south wall with the eastern post, F467. The post-pit was as wide as the wall and was almost certainly a deepening of the foundation trench to form a regular circular pit with a slightly dished bottom and very steep sides. The post was carefully set so that its north edge was practically on the centre-line of the wall and then it had been packed round with stone as part of the wall foundations. The post was rectangular, 0.18m by 0.15m, and aligned north-to-south. To the west lay the lower part of a hole, F353 (Fig. 46B), containing a closely matching, but slightly smaller, post. The plan relationship of the two is such that there can hardly be any doubt that both belonged to the same structure. The distance between the two was 2.3m and should indicate a double-leaved door. The posts of the frame ran into the ground, were probably tied at or beneath the planned threshold and at the top by

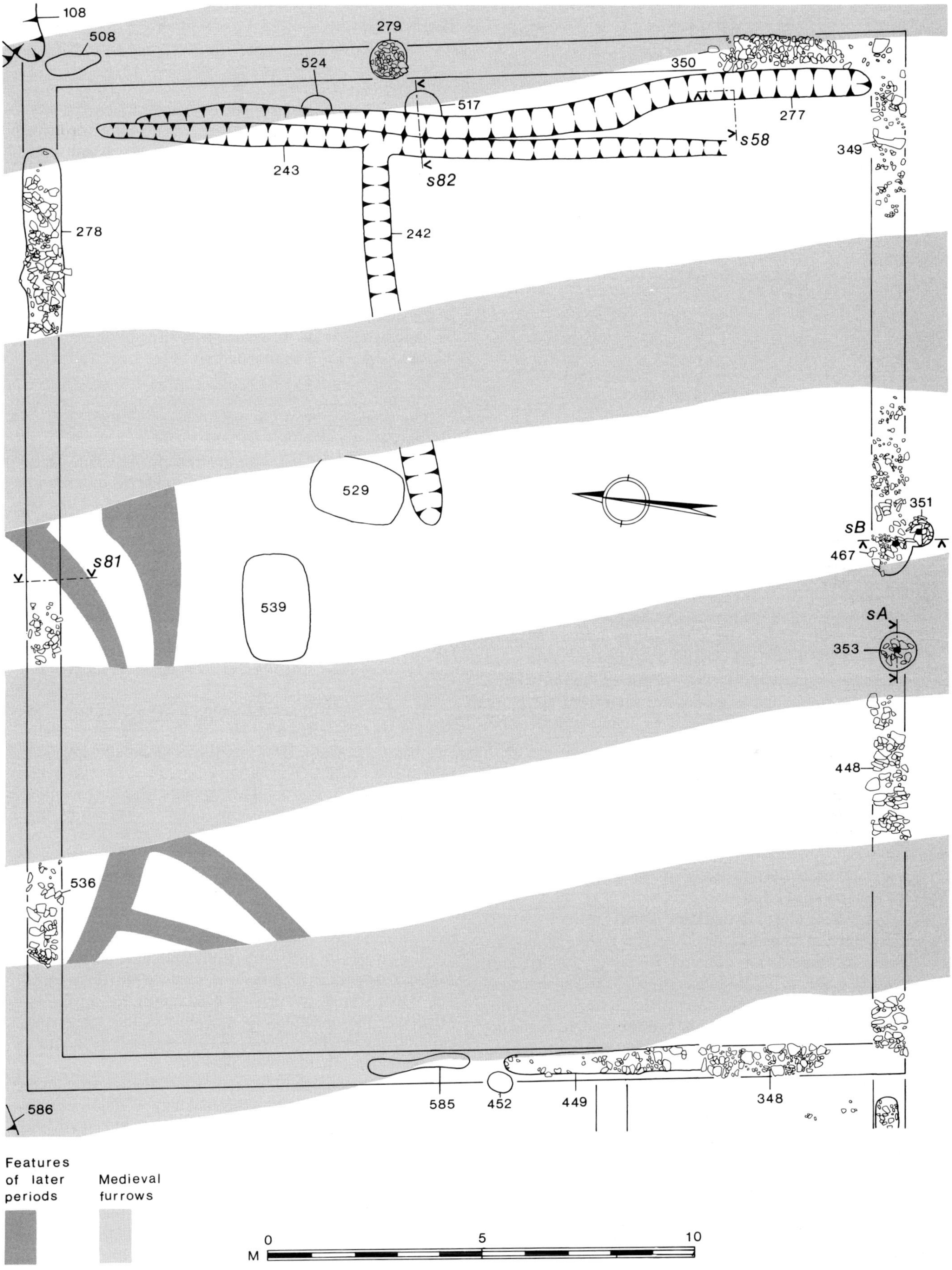

Figure 45 Plan of the rectangular building. Scale 1:125.

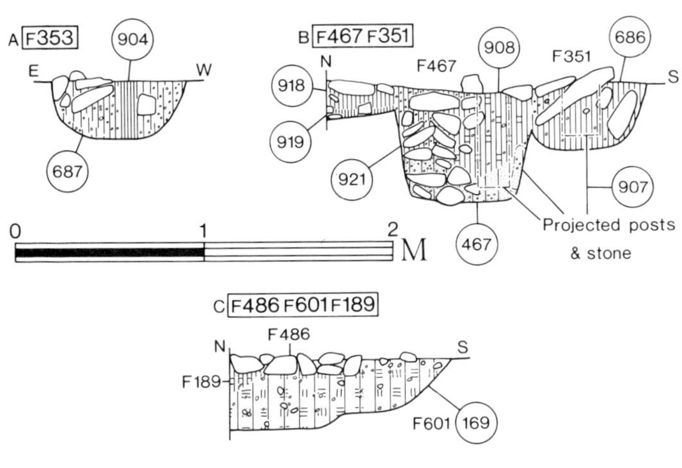

Figure 46 Sections A–B, in the rectangular building; C in the Mill-house.

a cross-member keyed into the wall. There may also have been ties into the wall nailed or tenoned lower down into the uprights to prevent whipping due to the weight of, and undue force on, the leaves.

Two features within the bounding walls, F529 and F539, both interpreted as possible tank bases, may have lain in the open. If so, a north range would have had a minimum width of 4m up to the outer wall. East and west ranges could each have been as much as 8.8m wide, and a range on the south side would have had its north edge where the suggested north wall of Period 5 may have been. No drain was found to take water shed from the roofs. Tanks may have been placed to take rainwater, but ground-water must have been a problem as the site here could not be excavated in winter because of standing water and there were no drainage ditches along the north side. Roofing schemes involving complete cover give impracticable results. An open yard is the most likely as it is easier to devise a structure and a plan which would allow reduction to the form in which it seems to have survived through part of Period 5.

The sheer size of the building makes the interpretation of its function more difficult. There was nothing to suggest living accommodation, although the pottery immediately south (Chapter 1, p. 23; Chapter 6, p. 55) showed that there had been a domestic focus nearby. The absence of effective drains is against animals having been housed. It should not have been a store as the ground was so wet, although the wall widths could easily have supported an upper floor used for this purpose.

The doorway may have had a second period as there was an equally carefully placed post just south of the original eastern one: F351 (Fig. 46). The pit for the post was rectangular, *c.*0.6m by 0.5m, and the post, 0.25m by 0.15m, was aligned north-to-south as before. The area of contact between the two post-pits was not enough for it to be absolutely certain that F352 was secondary, but that was the impression on excavation. The absence of a similar post to the west would have been due to the furrow there: F351 was only 0.35m deep, and only 0.31m survived of the depth of F353 (Fig. 46).

IV. The Mill-house, west of the Rectangular Building

(Figs 46C and 47)

Periods 4–5

Its remains have been described in Chapter 1 (see p. 23–4), but it is worth repeating that there was no sign of a north wall, that most of the south wall was missing and three furrows cut through the site. It was the flooring, which had sunk into the hollow of the earlier boundary ditch of the Main Yard, which showed that the structure had run through to the rectangular building.

The fragments of the south and west walls were each about 0.7m wide. The length of the structure from the south-west corner was about 17.7m. Using the limits given in the description of Period 4, the overall width of the building would have been about 6.75m, giving a span within of *c.*5.35m. The foundations of the west wall went to the base of the earlier ditch and this should be convincing evidence that stone was used up to the eaves.

The principal features inside were the three bases, Fs486–488. Their character was distinctive. Each was rectangular and all were approximately the same size, ranging from 0.9m to 1m in length and 0.7m to 0.8m in breadth. Each was made up of a carefully pitched bed of stone in a single course (Fig. 46, C) with remains of another laid flat across the top of selected stones fitted closely together. All showed that care was taken to make them uniform, flat and firm. Another can be fitted in to the east where there was a furrow, but this is discounted. Their make-up, spacing and construction does not suit the post-pads found in three of the barns. There is no obvious clue for what their function had been and there seem to be no parallels.

The discovery of several fragmentary millstones (see The Finds) raised, when first found, the problem of where they had come from. How a mill was mounted depended very much on how it was to be driven. One millstone (Stone 4, Fig.78) was complete enough to show how this was done: it had been driven from above. Therefore, the mill should have been at floor level and well bedded. Only this building produced suitable features and it is proposed that it had been a Mill-house. In which case, there would have been three mills in it (*cf.* Boon 1974, 289).

The pads were 2m and 2.4m from each other and the reconstructed distance between their southern edges and the nearest face of the south wall would have been about 1.75m, ample room for people or beasts to turn the stones. The upper part of the driving mechanism was probably fitted into the underside of a beam running across the building at a height which would allow people or animals to pass under it: hardly more than 2m.

As the stones would have shed the flour around their periphery, it should be obvious that they would have had to be raised so that a tray or trough of some kind could be fitted under them to catch it. There also should have been enough depth for any operator to have cleaned the inner side without difficulty. Two blocks, of a limestone akin to Barnack rag, had been thrown into the well just outside Barn 4. The stones themselves were fairly evenly sized and box-shaped: 0.62m by 0.62m by 0.27m, and 0.63m by 0.59m by 0.32m. The diagonal of each was 0.87m which would suit the millstones themselves. Each had a hole roughly made in the centre of one side which could have held a metal seating set in lead for the bottom end of the

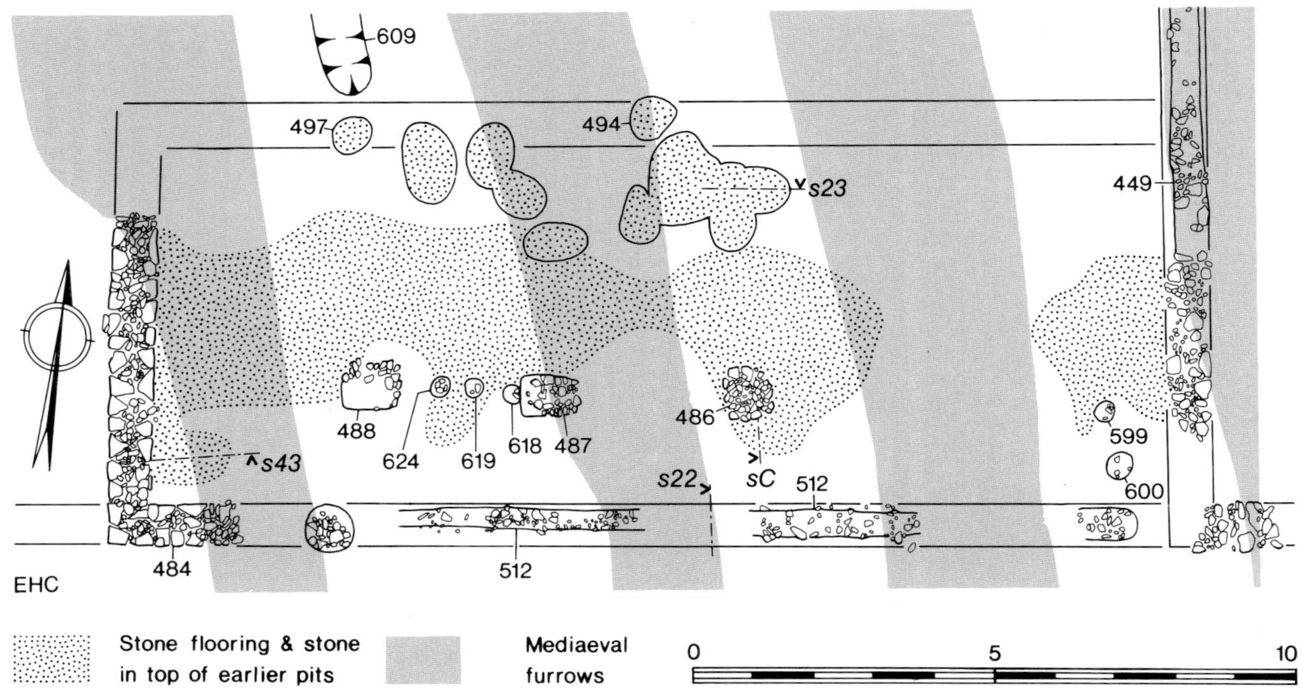

Figure 47 Plan of the Mill-house. Scale 1:125.

spindle running through the mill itself. If the period when the well was filled has been correctly assessed, it could argue for the Mill-house having survived into Period 5, even if it had been no longer in use.

In the reconstructed plan of the building, the line of pads is set towards the south side. Allowing 1.5m from the centre of the pads to the north, there would have been a further 2.35m to the estimated face of the north wall available for access, and perhaps for standing sacks of flour or equipment. The space at the east end of the building, beyond the 1.5m maximum working radius of the nearest stone, would have been about 5m. There should have been a door in the south wall. If this had been 2m wide, there would have been 2m–3m spare for storage and stacking against the end of the building, assessing a door set towards the nearest mill.

One important aspect of the production of flour is the need to sieve it to remove bran at least, to grade the flour and return coarse elements for a second milling (Moritz 1958, 156–7). In this case, the sieving could have taken place along the north side of the room with the flour being bagged and stacked at the east end. If wholemeal was required, a single milling should have been enough and there would have been no need for a sieving area (*ibid.*, 158).

The Mill-house should have been of stone throughout. Not only do the foundations speak for this, but also any mills in it would have been expensive and vital installations. On the grounds of security alone, the building would have had to be more durable and more fire-resistant than wood and clay. Cross-beams to hold the upper end of the mill-spindles could have formed part of a floor support system for a granary above. A granary here would have had a floor area of not less than 90m^2, one of the largest of the suggested granaries on Roman rural sites, and would complement three mills beneath.

There seems to be general agreement between ancient authorities and the archaeological evidence from Britain that there was no standard plan or system of flooring in non-military granaries (Gentry 1976, 5–6; Morris 1979, 113–19). In fact, the lack of uniformity makes it very hard to identify a granary in the first place, the most recent review (Morris 1979) demonstrating this admirably. Perhaps the only identified building which was unequivocally built to store produce in bulk is the granary at Lullingstone (Meates 1979, fig. 27).

Elsewhere, the presence of dwarf walls or ledges around a room have been interpreted as having been for raised floors in granaries. Morris suggests that Room 1 in Building F at Winterton was a granary as, in a second stage, it had joists run across the ground surface (Morris 1979, 114: Stead 1976, 29, fig. 16), but this could have been a domestic room and, as such, would suit the rather mixed character of the occupation in the Winterton barns. Building K also had ledges, the door threshold being set above these (*ibid.*, 73–5, fig. 37). The door was 1.6m wide in its second stage. Stone sleeper walls may be a better sign (*e.g.*, Brodribb *et al.* 1968, 18, fig. 4). To use these specific features to identify granaries is not always advisable, as Iwerne shows. A room attached to one side of Building B had ledges, or an inner wall, around it with the natural left unremoved in the centre. The argument was that this was to take a timber floor at a high level in the building and that it had been a granary, the assumption being that the main building only had a ground floor and the granary was placed in a kind of tower. Yet what was excavated was well preserved precisely because the building had been set in a cut in the hill-side. There is no reason why what was found should not have been a basement for domestic quarters opening out on to the uphill side. The timber floor in the projecting room would then match those in the rooms behind (Hawkes 1947, fig. 10B). However, the similar inner wall plan at Stroud could have served as a granary as there was a proper house elsewhere and the rest of the range looks very much as though it had been a functioning farm building (Williams 1909, pl. I).

An examination of the examples of granaries given by Appelbaum (1972, *passim*) shows that there were no good grounds for practically all of them. The problem remains: there must have been granaries and they should have been common. The wish to identify them is usually coupled with one to provide a functional interpretation for the parts of a site in terms of the activities which it is imagined took place. This applies to Orton Hall Farm where it would seem that a major grain store is necessary to serve the mills which are evidenced by the actual stones. These were in the north-east part of the site suggesting a source not far away. The grain should have been kept relatively close at hand. Barns 2 and 4 were closer to the main find-spots, but neither had any evidence for a special base and another function is proposed for these (Chapter 9, p. 230–1). The only candidates are the rectangular building, the present one and the one which stood immediately south of that.

The first had no signs of sub-division or any special feature and the damp site was ill-suited for storage, unless an upper level had been used. The last was only identified by the remains of one wall and there was no evidence for its character. The obvious candidate is the present building which, as has been pointed out, could have had an upper floor. This is one of the recommendations in the classical sources (Gentry 1976, 5) along with a north-easterly aspect so that the effects of high temperatures could be counteracted. The chief advantages of having a granary above mills would have been a proper gravity feed into them from it, the grain being stored well away from rising damp.

The available floor area would have been about 93m^2, about half the maximum area of the granary at Lullingstone, but a good deal more than the suggested stores at Iwerne, Stroud, Winterton and Shakenoak. These ranged from *c.*16m^2 to 47m^2. Gentry gives calculations for the thickness of walls to withstand the lateral thrust of grain. These show that the 0.7m of the walls at Orton Hall Farm would have accommodated a free height of 3m of grain (Gentry 1976, 35–6). When it comes to the load on the floor, the transverse beams suggested for the mills below may well have been substantial and, quite possibly, part of a strengthened floor put in to cope with the weights to be imposed on it.

V. The building south of the Mill-house

The presence of this is inferred from the single length of wall found running north-to-south. It was hardly more than 4m long and only about 0.5m wide.

Endnote

1. I am grateful to Mr R. Zeepvat for sending me details prior to publication.

Chapter 3. Features Associated with Fire

I. The Driers

Five were found, two (Nos 2–3) were linked in plan, and No.4 was replaced by No.5. Three had an 'H'-shaped flue plan with the main flue and stoke-hole running away from the middle of the cross-bar of the H. The fourth had, firstly, a rectangular basement with two piers inside and, secondly, a main flue which returned on either side. The driers are described in their probable sequence of construction.

1 F156, in Barn 2 (Period 3)
(Figs 48 and 50)
Like the double arrangement in Barn 4, this was of the common 'H' pattern in plan. The remains had been shaved off below the original ground level and were further damaged by a Period 4 ditch dug more or less through the centre. The surviving walls were of large uncoursed rag laid dry against the sides of the trench dug to receive the feature. The flooring was stone slabbing in the stoke-hole and running into the main flue. There were no signs of vents or floor over the structure. The siting of the drier against one of the main posts in the barn shows it was not thought to be a fire hazard. The stoke-hole was partly lined in stone and, on the east side, a possible hearth was intimately associated with it (see **16**, below), suggesting that both were in use together.

Overall plan size: 2.7m by 2.6m = 7.02m^2.

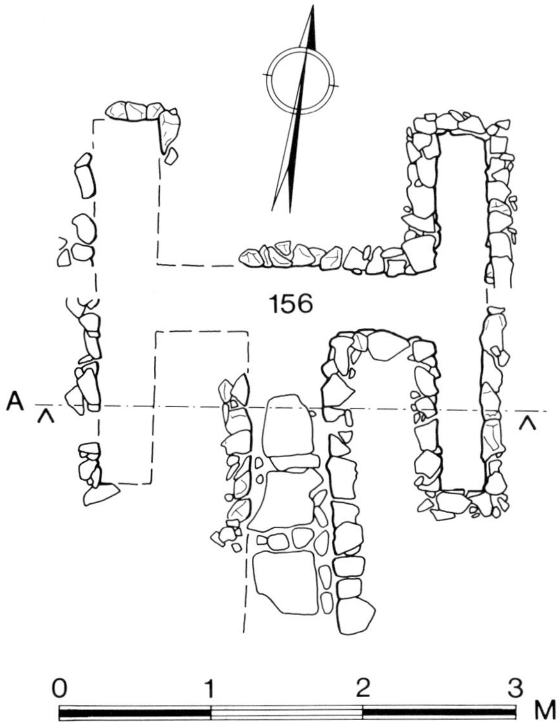

Figure 48 Plan of the Period 3 drier, F156, in Barn 2. Scale 1:50.

2–3 F168–F169, in Barn 4 (Period 4)
(Figs 49 and 50)
Both had been cut through by Period 5 gulleys and medieval furrows. After being uncovered in 1971, the remains were left open to all weathers and the destructive effect of growing vegetation. Both driers were of the same basic H pattern as drier 1. The two driers had the same stoke-hollow and were joined where one arm of each H impinged on the other.

F168 was the better preserved of the two and showed that the main flue and cross-flue of each were about 0.6m wide, the outer limbs being some 0.2–0.25m wide. The sides of the trenches dug for the feature were lined with limestone in a mixture of uncoursed rag and pitching with, here and there, flat coursed stones and all apparently laid dry. Earth was used to backfill behind this walling. The maximum surviving height of the walling was 0.4m. No closing walls were found at the ends of the smaller flues. A stone set at a slope at the end of one is taken to be a sign that each had had a vent at the corner of the flooring covering the whole.

Overall plan size: F168, 3.4m by 3.0m = 10.2m^2
F169, 3.2m by 2.6m = 8.32m^2.

The smaller size of the latter was almost certainly because it had to be fitted between two main posts and was further limited by the barn's west wall.

The western drier, F169, went out of use before the other; the main flue and the flue joining the two was blocked, but the latter may have been put in earlier to improve the draught. The damage to the main flue of F169 might have been done by a Period 5 gulley. The main flue leading into F168 was narrowed. Final disuse occurred when the barn was reduced in size, though the main flues were probably visible when the Period 5 ditches were dug.

In Barn 2, the drier was put as close as possible to one of the posts in the west row so that the eastern part of the nave could be occupied by other features using fire, but no Period 4 feature of a similar kind was identified in Barn 4. F168 was placed in the centre of the nave thus pushing any other features into the east aisle where there was a 1971 trench. The drier was almost exactly bisected by a line joining the two posts in the barn. F169 was equally carefully placed in the northernmost bay on the west side.

4–5 F159–F160, in Barn 4 (Period 5)
(Figs 51 and 52)
The original structure, F159, was extensively remodelled to form F160. The area of the upper platform or floor remained the same, the outer walls not being moved when the changes were made. The first version, F159, had a short main flue between the stoking pit and a nearly square chamber with a pier on each side, continuing the line of the main flue to support the floor. The whole was built in a rectangular pit and the outer walls were extensively rebuilt to form F160. The original stoking pit was done away with, a wall being built across the end of the main flue, the new stoking pit being that part lying outside the new floor. The new main flue was 1.2m long and 0.75m

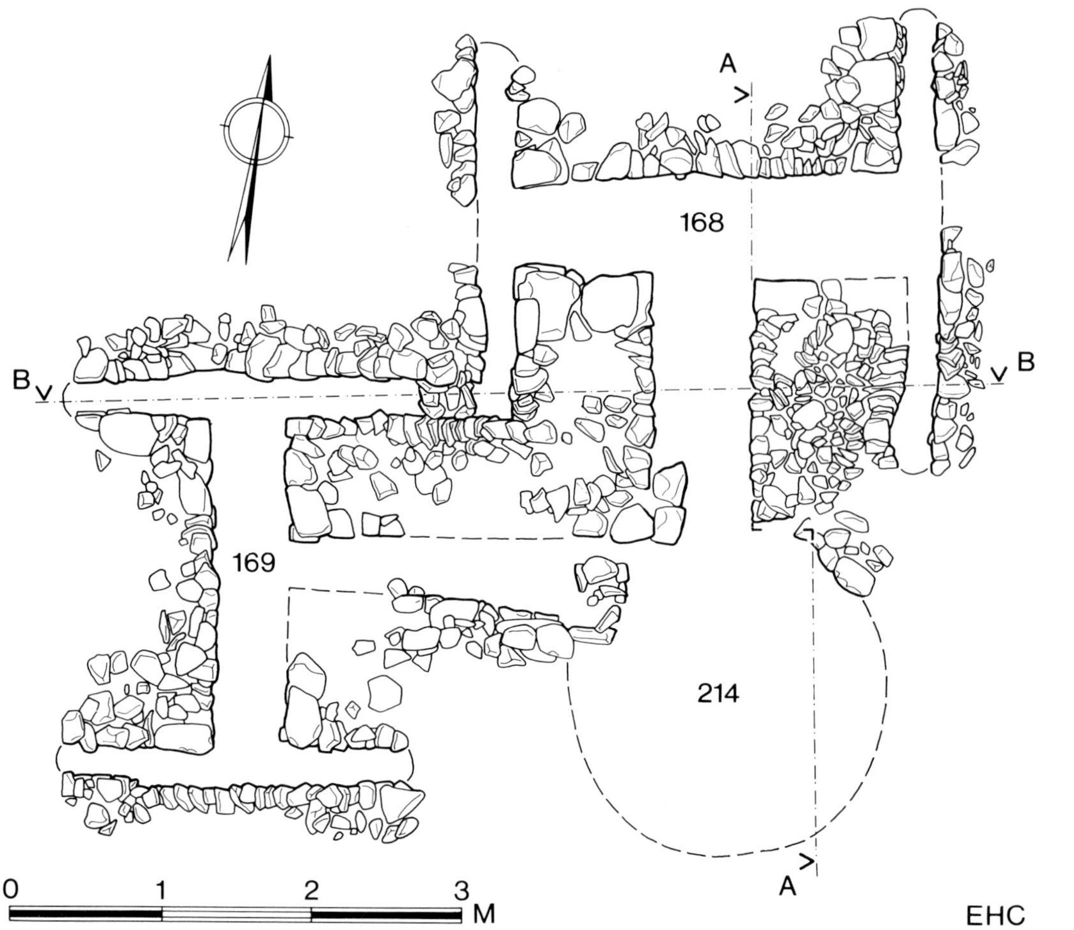

Figure 49 Plans of the Period 4 driers, F168 and F169, in Barn 4. Scale 1:50.

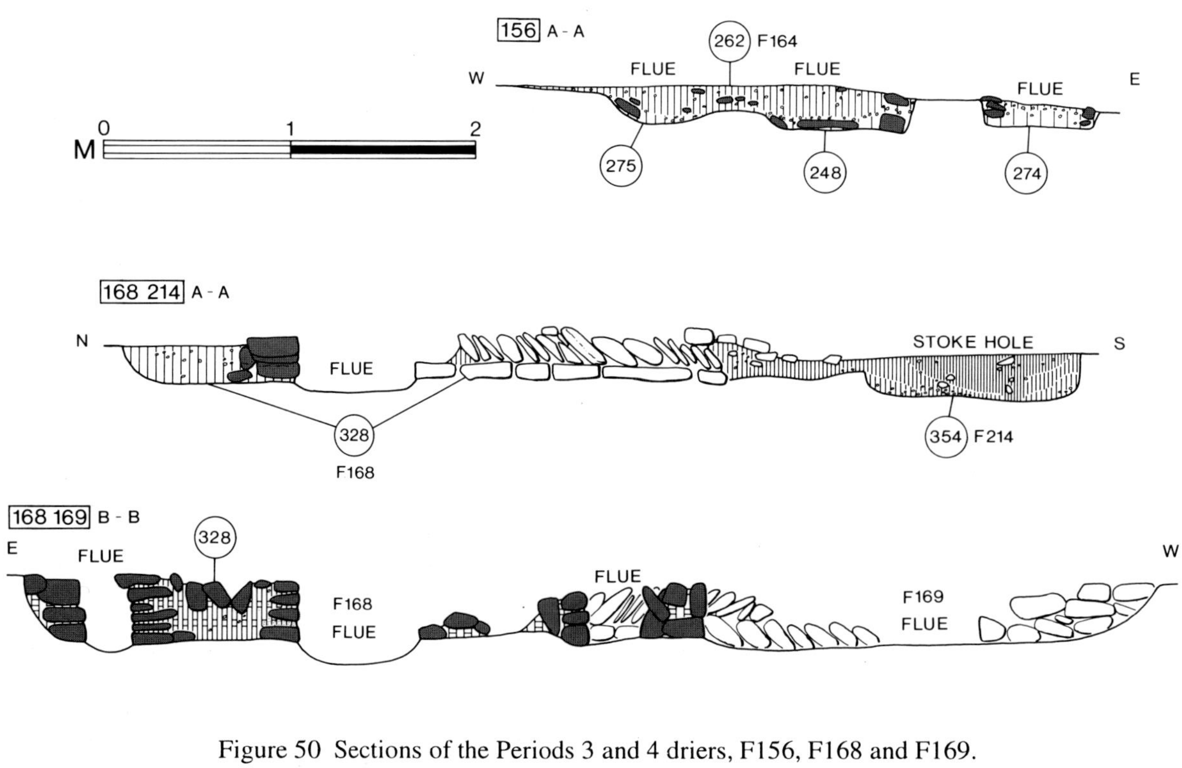

Figure 50 Sections of the Periods 3 and 4 driers, F156, F168 and F169.

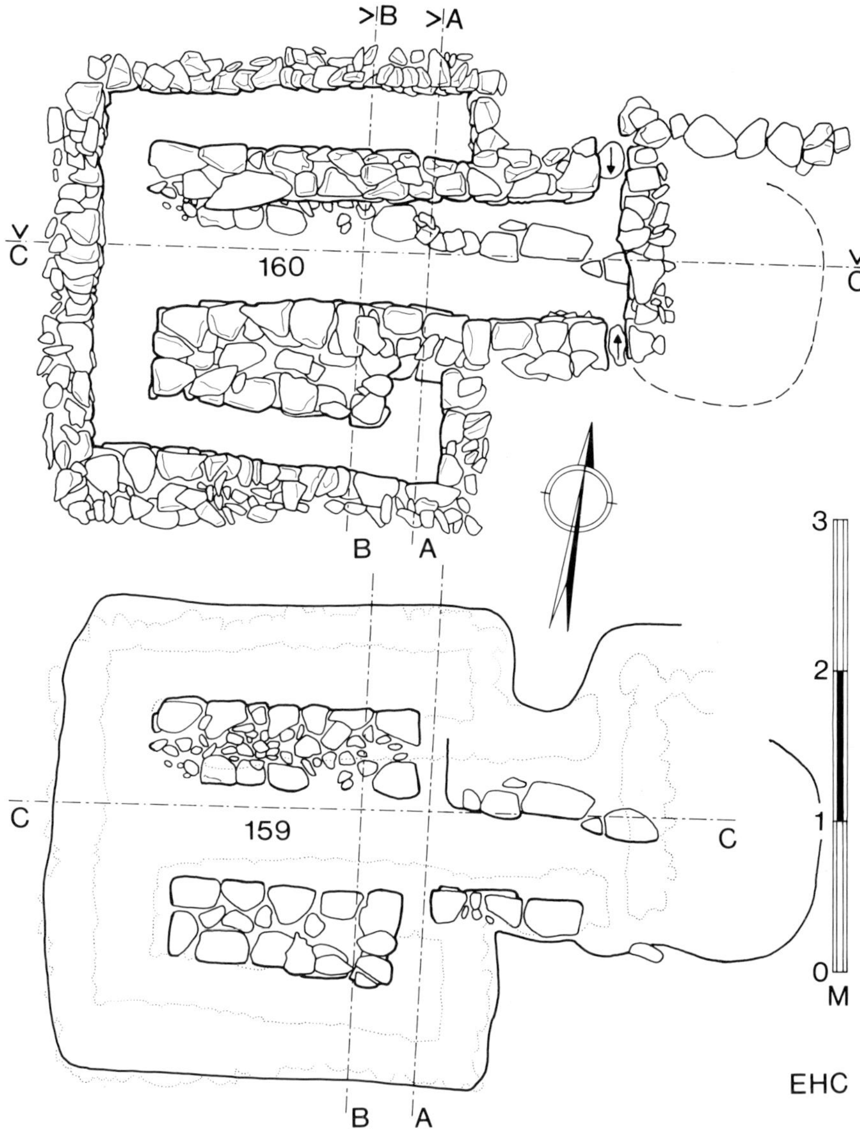

Figure 51 Plans of the Period 5 drier F159/F160. Scale 1:50.

wide. The freestanding piers inside were rebuilt as walls running from the east side to stop short of the end wall.

The surviving structure was of the same mixed limestone rag construction as that in the driers to the north. The material found between the stones, a clayey, gravelly silt, may have been deliberately selected to act as a mortar. When the drier was rebuilt, the new east wall was built on deposits partly filling the earlier pit so making the floor of the main flue slope down to the west.

The side walls of the main flue at the east end had breaks in them next to the end wall (Fig. 52, [159]C–C). The southern one was better preserved having a vertical stone at the south end with another inside that sloping down into the flue. While these could have been vents, this should imply that the floor over the whole had been run to the end of the main flue. On analogy with the other driers, this suggests that this one would have been fired at the opposite end, but there was no good evidence that there had been a stoke-hole here. The purpose, therefore, of the sloping stones is obscure.

Overall plan size: 2.6m x 2.4m = 6.24m^2.

F156	2.7m × 2.6m	=	7.02m^2	1:1
F168	3.4m × 3m	=	10.2m^2	c.1:1
F169	3.2m × 2.6m	=	8.32m^2	4:5
F159–F160	2.6m × 2.4m	=	6.24m^2	1:1

Table 3 Floor areas and ratios of the driers.

Discussion

The purpose of driers is to some extent in doubt, and for this reason the usual qualifying word 'corn' has been dropped. The possible uses of these features have been set out by Morris (1979, 6–8). We know very little about the growing, processing and consumption of either pulses or legumes and, apart from drying corn for storage or threshing, the alternative uses involving corn, preparing malt for brewing, and part-roasting for milling, may at times have been more important than the others. In the case of the Orton Hall Farm driers, one specific use is preferred (Chapter 9, p. 230–1).

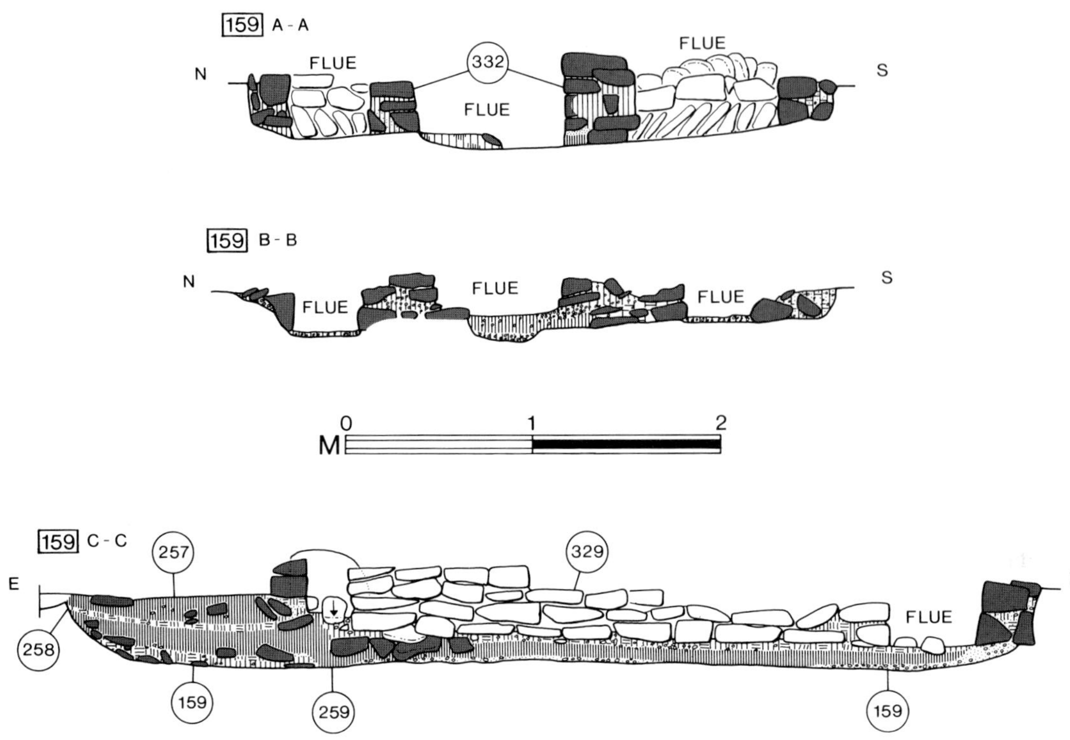

Figure 52 Sections of the Period 5 drier F159/F160.

No drier has been excavated, and published, in a sufficiently good state for it to be completely certain how it should be reconstructed. The remains have usually been badly damaged and shorn off at the under surface of the topsoil. Of the seventy-five examples gathered for comparison (MF5), only four provide significant information for what lay above the basic flues and even these were badly preserved. Therefore it is not surprising that most attention has been paid to the plans of the flues.

The comparative figures do not suggest that driers were built to standard sizes; if anything, the size should have been related to expected usage. The only major variation at Orton Hall Farm was the provision of two driers in the first instance in Period 4. However, the fact that one went out of use suggests that there was either no need for extra capacity, or there had been a drop in the demand for the processed material.

The comparative examples are arranged in increasing order of size, including those from this site, and are in MF5 along with their bibliographical details. In the following discussion, specific examples are cited by their number, in parentheses, in the list which is ordered usually by the overall area covered by the flues or, less often, by direct evidence for the floor area once heated by these. This was larger in some cases than the flues themselves might have indicated. In addition, there are a few examples of highly specialised plan or of great size: (80)– (82). These are the drier from Old Sleaford, the circular floor at Great Casterton and, finally, the late and very large installation at Verulamium. It is not certain that either of the first two had been in the open air, each lying in excavations hardly extending beyond the limits of the feature itself (*e.g.*, Corder 1954, pl. V).

The most recent discussion of driers (Morris 1979, 88–103) described ten major types before dealing with completely enclosed forms such as that from Great Casterton, or with features which were in a single line only. Of the seventy-nine listed in MF5, only three belong to the latter pattern, (4), (10) and (28), and these showed evidence that a floor over them was larger than the flue arrangement itself. There is no need to define in detail a large number of types as, with the exceptions of (7) and (26), there are essentially only two basic patterns, variations being minimal.

The simple 'T'-shape is the commonest form, and there are forty-eight examples listed. The type usually occurs by itself, but is sometimes combined to form a larger heated floor area, *e.g.*, (70) and (77), and is simplified to form the 'H' pattern like three of those at Orton Hall Farm. The other plan has a chamber containing piers reminiscent of a hypocaust of which (59) is one of the best known. Variations are to be found in three of the four chambers of two of the largest driers at Hambledon, (74) and (75). Generally speaking, as the listing makes clear, it is only when the floor area exceeds *c*.5m^2 that mutants of the basic 'T'-shape appear, and the fully blown hypocaust seems not to be smaller than *c*.7m^2. In other words, the need for a bigger heated surface gave rise to changes in the design of the flues beneath.

This conclusion seems unexceptional but for the great drier at Verulamium as, whatever its use may have been, it lacks a flue pattern commensurate with its size. The only evidence brought forward for it being later than the second period of the house in which it lay is that one of the flues cut the chalk foundations of the second stage of the building, it being thought that this could not have happened when the enlargements took place, as both walls and an ordinary room-heating hypocaust would have been built together (Frere 1983, 223). This may be doubted. The actual stoke-hole opening in the wall would have been built with the main structure, but the flues may well have been put in later in the same construction programme. As the flues and the stoke-hole through the wall had all been grubbed out (*ibid.*, pl. XXX, b), the structure itself could not be checked. Similarly, the first stage of the wall of the stoking area would almost certainly have been secondary in construction and, in this, would conform with building practice found elsewhere in Roman Britain. The burning or sooting on the foundations where the ends of the flues ran up to them was almost certainly because the flue gases discharged into flues set into the walls of an ordinary heated room. The fragments of mosaic found in the destruction of the system came from the floor laid over the flues of a channelled hypocaust: they could hardly have been the product of the *construction* of a drier through such a floor (*ibid.*).

Returning to driers in general, (4), (10) and (28) are only included because there was evidence for the floor area of the complete feature. The first two were small, hence neither needed an elaborate flue pattern. The one at Catsgore (28) may have had a sub-flue, but not at the low level of the main one. (22), (35), (39) and (42) are superficially of the same pattern: an 'H' with two parts of the sub-flues removed. That there can be a development from one 'design' to another is shown by the alterations made to drier 4 at Orton Hall Farm to create drier 5. In its earlier form it was like (63) and the first stage of (40) which could be described as simple hypocaust types, just as the others are basically modified 'T' patterns. The changes in both (35) and (40) point to what was probably a relative commonplace at the time: some installations needed to be altered to make them work properly. The principal tendency was to reduce the number of flues, and the blocking which separated driers 2, (62), and 3, (54), at Orton Hall Farm, may reflect this rather than the plain fact that one had passed out of use, although the evidence is that one ultimately did. The remaining plans, those classified in the listing as 'odd', are variations of both the basic 'T' and hypocaust design, only (7), (13), (26) and (31) defy easy allocation and (7), at least (see below), may perhaps not belong to the general class being considered here.

Virtually all the driers listed are rectangular — even (26) is best reconstructed as such — and less than 20m^2. All those over 12m^2 have double-flue plans. The proportions either deduced or seen directly show that all tended to be square. Only in double plans does a 1:2 ratio seem to be intended and is obviously a straight doubling of 1:1 for a single drier: (70), (72), (75)-(77). There are exceptions, *e.g.* (3), if the flues only are taken into account, but even here the report (Mellor and Goodchild 1942, pl. II) suggests that the floor was probably square. In most cases, only the flues remain but a notional wall thickness of 0.3m on the stoke-hole side reduces many an elongated plan to either 1:1 or nearly so. Those whose flue length can be estimated projecting beyond the definite or probable floor edge ((28), (37), (39), (40), (45), (63), (72), (74), (75), (78) and (79)) show that these are *c.*1m-1.7m in length. If this was general, then those with no surviving wall at the stoke-hole end can be reduced to 1:1 without any difficulty. This would, of course, alter the estimated floor size. The point of considering the proportions of driers is that none should have been awkward to operate and the smaller they are, the greater the likelihood that there had been only one person working them. It should follow that any which could not be easily managed may have been designed for a function differing significantly from the rest.

Forty-one of the assembled examples were definitely within buildings, and a further eight may have been. In the remaining cases, there is no certainty that there had been no surrounding building: many of the records are of old excavations or of those which were too limited in extent to establish the point one way or the other. Only complete stripping of the site around F156 (43) at Orton Hall Farm revealed the barn. Therefore, the reconstruction of the one at Foxholes Farm (33) as an independent building may not be advisable (Reynolds 1979). An experiment at Welton Wold showed that cover was needed to mask the effect of weather on an open site (Frere 1977, 383).

Those sited on chalk are at a disadvantage as natural wastage of the surface of the subsoil coupled with ploughing could have removed all trace of a building except for the most deeply founded elements. Durrington Walls (17) may be cited as an example. Here, a solidly built drier lay in an area with gulleys and post-holes, the latter not adding up to a building around the drier. However, ditches 6 and 22 may have run round one, their general alignments agreeing with the drier's axes (Wainwright 1971, fig. 5). Had this drier been open to the skies, it would have been a trap for rainwater. The lack of definition round some of the driers at Hambledon (Cocks 1921, pl. XIII) was almost certainly due to the excavation method used, and both these comments can be applied to those found on Rockbourne Down, Hampshire (Sumner 1914). Bearing these points in mind, and considering the number which were certainly fitted into buildings, it seems unlikely that any had been placed in the open air.

If sophistication in the upper construction and use of driers need not be looked for (Morris 1979, 10, 12–13, 15), insistence that driers should have been properly housed would be out of place. But, if even heating and a solid floor, whether of mortar over stone or tile, or plain or mortared wood, was a principal consideration, a building should be expected in practically every case. Where driers are found fitted into buildings, problems of roofing do not arise. Reynolds, in discussing the possible reconstruction of the one at Foxholes Farm, (33), arrived at a simple structure rising straight from the outer walls of the drier itself (Reynolds 1979, 30–2, pls IV and V). These were less than 0.25m thick (*ibid.*, fig. 1), but the reconstruction used thicker walls of 0.4m or more (*ibid.*, pl. IV). The experiment of drying corn straight from a field was counted as a failure (*ibid.*, 38). What the effects of a proper ambient atmosphere would have been cannot be known (*ibid.*, 32). A single-roomed structure would make it difficult for a worker to move around without crushing the very thing being treated. It is for this reason that emphasis was laid earlier on the number of sides from which a drier could have been managed.

The microfiche list briefly mentions the kind of building which can be identified and an assessment of the number of sides, excluding the stoke-hole side, from which each was approachable. The only exception is the installation at Highdown Hill (78). Here, all of the stoke-hole side was available to a worker as the main flue had been roofed over inside the building. In general, the larger the drier, the more sides were encumbered by such as walls or tanks and large driers tended to be in the middle of a building rather than in a corner. At two sites alone was the drier apparently approachable only from the stoke-hole side: Catsgore (7), and Darenth (73). In the first, the flue plan is distinctly unusual and may not have been a normal drier and, unlike the other three from the same site, no carbonised grain was noted in association with it (Leech 1982, 64–9, F275, F444, F421, F316). The remains at Darenth were of big installations, each of the order of 13.2m^2 and each fitted into adjacent rooms at the end of a barn. The remnants of the flues yield little sense and so perhaps they should be removed from the list.

The classic interpretation of the superstructure of driers was based on one of two found at Atworth (8) (Goodchild 1943) and this was used to reconstruct the one at Park Street (4) (O'Neil 1947, 46–7, figs 3 and 24, pl. II, B). Morris (1979) considered how driers were used and the bias of the study led to the conclusion that open flues for small sheaves or bundles of grain on the stalk were more likely: the evidence for double floors was not found convincing (*ibid.*, 9–10, 13–15). The plans of driers suggest that even heating was a pre-requisite, and it should follow that their use should have been efficient and uncomplicated. The evidence as to how driers were completed is good, given the conditions under which most were found.

The flooring of the common 'T'-shaped drier can be arrived at fairly simply. Both the Atworth and Park Street reconstructions follow the same model: the draught took the flue gases to the far end of a covered main flue where they were drawn back under an upper floor to exhaust from the system by means of side flues, one in the case of Atworth, (8), and two at Park Street, (4). This design demands a double floor. The evidence at Durrington Walls, (17), suggests that the main flue was vaulted right to the cross-flue at the back. Each end of the latter flue was rounded and then sloped outwards towards the top, so much so that the outer walls were cut into; there was no sign of a vent at the end with the stoke-hole (Wainwright 1971, fig. 6, pls Ib and IIa). Side walls were provided to carry an upper floor. If near-modern peasant practice was to have no bar between fire and crop (Morris 1979, 8) or only a temporary one (*ibid.*, 11), it should be asked why such sturdy driers were needed in Roman times, their construction and design being much more substantial than subsistence or marginal farming methods would need.

The length of between *c.*1m and 1.7m of flue provided on at least eleven driers not only removes the fire from the floor, but shows that a draught was needed to operate the system. The principle behind Goodchild's restoration was the distribution of heated air in order to heat the surface above evenly. The upper floor of the Atworth drier was, apparently, formed by stone flagging, probably supported at intervals by *pilae*. The Park Street reconstruction has a wooden floor (O'Neil 1947, fig. 24). The Durrington Walls drier would have needed another floor to even out the heat from the main flue beneath and the nails and charcoal found in the flues could have derived from a wooden floor (Wainwright 1971, 84).

The result seems to be a choice of design: the return of the flue gases over the main flue or the quick emission of these at the ends of the cross-flue without there being a full circulatory system at all. In other words, the radiant heat from the covering of the main flue was enough. This could be important. The Atworth drier (8) seems to have been of stone, probably with a dressing on top, and could have been fire-proof; a wooden floor may have been more vulnerable, but would have conducted heat more readily.

However, there is other evidence for superstructures. One at Thundersbarrow Hill, (5), had a projecting shelf along the back about 0.4m above the surface over the main flue. The same feature at about the same height occurred in one of the Rockbourne Down driers (53). Both point to a support for an upper floor and the need to provide, possibly, a low wall running round this. The double 'T' drier at Downton (72) had fallen slabs in one sub-flue and the photograph illustrating them (Rahtz 1963, pl. 1) suggests that these had not ceiled the flue, but had fallen from a higher level and could have come from another shelf. This may have acted as a deflector for rising flue gases and as a protection for a wooden floor against sparks. The alignment of boards resting on such a shelf would have been from back to front and the absence of a shelf at the back elsewhere may only mean that boards were run the other way. Most driers were less well preserved than these, but the occurrence of vaulting or slabbing over the main flue as far as the cross-flues of 'T' driers points convincingly to an overall design like that of either Atworth or, if it was different, Durrington Walls: (10), (14), (15), (16)?, (21), (24), (32), (36), (45), (47)–(49), (53), (63), (68), (72), (77), (78)?.

The behaviour of the ends of the cross-flues at Durrington Walls is matched at Brading (14) where one drier had ends sloped outwards and a main flue once covered with tile and stone slabs (Price and Price 1884, 17–18). Other cross-flues with sloped ends occurred at South Malling (16), Gadebridge (50) and Downton (72). The two inner sub-flues at Downton probably opened directly into the space under an upper floor. In this case, it may be that common practice for building single 'T' driers was unthinkingly applied here in a double one. The remaining evidence for flues may be divided into two: most, (21), (25), (28), (61), (66) and (76), had flues which opened directly out of the drier at the ends of sub-flues and through the side walls. Only one had evidence for a flue opening upwards into the space between the main flue and an upper floor: (45). In all these cases, the vents were at the end away from the stoke-hole. It looks as though a proper circulatory system was not the prime need in most 'T' and 'H' pattern driers and those at Orton Hall Farm fit in with this.

In other words, heating of an upper floor was intended to be even and gentle, conditions which admirably suit the preferred function of the Orton Hall Farm examples (Chapter 9, p. 230–1).

No example of the hypocaust pattern and its variants seems to have survived sufficiently for details of an upper floor to have been noted. However, it is possible that a primitive distinction can be made between these and the rest. If an evenly heated floor was of first importance, then none of the hypocausted group needed to have a double floor arrangement. The *pilae* were small enough for the floor supported by them probably to have been heated satisfactorily.

II. Other Features Associated with Fire

(Figs 53 and 54)

Introduction

All the remaining features associated with fire are gathered here. The interpretation of many is open to question and, rather than separate them into subjective categories, they are arranged in the order of period and in the order of context group within each. Hardly any had evidence for burning on their surfaces, but there can be no doubt that a fire had been in **18**, the evidence being confined to the structure alone.

The correct periodisation of **2** and **3** is in doubt. They have been put in the earlier period, and their possible true date has been discussed in the descriptions of Periods 1 and 5 (Chapter 1). The third feature of a similar character, **27**, F999, lay almost in line with **3** and contained Anglo-Saxon pottery.

Period 1

1 F901 Phase f (Fig. 16, [19]), ((39) and (51)). Sealed beneath the final fill of F969, the dating was second century and the feature may have been associated with the house thought to have been nearby (Chapter 1, p. 8). The shape was that of a long thin trench with a bowl at one end. The length of the trench is not known as it ran out of the excavation, but was 0.25m wide. The bowl, an elongated oval 0.46m wide and 0.7m long, was 0.35m deeper than the trench. On each side of the bowl was a thin limestone slab on edge. Running away from the back of the bowl for a distance of *c*.0.8m was a 'tail' consisting of a layer of limestone pieces, none overlapping. There seems to have been no lining to the trench and, although it was filled with a black ashy silt, clearly the product of burning, there was little evidence on the surfaces of the feature itself. There was no evidence for repairs to, or for structure over, the feature.

2 F1081 Period 1 or 5 (Fig. 53), (38). The shape was the same as that of **1**, the overall length being 1.56m with the 'bowl' or chamber having a diameter of 0.74m. The trench was 0.44m wide. The latter seems to have been purposefully filled with green clay (L2622), but only after the rest had been choked. Again, it betrays signs of burning without being burnt itself. The feature seems to have fallen into disuse. The basal layer was burnt (L2624) and had a sandy-clayey layer sealing it in the bowl. This may have been the remains of structure over that end. No superstructure or lining remained, but the intermediate fill contained burnt daub.

3 F1082 Period 1 or 5 (Fig. 53), (38). Having the same shape as **1** and **2**, the feature was deeper than **2**. The overall length was 1.65m, the 'bowl' being 0.65m in diameter; the trench end was 0.42m wide. There was no sign of any lining. Two main fills in the lower part suggest two phases. The one against the back of the bowl was a use deposit (under L2629) on a sandy layer which had not been cleaned out before either reuse or repair for a higher floor. The thin use-deposit was succeeded by a bed of dark clayey and sandy loam liberally speckled with charcoal (L2629) which probably accumulated during a period of disuse. It was then apparently mostly dug out and replaced with a much cleaner earth (L2628), but still containing charcoal. This formed a bed for new firings, possibly dressed with gravel which was then disturbed and acquired a near-black ashy content (L2627). Again, at the back of the bowl was a build-up containing a lot of comminuted charcoal (L2626). All were sealed by a fill (L2625) probably representing the abandonment of the feature. The way in which the layers accumulated inside the feature suggested that the 'bowl' had been covered and had had such a narrow entrance that cleaning out the unwanted fill would not have been easy. There was no trace of a superstructure.

Period 2

4 F1008, Barn 1 (66). A shallow nearly circular hollow 2.1m by 1.9m and only 0.08m deep. The interior was not burnt, but the fill contained the products of burning including burnt limestone. The feature lay in the middle of the barn and this suggests that it may have been a hearth, possibly like **24**.

5 F1011, Barn 1 (66). When excavated, the feature was 0.69m in diameter and 0.12m deep. However, when first exposed, there was an extension on the surface running south for about 0.5m giving the whole an oval plan. The fill had a high density of charcoal and a little burnt clay. It was the plan form as it was first uncovered which suggests that this had once been a feature of the same type as **1, 2** and **3**.

6 F1019, Barn 1 (66). The whole feature lay in an elongated hollow 1.95m long and 0.67m wide at most. At the west end were the remains of a stone structure consisting of thin, burnt, limestone slabs forming side walls. These may once have run across the west end and they seem to have been bedded against a clayey lining to the pit. The length of the chamber would not have exceeded 0.5m-0.6m and its depth was not less than 0.4m. The stoke-hole sloped down from the east, where it was barely 0.1m deep, to a depth of 0.25m just at the step down into the chamber. The fill was a darkened silt containing a lot of charcoal.

7 F1034, Barn 1 (Fig. 53), (66). Again slight, the 'bowl' was 0.7m in diameter and 0.15m deep, with the stoke-hole being 0.65m long and only 0.12m deep. The surfaces were not burnt and there was no trace of structure. The section shows a basal layer (L2254) whose hollow top had a layer of burning on each side (L2254) with an ashy charcoal deposit (L2252) in the middle. This suggests a restricted access between two parallel walls with a narrow slot between.

8 F1039, Barn 1 (66). A 'furnace' consisting of a stoke-hole and chamber with no evidence for a structure. The feature was slight, being only 0.11m deep. The whole was 1.6m long, the 'bowl' 0.8m across and the width of the stoke-hole 0.34m. The evidence for heat consisted of a burnt rim to the 'bowl', otherwise there was only charcoal flecking.

9 F1040, Barn 1 (66). The form was not quite like that of the preceding examples, the whole being more like a shallow trench with a shallow curved profile. The length was 1.55m, the width 0.4m and the depth at most was 0.2m. The fill contained a variable quantity of charcoal becoming dense in places with patches of burnt sandy clay and burnt earth. The hollow could have been a raking slot under a structure which had a fire raised on a grid of some form, thus accounting for the lack of burnt surfaces. The hollow may have resulted from repeated cleaning out.

10 F1041, Barn 1 (66). This was 1.15m long, 0.55m wide in the middle, and 0.4m wide at the ends. Its depth was about 0.1m. The fill had the charcoal flecking, but it was the form of the feature which places it here.

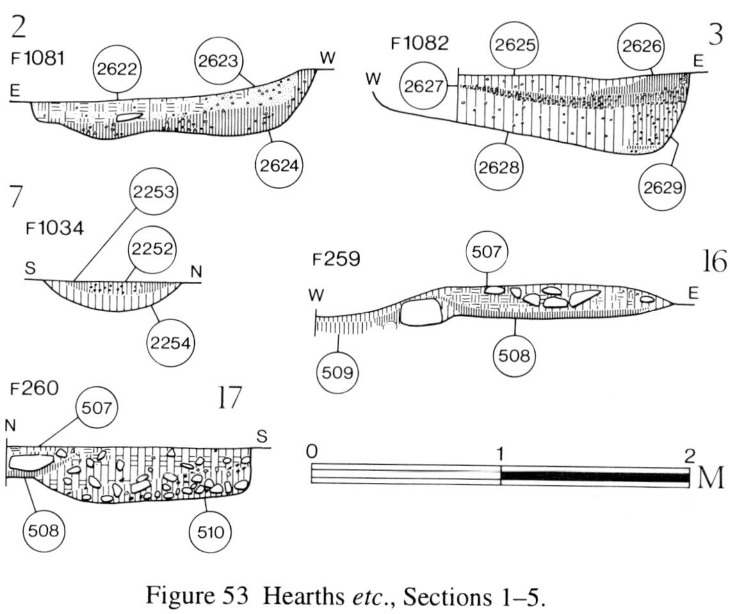

Figure 53 Hearths *etc*., Sections 1–5.

11 F1042, Barn 1 (66). This was very close to **10**, but the state of preservation prevented the relationship from being established. When first uncovered, they were thought to be one feature. Of the two, this was the slighter, being 0.57 in diameter and 0.15m deep. The fill contained not only charcoal, but also lumps of re-deposited material and burnt earth, suggesting that it had been dumped here.

12 F1066, Barn 1 (66). Of the same form as **7**, the 'bowl' was 0.57m in diameter and 0.2m deep, the trench being 0.75m long by 0.1m deep. The fill had some small fragments of burnt clay.

13 F1067, Barn 1 (66). A repeat of the last, the remains were slight, but represented a larger version: the 'bowl' was 0.84m in diameter and 0.1m deep. The trench was 1.09m long running from 0.1m to 0.05m deep. The fill contained highly comminuted charcoal and pieces of burnt clay.

14 F320 (88). A pad of clay, once *c.*1m in diameter containing large blobs of charcoal, it cut F227 and was cut by F162, the bounding ditch of Barn 4 in Period 5. It was also cut by F321. The feature may have been a hearth, but there was no sign of burning. However its surface had been planed away by ploughing.

Period 3

15 F258, Barn 2 (105). A hollow 0.64m by 0.58m, not obviously burnt, it contained burnt clay with plentiful charcoal both comminuted and in flecks. It may have been a raking hollow in front of an oven built on a lost floor.

16 F259, Barn 2 (Fig. 53), (105). Cutting into **17**, this feature started as a hollow 1.2m by 0.6m and about 0.18m deep. The lower fill (L508) lapped over the side wall of F156 drier flue sealing its construction trench. L508 was derived from burning, its hardness pointing to fire on its surface. This was sealed by L507, a hearth made of fired clay with burnt stone, previously burnt clay and the occasional lump of material similar to L508.

17 F260, Barn 2 (Fig. 53), (105). Later replaced or extended by **16**, this was a hollow 1.2m by 0.94m and *c.*0.27m deep at most. The fill, while the burnt content was not high, contained charcoal, fire-shattered pebbles, burnt stone and a little burnt clay, mainly at the south end. The hollow had some clay capping at the north end which may have been all that survived of a bedding for a hearth.

18 F105, Barn 2 (Fig. 54), (106). All that survived was the roughly circular hollow *c.*1.58m in diameter, in which the feature had been built, and the stoke-hole associated with it. The structural elements were the stone cheeks of the opening, 0.6m wide, and a stone pedestal or tongue inside. This was 0.7m long by 0.45m wide and projected from the back of the hollow. The burning on the inner corners of the cheeks, and on the front end only of the pedestal, suggests that there had been no up-draught. Pieces of burnt clay in the demolition deposit may have come from the superstructure. This feature is interpreted as having been a vat base, the pedestal providing support for the bottom of the vat (see Chapter 9).

19 F707, Droveway (Fig. 9), (111). Oriented east-north-east by west-south-west, the feature was 1.2m long and 0.5m wide and showed as two conjoined circles when first revealed. The west end was 0.14m deep and had burnt sides and base, and was filled with an ashy layer. The east end was 0.16m deep and had not been burnt, and its fill contained baked clay lumps and ash. The whole had been sealed with a bed of clay.

20 F1092 (130). The bowl was 1.1m long, 0.9m wide and 0.15m deep. The trench was only 0.7m long, *c.*0.6m wide and 0.11m deep. There was a ridge between the two parts. An ashy deposit lay around the northern side of the bowl sealed by demolition products containing limestone pieces, lumps of burnt clay and discoloured, almost black, earth. The limestone and clay may have formed parts of a superstructure.

21 F1047 (134). A shallow and large hollow some 4.5m north-to-south and at least 2.75m east-to-west, with no evidence for structure in or around it. The lower fill was basically a black ashy soil containing lumps of burnt clay and many limestone fragments. The upper was redeposited topsoil.

22 F1135, Barn 3 (138). A narrow stone-lined trench, 1.5m long, 0.32m wide, only survived for a depth of 0.1m, its west end lay *c.*0.35m from the west wall of the barn. The east end opened into a hollow 0.4m in diameter and 0.05m deep. There was a rough stone floor for 0.65m from the west end and here, for a distance of 0.7m, was evidence for intense heat on the side walls. Above this part there had probably been an opening into a superstructure. The limited extent of intense burning points to a raised platform walled across at both ends, in which case, the channel would have formed a flue and raking access, the bowl being a wear hollow rather than a stoke-hole as such. The original feature was therefore probably a raised hearth. The absence of any trace of slags or iron scale is the best sign that this had been a cooking hearth.

23 F1163, Barn 3 (138). Another 'furnace' sunk into the subsoil without any trace of structure. The bowl was 0.7m long, 0.48m wide and 0.17m deep. The trench was 0.67m long, 0.35m wide and about 0.1m deep. The 'bowl' lay at the east end. The fill of each part had no particular characteristic and the fired end had been deliberately backfilled.

Period 5

24 F137 (Fig. 54), (233). Originally showing as an oval 1.75m long and 0.95m wide, it was aligned south-east/north-west. It had irregular sides down to about 0.25m and then became a near-vertical-sided slot 0.36m wide with a rounded bottom. At the bottom and filling most of the slot was a grey-brown slightly clayey, silty

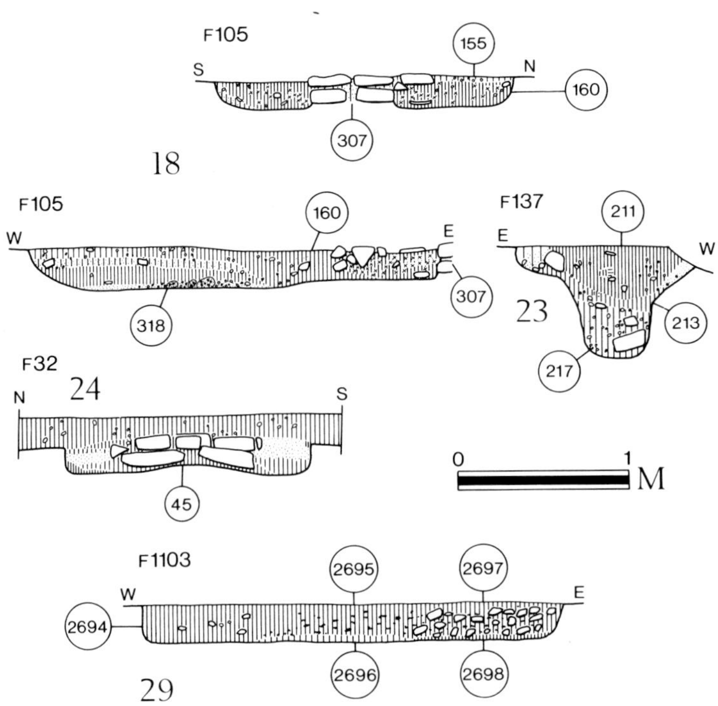

Figure 54 Hearths *etc.*, Sections 6–10.

loam with some slag, gravel and small, almost chips, of limestone (L217), but no charcoal. Above this and spreading out to either side of the upper part was a bed of slag set in a matrix of very dark silty loam with burnt pottery, charcoal and, at the base of the layer, flecks of displaced natural and pieces of limestone (L213). The top fill was very much the same as the bottom one, but with more burnt clay and a sherd of Anglo-Saxon pottery. A ledge on one side *may* have been used to seat part of a structure. The association of this feature with a possible Anglo-Saxon building has been discussed above (p. 32).

25 F32 (Fig. 54), (248). A circular shallow pit, only partly exposed in the excavation, with near vertical sides, some 1.25m in diameter and *c*.0.15m deep around the edges, the centre being very slightly domed. In the middle was a limestone bed in two courses filling the hollow. In the bottom, there was a layer of charcoal around part of the north perimeter and the feature could have begun as a fire-hollow later formalised by putting a square bed set with its corners to the cardinal points. There were other pieces of limestone set on a bedding of burnt earth in the upper surface of the hollow around the square. Only the edges of that were burnt and there may once have been a top dressing of clay.

26 F83 (Fig. 31, [101]), (249). The hollow was sub-rectangular, just under 1m across in either direction and hardly more than 0.1m deep. It had a flat bottom and gently curving sides which showed no signs of having been burnt. The fill and the matrix were heavily charged with charcoal, mainly comminuted and darkening the silty fills. At the east end were pieces of burnt clay in the top and bits of burnt tile. Again, this might have been a rake-hollow in front of an oven built at ground level, the burnt clay suggesting that it would have been to the east over the Anglo-Saxon ditch, F55, which was cut by the hollow.

27 F161, Barn 4 (258). Lying east of the drier F159–F160, the remains consisted of an oval of stones, about 0.5m by 0.35m inside, with intense burning on the inner face. Two stones sloping outward, about 0.2m apart, formed the narrow entrance. There was a hollow both inside and out with a ridge across the entrance.

28 F999 (289). The plan form and alignment recalled those of **2** and **3**. An elongated 'bowl' about 0.5m wide and with a rounded end lay at the east end. It was only 0.18m deep and it narrowed to the west to become only 0.4m wide. The overall length was about 2m. The lower layer of fill was a grey-brown loamy silt with pieces of limestone, the upper being darker with a fair amount of charcoal flecking and more and larger fragments of limestone.

29 F649 (311). A small shallow hollow, with a slight burnt rim, about 0.55m in diameter and no more than 0.27m deep, its fill was a near black silt heavily larded with charcoal blocks and twigs.

30 F1103 (Fig. 54), (322). Basically straight-sided with rounded ends and lying east-to-west, the feature was 2.07m long, 1.26m wide at the east end and 0.95m at the west. The bottom was essentially flat, mainly 0.25m deep and rising to 0.21m at the east end. The eastern third had a thin layer of near-black soil containing much charcoal sealed by a bed of selected alluvial cobbles (L2698) running up the sides. This, and the charcoal-filled layer below, may point to there having been two phases of use. The cobble bed was heavily burnt on top and most of the cobbles were heat-cracked, many with spalling and some were reduced to a friable state. On the bottom in the middle was a bed of sand (L2696) with evidence for a trail of charcoal-filled soil and limestone pieces (L2695) thinning out to the west. The hollow was finally filled with undistinguished silty soil merging with the layer below. L2696 contained Anglo-Saxon pottery.

31 F954, undated. No description of the feature survives and its relationship with F969 is unknown: it could have been in the top of that, or cut by it. The content, apparently, had heavily burnt limestone, patches of burnt grass or bracken as well as burnt pea gravel and clay.

Discussion

The features listed fall into three basic types: 'furnaces', 'hearths' and 'ovens'. With the single exception of **24**, F137, with its iron slag and Anglo-Saxon pottery, none had any direct evidence for its function. Most were filled with soils having evidence for fires nearby. Only three, **1**, **6** and **22**, had some evidence for above-ground structures penetrated below ground level. Only **22** had better evidence for an actual above-ground structure. As all, with the exception of **22**, only survived below the stripped surface of the site, all should have been high enough above any ground or floor level to have been used without undue effort. If so, the 'bowls' and 'trenches' then become parts cleared out to give draught to the fire above. Therefore, a general lack of evidence for great heat in these elements is to be expected, along with little sign of the actual use of the complete installation.

The restricted area in **22** which had been subjected to intense heat points to a substantial above-ground structure. The stone coursing on either side directly implies walling above ground. **1** and **6** only had thin slabs placed on the long sides to define a narrow channel. The slabs may have been put in to stop further erosion of the sides below a built structure, or put in to support the inner edges of such.

As for location, most lay inside buildings: **5–13** in Barn 1; **22** and **23** in Barn 3. This accounts for eleven of the seventeen, and in addition, **24** might have been housed. Of the other six, the possibilities that **2, 3** and **28** may have been covered has been discussed. Of the rest, **30** is unusual for the bedding of cobbles at one end and for the presence of Anglo-Saxon pottery and may have lain in the pentice structure along the south yard wall, if that survived: its dating indicates Period 4, unless the pottery was residual.

Only **1** and **20** appear to have been unequivocally in the open air. **20** lay outside the area of formal excavation and traces of a structure may not have been seen. **1** was in a ditch, possibly for shelter. It was the only one of all these features aligned north-to-south. **24** ran south-east/north-west, and the rest were east-to-west which was possibly the preferred direction, but probably dictated by the alignment of the buildings in which they either lay or with which they were associated. As for **24**, its alignment may have been the result of deliberately selecting a direction for the optimum draught for a blacksmith.

In common with most Roman sites, the evidence for cooking was very limited. Apart from the area in the north-west corner of the excavation which included features with pot bases still in position ((9)–(15)), which may have been used for food preparation (Chapter 1, p. 3), no obvious candidate offers itself. The character of a typical Romano-British cooking area has yet to be established and it could be that a formal raised hearth like those at Pompeii should be expected. In such a case, there is a possibility that one or more of the 'furnaces' may have been cooking hearths, especially **22**. The number of 'furnaces' in Period 2 in Barn 1 *may* reflect a domestic presence: this is the only period for which no domestic focus can either be identified or suggested. If this had been the domestic centre of the site in this period, it would suggest a fairly lowly status for the inhabitants. The analysis of the distribution of mortaria, assuming that they were for food preparation, does not lay emphasis on any of these features.

The remaining listed features were either hearths — **16, 17** and **25**, possibly **14**; or hollows — **4, 15, 21, 26** and **29**; or ovens **18** and **27** and these two were probably vat bases. Only **16** and **17** can be described as having been hearths proper. Neither had restricted signs of burning suggesting the presence of the side walls of an oven. **15** and **26** were unburnt in themselves and both may have been raking-out hollows in front of ovens. As for **29**, it was small, had a burnt rim and its fill contained plentiful charcoal. The evidence for fire in it points to it having been more like a hearth than anything else. It was only the evidence for fire having been associated with the fills of **20** which allows it to be included here.

However, the siting of **4** is of considerable interest: it was in the middle of Barn 1. The dating evidence was confined to some basically undatable shell-tempered pottery and the feature was only given to Period 2 because all the other features using fire in the barn seem to belong to that. The isolated nature of **4** is unparalleled in the other barns and may suit the projected Anglo-Saxon use of this one.

Neither **18** nor **27** suit plain ovens. The pedestal in **18** and the small size of **27**, coupled with its strong construction, point to another use. Both may have been vat bases. The pedestal in the first would have supported the bottom of something like a lead vessel. The small chamber in the second may have been a different way of dealing with the same problem by heating a smaller area of the base intensively. The function of vats in these contexts is discussed in Chapter 9.

Chapter 4. The Anglo-Saxon Buildings

Description and discussion is confined to the Anglo-Saxon House at the east end; the Sunken-featured Building; the structure over the Roman House; and the Granary with its fence; and, briefly, the two probable rectangular buildings represented by flooring. Post-rows belonging either to other houses or forming lines, have been covered in the description of Period 5 (Chapter 1), but there is a short note concerning the possible building in the middle of the Main Yard.

I. The Eastern Building

(Fig. 55, (225))

This, the best-preserved house, lay in the eastern central enclosure and was cut through by one furrow. The details of post-hole series which made up the structural traces are given in the microfiche (MF4).

The building was approximately 9.7m by 4.15m giving an area of 40.25m^2. Each end wall had six posts, the spacing in the east wall varying between *c*.0.75m and *c*.0.9m while in the west wall the difference was *c*.0.65m–*c*.0.9m. In the west wall, F291 was later than F356 suggesting that there had been at least one repair. The south wall had seven surviving posts between the gables, the north one may have had as many as nine. The basic spacing revealed varied between *c*.0.62m and *c*.0.87m and *c*.0.77m will fit virtually all. Apart from Fs155 and 152 cut by F154, F285 in the south row may have had two post positions pointing to a repair and the shape of F118 is suggestive of one as well. The north row was too shallow to betray such signs and the wall line seems not to have been fitted in directly between the end posts of the gables. However, the more variable spacing, and the poor alignment of the posts generally, could point to a history of repair: those not fitting easily into a single reasonable row, as those on the south generally did, were possibly part of a secondary set.

The lack of signs of doorways should show that any had been in the middle of a long wall where a furrow had removed the evidence. Unless F152 and its successor, F154, had been one side of a door, in which case it would have been off-centre and without a matching one to the north. Parallels suggest that this was not the common form. In Chapter 1 (p. 31) it was suggested that F408 may have been a drain through a door, but it would have passed through the site of another post-hole, had the door been further west.

Evidence for the size and shape of the posts was meagre. In the east wall, the four in the middle were all essentially rectangular and aligned east-to-west. Two had post-pipes measuring 0.11m and 0.15m north-to-south. The west wall had less definite evidence for a similar layout. All six were essentially east-to-west, but only Fs288 and 289 were markedly so. Traces of three posts were found and these do not offer much hope that there had been a simple well-managed carpentry practice: F288 had its post aligned with its pit; Fs289 and 356 were aligned north-to-south, the latter being a rectangular post. In the north wall, only F294 had evidence for a post, this time round. In the south row, three had signs of posts, Fs118, 132 and 285, all round. If these traces point in any direction, it is that the construction techniques only needed squared posts in the central run of each end wall. The layout of the posts used in the end walls shows that there had been no support from the ground for a ridge-pole.

The varying depth of the post-holes should be mentioned. The impression on excavation was that the north side was less well-founded than the south. If the building had been basically levelled, the topographical fall from south to north could account for the difference. Primitive analysis of the depths of the posts reveals two points. Firstly, all four primary posts in the middle of each end were deeper than 0.1m, both sets showing the tendency for the post-pits to become shallower towards the north. This supports the view that the building had been levelled: the length of the posts used may have been standard with a very small tolerance. Secondly, the end posts of each end wall conformed much more with the side walls both in post-pit character and in their variable depth. The south wall, depending on which posts in the area of F152 are included, varied between 0.18m and 0.05m in depth with F118 going down to 0.26m. There was no even drop in depth from one end to the other although the two shallowest posts lay in the western part. In the north row, the depth varied between 0.11m, F151, and 0.03, F109.

The conclusion is that the middle runs of the end walls were of selected timbers set deeper than the posts in the side walls which were probably all round with less care taken over how deeply they should be founded. This suggests that the roof had been gabled and not hipped.

A study of the general proportions and evidence for structure of what was called an early medieval tradition (James *et al.* 1984) drew attention to a system of proportions generally applicable to the selected examples. The basic unit was the square. One layout had two squares next to each other. A second had a square set out from either side of central doors and a third had two squares overlapping by the width of the central doorway (*ibid.*, 187–8, fig. 4). Using this principle here, each square would have had a side of 4.15m, leaving a gap 1.4m between which implies a central door or two opposed ones. Allowing 0.2m diameter for the door posts, the door would have been about 1.2m wide.

The scheme works for many sites, and Cowdery's Down (Millett and James 1983) and Chalton (Addyman *et al.* 1972; Addyman and Leigh 1973; Champion 1977) provide excellent examples, but it is not so successful at West Stow (West 1985), where the evidence for central doors was lacking in the best preserved plans, or at Bishopstone (Bell 1977). The reasonable fit of the Orton Hall Farm building with the scheme encourages a further look at the meagre structural evidence, but without going to extremes (Millett and James 1983, figs 67 and 70).

The use of cruck-framing at an early date has been argued, the evidence consisting of posts set within the line of a wall and forming one or more pairs across the width

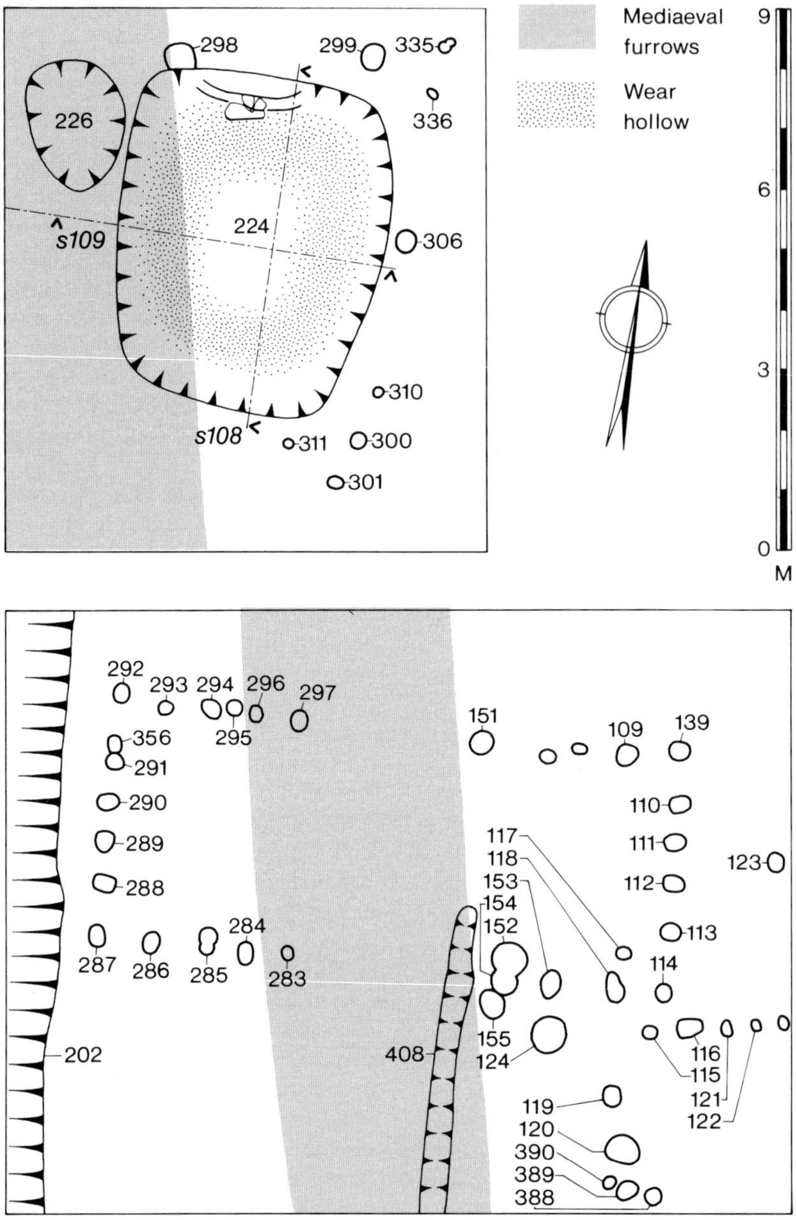

Figure 55 Plans of the eastern Anglo-Saxon House and the Sunken-featured Building. Scale 1:125.

of a building (James *et al.* 1984, 191–4, figs 7–8). F152 and the mark opposite, planned on the site but too slight to excavate, make an appropriate pair and occur half-way between the east wall and the west side of a central door defined by using squares. A further 'bay' set out from the east side of such a door brings us to the position where F285 lies and the apparent later version of this lay inside the line of the wall. The post-print was larger than the others in the side wall here. There was no sign in the north wall opposite this point, unless it is F295, the other posts here being replacements of an earlier line. The case is unproven, but there is the possibility that the roof structure had major supports more or less independent of the walls themselves. The result is that the building could have had simple proportions and a simple bay system supporting the roof, even if the resultant bays are not satisfyingly equal in width.

A disappointing result of this analysis is that none of the conclusions can be used to define any other in this part of the site. There is a marginal case in the north-west corner of the main eastern enclosure. If a clutch of post-holes running in a band from north-north-west to south-south-east are considered to be parts of an end wall, it is possible to bring in F33 or F41 as parts of a 'south' wall. Such a building would include F137, the probable blacksmith's hearth (Chapter 3:II, **24**).

II. The Sunken-featured Building, F224

(Fig. 33, [108], [109]; Fig. 55, (284))

This lay at the east end of the Roman Main Yard. There were two main elements which made up the plan of the building: a large and, relatively speaking, shallow pit and remnants of lines of posts around it.

The plan of the pit was approximately rectangular with rounded corners, with slightly bowed sides and steep sides. It measured 5.25m from north to south, 4.4m from east to west. Its area was of the order of 22m^2. It was 0.9m deep at most, the bottom being above the water-table. The evidence for a floor in the bottom was, firstly, a very thin trodden soil and, secondly, three rough steps about 0.75m wide formed of limestone slabs. The stones were laid on deliberately cut ledges in the natural. What was taken to be a hollow due to wear formed a wide and shallow 'channel' round the centre.

There was no evidence for roof supports either in the sides or bottom of the pit. The only traces of walling lay outside, there being a good line of post-holes down the east side. The west side had been cut away by a furrow. Only one post remained in the south wall and two were found in a north one. Assuming that side walls had been laid out at about the same distance from the edge of the pit, the building would have been about 5m wide. The surviving post-holes of the end walls suggests that these had been recessed; the effective length was not more than 6.4m. The area defined would have been about 32m^2. Three post-holes are not much to argue from, but the two at the north end did not lie on a right-angle with the east wall thus the whole may have been either askew or the area was slightly less than estimated (but see below).

The length of the east wall, centre to centre of the end posts, was 7.5m. The spacing of the three posts at one end and the pair at the other would suit a row of ten set *c.*0.83m apart. None of the holes had any post-pipes and no prints were noted. Apart from two, all were shallow, ranging from 0.03m to 0.07m. F300 was 0.2m deep and F306 was 0.4m deep. F306 lay near the centre of the wall and may have been more deeply founded to provide extra anchorage. Most of the posts could not have exceeded 0.15m across and all may have been less.

F311, the single post assigned to the south wall, did not differ significantly from most in the east wall. However, the two given to the north wall were larger in plan, both being over 0.4m across. Neither was very deep, not exceeding 0.11m. The lack of a 90 degree angle between the north and east walls may be illusory as F298 was so close to the edge of the pit that, at first sight, the fill appeared to cut it. While this may have been the case, the same effect would have been produced if the side of the pit had been eroded back into the fill of the post-pit.

The pattern of wear in the bottom of the pit is of interest as it would be thought that general activity would have been more in the centre rather than round the edges and this suggests that there had been a large obstacle in the middle. A table might produce this effect. The wearing down of the floor did not produce any accumulation of residues of natural mixed with tread dirts suggesting that it had been kept clean. The building seems to have been taken down and the hole used for a purpose which produced a green stain in its main fill. This indicates that the building went out of use before the site was abandoned.

The pit was, at 22m^2, of great size. The general area of those recovered at West Stow ranged from 2.79m^2 to 19.32m^2 (West 1985, 116). Only eight of the 69 found were larger than 12m^2. At Mucking, the sizes of the 113 examples recovered ranged from 10.5m^2 to 24m^2 (Jones 1974, 196) and most were smaller than the Orton Hall Farm example, especially when the actual structure tended to be confined to the edges of the pits themselves (*ibid.*, figs 3 and 6). If the estimated area of the actual building around F224, *c.*32m^2, is taken into account, the disparity becomes clearer. One with an area of 46.5m^2, so large that it needed roof supports, was found at Chalton, a site on which Sunken-featured Buildings were uncommon (Champion 1977, 365). An example at Upton, Northamptonshire, was 50m^2 in area. Its structure lined the edges of the carefully cut hollow (Jackson *et al.* 1969, 213–14, fig. 4). Both these large and more carefully designed features are later (Champion 1977, 367; Jackson *et al.* 1969, 214–16) than the present example which had passed out of use before the abandonment of the site, and this should make it at least late fifth century at the earliest.

III. The Western Building

(Fig. 56, (306))

This was built over the demolished Roman House and was cut by three furrows. It was not a perfect replacement of the Roman one in either alignment or position and seems to have been fitted into what was kept of the walling of the Roman Small Yard (see Chapter 1, p. 38). The structure did not belong to the primary Anglo-Saxon occupation as it straddled the robber trench of the Roman House which contained an Anglo-Saxon sherd. The building seems to have had a fairly long life as the evidence for repair is both consistent and widespread.

At least eight of the posts had been repositioned and no fewer than five positions were found for the one at the north-east corner. If there had been a double post here, the history of replacement and repair would be more like that of the other posts. However, a double-post system here would need to be repeated at the south-east corner, but was not. Most of the posts had been packed round with stone and there were enough traces of similar packing in earlier post-holes for this to have been a consistent feature from the beginning.

The post-sizing recovered showed some consistency as most were either 0.14m in diameter or 0.1m (two cases) more or less. Nearly all were round, only one rectangular pipe being recorded: F820. The largest post was in F561 where both versions were 0.23m and 0.24m in diameter respectively. Whether this was matched on the south side could not be checked due to damage caused by a furrow. The possibility that crucks had been used in the eastern House has been discussed, but there is insufficient evidence here: the anomalous siting of F764 is dealt with below.

The overall plan is difficult to arrive at; its area was *c.*95m^2, in other words, about two and a half times as big as the Anglo-Saxon House, without counting the annex. The problem, however, is to decide what its shape would have been and whether these primitive determinations are valid.

The north side was very badly affected by the digging of trenches along the sides of the walls of the Roman House in the 1960s which were later extensively robbed by the farmer to improve his field. Further ploughing did more damage. The survival of the group of Period 1 features north of the Roman House highlights the absence of Anglo-Saxon post-holes there. The choice is either to run the north wall of the Anglo-Saxon building more or less in line with the Roman one, which creates a plan tapering from one end to the other, or to argue for a double taper and make the building 'boat'-shaped. The plan (Fig. 56), shows that this may not be advisable.

The definition of the west end is difficult. A deep furrow left only the bottom parts of Fs762 and 763. The variation in the depths of the posts surviving further east

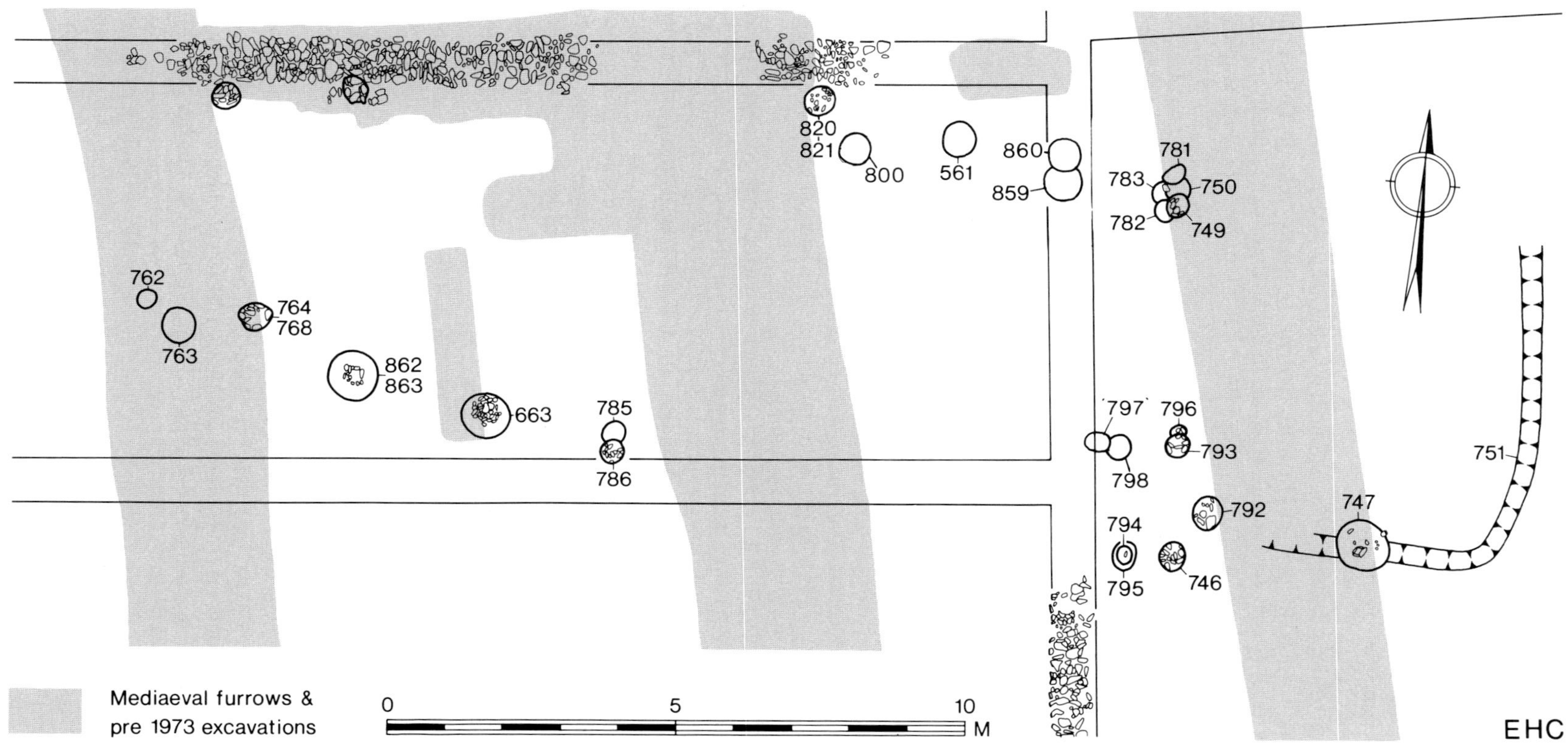

Figure 56 Plan of the Anglo-Saxon Hall and annex superimposed on the Roman House. Scale 1:125.

ran from 0.5m to 0.15m which shows that the two in the furrow can be seen to have been deeper than usual and others *c*.0.2m or less would have been destroyed. This applies to any in the furrow further east and for the ending of the north row at Fs820–821. If the east wall is looked at, only one post, F795, exceeds the 0.26m depth of F782, the rest being less than 0.2m hence the whole of a west wall could have been lost in the western furrow. As for the north wall, if it had been partly based on the footings, no trace would have remained and the trenches dug in the 1960s would have removed all shallow ones. However, if there had been a levelled structure, on analogy with the Eastern Building, the northern post-row would have been shallowly founded anyway. The unnumbered hole opposite F763 *may* have formed part of the west wall, but probably not in the line of the north wall. While this may not be the best interpretation, further speculation is unprofitable.

Bearing these points in mind, we can look at the theory of the use of a square (James *et al.* 1984). Using the length between Fs795 and 783, a square 6m by 6m can be used, but no simple ratio of complete squares emerges. There would have to be an overlap and that might be where doors were placed. In which case, they would not be central. On the whole, the building was too poorly preserved for further discussion of its structure to be justified. In any case, the examples which have been given for this type of plan are not helpful (James *et al.* 1984, 16–24, fig. 4) and some do not fit the proposition when the excavation details are looked at (*ibid*., fig. 4, 16, 19, 20). The remainder are examples from Cowdery's Down (Millett and James 1983, figs 30, 37 and 39). The area of these runs from 86m^2 to 101m^2, bracketing the *c*.95m^2 which roughly applies at Orton Hall Farm. One other model may be appropriate in which the use of squares as before seems hardly appropriate (James *et al.* 1984, 186, fig. 4, 27–8). The examples are again from Cowdery's Down (Millett and James 1983, figs 46 and 51) and each is large. One had a main area of 139m^2 with a room at each end, the other had an overall area of 156m^2, but there seems to have been a partition which would have reduced the area. The point about both of these plans, as well as those of other large buildings, is that they have ancillary spaces.

East of the main structure lay F751. This may have formed an annex, and the line of the east boundary suits a line drawn between Fs795 and 783 and its cluster. If so, the main building may have had a skewed end roughly parallel with the old Roman boundary here. An annex here would have measured *c*.6.4m east-to-west from the end of the main wall. There was no trace of a northern boundary but the width could not have been less than 5.5m. There is no guarantee that the structure was actually attached to the Hall: the isolated nature of the aberrant post-holes, Fs792, 793, 793 and 796, offers no clue. The area of the annex at best would have been about 32m^2. That of the Anglo-Saxon House at the east end of the site was 40.25m^2: the space offered by the annex is far from negligible. Its area added to that of the main structure yields a total of *c*.140m^2 which places the whole into the largest category of Anglo-Saxon building types known save for royal halls. The Orton Hall Farm building ought to have been a hall and is discussed from now on as though it had been.

The relationship of the Hall and its probable annex to the two yards of the Late Roman farm is potentially of interest. The east end of the Hall and the line of the east wall of the Small Yard are more or less coincident, the annex lying in the old Main Yard, a relationship which looks odd to modern eyes.

The same basic arrangement can be seen at Chalton where Building AZ1 and Building A1/B1 are each attached to a fenced yard, the former to the centre of a long side, the latter towards the corner of a short side (Addyman and Leigh 1973, fig. 3; Champion 1977, 366). The former also has a room between its main part and the yard. The same effect can be seen at Cowdery's Down where Building A1 lies outside but joined to the centre of one side of a yard. In the succeeding period, Building B4 lies half in and half out of an enlarged enclosure and probably again in the centre of a side. Building A1 has a square annex lying inside the fenced area while Building B4 has immediately beyond its end, and in the yard, a subsidiary building, B5 (Millett and James 1983, fig. 27). These two arrangements would suit either of the possibilities which may have been present at Orton Hall Farm. These examples suggest a repeatable plan unit which should reflect something of the manner in which each was used, and involve a major building which probably had the highest social status on this site.

If these observations are a correct reflection of what actually was the case in the Early and Middle Saxon periods — the span covered by Orton Hall Farm, Cowdery's Down and Chalton — and if it is a distinction valid for an area greater than a region, it would suggest that, for a while, there was a relatively important seat here. That the west end was the more important within the excavation is shown by the presence to the south of a granary, the most important demonstrable economic indicator on the site.

It has been argued that the Hall and the Granary were placed deliberately in the Roman Small Yard (Chapter 1, p. 42) and there are signs that, when that began to break down, new bounding limits were introduced along the western side of the site. It has also been argued (p. 40) that Barn 1 survived into Anglo-Saxon use of the site and if F1008 (see p. 84) in its middle was also of Period 5, the whole begins to look like a hall. There would then be two potential halls, but not necessarily coeval. The distribution of Anglo-Saxon pottery shows that there was remarkably little in the vicinity of the Hall, but there could have been a well controlled system for rubbish disposal at all times (Millett and James 1983, 249–50). The logical sequence would have been for the Hall to replace Barn 1: it is easier to imagine that the barn would not have survived to the end of the fifth century, while the Hall itself was not a primary structure in the Anglo-Saxon phases of the site.

IV. The Granary

(Fig. 57, (313))

The main structure had been cleared by Mr Dakin. Fortunately, the excavation technique used was to leave the stonework standing like a tube by taking away the surrounding earth. This ensured the basic preservation of the post-holes but without any dating evidence, and the sizes of the original post-pits were lost. The brief report prepared by Mr Dakin does not mention any dating material (see MF6). None can now be identified.

The main structure consisted of three rows of three posts each set in a pit and packed round with stone. A further post-hole lying just to the south was closely

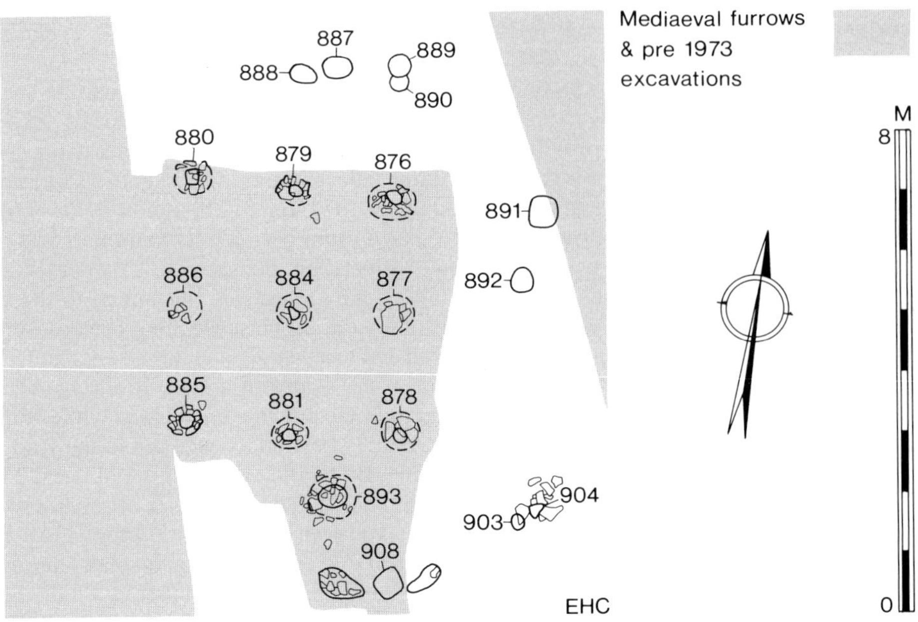

Figure 57 Plan of the Granary. Scale 1:125.

associated with the set. Around the posts were traces of a bounding fence made up of three posts on the north, another three on the south and four on the east side, but in pairs. A furrow had removed any trace on the west side. The best direct evidence for an Anglo-Saxon date was the presence of Anglo-Saxon pottery in F893. Apart from that, there was its location in a corner of the Small Yard and the difficulty of conceiving a Roman date for it when there was an abundance of stone for building and when the technique is so similar to the Hall to the north.

The character of the fence is also suggestive even if ten in four groups does not look promising. The basic spacing of the north and south group was 0.7m between the posts. The north, south and east lines were *c.*2.1m, 2.55m and 2.3m respectively from the nearest row in the set of nine. As the post-holes in each group did not form straight lines, the measurements are from a mean line running through them. A completed arrangement would have been of posts alternating on either side of a central line. Fences are not common on Roman sites and such a distinctive layout would not normally be expected, the Period 2 fence on the east side of what was to become the Small Yard being a good example.

However, such fences are known on Anglo-Saxon sites and well-preserved remains were recovered from Cowdery's Down where the phase C fencing was markedly of this form, where the spacing was about 0.4m (Millett and James 1983, fig. 28). The other style of fencing found there was made up of pairs of posts across the line and alternating with single posts directly on it (*ibid.*, phase A1). The distinctive pattern of the fences at Cowdery's Down enabled fragmentary buildings to be separated from the mass of post-holes.

The reconstructed form of the Cowdery's Down fences was, for phase A, wickerwork panels tied into the single posts with angled supports rising from the pairs (*ibid.*, fig. 32). The phase C fence was reconstructed with planks trapped between the alternating posts which were tied across the top (*ibid.*, fig. 38). The same pattern was found at Chalton where the spacing was similar (Addyman and Leigh 1973, fig. 10), but one had wider spacing at approximately 0.6m (*ibid.*, fig. 9), much closer in character to those at Orton Hall Farm. The alternating pattern is detectable at Catholme (Losco-Bradley 1977, 358), but no fence, apparently, was found at West Stow (West 1985, figs 6 and 7), or coherently at North Elmham (Wade-Martins 1980, but see fig. 159) and site conditions at Bishopstone were not conducive to good preservation (Bell 1977, 197).

The enclosure which can be developed around the central set of post-holes measured approximately 8.6m north-to-south by 8.1m east-to-west. The neat pattern of post-holes making up the three rows measured overall, from centre to centre, 3.75m north-to-south and 3.5m east-to-west. This partly accounts for the elongation north-to-south of the fenced enclosure.

The area covered by the nine post-holes is about 13m^2. The posts themselves were generally between 0.12m and 0.18m in diameter, the exception being the central one of the north row which was 0.26m in diameter. A building with a span of hardly more than 3.75m is too small for it to need a central pole to carry the roof. And a raised floor would only really need the central post to support it if there had been a heavy load on the floor.

A raised floor should mean that there had been, at best, steps up to it, a ladder at worst. F893 probably belonged to a set of steps with a door above in the south wall. The post was sited more or less in the middle of the bay opposite it. It lay 1.2m away from the centre-line of the row and this would give 3ft for the width of any steps laid out against the side of the building.

A scrutiny of the plans of West Stow and most other sites fails to reveal an adequate parallel for the nine-post structure. However, there is a six-post building at Catholme the span of which was comparable with the overall width of the Orton Hall Farm structure. The timbers were much larger and out of all proportion to the floor area (Losco-Bradley 1977, 358, 360: blg 39). There may also have been another (*ibid.*, blg 45). The site is difficult to assess, as there are only small-scale plans available and none periodised. Structures with similar characteristics are found in north German contexts of both the late Roman and Migration period, for example, at Flögeln (Zimmermann 1974, fig. 10). None seems to have been identified in an exclusively Romano-British context and, taking the style of fence into account, the Orton Hall Farm plan is assumed to belong to a solely Anglo-Saxon phase of the site.

The specialised nature of the construction and the care taken to separate it from normal traffic, as well as having the most securely seated posts of any Anglo-Saxon building for what was almost certainly a raised floor, are all features which would be expected for the storage of something as economically important as grain. In default of any more likely explanation or crop, the nine posts are taken to be parts of the substructure of a granary.

The building was almost equidistant from the wall of Barn 1 and the projected face of the west wall of the Small Yard. It makes sense to see the Small Yard as still in being along with Barn 1 and remaining the most secure part of the site. But what was present at the north end of the Small Yard is not clear. The discussion of the Hall suggested that it replaced Barn 1 and, either way, the Granary would have been in the most important part of the site and the Hall should be a guarantee that the west end was just that.

V. Other Buildings

Structures F865 and F871 (312) (316)

Of these two, F871 was poorly preserved and lay outside the area of formal excavation. It consisted of a spread of limestone, fairly densely placed, but with no sign of its having been pitched or otherwise packed, set in a matrix of dark earth. It contained some Anglo-Saxon pottery and it was the straight edges to east and west which suggested that it might have been the floor of a building analogous with F865. Its width would have been about 5.2m and not less than *c*.5.5m in length. This would give a minimum area of *c*.29m^2.

It was the other structure, F865, which showed that F871 may have been part of a building. F865 was well marked and rectangular although the eastern end was partly or wholly removed by a furrow. The width was 3.9m and it survived for a maximum length of 4.6m. The minimum area would have been about 18m^2. The apparent south-east corner is in doubt: the surviving trace could have been the result of medieval ploughing. A post-hole was found at the north-west corner with its centre about 0.2m to the north. There was no sign of a post-pipe, but its size at 0.4m diameter would fit in well with other demonstrable post-holes on the site. The 'floor' itself was a distinctive mix of loam, gravel and clay. There was a sherd of Anglo-Saxon pottery incorporated in it.

If the post-hole was the only survivor of a set making up a bounding structure, the width of that would have been about 4.3m. This is close to the width of the Anglo-Saxon House at the east end of the site. If it had had a double-square plan (James *et al.* 1984), it would hardly have carried across the furrow. But had it been like the other, with a square on either side of opposed doors, some sign of the floor on the east side of the furrow should have been present. However, the attested floor area is too great for the building to be called a hut.

The alignment and siting of the building call for comment. It was approximately parallel with the Hall to the north rather than with the Granary and Barn 1 to the south. Its west end butted the side of the apparent track running round the west end of the Barn to the Hall. The track cut across the west side of the Small Yard and may have been a secondary stage in the development of the Anglo-Saxon site at least. Therefore, F865 may also have been secondary.

The possible building in the Main Yard

The evidence for this has been reviewed in the description of the site (Chapter 1, p. 37; Pl. II). It consists entirely of the Anglo-Saxon pottery from one of the 'furnaces', F999 (Pl. VI), and the possibly dubious association of two other similar features, Fs1081 and 1082 (38) (Pl. II) assigned to Period 1. Whether all three should be given to Period 5, or not, there would have been some kind of building sheltering the activities here. It is the lack of coherent evidence for one or more of these which prevents a satisfactory picture from emerging.

However, the size of a single building is a problem. The maximum length from east-to-west could be between Fs998 and 1044: *c*.12.5m. The width is less easy to estimate but could have been from F1026 to part of the curious arrangement at the south-west end of F1000: *c*.6.75m. The area would, therefore, be 83m^2 which seems large for a plain, utilitarian building, but is not excessive in itself.

Chapter 5. The Finds

In the interests of economy, the reports on all finds except the coins, the Anglo-Saxon objects and the millstones have been placed in microfiche, MF7 and MF8. This also contains the list of all small finds from the site along with brief identifications, layers, periods, *etc*.

I. The Worked Flints and Other Prehistoric Finds

by Dr Helen Bamford

Microfiche MF7; MFFigs 58–60; MFTable 4

II. The Coins

by Dr Simon Esmonde Cleary

The sixty-three Roman coins recovered are all of types entirely usual on British sites. In terms of dating for the archaeological features all they can provide is a *terminus post quem* for the deposits in which they lay. Forty-eight of the coins were single, only six layers having more than one coin. Of these six, four were associated with the stripping of the site and so are of questionable stratigraphic integrity. The other two (L313 and L574) were both of Period 5 whose proposed date bracket is entirely consonant with those of the two coins in each of the two layers. Nor were there any sequences of layers dated by single coins. Clearly, the dating established by the stratigraphic and pottery evidence must take precedence over the feeble contribution from the coins.

While the date range of the coins is broadly comparable with that of the site as a whole, it should be noted that the majority of the first- and second-century coins were in much later deposits, probably through re-deposition. The coin-list goes down to the latest issues to reach Roman Britain, though again only providing *termini post* rather than an end date. There are too few coins overall to attempt to trace whether there is any significant fluctuation at this or any other period.

Distribution of the coins by site period

Period 1	*c*.50–*c*.175	none
Period 2	*c*.175–225/250	none
Period 3	225/250–300/325	3, 6, 14, 25, 37, 62
Period 4	300/325–*c*.375	41
Period 5	*c*.375– early sixth century	2, 10, 12, 15, 20, 26, 29, 34, 35, 39, 42, 45, 48, 50, 51, 61
unstratified including post-5		1, 4, 5, 7, 8, 11, 16, 17, 19, 21, 22, 23, 24, 30, 31, 32, 33, 36, 40, 46, 47, 52, 53, 54, 55, 56, 57, 58, 59, 60
unprovenanced		9, 13, 18, 27, 28, 38, 43, 44

III. Roman Finds

by D.F. Mackreth

The objects are arranged in categories of use. As the collection is not large, these are broad in definition. Those defying immediate identification are gathered together. All obvious Anglo-Saxon finds are placed after the Roman finds and are followed by the few medieval objects. There is a single number sequence through the whole series. Roman finds after 107, and medieval finds will be found in MF8.

Personalia

Brooches

(Fig. 61)

All are of copper alloy, unless otherwise stated.

Colchester Derivatives

1 The axis bar of the hinged pin was housed in a slot behind the wings. Each wing has three relieved mouldings, with traces of beading, separated by flutes. The bow has a central ridge down the upper part ending at a pair of lenticular bosses, each with a median moulding and bordered by punched dots. Lines of punched dots on the sides of the lower bow give the effect of concave-sided lozenges lying across the bow. The foot is a simple projecting moulding.
sf.108, Period 5.

2 The pin is hinged. Each wing has a small moulding at the end separated from the rest of the wing by a flute. The upper bow has double repeat of the single elements on the wings of brooch 1. The lower bow has a flat face on either side of a central arris. The foot is a projecting moulding.
sf.364, unstratified.

Both brooches are marked by the kind of moulded ornament which can be continuous down the bow. This style of decoration is commonly found in the East Midlands. The indicated dating is from the later first century to *c*.125/150, but may be largely before AD 125.

3 The pin is a piece of wire wrapped round an axis bar in a slot behind the wings. Each wing has a groove at the end. The bow, broad at the top, tapers to a narrow foot-knob. The top of the bow has a triangular boss defined grooves with cross-cuts on the outer ridges.
sf.996, unstratified.

The brooch belongs to a small family having a limited repertoire of decorative motifs combined to produce different effects. The dating has recently been considered and the conclusion was that the *floruit* was later first century to late second (Mackreth in Jackson and Potter, forthcoming).

4 Not illustrated. A fragment of a brooch whose pin had been held in the Polden Hill manner: an axis bar passed through the coils of the spring and pierced plates at the ends of the wings, the chord was held by a rearward-facing hook.
sf.538, unstratified.

Enough survives to show that it belonged to the type known to the writer as a Dolphin and a recent discussion of the whole family showed that its date-range is *c*.75–150/175 (Mackreth in Timby, forthcoming).

5 Not illustrated. A catch-plate and plain lower part of the bow are all that is left of a Colchester Derivative. The brooch should not have been in use after *c*.150/75.
sf.756, Period 1.

Late La Tène

The following four brooches have or had four-coil-internal-chord springs integral with their bows.

6 Only the oval-sectioned bow and the top of the catch-plate survives. The catch-plate seems to have had a central piercing.
sf.709, unstratified.

7 The bow has a flat back, a swelled front with a groove down each side and a recurved profile.
sf.760, Period 1.

8 The distorted bow has a thin rectangular section tapering to a plain foot. The catch-plate was forged.
sf.854, Period 1.

9 The distorted bow is of sheet metal, lanceolate in shape, and has lost the lower part with the catch-plate. Four lines incised lines on the front form a concave-sided lozenge.
sf.200, Period 5.

Only 6 is likely to be pre-Conquest. Had the catch-plate been open-framed, the brooch would basically have been a *Drahtfibel*. There

No.	Emperor *etc.*	type	date	comment	sf. no.	Layer	Period
C1	Claudius I	*as*	41–54	illegible	691	999	post-5
C2	Domitian	*dupondius*	81–96	illegible	201	472	P5
C3	Hadrian	*as*	117–34	illegible	407	885	P3
C4	Antoninus Pius	*sestertius*	145–61	RIC 760	813	1744	-
C5	Faustina II	*as*	175–plus	illegible	381	347	-
C6	"	*as*	175–plus	illegible	392	853	P3
C7	Commodus	*denarius*	180–92	uncertain	524	1000	-
C8	Gordian III		240	RIC 319b	382	247	-
C9	Gallienus		260–8	RIC 283	80	-	-
C10	"		260–8	RIC as 309, but mm.zL	288	574	P5
C11	"		260	RIC 297	894	295	-
C12	Claudius II		268–70	RIC as 79	152	109	P5
C13	Radiate		260–80	reverse: incuse	86	-	-
C14	Carausius		286–93	RIC 300	1013	2372	P3
C15	"		"	reverse illegible	1015	2699	P5
C16	"		"	reverse illegible	1056	2740	post-5
C17	"		"	RIC 835	1057	2825	-
C18	Allectus		293–6	RIC 124	81	-	-
C20	Barbarous Radiates		270–90	reverse *Pietas*	12	86	P5
C21	"		"	reverse *Spes*	469	1049	post-5
C22	"		"	reverse *Pax*	695	1963	-
C23	"		"	reverse *Pax*	823	1743	-
C24	"		"	reverse *Spes*	932	2503	-
C25	"		"	reverse *Salus*	1000	2356	P3
C26	Constantine I		323–4	RIC VII as London 289	84	-	-
C27	"		330–4	LRBC I 72	15	89	P5
C28	Constantine II		330–5	LRBC I 81	85	-	-
C29	"		337–41	LRBC I 132	159	330	P5
C30	"		337–41	LRBC I 124	438	938	-
C31	"		341–8	LRBC I 158	609	1008	-
C32	"		341–8	LRBC I 140a	772	1962	-
C33	Constans		348–50	LRBC II as 30a	434	785	-
C34	Constantius II		330–5	LRBC I as 70	257	575	P5
C35	"		341–8	LRBC I 145	1060	2866	P5
C36	Urbs Roma		330–40	LRBC I as 51	541	1066	-
C37	Constantinopolis		330–40	LRBC I as 52	641	1273	P3
C38	House of Constantine		335–45	LRBC I copy as 87	87	-	-
C39	"		337–41	LRBC I as 126	271	574	P5
C40	"		341–8	LRBC I as 161	353	785	-
C41	"		341–8	LRBC I as 158	635	1317	P4
C42	"		350–60	LRBC II copy as 25	974	2597	P5
C43	Magnentius		350–1	LRBC II 54	82	-	-
C44	Valentinian I		364–75	LRBC II as 280	83	-	-
C45	"		"	LRBC II as 96	151	313	P5
C46	Valens		364–78	LRBC II as 504	236	295	-
C47	"		"	LRBC II 705	326	615	-
C48	Gratian		367–75	LRBC II as 517	162	313	P5
C49	"		"	LRBC II 505	505	1045	-
C50	"		"	LRBC II 1331	678	1980	P5
C51	Arcadius		393–5	LRBC II 2578	980	2593	P5
C52	House of Theodosius		383–8	LRBC II as 782	362	295	-
C53	"		388–402	LRBC II as 796	433	817	-
C54	fourth-century			illegible	337	37	-
C55	fourth-century			clipped illegible	399	785	-
C56	fourth-century			illegible copy	406	825	-
C57	fourth-century			illegible	767	1963	-
C58	third- or fourth-century			illegible	334	295	-
C59	third- or fourth-century			illegible	378	295	-
C60	third- or fourth-century			illegible	838	1896	-
C61	third- or fourth-century			illegible fragments	1043	2762	P5
C62	third- or fourth-century			illegible	1078	2864	P3
C63	one coin stolen before being recorded				318	670	-

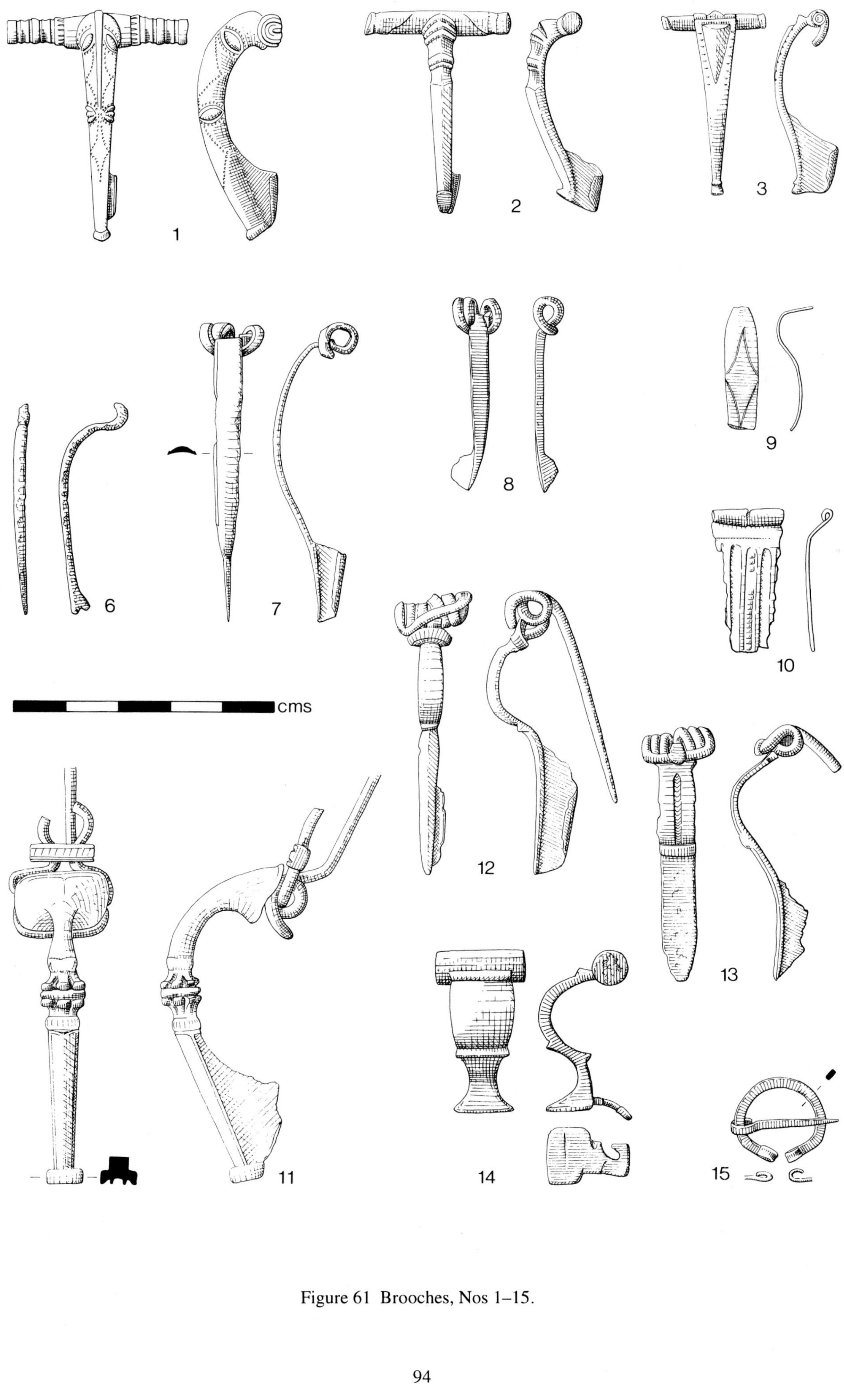

Figure 61 Brooches, Nos 1–15.

is not enough of the spring present to show whether the chord had been external. An open catch-plate would indicate a date in the first century BC, and the profile suggests a date before *c*.AD 50. The other brooches have no early trait and, as the type is common after the Conquest, a general *floruit* should apply: mid-first century to *c*.AD 80–100. However, 7, with its trapezoidal catch-plate and slack profile, should not be a survivor in use as late as *c*.AD 75.

Hod Hill

10 The axis bar for the hinged pin was housed in the usual rolled-over head. The broad bow is thin with a slight moulding across the top. The upper bow has a sunken bead-row down the centre with a wide and shallow flute on either side. The lower bow is missing.
sf.757, Period 1.

Recent discussion has shown that no Hod Hill has been published from a pre-Conquest context and that the type was in common usage into the 60s of the first century, but had almost completely passed out of use by AD 70 (Mackreth in Jackson and Potter, forthcoming).

Trumpet

11 The spring is mounted to a loop behind the head of the bow by means of the wire carried over the head to form a loop. The waist is bound by a collar with a central beaded moulding. On the head is a nib which prevents the loop moving over the front of the brooch. The plain trumpet has a groove around the upper end and a slight median arris. The knop has a central 'petalled' moulding with, above and below, a double moulding, the inner one being wide and beaded. The lower bow has a central arris, a groove down each border and a triple moulded foot-knob with, underneath, a central recess with a groove round it.
sf.962, Period 3.

The dating of Trumpet brooches has recently been reviewed and the *floruit* can be seen to run from the later first century to *c*.150–175 (Mackreth in Garrod and Atkin, forthcoming).

Unclassified

12 The brooch has a Colchester-style sprung pin. The wings are short and stubby. The upper bow proper is straight-sided with a flat back and a swelled front. The profile has a recurve, the point of inflection being marked by a small cross-moulding, and the lower bow is nearly straight with, at the top, an arris.
sf.624, unstratified.

13 A repeat of the last, but with a sunken bead-row down the upper bow.
sf.420, Period 3.

Both belong to a distinctive type of continental import usually dated to the earlier first century AD (Böhme 1972, 10). None comes from a British pre-Conquest context. It has, however, been shown that developed forms of *Augenfibel,* which these essentially are, but with 'eyes', are not to be expected before the reign of Claudius and would seem to continue to *c*.AD 70 in use (Kunow 1980, 157–9).

Knee

14 An open-backed cylinder houses the spring. The bow has an S-shaped profile with a step just below the head and another, both in front and behind, at the point of inflection. The bow narrows downwards then splays out to form a flat foot, the catch-plate being in the same plane.
sf.373, unstratified.

Discussion of the large collection of Knee brooches from Catterick has shown that all types of Knee have the same *floruit*: essentially the second half of the second century and into the early third (Mackreth inWilson, P., forthcoming).

Penannular

15 Complete, the ring is only 17mm in diameter and made from a thin rectangular-sectioned bar. The terminals are roughly folded back over the ring. There is no decoration. The pin is straight.
sf.597, unstratified.

The type was well established by the middle of the first century AD, but the present form seems to be more Late Roman than anything else (Mackreth in Ellis, forthcoming).

Fragments

Not illustrated

16 Part of a pin and a coil of a spring.
sf.677, unstratified.

17 Part of a pin and two coils from a spring.
sf.711, Period 4.

18 Part of a pin from a brooch.
OHF 71 sf.21, unstratified.

19 Iron. Possibly part of a brooch pin.
sf.714, Period 1.

20 Iron. Two pieces possibly from a brooch pin.
sf.786, Period 1.

21 Iron. Possibly part of a brooch pin.
sf.456, Post-5.

22 Iron. Possibly part of a brooch pin.
sf.119, Period 5.

The high number of iron pieces, when iron brooches themselves seem to be absent, is worthy of note. Iron brooches die out in the first century AD and go back well before the Conquest. The Period 1 dating of 19 and 20 would suit this, but 22, from Period 5, and 21 which is later, possibly come from Anglo-Saxon brooches whose springs and pins were almost invariably of iron.

Belt Fittings

(Fig. 62)

23 Copper Alloy. The outer part of a D-shaped **buckle plate** wrapped round the cross-bar of the buckle has part of the slot for the tongue. The outer edge is turned down to grip the belt leather. Three holes round the curved periphery once held rivets tying the front and back together.
sf.525, Period 4.

Similar belt-plates from Lankhills, Winchester, point to the middle and later part of the fourth century (Clarke 1979, 270–2, fig.34, 27, 70, 481, 533).

24 Copper Alloy. **Strap end**. Enough remains to show that it belongs to the 'amphora' type, although lacking open-work 'handles'. The absence of decoration may mean that this was the back-plate and has one hole to attach the item to the strap.
sf.1086, unstratified.

The parallels suggest a fourth-century date with an emphasis on the middle and later parts.

25 Copper Alloy. Probable **buckle loop**. A band 2mm wide and 1mm thick, flat on one face and curved on the other. One end expands and has the remains of a piercing, the other seems once to have been the same.
sf.344, Period 1.

The item could have been a movable buckle loop even if of slight section, but one from Skeleton Green serves as a parallel: mid-first century (Partridge 1981, 105, fig.54, 2).

26 Copper Alloy. **Spectacle buckle**. Two conjoined 'D's with a slight waist at the cross-bar.
sf.1034, Period 5.

Roman spectacle buckles are rare, most being more substantial and elaborate. The British dating evidence covers the first to fourth centuries. (Stead 1976, 212, fig.111, 107; Crummy 1983, 52, fig.66, 1819; Wedlake 1982, 207, fig.85, 7).

27 Copper Alloy. **Buckle tongue**. The wrap-round is lost and the other end is has a worn zoomorphic form. The snout is now rounded and each ear has a groove across it.
sf.1044, Period 5.

This style seems not only to be fourth-century but also late (Clarke 1979, 270–2, fig.34, 126, 481).

28 Iron. **Buckle tongue** (not illustrated). Identifiable by the beginning of the wrap-round, the dip in the profile and the 'kick' at the free end which sat on the buckle loop.
sf.926, Period 5.

29 Iron. **Buckle tongue?** (not illustrated). Corroded, the start of the wrap-round is present and the rest is straight: 24mm long; 4mm by 3mm in section tapering to a point.
sf.1051, Period 5.

Finger Rings

(Fig. 62)

30 Copper Alloy. The band is 3.5mm wide and 1mm thick, with a decorated zone which is flat, not following the curve of the hoop, and moulded with three shallow domes.
sf.454, unstratified.

Such thin rings, usually with ornament, tend to be Late Roman (Crummy 1983, 49–50, fig.51, 1787; fig.52, 1791, 1793).

31 Copper Alloy. The section of the fragment is rounded inside and out, but flat on either side.
sf.943, unstratified.

While possibly a ring of late fourth-century date (Clarke 1979, 218–9, fig.75, 146; fig.87, 389, 401; fig.90, 337; fig.98, 559, 565, 567, 570–1, Crummy 1983, 47, 50, fig. 50, 1768; fig.52, 1789), it could also be part of an earring of Allason-Jones' (1989, 142–6) type 1 and so be first to fourth century.

32 Copper Alloy. Complete and plain. The bascially circular section has flat sides like 30.
sf.155, Period 3.

The date-range is mid-second century to the end of the Roman period. However, the flattened sides may point to a fairly late date in the overall *floruit*: second and third century.

33 Copper Alloy (not illustrated). A piece of wire, just over 1mm in diameter, whose curvature is too marked and regular to suit a pin.
sf.583, Period 5.

Such simple rings appear to be Late Roman.

34 Iron. The X-ray shows this to have been made from very fine wire wound about three times.
sf.842, Period 1.

No parallels have been noted, such insubstantial iron rings seldom survive to be recognised.

35 Iron (not illustrated). Made from wire *c*.2mm in diameter and distorted, the internal diameter had been *c*.15mm. No bezel emphasis is obvious, however, the object may have been a finger ring.
sf.682, Period 1.

36 ***Intaglio*** (Pl.VIII). Dr Martin Henig comments: 'The stone is a red jasper with a flat upper surface and is in excellent condition apart from some chipping around the edges. It is very highly polished except within the area of cutting. Dimensions: 16 x 13 x 3mm. Shape: F1.
sf.171, Period 2.

The device portrays a youthful male figure seated in a light two-wheeled trap (*cisium*) and driving the pony to the right when the impression is looked at. There is a ground line. The personage was previously identified as a Cupid (Henig 1974, 20; Henig 1978, 293, pl.XXVI, App.74), but the 'wings' may well be the hood of a countryman's cape as is the case with two composite examples (Krug 1980, 182, pl.72, 53; Henig 1978, 289, pl.XXIV, App.26). Date: the use of red jasper and the bold competent manner of execution suggest that the gem was engraved in the Antonine period.'

Plate VIII The *Intaglio*, No. 36.

Bracelets
(Figs 62–63)

37 Copper Alloy. Part of a penannular ring of *c*. 40mm diameter, worn and now 50mm long, the broken end is 3mm by 2mm expanding to a flat-faced terminal 7mm by 6.5mm across. At the terminal is a moulding, then a shallow swell, with longitudinal striations, followed by another moulding, and then a swelled face separated from the rest of the ring by a groove.
sf.368, Period 2.

Two specimens, much thinner in section, appear to be second century (Bidwell 1979, 235, fig.73, 17; Wedlake 1958, 251, fig.57, 12B).

38 Copper Alloy. A curved strip, distorted at the broken end, with a shallow ogee outer profile has a blunt curved end 7mm at the break, then a waist 6mm across expanding to 8mm. No more than *c*.2.5mm, the fragment had an internal diameter of *c*.80mm.
OHF 71, sf.2, unstratified.

Though worn, the object is probably the terminal from a snake-headed bracelet. None seems to be first century, the dating suits the third and fourth centuries: (Stow 1982, 124, fig.61, 31; Wheeler and Wheeler 1936, 210, fig.45, 44; Whiting *et al.* 1931, pl.LVII, fig.1: pl.XVI, 142–3).

39 Copper Alloy. Almost complete, the band 43mm, diameter, 1.75mm wide and *c*. 1mm thick, has a continuous series of cross-grooves giving a beaded appearance. Each end is slightly expanded and pierced. The broken terminal had a hole for a rivet.
sf.1083, Period 3.

The variable dating ranges from after *c*.AD 150 (Neal 1976, 21, fig.XI, 8), through the third century (Allason-Jones 1983, 119, fig.77, 174), to after AD 350 (Webster 1975, 205, fig.111, 28).

The following five items are fragments of cable bracelets.

40 Copper Alloy (not illustrated). Three strands, *c*.35mm surviving length, 2.2mm diameter wire, internal diameter no less than 40mm.
sf.153, Period 3.

41 Copper Alloy. Three strands, *c*.45mm surviving length and 1mm diameter wire. One strand has been wrapped round another to form part of a terminal.
sf.527, Period 4.

42 Copper Alloy (not illustrated). A strand from a bracelet, *c*.57mm long and of *c*.1.5mm diameter wire.
sf.548, Period 5.

43 Copper Alloy (not illustrated). Three distorted strands *c*.40mm long of 2.2mm diameter wire.
sf.350, unstratified.

44 Copper Alloy (not illustrated). Two fragments 34mm in length of three strands hammered to form a nearly square section *c*.2mm by 2mm.
sf.874, unstratified.

None seems to be indisputably earlier than fourth century, the bulk of those from Lankhills came from graves earlier than *c*.AD 370 (Clarke 1979, 313–4) suggesting that the *floruit* was over by the end of the fourth century and this is supported elsewhere. 44 may be an example of Allason-Jones' ear-ring type 6 (1989).

45 Bone (not illustrated). Flattened oval section 3.5mm by 2.5mm and 47mm.
sf.217, Period 4.

46 Bone (not illustrated). Flat inside and out with rounded sides, 27mm long by 5mm x 2mm.
sf.607, Period 4.

47 Bone (not illustrated). Two pieces, 179mm in total length, tapering from the fracture to a rough end with a rivet hole 1.5mm of diameter 15mm from the end.
sf.713, Period 4.

48 Bone (not illustrated). As 45, 17mm long, 4mm by 2.75mm.
sf.742, Period 4.

49 Bone (not illustrated). Flat inside, rounded outside, one rounded edge, the other flat. 26mm long, 4.5mm by 2.5mm in section.
sf.54, Period 5.

50 Bone (not illustrated). As 46, 105mm long, 6mm by 2.5mm in section.
OHF 71, Trench VI, +, unstratified.

51 Bone. Flat on the inside, the curved outer face has two longitudinal grooves. 58mm long, 6.5mm by 2.5mm. Diameter: *c*.120mm.
sf.1053, Period 5.

Thirty-eight of the forty-two bone bracelets from Lankhills dated between AD 310 and 370 (Clarke 1979, 313–4) and most of the other dated examples noted would suit this date-range which fits the evidence from Orton Hall Farm. Decoration is uncommon, apart from grooves on either side of the metal sleeve over the join and these probably result from crimping the sleeve to fix it.

52 Copper Alloy. A broken strip whose axis is at right angles to the wrist, 83mm long, 1.5mm by 1mm in section. The outer edge is 'castellated' with four, possibly hand-cut, cross-ridges in the hollows. One end is tapered to form part of a riveted lap joint, part of the rivet surviving. Internal diameter: 80mm
sf.736, Period 4.

A common design south of Hadrian's Wall, few are dated. Only one seems to be as early as the late third century (Brodribb *et al.* 1971, 114, fig.49, 104), otherwise the evidence favours the second half of the fourth century.

53 Copper Alloy. A fragment 12mm long and 2.5mm by 0.75mm, the outer surface has a raised panel of four ridges with a concave face on each side.
sf.1033, Period 5.

No dated parallel has been noted, but, like most similar bracelets, probably fourth-century date and possibly late.

54 Copper Alloy. Part of a broad flat band, 8.5mm wide by 1.5mm thick and only 25mm long, the outer surface has a central hollow, with deep punch-marks along it, a shallower one on each side; at one end are three cross-grooves. Each edge has triangular indents leaving

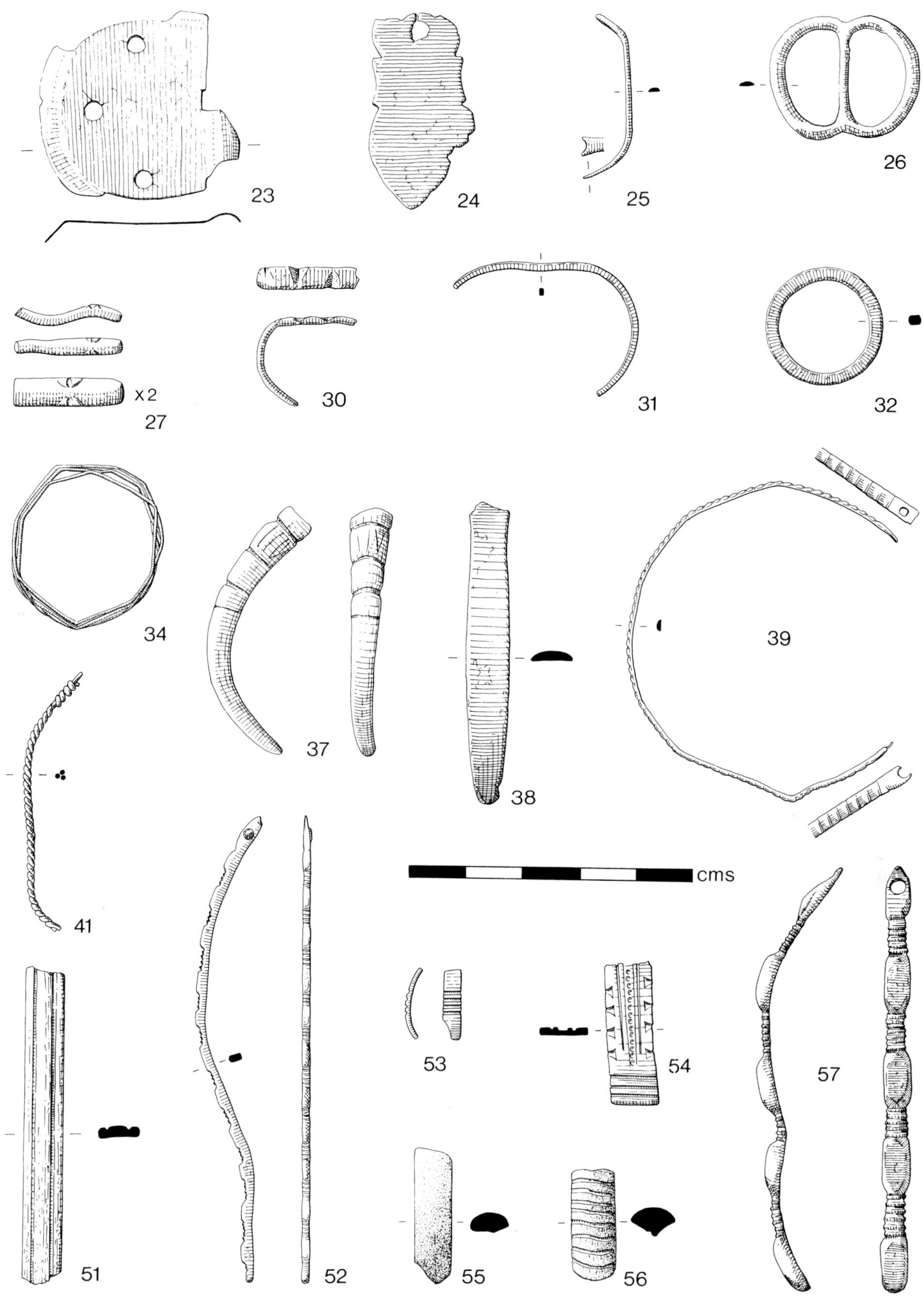

Figure 62 Belt fittings and buckles, Nos 23–29; finger rings, Nos 30–35; bracelets, Nos 37–57.

between wedge-shaped platforms tapering back from the edges. Internal diameter: 80mm.
sf.177, Period 5.

Recalling some Crossbow brooches, the decoration is unusual on bracelets and marks a group characterised by variable designs. The Lankhills examples are mostly late fourth century (Clarke 1979, 307–9, fig.37) and the bands of the late variety tend to widen towards the centre. The parallels noted are either specifically late fourth century, or come from contexts spanning that time.

55 Shale. The smooth elongated oval section has a ridge inside caused by cutting the ring from the core. 27mm long, 8mm by 4mm in section, internal diameter: 60mm.
sf.248, Period 5.

56 Shale. Curved outside and triangular inside the section displays turning marks and the usual ridge. The band has a series of badly made cross-grooves at an angle to the edges. 21mm long, 8mm by 5mm in section, internal diameter: 60mm.
sf.238, Period 5.

Although found from the first to third centuries, shale bracelets are especially common in the fourth, but few appear to be after *c*.AD 375. One parallel for the present piece is dated to the fifth century (Goodburn and Grew 1984, 77, fig.33, 305).

57 Copper Alloy. 75mm long with part of a 'hook-and-eye' fastening, the design simulates stone beads threaded on a metal chain (*e.g.* Neal and Butcher 1974, 133, fig.58, 75).
sf.7, unstratified.

Few come from dated contexts and they indicate a date after *c*.AD 350 (Wheeler and Wheeler 1932, 82, fig. 17,Q; Goodburn 1984, 33, fig.10, 65–7).

58 Copper Alloy. A thin band whose section, 3mm by 1mm, is at right angles to the wrist. The worn outer edge has hand-cut V-shaped notches. 25mm long, internal diameter: 60mm.
OHF 71, sf.34, Period 5.

Only one precise parallel has been noted. Its context was dated AD 350–450 (Crummy 1983, 40, fig.43, 1657).

59 Copper Alloy. Made from wire 1.5mm in diameter, the complete end is a simple hook. Corroded onto the ring is part of the other end which was wound round the main ring. Internal diameter: probably 60mm.
sf.418, unstratified.

Normally, the ends of the ring overlap and wrap round each other (*e.g.* Clarke 1979, fig.73, 81, 88). The present system is unusual. The dating is: *c*.AD 100–280 (Cunliffe 1971, 107, fig.41, 47) and after 350 (Webster 1975, 203, fig.111, 24).

60 Copper Alloy (not illustrated). A flat strip, corroded and apparently undecorated, 16mm long, 2mm wide and 0.5mm thick.
sf.989, unstratified.

Although not certainly a bracelet, the fragment suits this class better than any other. It is probable that the use of thin and narrow bands is a more important factor than the presence or absence of ornament. If so, the date-range offered by the general parallels is, again, fourth-century.

Beads etc.
(Fig. 63)

61 Glass. A pale translucent blue bead of five segments 15mm long. The oval section, with a 3mm maximum width, was made by winding a glass rod round a former.
sf.395, unstratified.

Belonging to Mrs Guido's small segmental class (Guido 1978, 91–3, fig.37,1), most specimens are fourth century (*ibid*., 92).

62 Glass. Opaque sage green, striations show that the bead was made by winding a cane round a former; 5mm outer diameter, 2mm inner diameter, and 2.5mm long.
sf.1077, Period 5.

The type had a long life, but was particularly common in the third and fourth centuries (*ibid*., 95, fig.37,5).

63 Copper Alloy. A wire loop not likely to be from a brooch and, at 12mm high by 9mm wide, too small to be from anything other than a trinket and not part of typical necklace fastenings (*e.g.* Clarke 1979, fig.75, 140; fig. 87, 405; fig.90, 363), but could have been used in a bell (Crummy 1983, 51, fig.54, 1811).
sf.181, Period 2.

Pins
(Fig. 63)

All the pins found belong to types identified by Mrs Crummy (1979). 64–68, and possibly 69, are her Type 2; 71–77 are Type 3 and 78–9 are Type 4. The two Type 1 pins, 80 and 81, are placed last as there is some doubt about their function. 82–93 are fragments of shafts and, apart from signs of a swelling, could have come from needles. However, the ratio of these to pins on this site is so low that they are better placed here. The cited parallels are all close: in the case of Type 3, only those with swelled shanks have been given.

64 Copper Alloy. 118mm long with a diameter of 1.2mm at the top. The head consists of a cone above a reel and is 3mm long.
sf.700, Period 1.

65 Copper Alloy. The upper 56mm survive. The top is 1.4mm in diameter. The head is 2mm long and of the same form as that on 64.
sf.806, Period 1.

66 Bone. 101mm long, 2.5mm in diameter at the top.
sf.918, Period 1.

67 Bone. The upper 55mm long survive, the head is 3mm in diameter.
sf.306, Period 2.

68 Bone. 37mm long down to the break, the top is 6mm in diameter.
sf.263, unstratified.

69 Bone (not illustrated). Only 19mm long, The shank is 3mm in diameter and almost straight-sided. What is left of the head has half a groove.
sf.706, unstratified.

Of the forty-two dated specimens noted, thirteen were mid-first century into early second, nineteen ran on to the late second century and only ten dated from then until the middle or the latter part of the third, suggesting that any in fourth-century contexts were residual. Mrs Crummy's dating (1979, 160–1) could possibly be refined as those after the mid-third century could also be residual.

70 Bone. Crudely trimmed and highly polished, the top is 75mm long, 6mm thick widening to *c*.7mm. The point is lost.
sf.1090, Period 5.

71 Copper Alloy. The overall length is 55mm, the top is 1mm in diameter and the swelling is 1.6mm across. The head is fat and biconical.
sf.383, unstratified.

72 Bone. 68mm in overall length. The point is lost. The top is 3mm in diameter swelling to 5mm. The roughly trimmed head has a flattish top.
sf.412, Period 3; sf.242, Period 5: residual.

73 Bone. 65mm long without the point. The upper shank is 2.5mm in diameter and increases to 3.5mm.
sf.967, Period 3.

74 Bone. 52mm survives, the well-rounded head is 2mm in diameter, swelling to 4.2mm.
sf.963, Period 3.

75 Bone. 67mm long, the point is lost, the top of the carefully finished shank is 3mm in diameter, the swelling 4mm.
sf.567, Period 5.

76 Bone. Most of the head is missing, 62mm survives; the shank is 2mm diameter at the top swelling to 4.5mm.
sf.977, Period 5.

77 Bone. Probably of the same type as the others here, 40mm survives, most of the head is missing as is much of the rough shank, 4.5mm diameter at the top and 6.5mm at the swelling.
sf.241, Period 5.

Type 3 pins are given a *floruit* running from *c*.AD 200 to the end of the Roman period (Crummy 1979, 161). Few appear to be earlier and may have been wrongly dated. The dating of those both published by Mrs Crummy and later point to a beginning in the third century, by the end of that century and for most of the fourth it was relatively common. The three earliest pins at Orton Hall Farm, from Period 3, fit this pattern.

78 Bone. The surviving length is 92mm; the top of the shank is 3mm in diameter swelling to 4.8mm. The head is a cube with the corners removed so that the main faces become lozenges.
sf.210, unstratified.

79 Bone. The point is lost and 60mm survives with a top diameter of 3.2mm; 3.7mm at the swelling. The head is a repeat of 78.
sf.510, Period 4.

Both belong to Type 4 and date from the mid-third century to the end of the Roman period (Crummy 1979, 161–2). Never as common as the preceding two types, the date range is not altered by later published examples, unless the type started later in the third century.

80 Bone. The head is a very shallow cone 6mm in diameter and the pin tapers away immediately from it. The surviving length is 49mm.
sf.701, Period 1, joins sf.905.

81 Bone. As the last, 5mm diameter at the head, 54mm surviving length.
sf.881, Period 2.

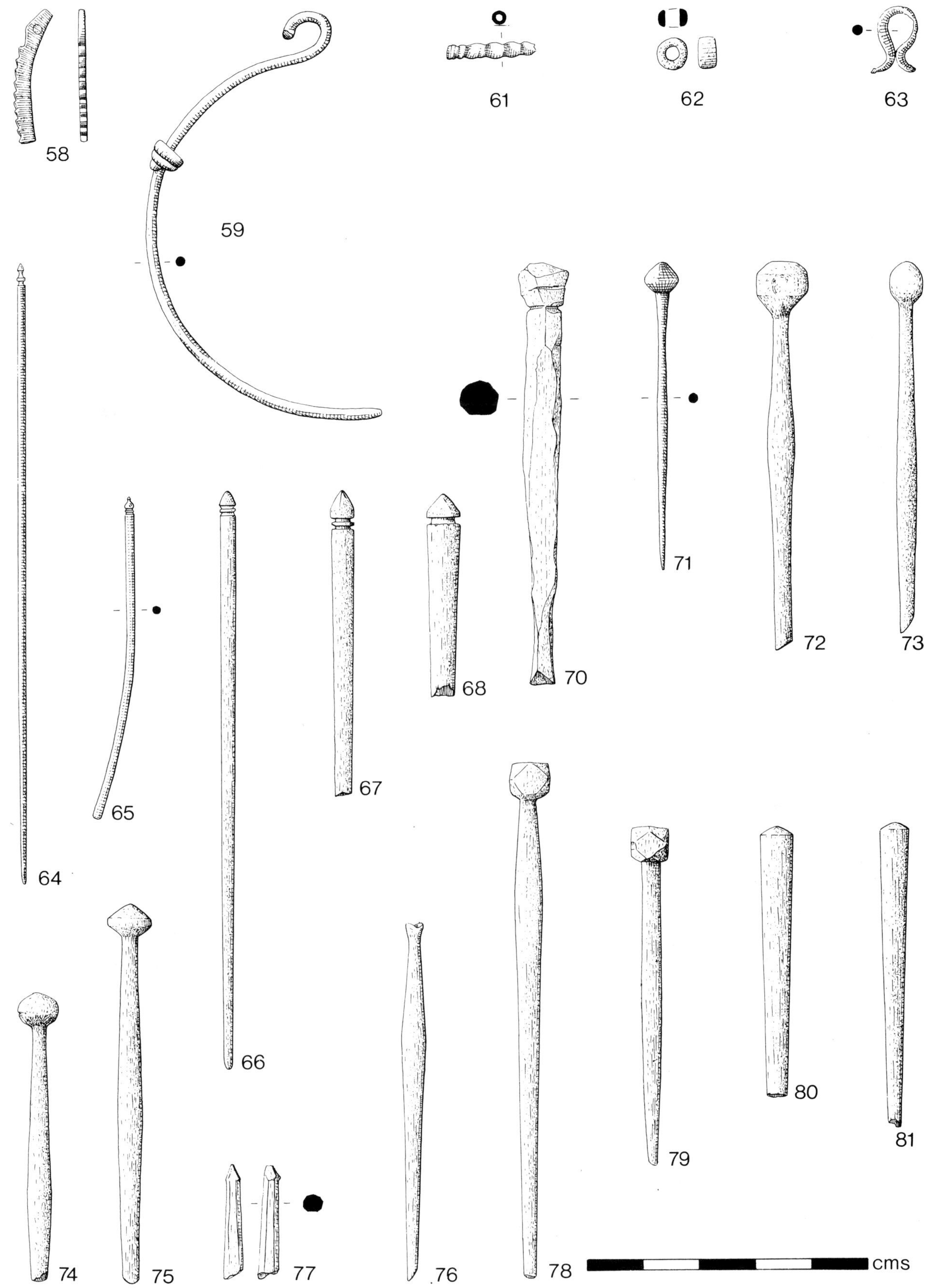

Figure 63 Bracelets, Nos 58–59; beads *etc.*, Nos 61–63; pins, Nos 64–81.

Both are Type 1 items dated from the later first to the mid-third century (Crummy 1979, 159–60), and are well-polished and could have been used as pegs.

Fragments of shanks
not illustrated

82 Bone. Swelled fragment, 3.5mm diameter at the top increasing to 4mm, 60mm long.
OHF 71, sf.19, unstratified.

83 Bone. Swelled fragment, 2.5mm diameter at the top increasing to *c.*3.5mm, 58mm long.
sf.966, Period 5.

84 Bone. Swelled fragment, 3.5mm minimum diameter, 4mm maximum, 28mm long.
sf.502, Period 3.

85 Bone. Point only, 36mm long.
sf.185, Period 5.

86 Bone. Point only, 28mm long.
sf.1058, Period 4.

87 Bone. Blunt point, 25mm long.
sf.281, Period 5.

88 Bone. Point only, 27mm long.
sf.481, unstratified.

89 Bone. Plain tapering, possibly intentionally stained green (MacGregor 1978, 35), 35mm long.
sf.477, Period 4.

90 Bone. Plain, 33mm long.
sf.307, Period 2.

91 Bone. Plain, 19mm long.
sf.1131, Period 4.

92 Bone. Plain, thicker than usual, 5mm maximum diameter.
sf.363, unstratified.

93 Bone. Crudely trimmed, highly polished, the presence of the head is marked by a constriction. 8mm diameter at the top tapering to 4mm at the break which is worn.
sf.449, unstratified.

Combs and Cosmetic Items
(Fig. 64)

94 Bone. **Comb.** Part of one end of a double-sided comb 58mm wide. The toothed segments are held in place by rivets through a central rib on each side. In section, each segment is flat in the middle under the ribs and then tapers to the rounded ends of the teeth. Saw marks on each side of the ribs show that the teeth were cut after assembly. Those on one side occur at 6 or 7/10mm and about 4–5/ 10mm on the other. Each rib has bordering grooves and rounded corners. The teeth become progressively shorter as they approach the end of the comb, leaving a solid triangle rising from the ribs, which is carefully cut to form two adorsed beast's heads separated by a concave curve. The snout is indicated as well as the brow; the eye is a hole through the plate; the ears are formed by two small nibs.
sf.707, Period 3.

Such combs are not common finds and the best parallels for the present piece come from cemeteries: second half of the fourth century, possibly more popular towards the end (Crummy 1983, 56–7, fig.58, 1855, fig.59, 1860; Clarke 1979, fig.31, 323, 479, 610). The date of the Orton Hall Farm comb is earlier than these which could result either from conservative dating of the pottery, or because the dated items only entered the ground at the end of their period of popularity. One from Lankhills, in which a mouth is crudely indicated by a simple cut, shows the evolution of more developed zoomorphic terminals (Clarke 1979, fig.93, 473).

95 Copper Alloy. **Tweezers.** Corroded and undecorated, the whole formed from a strip of metal 5mm wide and at least 105mm long.
sf.568, unstratified.

The dated examples seem to show that such tweezers were most common in the first and second centuries, but there is not a sufficient fall in numbers to show that fourth century items must be residual.

96 Copper Alloy. **Cosmetic scoop.** The scoop, damaged, was once 5mm in diameter. The shank is plain, 96mm long ending in a slightly blunted point. Its maximum diameter is 2.5mm.
OHF 71, sf.17.

The object could have been used as a stylus: the straight edge just short of the bend would have been effective as the erasing end, but it almost certainly began as a cosmetic accessory. Found in the first century, the type belongs more to the second century and continues into the third. The few which can be dated to the fourth century were probably residual.

97 Copper Alloy. **Scoop**. One end missing, the scoop, 5mm in diameter, is mounted on a tapering shank 11mm long, rising from an elaborate four-element moulding, 15mm overall. The outer ones have beaded edges. The rest is 65mm long reducing to 2mm in diameter.
sf.762, Period 1.

Mouldings at the base of this kind of scoop seem to be rare, the bowl is usually larger, dished, and elongated. A pin from Leicester with an exact repeat of the mouldings here shows that single workshops produced a whole range. The pin was dated *c.*125–30 (Kenyon 1948, 262, fig.89,3). The main bias of the dating of articles similar to the present one is towards the late first century and the earlier part of the second, possibly as far as the middle.

98 Iron. Only the longer of two pieces is illustrated. It is a carefully squared rod, 2.2mm by 2.2mm, and 271mm long. One end is broken, the other is swelled to form a rounded-ended elongated thickening 13mm long and 4mm in diameter.
sf.345, Period 5.

99 Iron. A repeat of the last, only 62mm long, one end is thickened, the other may be broken and may have been flattened.
sf.1062, Period 5.

No parallels have been found, but there is no reason why they could not have been used for cosmetics. The profile of 98 is reminiscent of that of a latch-lifter. Both come from Period 5.

Footwear
(Fig. 64)

100 Iron. **Hobnails** (not illustrated). 522 were found and their distribution by period is: Period 1, 19; Period 2, 16; Period 3, 18; Period 4, 154; Period 5, 127. 188 were recovered from unstratified contexts.

Although the large number in Period 4 could reflect an increased use then, there are too many imponderables to allow such a straightforward interpretation. However, only in Period 4 did the whole of the Main Yard become a general concourse area compared with earlier periods. Some hobnails were found corroded together, or lying in the formation to be expected on a sole, but most were detached and usually very corroded. That some variation in size and shape can be expected is shown by those found at Lankhills (Clarke 1979, fig.38), but there are not enough here for close analysis. It is worth bearing in mind a *c.*300mm long strip of thirty-four in a single row found at Brough-on-Humber (Wacher 1969, 94) which should be a warning that not all were used in footwear. The standard literature reveals few shoes preserved well enough for their complement of hobnails to be counted. Much depends upon the size of shoe as well as the kind of sturdiness required. A rough assessment suggests that most shoes with hobnails had between about 20 and 65. However, there is possibly another group at around 90 and there are two which may have had more than 110 (Keppie 1975, fig.25, 57: half-shoe had *c.*60, MacGregor 1976, 14, fig.11, 146).

101 Iron. A slightly curved plate, 62mm long by 16mm wide with, at one end, an upright section *c.*15mm long and 8mm high curving back over the plate.
sf.402, Period 3.

102 Iron (not illustrated). A more fragmentary plate of the same kind, 62mm long by 13mm wide.
sf.402, Period 3.

103 Iron. A fragment only 30mm long by 8mm wide. One end is broken. At the other is a turn up 10mm high with a slight bend to one side.
sf.965, Period 3.

All three are **boot-plates**. 103 is the more common pattern, only one resembling 101, published as a spoke-shave (Leach 1982, 255, fig.124, 14), has been noted. Most parallels for 103 are oval in plan suggesting that they reinforced toes or heels. Few are straight like 103 and the writer has not seen a published shoe with a boot-plate in position. The longer, straight-sided ones might have been in a row along the full length of a sole allowing it to flex in use. Very few long enough to span the foot have been published (*e.g.* Frere 1984, 103, fig.44, 131).

Counters
(Fig. 64)

104 Bone. A counter or gaming piece turned on a lathe, 20mm in diameter, 9mm thick, one side is flat with a deeply scored cross, the other having a cone-shaped hollow *c.*5mm deep. The flat face has many small scratches running at right-angles to each other suggesting that the object may have been pushed along set courses on a gaming board.
sf.939, unstratified.

105 Clay. Not illustrated. Trimmed pre-fourth century pot base with worn edges and upper surface, *c.*36mm in diameter.
sf.157, Period 4.

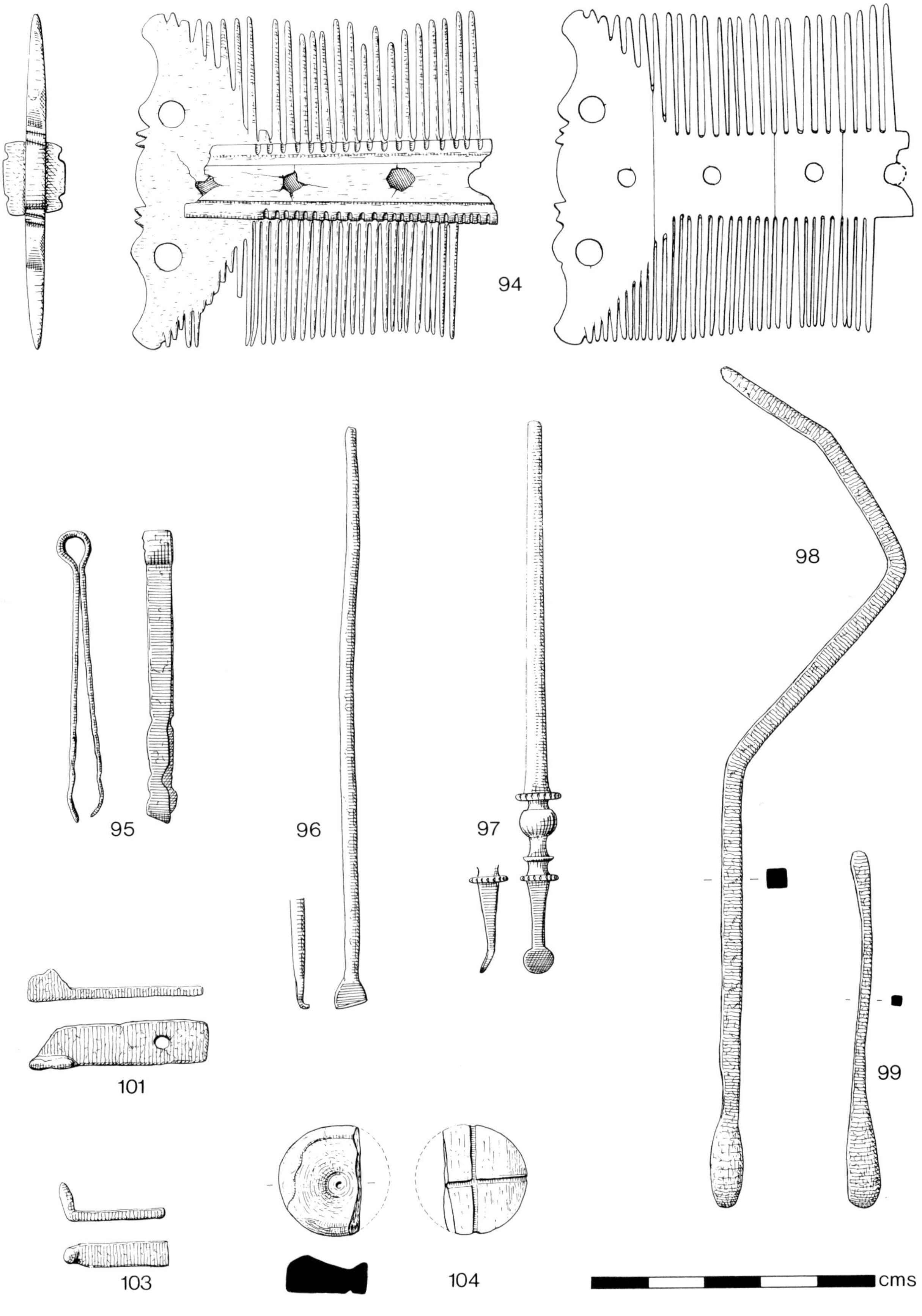

Figure 64 Combs and cosmetic items, Nos 94–99; boot-plates, Nos 101 and 103; counter, No. 104.

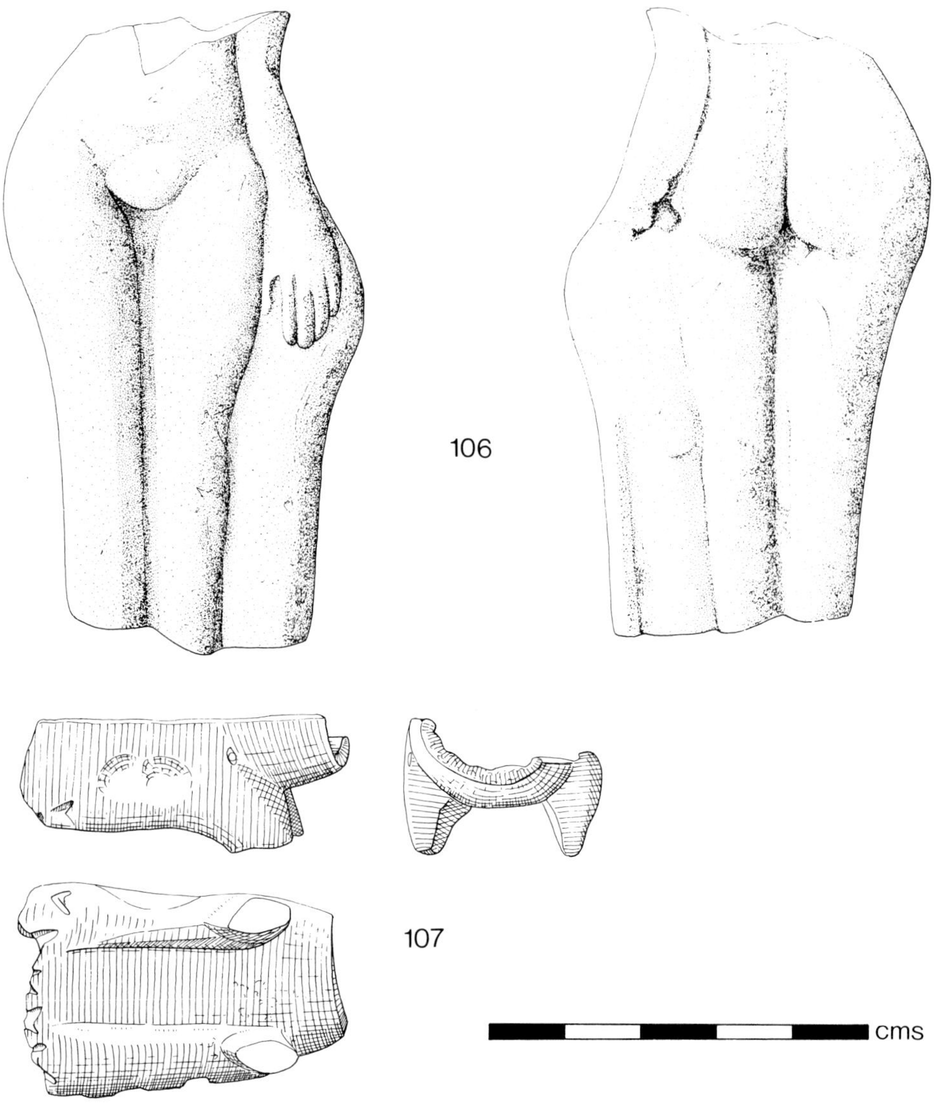

Figure 65 Cult objects: pipe-clay figurine, No. 106; copper/alloy animal, No. 107.

Cult Objects
(Fig. 65)

106 Clay. A pipe-clay figurine of Venus, all below mid-shin and above the waist is missing. She stood with right hand raised and the left lying on the top of a vertical element against which her left leg rests.
sf.549, Period 4.

107 Copper Alloy. The lower fore-quarters with the very top of the legs survive of a hollow-cast miniature animal. The neck seems to thrust forward and to have been cut off and the upper edge of one flank shows similar treatment. The other edges are irregular and roughly broken. On the belly of the beast, and running along the break is a series of cells which, as the complete example shows, were V-shaped and were probably intended to represent dugs. No trace of enamel survives. The type of animal is unidentifiable. The casting has a flaw just above the right foreleg.
sf.91, unstratified.

Both pieces should be connected with religious practices: the Venus, probably made in the second century (Jenkins 1967, 19–20), obviously was and animals as *objets d'art* seem to be unknown in Roman Britain. Not having been subjected to normal wear and tear, both items should have lasted a long time. As cult objects, their disposal in a mutilated form can be considered as desecration and the deposition, about the middle of the fourth century for the Venus, may have been related to a major change in religious practice on the part of their worshippers.

Microfiche MF8; MFFigs 66–73

108–110 Vessels.
111–120 Knives and shears.
121–123 Knife handles.
124–131 Styli.
132–139 Needles.
140–169 Tools.
170–178 Weaving equipment: loomweights; spindle whorls; cheese-press(?); pin-beaters.
179–189 Whetstones.
190–194 Locks and keys.
195–204 Chains and general suspension items.
205–231 Fittings.
232–242 Miscellaneous.

IV. Anglo-Saxon Finds

by Martin Howe and D.F. Mackreth
(Figs 74 and 75)

Only those items which are recognisably Anglo-Saxon have been gathered here. Those which could be Anglo-Saxon, *e.g.* amongst the brooch fragments and tools, are on microfiche. M.D. Howe commented on Nos 243–250 inclusive.

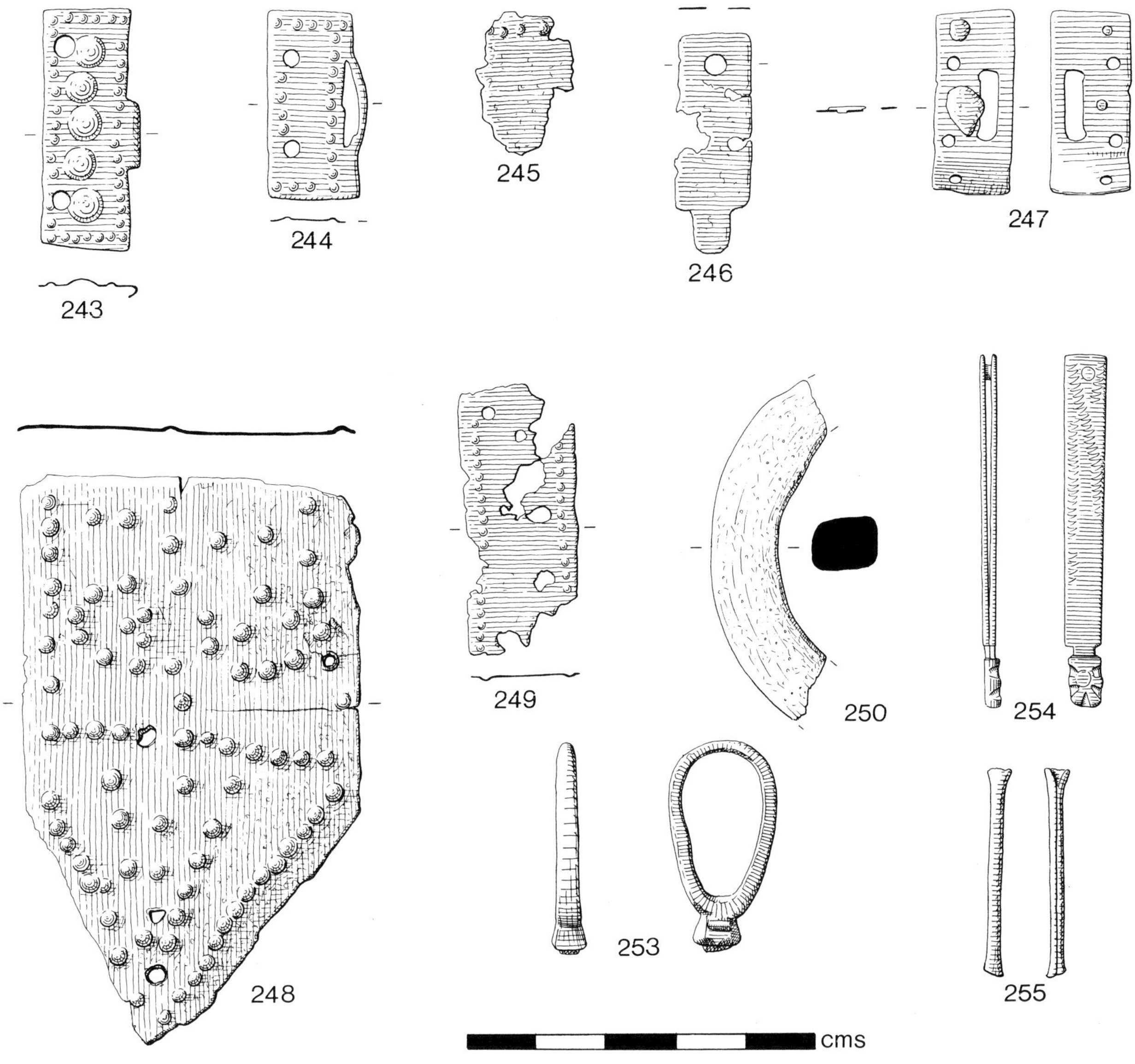

Figure 74 Anglo-Saxon objects, Nos 243–250. Medieval objects, Nos 253–255.

Copper alloy

243 Complete **'hook' half of a pair of wrist-clasps**. The plate is cut from sheet-metal and measures 35mm long, 12mm wide and is 0.8mm thick. It is decorated with five bosses, each 4mm in diameter, placed in an irregular line across the centre of the plate which is bordered with forty carefully placed bosses, each 1mm in diameter, two of these are pierced by holes used to stitch the clasp to the garment. Hines' Form B7 (Hines 1984, 380).
sf.401, unstratified.

244 **'Catch-plate' from a pair of wrist-clasps**, 22mm long, 11.5mm wide and 0.3mm thick. The incision forming the catch was crudely made using a cold chisel. The catch-slot has bowed out. The plate is decorated with twenty-two fairly carefully placed raised bosses 1mm in diameter and allowance has been made in the design for the attachment holes. Hines' Form B7 (*ibid.*).
sf.1052, unstratified.

245 A piece of a **wrist-clasp** too incomplete to permit a detailed analysis. The fragment has three raised bosses 1mm in diameter marking one of the edges of the plate which is 0.3mm thick. The raised bosses suggest that the clasp may belong to Hines' Form B13b, those having raised-boss decoration at the edges and soldered decorative strips.
sf.341, unstratified.

246 Three fragments of a much broken **wrist-clasp or strap-end**. Two of the pieces show some of the original edges of the plate, the remains of one raised boss and an attachment hole.
sf.1039, L2755, (280), F254, Period 5.

247 The **catch-plate from a pair of wrist-clasps**, well preserved but badly bent. It measures 25mm long, 10mm wide and 0.4mm thick, and has been carefully made. Two holes pierce the plate for the stitching and a further three holes are present, one of which retains the remains of a rivet or stud. The central hole retains a corrosion pattern which suggests that the other holes also housed rivets. The presence of the rivets indicates that this example belongs to Hines' Form B8 (Hines 1984, 380).
OHF 71, sf. 18, Period 5.

Wrist-clasps are frequently found in cemeteries in East Anglia and the Midlands. They are held to be peculiar to the Anglian group of early Germanic settlers. The clasps are normally items of female costume.

The Orton Hall Farm clasps are of two types: those with raised-boss ornamentation of Hines' Forms B7 and B13b, Nos 243–246, and the single example, No. 247, which once had rivets of Hines' Form B8. These types were made from sheet copper alloy. Those with raised bosses were made in the following way: the plate was cut using shears or a cold chisel and on the 'hook-plate' a process was formed on one of the long sides ultimately to form the hook. If the plate was to have a curvature (to assist attachment and make it more comfortable to wear), it was formed at this stage and the decoration and attachment holes were then punched in. The bosses were raised using a punch with a rounded end to strike the metal into a depression of corresponding size in a block. Lastly, the hook was formed by bending over the process on the long side. The object may have been annealed during this operation to prevent the hook breaking off. The construction of the corresponding clasp was in a like fashion, the 'catch-slot' being formed by piercing two holes and cutting away the metal between with a cold chisel.

With the exception of No. 247, the Orton Hall Farm wrist-clasps are closely paralleled from local cemeteries at Woodston, Huntingdonshire (Walker 1899, 343–9; Wyman-Abbott 1920, 34–40) and Nassington, Northamptonshire (Leeds and Atkinson 1944, 100–28). The 'hook-plate' of a pair of clasps from Woodston (Peterborough Museum, L524) measures 27.5mm long and 11.5mm wide. They are decorated in similar fashion to No. 243 and, although differing in precise dimensions, show a broad similarity. No. 247 also has dimensions which are in accord with these examples and it is possible that all these objects were produced in the same manufactory.

It is unfortunate that, for the few stratified items, none of the associated material aids the dating of the Orton Hall Farm wrist-clasps. The Nassington cemetery is the only reliable local site that can produce dated parallels: the Woodston material was nearly all collected from nineteenth-century gravel workings and lacks any detailed record. Although the Nassington cemetery was excavated under far from ideal conditions, a number of wrist-clasps were found which, from their associated grave goods, can give a relative date. Grave 20 (Leeds and Atkinson 1944, 107) was partially disturbed and contained the remains of two skeletons one of which had a pair of raised-boss decorated clasps at each wrist and two Small-Long brooches of Leeds' Square-Headed Type ii at the left clavicle and pelvis (Leeds 1945, 26). Both brooches have a crescentic foot and are decorated with punched triangular designs (*ibid.*, pl. XXVIII, b). This type of brooch dates to the middle years of the sixth century. Wrist-clasps found in grave 16 (*ibid.*, 107) were associated with two Swastika brooches of mid sixth-century date and grave 28 (*ibid.*, 109) produced clasps associated with Cruciform brooches of Åberg's Group IV, again dating to the middle years of the sixth century. Hines dates his Form B7 and B13b clasps to the sixth century (Hines 1984, 74), and they are amongst the most common forms encountered.

The Form B8 clasps have a late fifth to early sixth-century date. They are, however, rare in England, deriving from Hines' Form B1 which are common in Scandinavia (*ibid.*, 75).

Because of the way in which they were attached, wrist-clasps were easily lost. The copper-alloy clasp would quickly wear through the thread joining it to the garment and the wearer would only become aware of the loss when the sleeve hung open. It is likely that wrist-clasps were replaced fairly frequently and are not likely to have been regarded as heirlooms. Thus it is probable that when they occur in grave assemblages, they are contemporary with the other grave-goods.[1]

248 Copper alloy. **Plate** with a rectangular body and triangular foot measuring 80mm and 47mm overall and 1mm thick. The whole is extensively decorated with raised dots. The outer edges of the plate are bordered with 25 raised dots which do not extend across the upper margin of the plate. The surface condition of this edge shows that the plate was not cut down from a larger one. The fitting is divided by a longitudinal line of twelve raised bosses which merge at the foot and the rectangle is divided from the foot by a line of ten raised bosses. The foot has two further lines of four and six raised bosses radiating from the point. The rectangle is divided by the longitudinal line of bosses into two fields each of which contains a 'rosette' of nine raised bosses. One 'rosette' was clumsily formed showing that insufficient room was left for it when the field was formed. The decoration of the rectangle is completed by an upper border of four raised bosses maintaining the rectangle and respecting pre-existing lines. The lower margin of the rectangle is dented and split. The damage occurred after manufacture and looks as if it was caused by a direct blow from a sharp edge. All the edges of the plate are slightly bent up and it is suggested that this resulted from the plate having been fitted with binding strips which pinched it in.
sf.853, L1821, (314), F898, Period 5.

A plate of similar dimensions with raised-boss decoration was recovered from grave 48 in the Wakerley cemetery (Northampton Museum, no. 690569). This female burial contained a pair of Small-Long brooches of Leeds' square head and lozenge foot type (Leeds 1945, 36). The brooches also have moulded finials which are held to be a 'mark of early or comparatively early date'. The presence of punched decoration does, however, suggest that these examples belong within the first half of the sixth century rather than the late fifth. The excavator, from the position of this item, interpreted it as a strap end. If the Orton Hall Farm plate is considered as such, it may have been attached to a fairly wide belt end, but lacks any obvious means of joining it to the belt. Four irregularly placed holes on the Orton Hall Farm example superficially suggest that the plate was riveted in position, a means of fastening widely used on belt fittings. However, on closer examination, the 'holes' are merely over-struck raised bosses, the punch piercing right through the metal. Thus, unless binding strips were used to attach the plate to a belt — which seems unusual and unnecessary — it is doubtful if the Orton Hall Farm plate was a strap end.

Although the form of the object might suggest that it was part of a scabbard chape, especially if it had been fastened into position using binding strips, the object is very difficult to identify positively as one. The chapes of many Anglo-Saxon swords of the pagan period are in the form of a U-shaped binding strip which does not cover the whole foot of the scabbard (Neville 1852, pl. 34, 96). Thus if the present plate was one, it would appear to look back to a type used by the Roman army. However, the surviving chapes from the early and later periods of the Roman occupation do not support this hypothesis. Scabbard chapes of the Conquest period are generally triangular in shape and smaller than the Orton Hall Farm piece. Also the later Roman chapes become more tubular, pinching in at the foot of the scabbard. The Orton Hall Farm plate is attributed to the Anglo-Saxon period because its raised boss decoration closely resembles that on the wrist-clasps. Also, its similarity to the Wakerley example suggests that a date within the first half of the sixth century would be appropriate for the Orton Hall Farm plate.

249 Copper alloy. A much corroded **strip** broken away at its left-hand end. The strip measures 17mm in width, is 0.1mm thick, decorated along its upper and lower borders with raised boss punches which, where visible, measure 1mm in diameter. At its right-hand end the strip is pierced with two punched attachment holes which show marks suggesting that the strip was held in position by small nails. There is no evidence for hooks or catch-slots and thus the strip is most probably the remains of a bucket or casket mount.
sf.559, unstratified.

The object is similar to a copper-alloy strip from Spong Hill, North Elmham, Norfolk (Hills and Penn 1981, 58, fig. 143, pl. VII, A2164) which was associated with a bossed urn with stamped decoration dating to the earlier part of the sixth century. It is suggested that the strip was a bucket mount. The same identification and dating seem appropriate for the Orton Hall Farm example.

250 Ivory. The broken segment of a **ring** clearly showing from the grain of its section that it was sliced from the tusk at a right-angle to its length, thus forming a natural circle and reducing the labour involved in manufacture. The ring shows the marks of having been sawn and rasped or sanded on its underside. The upper surfaces have been well-rounded by initial working and wear resulting from use. The ring's sub-rectangular section measures 10mm by 8mm. It has a reconstructed external diameter of 52mm and was thus cut from near the tip of the tusk. Such a diameter does not permit the passage of even the smallest hand and thus this object was not a purse-ring or a bracelet. There is a constriction on the upper face and it is suggested that the ring served as a belt buckle.
sf.1114, L483, (213), F31, Period 5.

Ivory rings are fairly frequent finds from Anglo-Saxon female burials where they occur as one of the fittings on a chatelaine. Most examples are, however, much larger than the Orton Hall Farm fragment. Rings from Caistor-by-Norwich measure 100mm and 108mm in diameter (Myres and Green 1973, 101) and a ring from grave 31 at Nassington (Leeds and Atkinson 1944, 110) measures 121mm in diameter. It has been demonstrated (Vogt 1960, 70ff. and figs; Green 1973, 100–3) that these larger rings once supported bags, but the Orton Hall Farm ring would have been too small for such a purpose. Similarly, none of these rings has been convincingly explained as a bracelet.

The constriction on the Orton Hall Farm ring was produced largely by hard wear, but may have originally been cut into the ring to form a convenient attachment point for a leather or textile belt. The ring fragment still retains a high surface lustre and this is again indicative of hard use and rubbing against other materials. The piece could have been part of a conventional ring-and-tongue buckle, but to the writer's knowledge, no ivory buckles in this form have yet been found in pagan Anglo-Saxon contexts although bone buckles are known from late Roman sites. It is suggested that the ring was part of a lightweight leather or fabric belt. The belt end passed through the ring and then wrapped-back around the secure belt to make an effective closure. The breaks at either end of the fragment are of interest. The right-hand break is reasonably fresh and probably the result of plough damage. The left-hand break occurs at the constriction and is much darker in colour. This break, at the weakest point of the ring, is ancient and was probably the cause of loss.

It has not proved possible to find an exact parallel for the ring fragment and thus, with its lack of context, dating is difficult. Barbara Green suggests a late fourth- to early fifth-century date for cremation urns N 52B and P 53B from Caistor-by-Norwich. Both of these pots contained rings and thus, by inference, she is attributing a correspondingly early date to the rings. However, she states (Green 1973,

102) that rings also occur in late fifth, sixth and seventh-century contexts in other parts of the country. The ring from grave 31 at Nassington (Leeds and Atkinson 1944, 110) was associated with a Cruciform brooch of Åberg's Group IV, a group which dates to the middle years of the sixth century. It is possible that the Nassington ring was regarded as an heirloom and may thus have been of some antiquity when buried. However, in the light of the date range of the other non-pottery finds from Orton Hall Farm it is more likely that the fragment is of sixth-century date.

251 Bone. **Comb** (Fig. 75; Pl. IX). Built up of three layers, the form is that of a Frisian Barred or Barred Zoomorphic comb. The central part is extended to form a regardant head at each end economically depicted by the shape of the plate and with a double ring-and-dot for the eye. The teeth are on one side only and the ends flare out. The handle is made up of three thicknesses and can be roughly described as being mushroom-shaped rising from a rectangular pedestal. The curved edge is bordered with a line of ring-and-dot ornament, heavily worn in the central section. The extra thickness of handle and the ribs on each side are fastened by means of iron rivets to the seven separate sections which make up the middle part of the comb. The other decoration is confined to three parallel grooves at the base of the handle and to the ribs. One of these is flat and has a repeat of the three grooves along each edge. The other side of the comb has two semi-circular-sectioned ribs with cross-grooves arranged on each in three groups. Although the spacings on each rib appear to match those on the other, the layout shows that there was no intention of having an exact repeat as the upper rib has a wider band at each end than the other, and a narrower one in the middle.

sf.1042, L2762, (328), F1114, Period 5.

Combs of this pattern have been discussed by Dr Hills (1981). The form seems to be a mixture of two types of comb common in the late empire and dating usually to the end of the fourth century into the fifth (*ibid.*, 98–100). One style has a semi-circular back while the other is triangular; both are single-sided combs. It is the latter which occurs with zoomorphic ornament, recalling that found here, and her conclusion is that such combs are less common than the other type and may also have been made by non-Roman craftspeople (*ibid.*, 100). While a start date for the present style in the early fifth century seems to be indicated by its obvious origins, the actual associations of such combs lead to a less precise date and the argument may be summarised by saying that the range is remarkably similar to that of the Roman types: late fourth-early fifth century (*ibid.*, 103–4). The largest published group of such combs comes from the large and prolific cemetery at Spong Hill, North Elmham, Norfolk. A comparison of the Orton Hall Farm specimen with those is instructive: there, there are five of which only one, from grave 1470, is close, the next nearest are from graves 1475 and 1227, while those from graves 1450 and 1556 can be described as crude and, on purely typological grounds, late in the sequence.

The pottery forms which accompanied these five combs show that it is only with the last two that patterning is present in the decoration: 1227 has simple linear and dot ornament and the other two are plain. It is hard to use the pottery from Spong Hill and compare it with that from the deposit in F1114 (329) from which the Orton Hall Farm comb came because the functions of the two sites were dissimilar and the domestic wares present here show the usual bias towards plain forms. However, it cannot be ignored as it was this same deposit which produced the mortarium in a Anglo-Saxon fabric (Anglo-Saxon Pottery, No. 15), the markedly biconical vessel (Anglo-Saxon Pottery, No. 1) in what could be taken to be a Romanising fabric, as well as the fragments of the hollow-based bowl (Anglo-Saxon Pottery, No. 13).

The Orton Hall Farm comb appears to be typologically earlier than any at Spong Hill in both having a generally better definition of parts and a basically more triangular form as the mushroom-shaped handle rises above the zoomorphic heads to a greater degree than in either of the relatively close specimens from graves 1470 and 1475 at Spong Hill. Thus the associations and parallels for the comb emphasise an early date which cannot be more closely defined than late fourth century into the fifth, a date which conforms with the earlier stages of Period 5.

252 Clay. **Loomweights**. Not illustrated. These are arranged in order of the small find serial number:

sf.21, unstratified.
sf.50, (244), L151, (243), F81, Period 5.
sf.126, (225), L180, (224), F118, Period 5.
sf.192, unstratified.
sf.193, unstratified.
sf.215, unstratified.
sf.360, unstratified.
sf.698, (278), L313, (277), F200, Period 5.

All formed parts of the ordinary annular loomweight belonging to the Early Anglo-Saxon period. Although they were damaged, it seem that there may have been three sizes yielding three markedly different weights:

sf.21, *c.*175gm
sf.50 and sf.698, *c.*250gm
sf.192 and sf.215, *c.*335gm

With such limited information it is hard to draw conclusions, but there may be elements here of a system of weights in which each larger size is half as heavy again as the preceding one.

V. Medieval Objects

by D.F. Mackreth
(Fig. 74) Microfiche MF8

253 Purse-frame suspension loop.
254 Strap-end.
255 Binding-strip.

VI. The Glass

by John Shepherd
Microfiche MF8; MFFig.76

1–60 Roman vessel fragments.
61–90 Window glass.
91–99 Medieval and post-medieval vessel and window-glass.

VII. The Millstones

by R.J. Spain
(Figs 77–9)

The numbers in [ ] brackets are the sample numbers in the Nene Valley Research Committee's sample series.

1 (L2885) [2273] Period 3. Fragment of a bottom millstone with part of the eye and rim showing. It is a segment of the original stone with two almost perfect radial fractures. Thickness at the eye is approximately 80mm and, at the rim, 55mm. The rim is very irregular and appears to be much fractured without any discernible dressed profile. The inclination of the grinding face is 10 degrees and its convex face is covered by a number of rough pick marks or indents each some 8mm to 15mm diameter and a maximum of 3mm deep. A radial section of the grinding face is convex but irregular and part of it is roughly flat. The back surface of the stone — the underside — is roughly dressed. The diameter of this stone was at least 680mm, but the heavily fragmented rim means that it could have been larger. Unfortunately, we have a similar difficulty in estimating the diameter of the eye. There is not enough of its perimeter to get a reliable figure, but it was probably some 80mm in diameter. The weight of the original stone must have been at least 32.5kg, possibly more.

2 (L309) [185] Period 2. Rim fragment from a bottom millstone which was 890mm in diameter when complete. No evidence of the eye appears on the fragment. The rim has a vertical, though slightly convex, face and a thickness of 82mm. Thickness of the fragment at a radius of 225mm from the axis of rotation is 93mm, suggesting that the thickness at the eye was probably close to 110mm. A close inspection of the grinding face does not reveal any evidence of furrows or pecking, although it is possible that any such dressing has been worn away. The inclination of the grinding face was probably close to 3 degrees and a radial section shows that it has a slight concavity of 2mm to 3mm across the specimen. Assuming an eye diameter of 115mm and a thickness at the eye of 110mm, the weight of the complete stone was approximately 126kg.

3 (L309) [186] Period 2. Large rim fragment from a lower millstone. No evidence of the eye appears on the fragment. The original diameter of this stone was 890mm and its weight, assuming an eye diameter of 100mm, was probably close to 130kg. A radial section of the fragment shows that the grinding face has a slight concavity which has a maximum depth of 2mm to 3mm, some 100mm from the rim. The inclination of the grinding face is 7 degrees from the horizontal. It is possible that the complete stone had a concave-convex radial section on the grinding face. The thickness at the rim is 75mm and at the innermost part of the fragment approximately 95mm, suggesting that, close to the eye, it was probably in the order of 100mm thick.

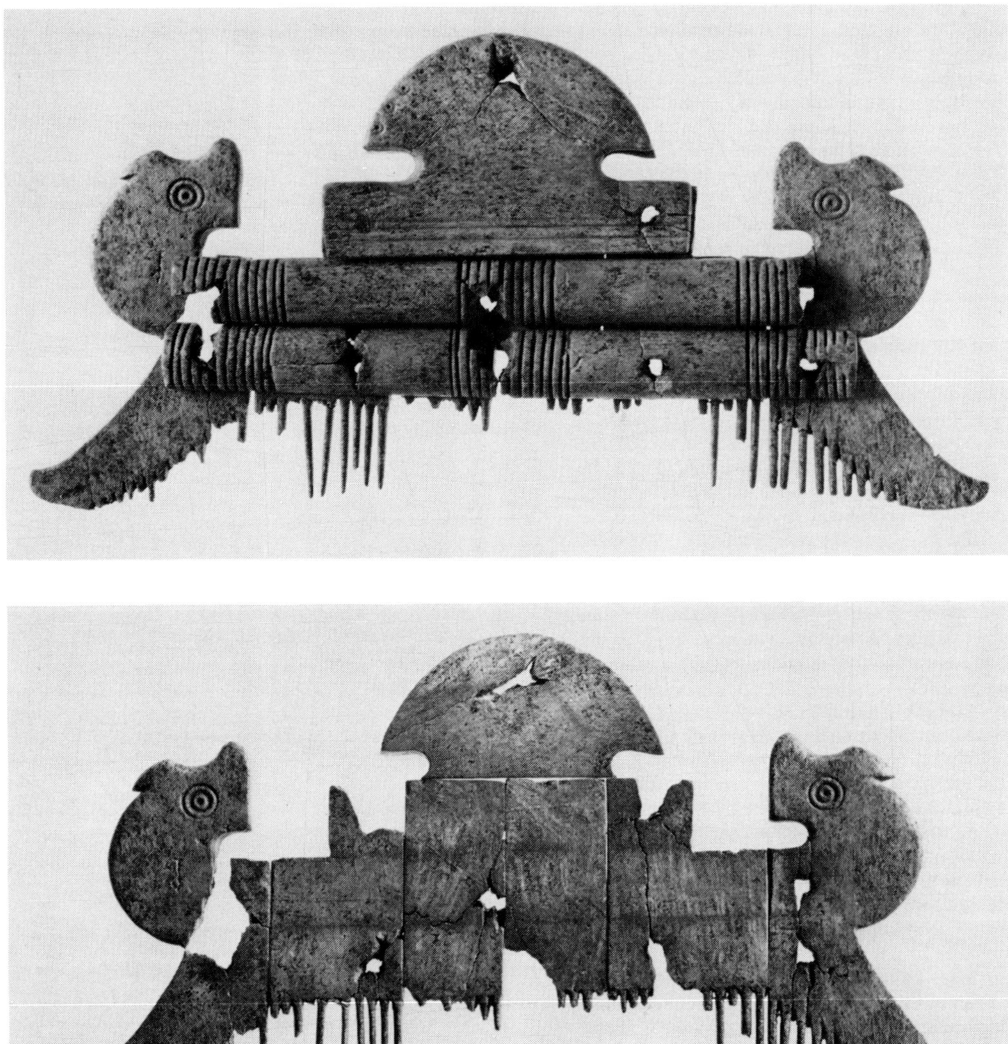

Plate IX Barred Zoomorphic comb, No. 251.

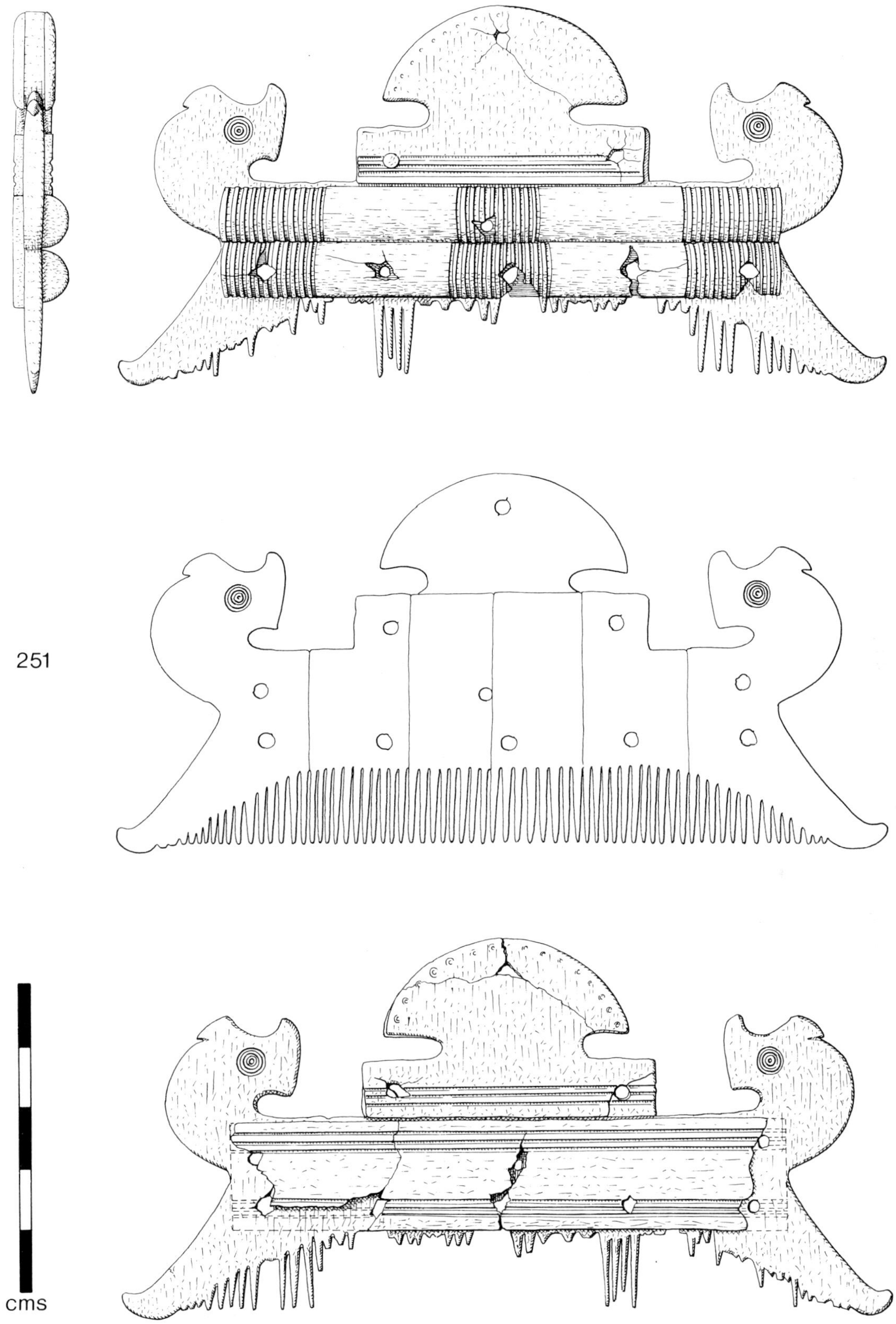

Figure 75 The Barred Zoomorphic comb, No. 251.

The milling face is relatively smooth with slight undulations, probably natural rather than the result of non-rotary grinding. There is a slight suggestion of circular furrows near the rim.

The rim section has a clean square section with a vertical face dressed smooth with faint vertical tooling and the other face of the stone is roughly dressed.

4 (L379) [182] Period 4. Fragment of a top millstone showing part of the rim, eye and rynd cavity. It has two roughly radial fractures, one of them passing through a lifting-eye cavity which passed through the stone to the grinding face. This hole is roughly tapered, narrow and nearest the grinding face, some 20mm to 25mm diameter, with a rounded edge at the top rim. The diameter of the complete stone was *c*.1m, with a slightly curved profiled rim dressed smooth. Thickness at the rim in the fragment is 35mm and at the eye 40mm–45mm. Its eye was approximately 130mm in diameter, rather irregular in shape and possibly enlarged at the grinding face. The most interesting feature of this stone and probably of the whole collection, is the remains of a rynd cavity on the top face of the stone. This is most unusual for, in Roman millstones of the disc type, the rynd cavities are normally on the underside of the top stone. The cavity shows that the diameter of the rynd was 250mm and the blade width not more than 25mm wide. It is not possible to ascertain if this was a two or three blade rynd although the general shape of the cavity shows that it was not a butterfly-shape rynd, but rather a parallel bar or blade type. A section of the emplacement shows that one radial edge is far more pronounced, with a definite shoulder, than the other which rises up towards the main surface producing an indistinct edge to the cavity. The probable reason for this is that the distinct shoulder was the leading edge of the rynd and if this was so, this fragment came from a left-hand pair of stones, *i.e.*, the top, or runner, stone revolved anti-clockwise when viewed from above. The depth of the cavity, a maximum of 13mm in relation to its width, suggests that the rynd was more likely to have been a blade type rather than a bar type, but as it was possible that it projected above the stone top surface, we should avoid drawing a conclusion on this point.

With the rim lying in a horizontal plane — the working plane — the centre of the stone was approximately 50mm above the rim, *i.e.*, the inclination was 6 degrees. On the grinding face there is evidence of crude circular or hoop furrows.

The fragment does not show evidence of furrow or 'peck' type dressing adjacent to the eye. A radial section of the grinding face shows that it is generally concave with a maximum depth of concavity of 4mm to 5mm, roughly in the middle of the face. The surface rises to some 3mm or 4mm at the eye over the last 60mm to 70mm, and there is a pronounced concavity in the last 30mm or 40mm adjacent to the rim. This might have been due to the bedstone being very slightly smaller in diameter than the top stone.

The top surface of the stone is of medium dress and smooth in parts, and the fragment shows no sign of any degradation due to uses other than rotary milling. The stone is of rough texture with gritty inclusions up to 3mm diameter with small incidental quartz flecks.

5 Unidentified. A rim fragment which was 800mm in diameter when complete. No evidence of the rynd or eye cavities exists on the fragment, but assuming an eye diameter of say 105mm, the weight of the complete stone would have been in the order of 42.2kg. Thickness of the rim is approximately 45mm and at the inner radius of the fragment 30–35mm. On the original top surface of the stone is a lead filled cavity some 30mm by 35mm in area with the remains of an iron insert, undoubtedly a lifting eye or hook. This cavity does not penetrate the stone. One of the radial fractures has passed through another, roughly circular cavity, some 25mm in diameter and at the same distance of 125mm from the rim, but this one, which is devoid of lead, passes through the stone to pierce the grinding face. There is no noticeable taper on this hole although its top edge is rounded and it is not undercut.

The outstanding feature of this fragment is the very pronounced circular furrows on the grinding face. It has the remains of nine deep-sectioned furrows some 23mm apart and from 6mm to 10mm and 3mm to 4mm deep. The lands between the furrows are irregular with some parts flattened and on those parts of the face where furrows are absent the dressing is very rough. As expected, the grinding face is concave and a radial section shows that it is more or less flat. The inclination of the grinding face is 7 degrees from the horizontal, which means that the projected grinding face meets the axis of rotation 50mm above the rim plane.

6 (L309) [197] Period 2. This fragment comes from a top millstone. Fortunately, enough of the rim remains to allow a calculation of its diameter — close to 1.06m — to be made, but we do not have evidence of the eye or rynd cavities. The grinding face, which is concave and fairly smooth, has no traces of furrows. A radial section shows that this face is flat. The rim is 70mm thick and has roughly vertical tooling marks on its face. Not enough of the rim remains to allow us to determine the inclination of the grinding face. The top face of the stone is dressed smooth with slight undulations. On one corner of the fragment adjacent to the rim is a small cavity on the grinding face. Its irregular area and shape and maximum depth of only some 4mm suggests that it was caused by a random fracture, probably subsequent to the main edge fracture.

The thickness of the fragment varies between 65mm and 73mm, and is of sufficient size for us to suggest that the whole stone was of more or less constant thickness.

7 (L387) [181] Period 5. A square-shaped fragment from an upper millstone. No evidence of the eye or rynd is apparent although part of the rim is present. There is no recognisable profile for the rim section and, as it is very irregular and well worn, it is not possible to ascertain the diameter of the complete stone. Although the diameter is indeterminable, it was at least 720mm.

On the grinding face there are the remains of some very pronounced circular furrows, almost a V-section. Their average distance apart is some 24mm and depth near to 5mm. The other face of the stone is convex, of fairly good curved symmetry slightly pock-marked towards the middle of the stone. Thickness of the rim is close to 50mm, but towards the eye the original stone must have been at least 70mm–75mm thick. This stone when complete must have weighed at least 52kg.

Discussion and summary

This small collection of millstone fragments is worthy of close study because among the stones are two unusual features. The outstanding feature of the collection is the rynd cavity in the upper face of the top stone, No. 4. I agree with D.F. Mackreth that this stone, and probably the others at this milling centre, was overdriven. This means that the stone was driven by a rynd attached to a vertical spindle projecting above the stone. Animal or manpower would have provided the necessary torque via a horizontal beam. The driving spindle would have been supported in a footstep bearing underneath the bedstone and steadied by a top bearing held to an overhead beam.

Many of the extant Roman disc-type millstones (as distinct from the Pompeii hour-glass type) must have been overdriven for it is unlikely that they were all associated with watermills. As time passes, more examples appear of Roman or Romano-British milling centres where water-power was clearly not used — the situation of the Roman forts on the Limes frontier in the Taunus range of mountains, in particular Saalburg and Zugmantel, are examples. In all of these stones, the rynd cavity is on the underside of the top stone which means that it had to be of sufficient depth to ensure that the rynd ran clear of the grinding surface on the lower stone. As the stones became thinner due to their working, the rynd cavity had to be deepened to ensure that the rynd did not interrupt the grinding action and deface the lower stone. With the rynd under the top stone, the spindle had to be supported and this was normally done by extending it through the eye of the bedstone to rest on a footstep bearing.

With an arrangement whereby the rynd was to view on the top surface of the upper stone, two important observations can be made: 1, the rynd was clearly not used for carrying the weight of the upper millstone; and 2, it was therefore impossible to adjust the gap between the stones by raising and lowering the millstone spindle with the rynd attached.

This would support the idea that the normal working conditions were for the grain and meal to take the weight of the top stone and the rynd was essentially used for transmitting the torque and turning the stone. It is worthwhile recalling that more modern millers, who of course were using very much thicker millstones (up to

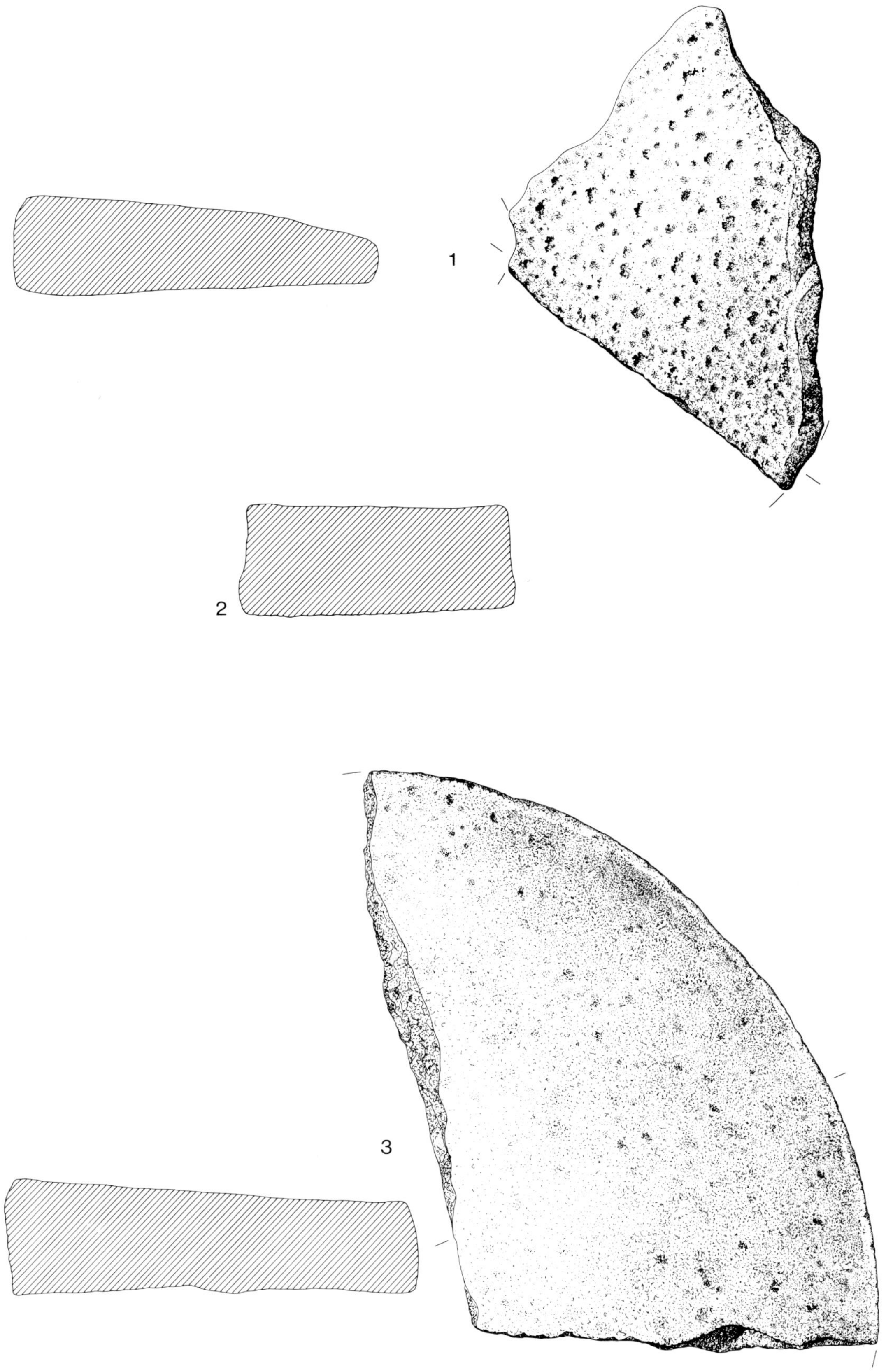

Figure 77 Millstones. Scale 1:5.

250mm thick when new), adjusted the gap between the stones to achieve the desirable standard of product. Obviously some weight was necessary for the proper reduction of the meal. It is a fact that when the water and wind corn-milling industry was in decline at the beginning of the twentieth century and many country millers found themselves in a difficult economic plight, they were, by force of circumstances, having to run their millstones to unusually thin sections. When the top stones became too light in weight they were used as bottom stones and it was not uncommon for them to have to pass the meal through the stones twice. The relevance of this evidence is that there must be a minimum pressure or top-stone thickness for satisfactory milling. To attempt to put a figure on this is impossible and we should remember that modern millstone dressing techniques were greatly superior to Roman techniques and that the standards of product were no doubt much different.

On three of the millstone fragments there is evidence of circular furrows on the grinding face. On two of them these furrows are very deep. All three of these stones are top millstones, and none of the three bottom millstone pieces exhibit this feature on their grinding face. A radial section of these furrows shows them to be more or less V-shaped, roughly the same distance apart and having some peak and trough symmetry, and although they appear to be perfectly circular in plan one must conclude that they were generated by dressing rather than being a product of natural rotary action. If these furrows had been produced by natural wear of the revolving stones, the bottom stones should exhibit the same profile (but mirrored or reversed so that they would fit together), but none of these fragments do. Of course, the sample is small but some significance should be given to the predominance of this feature in the fragments of top stones (three out of four) and its absence in the three bottom stone fragments.

Circular furrows are unusual on Roman or Romano-British millstones, and are in fact rare. Sometimes, one or two circular furrows can be seen on top or bottom millstones near the eye, but such marks are usually not a full circle and the grooves often 'run out' on the surface. These are caused by stones getting jammed in between the grinding faces and gouging the surfaces. The Orton Hall Farm specimens are clearly not in this category.

With true circular furrows there is no 'scissor' action when the stones are rotating, unlike all other types and styles of furrow dressing which are normally repeated on both grinding surfaces. Inclined grinding faces would tend to hasten the movement of meal from the eye of the stone towards the rim, which is undoubtedly the main reason for the feature during the period in question. It might also have some historical significance and be a natural development from those querns which had steeply inclined grinding surfaces, but such speculation on their evolution is beyond the scope of this report.

If both the upper and lower stones had these deep circular furrows, the movement of meal would be greatly impeded; indeed even with an inclined face of the order of 10 degrees from the horizontal it is difficult to imagine any practical through-put would be achieved.

Only one advantage can be tentatively identified for having circular furrows on one face: it would probably increase the pressure on the meal. This might improve the quality of product particularly where the top stone has become thinner consequently providing less pressure on the meal. One disadvantage is that the wear pattern on the lower stone would also be circular; each projecting 'peak' on the upper stone would tend to create a trough on the lower. Naturally all forms of furrow design cause more or less equal wear in both stones, especially when the layout is repeated on both stones, the profile becoming shallower until the peaks in the top stone fit the troughs in the lower stone. When this stage has been reached, we can speculate that two changes will have occurred: 1, the high pressure areas originally present under the peaks resting on the flat bottom stone would have disappeared, replaced by widespread normal pressure areas; and 2, the meal on its helical and outward migration towards the rim would now have to move uphill to overcome each successive furrow.

Allowing that the grinding face was inclined up to 10 degrees from the horizontal on all these stones, the furrow wear would have to be well advanced on the bottom stone before the meal had to travel uphill in absolute terms. If this was allowed to occur, we must conclude that the through-put of the stones would have been reduced.

Of course, any other form of furrow dressing would have no doubt improved greatly the effectiveness of the grinding. However, the use of circular furrows to improve the grinding effect would only have been achieved providing the furrows did not become choked or full of meal. If these furrows became full of meal during working — and it is difficult to see how this could have been avoided in the absence of any self-cleaning or ventilating channels — then the weight of the top stone would be, once again, spread over the whole of the bottom stone and no increase in pressure would be gained. Without the benefit of practical experience of circular furrows, we are not able to sustain such criticisms — they must remain speculative. We ought to give some credit or benefit in our doubts to the operators of these stones and there had to be some advantage to them adopting this style of dress. A reasonable conclusion would seem to be that an increase in pressure could be gained providing the feeding rate was controlled and did not exceed the rate at which the meal naturally migrated towards the rim and cause the furrows to become congested. With these pronounced peaks the area of contact is much reduced and such stones can be described appropriately as 'high pressure' stones. If these profiles were more or less maintained on the top stones when they were periodically redressed, their working life would have been less than that resulting from more normal types of dressing.

Among the finds from Orton Hall Farm were two stones which showed strong evidence of having been a footstep bearing. They are irregular roundish rather than flat stones with a small cavity in one of their faces which may have served to support one of the iron millstone spindles. Such stones would have been needed below each bedstone for the spindle to sit on. One might argue that the spindle could sit on the lower stone, an arrangement which sometimes exists in querns, but the main disadvantage with this is that a considerable point load would exist on the stone which, as the stone wore away and became thinner, could well cause breakage. Luckily, one of the lower stones found had evidence of an eye so that we can be fairly certain that footstep bearings were employed. With the rynd cavities on top of the upper stone, the weight of the rynd, spindle and any levers attached, had to be carried by either the lower bearing stone or the top stone. When initially setting up the stones, the rynd has to sit in

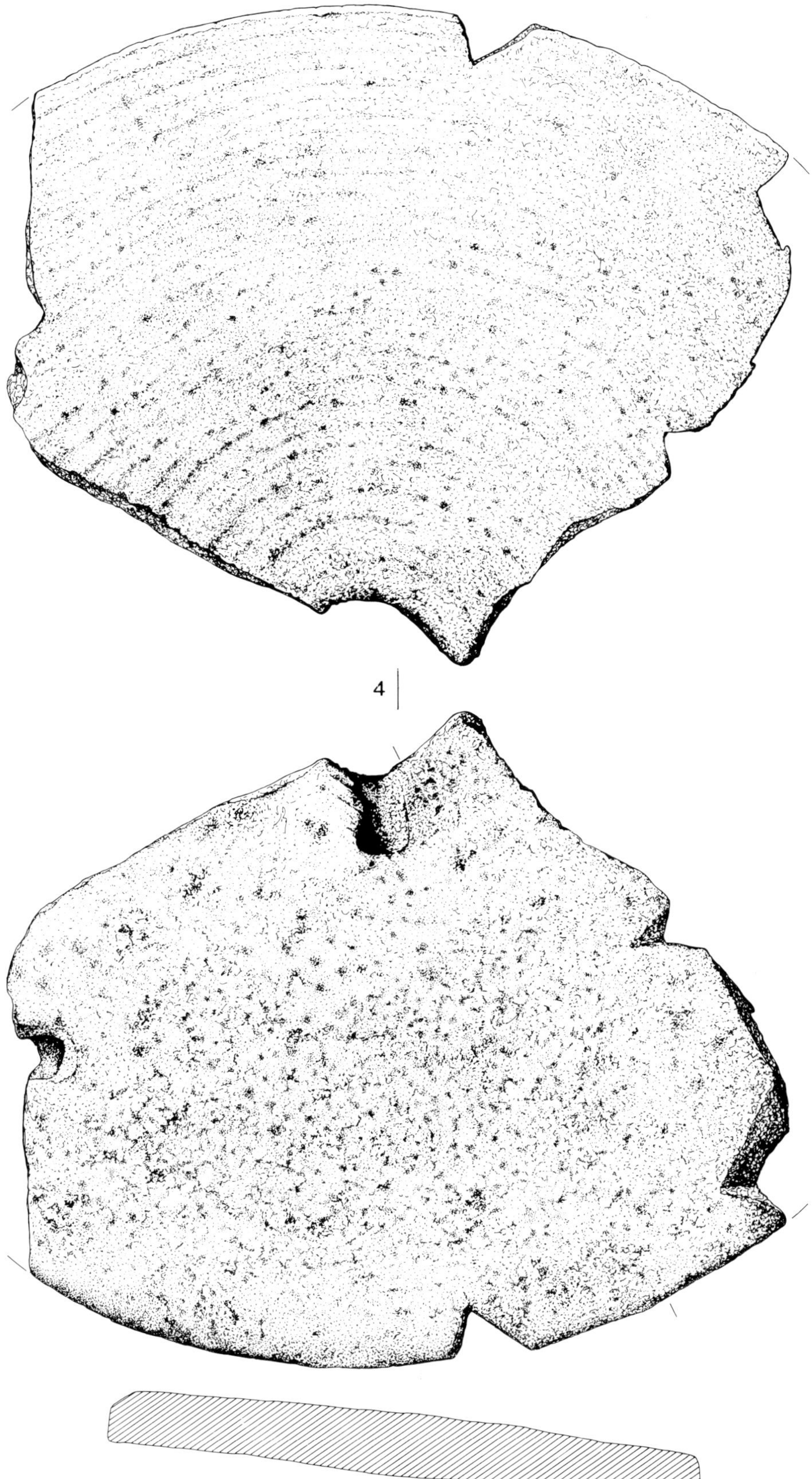

4

Figure 78 Millstones. Scale 1:5.

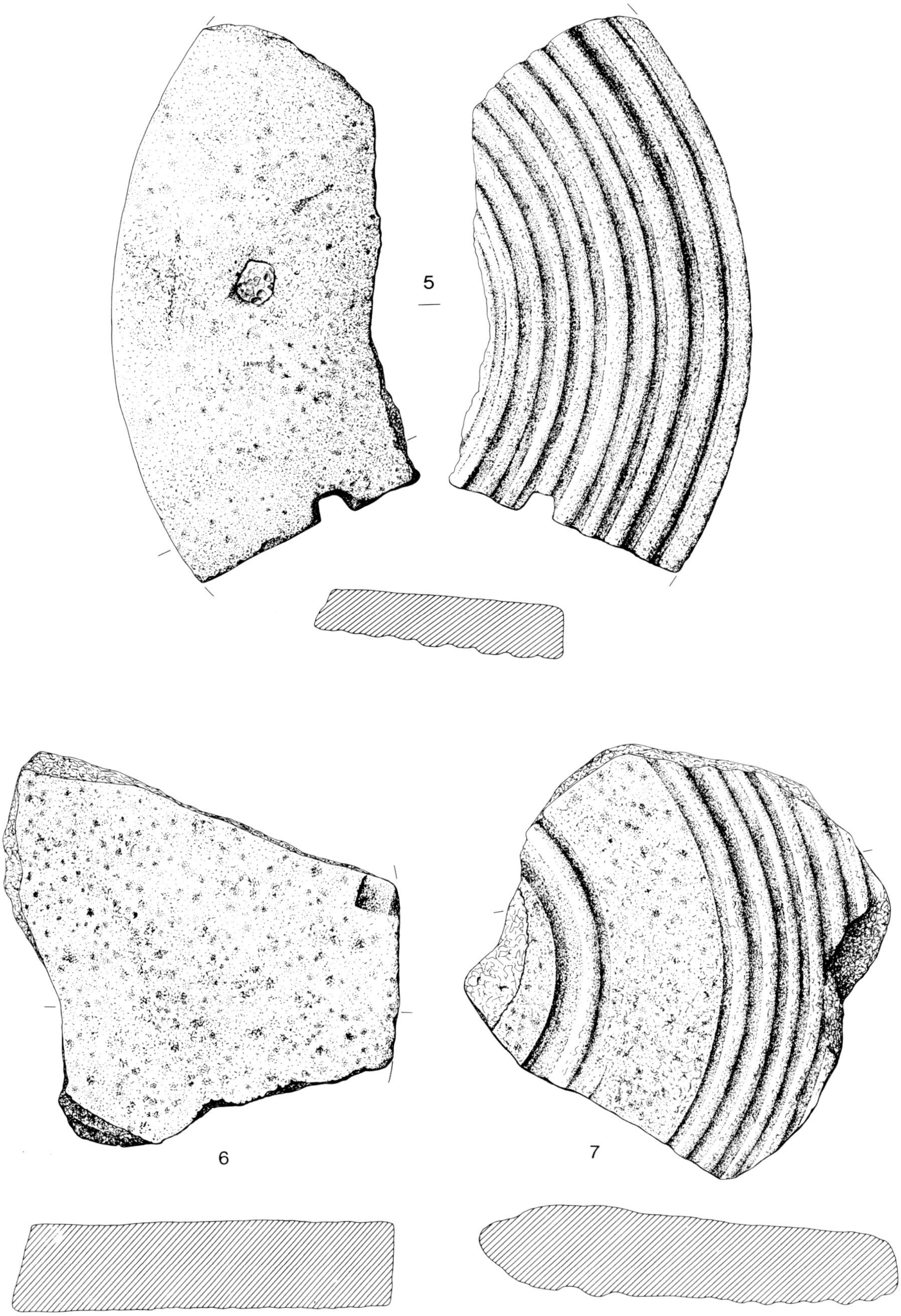

Figure 79 Millstones. Scale 1:5.

its cavity in order to transmit the torque and drive the stone, and its lower end has to be constrained so as to stop it wobbling while rotating. Even if some of the weight is initially taken by the bearing stone, it would quickly wear away under the considerable point load, so that for most of its working life it would be steady and provide stability for the millstone spindle.

Once the bearing stone had been positioned, the hole in which the spindle sat would deepen by wear only by the amount that the rynd lowered as the stones wore away. If the operators were in the habit of redressing the grinding faces periodically, and there is considerable doubt on this point, then they would probably re-adjust the bearing stone or, alternatively though less likely, the lower stone, otherwise the rynd would not sit in its emplacement and drive the stone properly.

With the spindle passing down through the bedstone it would have been necessary to close and seal the eye around the spindle so that the grain and meal did not fall through. This could have been achieved by a wooden collar which may also have acted as a steady, possibly with an iron sleeve to reduce wear.

Whether or not such stones could be readily adjusted for height without having to remove the millstones is difficult to decide. To have been able to do this, a horizontal cavity or slot would have been required across the top of the base on which the lower millstone sat.

In speculating whether or not a top bearing or 'steady' existed on the millstone spindle, the stability of the rotating elements must first be examined. Clearly, if there existed any instability likely to cause displacement or undue vibration of the top millstone during working, then the top of the millstone spindle would have needed restraining. To a limited extent disc millstones with an inclined grinding face were self-centring when rotating but probably not sufficiently stable to rule out displacement of one or the other. One might also argue that circular furrows gave some rotational stability, especially when they were deep in section and occurred on both top and bottom grinding faces. However, with the torque being applied direct to the spindle by one or more men or animals without gearing being employed, it is considered that a top bearing would be necessary.

Such a bearing had to be supported from an overhead beam and structure independent of the rotating levers, and to avoid a long iron spindle it is quite possible that a wooden shaft with iron inserts at each end was used.

One final part of the arrangement remains to be examined: was the feeding hopper supported by the rotating beams or was it static and carried by an independent structure? Whichever, it had to be positioned off-centre and beside the vertical shaft. Probably the most convenient arrangement was for the hopper to be carried by the levers, although the layout of this structure and the configuration of the levers is open to debate.

Other evidence from the site lends support to the idea that the millstones were overdriven. In one of the buildings the remains of three stone bases were found, made up of selected stones carefully arranged to provide a plinth of 0.75m by 0.6m. They stood at least two courses high. These were clearly not post-bases for they were quite different from those found in the barns of the site and we must therefore assume that they had some other agricultural/ industrial function. The most likely use for such bases is to support pairs of millstones. It is interesting to note the position of these bases. Examination of the three bases shows that their centres were 1.5m from the internal face of the south wall of the building and 3.05m and 3.25m between each other. The relationship between these dimensions may not be coincidental; as millstone bases they would need to be twice the distance between each other as they were from the wall. This would be necessary to allow adjacent millstones to be worked simultaneously. If this projected arrangement is correct, and there is a strong possibility that it is, these millstones were powered by men and not animals. Allowing for millstones of, say, 1m diameter, the clearance around them provided by the position of the bases is 1m. This is not enough for an animal, but sufficient for people walking around.

Provisional calculation, assuming:

1 Weights for a comparatively new top millstone, iron rynd, spindle and wooden levers at 200kg.
2 A radius of effort of 1.1m.
3 A speed of 7rpm.
4 A co-efficient of friction of 0.5.

Then this suggests that the necessary torque could be maintained by two people walking around. In the physical limitation of this layout, up to four people could have walked around each pair of millstones and the obvious lever arrangement for this would have been a cross configuration forming radial arms.

Endnote

1. We are most grateful to Dr John Hines for his help and advice during the preparation of this section of the report.

Chapter 6. The Roman Pottery

by J.R. Perrin, with contributions by F. Wild and K.F. Hartley

I. Introduction

In all, 1,173 features and 2,975 layers were identified and excavated either fully or in part. Those which could be assigned to the five main periods and their phases contained over 32,000 sherds of pottery weighing almost 560kg, with a combined rim length amounting to the equivalent of some 426 vessels. A possible 3,777 different vessels were noted by catalogue entries (see below). A further 12,000 sherds weighing more than 183kg belong to those layers and features which could not be assigned or were surface clearances, and in the interests of speed and economy of time, no rim percentages were noted for this pottery.

The dating used is based on evidence from both published and unpublished local material. Key sites were the various excavated kilns: Water Newton (WN), Chesterton, Sibson, Stibbington (ST) and Stanground, and pottery groups from Normangate Field and Chesterton. The latter was of especial importance for the occupation there spanned the second to fourth centuries and a number of independently dated sequences in each of the centuries was identified.

A note of caution is needed: reference to a particular fabric or form at a given kiln site should not be taken to mean that either or both were exclusive to that site. Only a small number of kilns has been excavated and there must be many more awaiting discovery, apart from those destroyed with little or no note. The 'local' evidence was supplemented by that from well-stratified and published sites such as Great Casterton, Verulamium, Godmanchester, various Fen and Fen-edge settlements, and others in Northamptonshire. The proportions of the various wares in each significant deposit were monitored closely, and were very often used to adjust dates already given.

Research into the full development and chronology of the sizeable Lower Nene Valley industry is still in its infancy. A lot is known of the more widely exported types of vessel and ware, especially to independently well-dated settlements and zones such as Verulamium and Hadrian's Wall, but the dating of these is often based on assumptions which may not be correct (see below). Even so, the evidence of these non-local sites is of paramount importance. The real problems are how the industry operated in its local setting and the possibility of typological and chronological variation from site to site.

The local 'utilitarian' grey ware and cream ware production, apart from mortaria, is only of local significance and Orton Hall Farm was the first major excavation on which the framework formed in considering sites such as Chesterton could be tested. But it was not the type of site which allowed typologies and chronologies to be refined. Although by giving broad date-ranges there is less likelihood of major errors, the precise effects that the nature of the site itself had on pottery supply and use cannot be fully appreciated until there is a greater body of comparative information. Some attempt is made to appraise these factors in Sections VI and VII. The uncertainties concerning the beginnings of the industry, the range of its early products, the varieties of wares and vessels, the decline of the grey ware production, and the final phases are of importance.

II. The layout of the report

The pottery is considered by period. This allows it to be more easily related to the archaeological and interpretative discussion but, equally important, it provides a sequence that can be followed in relation to the site itself rather than as a type-series for the industry as a whole, which it is not meant to be, nor could be. Each period discussion is based on the assemblages which can be dated reasonably well from contemporary features and layers. Little attempt has been made to omit residual pottery from many of these groups: firstly, it was not always obvious what may have been truly residual vessels (see Section VIII), and secondly, it was thought important to show the range of an assemblage. These main deposits are supplemented by others having pottery not represented in the main groups, additional examples of particular types, or because the feature itself was of interest in that period or phase.

An 'average percentage', **A%**, has been used experimentally when discussing the pottery, to try to minimise the inherent and unavoidable variations when using just one or even two of the available quantification methods. The A% (see Table 5) combines the percentages related to number, weight and rim length to give one figure. It is not thought that this is any more valid than the others, but it is certainly not less so, and allows easier comparison.

All vessels represented by rims or large rimless profiles were also classified according to form. Generally, no attempt was made, unless it was obvious, to identify the form of smaller sherds. For example, even if in very small sherds, beakers are easy to recognise. Others could also be identified by virtue of their highly individual fabrics in which only a few types were made: *e.g.* Horningsea ware jars. A more painstaking examination of the pottery would undoubtedly have revealed more of these and additional time spent on seeking joining sherds would have reduced the number of vessels. However, constraints on time, funding and personnel limited the work to basic card entries and the figures noted in the discussion must be viewed as the results of a reasonably thorough examination proportionate to the resources available.

The numbers of each form can be presented and compared in two ways. First there is the simple number of entries as outlined above, and second there are the figures for total circumference given as an estimated vessel equivalent (EVE). The latter might be considered as providing a 'minimum' number, while the former could represent a 'maximum' if it is thought that the number of unrecognised joins would be balanced by the unrecorded body sherds. Certainly the difference in the figures for the stratified pottery — 3,777 entries compared with the equivalent of 426 complete rims — is best explained in these terms. If only about 15 per cent of the site was excavated the vessel numbers can be increased to 25,180 and 3,240. Considering that the site was occupied for approximately 400 years, this gives an average annual pottery turnover of either 63 or 8 vessels, the former being the more likely. One of the shortcomings in the use of EVEs as a quantification method is the possible underestimate of the number of vessels, therefore, the discussion on numbers of different forms uses the figures provided by entries.

The numbers of sherds, weight and amount of rim, and their percentages, for the various sets of figures in the text, are given in microfiche (MF11).

III. Introduction to the catalogue

Presented in periods, and mostly by feature within each period, there are some entries of important or interesting vessels included under fabric alone. Where the feature contained sufficient pottery, this is arranged in wares which means that some of the corresponding pottery figures are somewhat mixed in character with different wares, vessels and dates in close proximity. This is an unavoidable consequence of the decision to deal with the pottery from the viewpoint of the site itself. The catalogue information is restricted to colour and find spot, occasionally with additional information thought to be significant.

Fabric and surface colours were matched with those of the 'Munsell Soil Color Charts' (1971 edition), and are given in a 'range' form because the colour of any vessels' fabric and surfaces were rarely uniform, there often being considerable variation. It was felt that too much concentration on the appearance of any one vessel should be avoided. The potter would have aimed at an overall colour range and used a variety of base clays, slips and firing techniques to achieve it. He would not have been too concerned about variation within this. The colours are given in numerical form prefixed by the letters CR and are listed below.

Colour Ranges

1	weak red	10R4/4, 2.5YR4/2.2
2	red	10R4/6, 2.5YR5/6, 2.5YR5/8
3	light red	10R6/6, 2.5YR6/6, 2.5YR6/8
4	reddish yellow	5YR5/6, 5YR5/8, 5YR6/6, 5YR6/8, 5YR7/6,
	yellowish red	5YR7/8, 7.5YR6/8, 7.5YR7/6, 7.5YR7/8, 7.5YR8/6
5	pink	5YR7/4, 5YR8/4, 7.5YR7/4, 7.5YR8/4
6	pinkish white	5YR8/2, 7.5YR8/2
7	pinkish grey	5YR6/2, 5YR7/2, 7.5YR6/2, 7.5YR7/2
8	reddish brown	2.5YR4/4, 2.5YR5/4, 5YR4/3, 5YR4/4, 5YR5/3, 5YR5/4
9	light reddish brown	2.5YR6/4, 5YR6/4
10	light brown	7.5YR6/4
11	brown	2.5YR5/2, 7.5YR5/2, 7.5YR5/3, 7.5YR5/4, 10YR5/3
12	dark brown	7.5YR3/2, 10YR4/3
13	dark reddish brown	5YR2.5/2, 5YR3/1
14	very pale brown	10YR7/3, 10YR7/4, 10YR8/3, 10YR8/4
15	light brownish grey	2.5Y6/2, 10YR6/2
16	greyish brown	10YR5/2, 2.5Y5/2
17	dark/very dark greyish brown	10YR3/2, 10YR4/2, 2.5Y3/2, 2.5Y4/2
18	light grey	5YR7/1, 7.5YR7/0, 10YR7/1, 10YR7/2, 2.5Y7/2, N7
19	light grey/grey	5YR6/1, 10YR6/1, N6, 5Y6/1
20	grey	10YR5/1, 5Y5/1, N5
21	dark grey	10YR4/1, 5Y4/1, N4
22	very dark grey	5YR3/1, 10YR3/1
23	light olive grey	5Y6/2
24	light blue grey	5B7/1
25	dark greenish grey	5BG4/1
26	white	5YR8/1, 7.5YR7/0, 10YR8/1, 10YR8/2, 2.5Y8/2
27	yellow	10YR7/6, 10YR8/6
28	black/ reddish black	10R2.5/1, 2.5YRN2.5/0, 5YR2.5/1, 7.5YRN2.5/0, 10YR2.5/1, 2.5YN2.5/0, 5Y2.5/1, 5Y2.5/2

The following points should be borne in mind:

1. When a number of different colours occur arbitrarily on the same pot, the ranges are linked with a + (*e.g.*, 18+24).
2. If the core varies from the surfaces the colours are separated by a / (*e.g.*, 18/14/18). In these cases, the internal surface colour is given first, then the core, then the external surface. The colours of the surfaces can vary (*e.g.*, 18/14/22).
3. Where there is a 'sandwich' core, the colours of this are separated from the surfaces by a // (*e.g.*, 21//11/19/11//21).
4. When the colour of a pot is uniform except for just one surface, a hyphen, -, is used (*e.g.*, 3–14). Sooting is not treated as a different colour.
5. If the surface of the pot has different coloured zones or areas, these are distinguished by a comma followed by text (*e.g.*, 18/5/18,21 patches).
6. Colour ranges are not given for non-local wares which are adequately described elsewhere.
7. The find spot is feature, F, followed by layer, L. If a layer did not have a separate feature it is termed unfeatured, UF. Some of these areas are named (*e.g.* Yard). A layer not assigned to a feature or area is said to be unassociated, UA. A number of vessels came from hand-stripping, HS, machine-stripping, MS, or 1971 excavation layers, 1971. The fabric abbreviations are given near the beginning of Section IV. Levels of uncertainty are indicated by queries: ? means probably; ?? means not certain; ??? means probably not.

Concordances of draw-sherds to layers and features, and of final report to initial selection are given in MF11.

IV. The wares and fabrics

The Nene Valley Research Committee Roman pottery fabric series and the computer programs and categories used to record the Orton Hall Farm Roman pottery on card and computer disk are described in detail on microfiche (M11).

Throughout, the pottery is designated by commonly accepted names rather than numbers. A full list of the various fabrics and the abbreviations used is given below. Where a particular fabric is well known and has been fully described elsewhere, reference is made to the appropriate source. The main local fabrics — those of the Lower Nene Valley industry, the shell-gritted wares, the 'belgic' wares and some of the grey wares, including 'London ware' type — are described in greater detail. Some aspects and problems relating to the origins and development of the major local industry are also discussed. Where the forms of a particular ware are more significant than the fabric, for example with shell-gritted wares, the discussion is biased towards the form.

The wares recorded

Lower Nene Valley grey ware	(LNVGW)
Lower Nene Valley colour-coated ware	(LNVCC)
Lower Nene Valley cream and white wares	(LNVC/W)
Lower Nene Valley, Stanground	(LNVS)
Late Iron Age shell-gritted ware	(LIASG)
Late Iron Age to early Roman, transitional shell-gritted ware	(TSG)
Roman shell-gritted ware	(RSG)
Roman shell-gritted ware. Bourne/Greetham (Bolton 1968)	
Grey wares	
'Belgic' wares	
Grog-tempered ware	
'London ware' type	
Central Gaulish colour-coated ware (Greene 1978)	(CGCC)
Lower Rhineland colour-coated ware (Anderson 1980; Anderson and Anderson 1981)	(LRCC)
Lower Rhineland white/cream ware (mortaria) (Haupt 1984; Richardson 1986)	
Colchester colour-coated ware (Hull 1963b)	(CCC)
Oxfordshire colour-coated ware (inc. mortaria) (Young 1977)	(OXCC)
Oxfordshire white/cream ware (mortaria) (Young 1977)	(OXW)
Mancetter-Hartshill ware (mortaria) (Hartley 1961; 1971; 1973)	
Verulamium region ware (inc. mortaria) (Frere 1972; 1983; 1984)	(VR)
Hadham region ware (Going 1987)	
Trent Valley ware (Todd 1968a)	
Horningsea ware (Hughes 1902; Walker 1912)	
Black-burnished ware, Category 1 (Gillam 1976; Farrar 1973; Williams 1977)	(BB1)
Black-burnished ware, Category 2 (Farrar 1973; Williams 1977)	(BB2)
'Black-burnished ware' type	

Local and probable local wares

1 Lower Nene Valley: the fabrics

At the time of writing, there seems to have been two main clay sources. One was used primarily by the potters working near Durobrivae, perhaps also at Great Casterton, and the other by those at Stanground, and possibly other sites such as Stilton and Yaxley.

1a Sibson, Stibbington, Water Newton *et al.*

Current evidence suggests that potters working in these centres always used clays of the Upper Estuarine series.[1] The fabric is usually hard and well-fired with a clean fracture which is occasionally laminated. It can contain voids, but is usually dense. The most noticeable inclusions are red, orange or black haematite which can be rounded, sub-angular or angular, and up to 2mm in size. These occur sparsely but can be more common. The haematite occasionally seems to stain the surrounding clay and sometimes appears as thin streaks up to 4mm long running parallel to the vessel wall. Use of a microscope reveals abundant minute grains of quartz, usually translucent and opaque, but occasionally pink in colour and sometimes as large as 1mm. Small amounts of fine mica and some white lumps are also present. These lumps vary in size between 0.5 and 4mm and can be rounded or sub-angular and do not react with hydrochloric acid. They are more noticeable when the basic fabric colour is other than white. Fabric texture is also variable and does not necessarily reflect the amount and size of the inclusions. Levigation to different degrees was obviously an important part of the clay preparation.

The clay was prepared and fired differently to produce grey, fumed, white or cream wares, and slipped to give numerous varieties of colour-coated ware. The iron-free clay gave white or cream ware in both reducing and oxidising kilns. The clays with more natural or added iron fired to varieties of pink to orange, and grey under oxidisation and reduction respectively. Most of the slips or colour-coats used appear to have been iron-rich. Vessels coated with slips rich in mica are also known.

1b Stanground

The potters working here seem to have used two different types of clay.[2] Most pottery was made from calcareous Oxford Clays from the underlying geology of the Stanground area. These contain sub-angular and slightly rounded quartz grains, together with minute quartz crystals, numerous small and large fragments of fossil shell and some form of gypsum. Voids are common. The colour of the fired fabric is usually a drab grey, but can be orange, and grey/orange 'sandwich' sections occur. Many of the vessels were slipped but the quality is generally poor and it is clear that the potters had difficulty in achieving the desired effect (Dannell 1973). Most of the pottery made from Oxford Clays were fired grey, though orange or red examples do occur. These probably represent mis-firings.

A relatively small proportion of the Stanground pottery was made from a clay similar to fabric 1a and seemingly reserved for vessels such as beakers and flagons, as well as some unusual forms. Not all pots in this clay found at the Stanground kilns may have been made there, some possibly being imported for use on the associated rural settlement. Research has shown that some mixing of the two clays occurred,[3] strengthening the belief that supplies of better potting clay were brought down river to Stanground.

The Dannell (1973) article attributed the colour-coated production to the potter who stamped some of the vessels. The stamp was read as 'INDIXIVIXUS' although, as an essentially illiterate stamp comprising vertical and diagonal lines, this was just one of a number of possible interpretations (Mossop 1960). Following the publication, similar vessels with or without a stamp, and especially those with grey colour-coated surfaces, have unfortunately been linked to Stanground (*e.g.* Jackson and Ambrose 1978, figs 45 and 50, nos 87 and 170–1), even when there is little or no evidence for this. The third-century date suggested by Dannell in 1973 is still thought to be valid (Dannell *et al.* 1993, 89–91).

The character of Stanground grey and colour-coated wares was not fully understood when the Orton Hall Farm pottery was first processed, and these were not classified separately. Unfortunately, limitations of time, finance and personnel prevented any re-examination of the pottery, therefore accurate figures for the amounts of Stanground pottery used on the site cannot be given. Individual vessels of definite or possible Stanground origin are noted in the various period discussions.

1c Other possible centres

There is now some evidence for pottery production south of the river Nene in the Stilton/Yaxley vicinity, in the form of fire-bars and kiln-dome material (Swan 1984, MF2.386).[4] Until kilns as such are located and excavated, their products cannot be identified. It is likely that the

Orton Hall Farm site, no great distance away, would have used pottery from these possible kilns.

The Lower Nene Valley industry: origins and demise

The earliest known Roman pottery production in the Lower Nene Valley is that associated with the Longthorpe fortress (Dannell and Wild 1987) and the shell-gritted and Gallo-Belgic type vessels apparently fired in kilns near Water Newton (Hartley 1960). Neither of these can be considered as part of the major industry for which the region is best known, and it is not certain when its first kilns started production, or where they were sited. Consequently their vessel and ware ranges are uncertain. The best guide available is probably the pottery found in three pits, one at Chesterton (Perrin, to be published) and two at Normangate Field, Castor (Perrin and Webster 1990), each dated by samian ware to the second quarter of the second century.

This included Roman shell-gritted (RSG), suggesting that there was some continuity in the production of this variety locally and, more significantly, grey and cream wares which, in basic appearance and fabric, were similar to those of the more well-known later local potteries. The precise date of the deposition of these groups cannot be ascertained: they might as easily be of *c.* AD 150 as *c.* AD 120. However, the samian ware and associated pottery like BB1 indicated a date around AD 130–140 rather than earlier or later. The first kilns of the Lower Nene Valley industry proper were probably in operation during the reign of Hadrian. It is tempting to link the start of the industry with other events in the area. The 'opening-up' of the Fens is attributed to the time of Hadrian (Salway 1970, 7–9) and, as a result, it is likely that what was to be the nucleus of the later walled town of Durobrivae received a major boost. As far as the Fens are concerned, Lower Nene Valley products seem to predominate early on (Hartley and Hartley 1970). Such interconnected events could have provided the stimulus for the development of the major local pottery industry.

All the kilns excavated in modern times are of late second, third or fourth-century date. Some found by Artis in the early nineteenth century might have been earlier but no evidence survives (Swan 1984, 95–7). The most likely location for the earliest kilns would seem to be the area adjacent to the fort at Water Newton and the later walled town of Durobrivae and its suburbs. The remains of a kiln at Sulehay, though the associated pottery was of mixed date, might indicate another production centre, and second-century occupation is certainly attested in the vicinity (Hadman and Upex 1975). A second-century kiln at Great Casterton (Corder 1961, 50–3) was perhaps another outlier. By the end of the second century the main kiln sites had become established where they were to be for the rest of the Roman period. Possible production at Yaxley and Stilton would most probably represent attempts to take advantage of the ease of distribution to markets in the Fens and along the line of the major Roman north-to-south road. Those at Stanground could have been connected with reorganisations of supplies within the Fens at the end of the second or early in the third century.

Although most of the early pit-group material already mentioned is similar in fabric to that of the later production, some is slightly different. This could be because the clay used in the earliest period was from another source, or prepared differently, or because some of the pottery was not made in the 'local' area. Sites such as Sulehay and Great Casterton show that it is not easy to define the geographical limits of the early 'Lower Nene Valley' potteries. There is considerable variety in the clays within this larger area but it is unlikely that these would appear sufficiently different under the kind of analysis presently available for separate production centres to be isolated. Once it is accepted that some of the grey and cream ware pottery in early second-century deposits in the Lower Nene Valley could have been made outside the 'local' area, reliance has to be placed on intangibles such as probability and expectation when attempting to assign particular vessels to a possible source. In any case, in attempting to identify the earliest products, the earliest pit groups may have been biased in not having contained a comprehensive range of what was being manufactured at the time.

Many of the vessels had broadly similar characteristics. The grey ware jars were generally fairly squat in form and had simple everted rims. Most vessels of all classes had carinated or angular profiles and were decorated with cordons and burnished vertical lines or latticing. One of the most easily recognisable types was the jar or beaker with one or more raised cordons around the shoulder or girth, itself often with diagonal incised cuts on it (*e.g.*, Fig. 87, No. 193; Guide, fig. 1, nos 1 and 2). The cream ware included flagons with ring-necks (Fig. 81, No. 38) and one akin to Hofheim types, as well as flat-topped and hemispherical flanged bowls, various jars with a number of rim forms including beaded and lid-seated, and mortaria. There was one lid.

The pits contained few roughcast colour-coated beakers and all apparently of Lower Rhineland origin. Some of the other apparently colour-coated wares may have been misfired grey wares (see below), but finds from other local sites, including Orton Hall Farm (*e.g.*, Fig. 83, No. 83), together with evidence provided by the Great Casterton kiln (Corder 1961, 50–3) suggest that colour-coated wares were also part of the early repertoire. The beakers from the early pit groups show, however, that imports were available at the time and secure identification of source can be difficult especially when dealing with small sherds. Use and soil conditions can also mask the differences in the slip quality and density of the products of the various centres, which might be more obvious on less abraded fragments. The decorative motifs used by the potters of each region are not likely to help either, as there could have been considerable imitation, especially if potters migrated from one area to another. Considerable care has clearly to be taken before particular sherds can be confidently assigned to one source or another. Little progress will obviously be made on any aspect of the early industry without further excavation and research, especially on kiln sites and fairly high-class settlements.

Once established, the potteries quickly grew into a large-scale industry providing the large population in and around the Fens with the bulk of their requirements. Four main types of ware were manufactured: shell-gritted, grey, white/cream, and colour-coated. Certain vessels made in the oxidised version of the latter were marketed throughout Roman Britain, and the trade enabled more utilitarian products, such as mortaria, also to be traded afield at various times.

The range of vessel types was widest in the second century; thereafter there was a tendency to standardise until the later phases of the industry when some new types appeared. Most of the pottery classes underwent some change during their production life. Not all, however, are represented at Orton Hall Farm. The most noticeable change in the later industry was the great reduction in, even cessation of, the making of grey-coloured pottery. Why this occurred is uncertain, but as the transition seems to have been in the later third or early fourth centuries, a response to provincial economic and administrative reorganisation cannot be ruled out, though demand and competition are perhaps more likely.

The similar and limited largely fourth-century range of vessels produced in the Lower Nene Valley, Oxfordshire, the New Forest and Swanpool near Lincoln, probably reflects increased competition for markets, and the need to minimise overheads. Though the local industry was still active, it no longer had the complete monopoly even locally, as the occurrence of Oxfordshire, Hadham area and Trent Valley type products at rural sites like Orton Hall Farm demonstrates. Finds throughout the province at this time indicate that there had been no significant decrease in the traditional market areas to which Lower Nene Valley pottery had been traded, though the amounts concerned might have been less than in the third century.

The demise of the industry is something which will probably never be fully understood. The pottery from the latest phases of any site will generally be very mixed and so full of residual material as to make certain identification of the latest types almost impossible. The late deposits from the Great Casterton villa (Gillam 1951; Corder 1961) and the well at Stibbington (Perrin 1981b, 448) indicate that production probably continued into the fifth century and, for the present at least, the pottery from these is the only comprehensive guide to that used by the area's inhabitants at the end of Romanised Britain.

Colour-coat or slip?

Although LNVGW and LNVCC are grouped and discussed separately, they are not completely different and, in fact, overlap in one important and significant way. The accepted definition of a colour-coat is:

> Pottery which has been dipped into a slip rich in iron compounds; the colour of the slip varies but is usually darker than the paste, and occasionally the surface has a metallic lustre (Webster 1976, 13).

In theory, therefore, this covers all the possible colours which could result from beige or light brown through to dark red or dark grey but, in practice, there has been a tendency to equate the term mainly with those vessels with orange, brown or red-fired slips. This has led, in turn, to the belief that no LNVGW was slipped when, in fact, a large amount was. The use of slip on some of the grey wares of other centres, for example Crambeck in East Yorkshire, has not been fully appreciated. This erroneous use of the term 'colour-coat' is not that serious in itself but it can cause problems when linked to date and typological development.

Current evidence suggests that certain colour-coated forms, primarily 'fine' wares such as beakers, flagons and 'boxes', were not part of the early Lower Nene Valley industry's product range, but were added subsequently. The practice of slipping or colour-coating was in use from the beginning, however, and it is important to stress that it is only forms that vary in date, not the process. The only difference between 'grey' and 'colour' slip is that the 'fine' wares tended to be fired in oxidising conditions. This distinction can also be false in practice, however, for bad firings or subsequent use can occasionally oxidise reduced slips and vice-versa, a fact confirmed by recent experiments.[5]

A further complication occurs in the later industry. There is no doubt that grey ware as such was no longer wanted. In fact, as far as slipped vessels were concerned, they were no longer fired in reducing conditions. The danger here is that, for example, a wide-mouthed jar or bowl with a red colour-coat (*e.g.*, Fig. 94, No. 352) will automatically be dated later than its grey equivalent, and this does not allow for variations due to misfirings or usage. Although, therefore, most of the late, fourth-century, colour-coated wares have surfaces in the oxidised range of colours, grey examples can still occur. Conversely not all the earlier examples of these types need be grey, and the few red colour-coated jars from the third-century Stanground kilns[6] provide a clear warning.

This point is also relevant to the production of imitation samian vessels. In common with other industries such as the New Forest, Oxfordshire and Swanpool, the Lower Nene Valley potters produced a range of colour-coated imitation samian types in the later third and fourth centuries. In the Lower Nene Valley, however, unlike in these other centres, colour-coated imitations of samian ware had been made from, probably, soon after the start of the industry and, at least, from the end of the second century. There is, therefore, a potential problem in dating specific vessels, and stratification and association are of especial importance.

Once it is accepted that LNVGW, and other grey wares, could be slipped, the remaining difficulty concerns the positive identification of a slip on a particular vessel, as this is not always obvious. Where the basic clay was iron-free, grey surfaces could also be achieved by mixing in iron-rich clay before firing, or using a process which is today variously described as fuming, sooting, smoking or smudging. This involved adding 'green' material to the fire of the kiln, usually towards the end of the firing, thereby producing smoke which penetrated the pores of the clay turning the outside of the vessels grey (Shepard 1974, 88). The appearance of the vessel surfaces depended on the quality of the smoking process. At best, this achieved an all-over even grey colour, similar in appearance to vessels that had been slipped. It could, however, be rather patchy, of different shades, or 'speckled'.

This is not as critical as the recognition of a slip itself, as it does not have the same dating and typological connotations. It is not clear why the fuming process was used, although, as it did not require a reducing kiln, or additional separate stages of clay mixing or slip preparation, it may have been more economical in terms of manpower, supervision, maintenance, and materials. It might equally, however, have simply been a method passed down to some of the potters. Some experiments on fumed medieval Dutch pottery[7] have shown that the process improves impermeability, and it might, therefore, be linked to the storage and handling of liquids or wet goods.

On the other hand, in the Lower Nene Valley, and at centres like Crambeck, many of the high-quality potting clays were basically iron-free and would not fire to colours other than white or buff without additional preparation or treatment. Coating the surfaces with an iron-rich slip, and firing in reducing conditions was another of the methods enabling grey ware to be produced.

In the Roman period generally, the popularity of red samian ware was a sufficient incentive for the potters of other industries to use slips to provide their imitation forms with the requisite red surface which otherwise would have merely been the colour to which the local clay fired. Even if the local clays could give red pottery, a slip was often used to give the vessels a more authentic 'samian ware-type' surface. Some of the dark grey and black surfaced pots might have been attempts at the imitation of pewter, and some of those with mica-rich slips were obviously designed to give vessels the appearance of metal. However, the bulk of the colour-coated vessels were not such imitations. The main reason for the use of the technique could have been connected, as perhaps with the fuming process, to permeability (see above, and Shepard 1974, 191). The greater range of final vessel colour that could be achieved and the smoother and more even surfaces that a slip provided may have been of incidental value.

2 Shell-gritted wares

There are six different groups of shell-gritted ware distinguished by fabric and/or form. Much of this pottery would have been made locally because it occurs in such large amounts in all periods, and as has been pointed out, the raw shell-bearing clays exist in the immediate vicinity of the site.

2a Late Iron Age (LIASG)

Almost all the LIASG from Orton Hall Farm was in a coarse open-textured fabric containing abundant large shell inclusions, together with limestone lumps up to 10mm in size and particles of ironstone. Colours ranged from orange through red and brown to black. A few sherds contained finer shell and some quartz, limestone and ironstone, but had a similar colour range. The vessels in LIASG from Orton Hall Farm can be paralleled closely in fabric and form on other Lower Nene Valley sites. A number of these had both larger amounts and range of forms, and the published reports on some provide a comprehensive introduction to, and description of, all aspects of this ware. The Orton Hall Farm material (*e.g.*, Fig. 80, Nos 1–5) can add little to these and readers are encouraged to look in particular at Fengate (Pryor 1984), and Werrington Enclosure (Rollo 1988, 112).

2b Late Iron Age to early Roman, transitional (TSG)

The fabric of TSG does not seem to vary significantly from that of LIASG except that the shell is often smaller and finer, and was perhaps deliberately crushed. There is a suggestion that the pottery was fired to a higher temperature, but it is not clear if this was the result of new technology. Vessels are characterised by the occurrence together of elements of Late Iron Age and Roman pottery traditions. The most common combination is a hand-made body with apparently wheel-finished surfaces and wheel-finished or thrown rims. The pots tend to be more symmetrical and the walls are of more even thickness. In addition, the forms are generally more stylised and standardised with definite rim, neck, shoulder and body sections (*e.g.*, Fig. 80, Nos 6–11). Decoration is also significantly different with an emphasis on horizontal, evenly-spaced and uniform grooving, rather than random, mainly vertical and diagonal, incised and scored lines over a large part of the vessel surface. Bands of simple decoration occasionally occur between the horizontal grooves.

2c Roman shell-gritted ware — general (RSG)

The basic fabric is laminated with abundant shell inclusions up to 10mm in size, and can also contain limestone, ironstone and quartz. Additional inclusions occur randomly in some sherds, but these have no apparent link to particular vessel type or date. The main difference from LIASG lies in its generally well-fired nature and the oxidised range of colours, orange, buff, reddish-yellow, reddish-brown, in which it invariably occurs, except where altered in use, especially by sooting. The types and range of vessels also vary from those made in LIASG and reflect the general traditions and styles of Roman pottery as a whole.

2d Roman shell-gritted ware — Bourne/Greetham products

This pottery is distinctive in form and fabric (*e.g.*, Fig. 80, Nos 12–14). The latter is hard and relatively thin, with jars not usually exceeding 5mm except at and around the base and rim, and contains a moderate amount of both fine and coarse shell up to 3mm in size. It can also have lumps of limestone and ironstone, together with a few grains of quartz. The surfaces appear smooth and the ware is usually fired to a dark grey or black, though there can be orange, red or pale brown patches resulting from either variable firing or use. The dark background colour makes the pieces of shell in the surface of the vessels stand out markedly. The fabric has been noted at Fengate (Hayes 1984, 610, ware 6b) and occurs on other local sites, such as Thistleton, Chesterton and Ashton and as far west as Leicester.[8] The majority of the vessels appear to be medium-sized jars with curved-over rims. As a type which combines hand-made and wheel-made characteristics, it is a continuation of the 'transitional' style into the full Roman period.

2e Roman shell-gritted ware — fourth-century non-local

The fabric contains abundant, mainly fine shell but with occasional larger pieces up to 5mm in size. Ground conditions can often make the ware seem vesicular. The main characteristic is a smooth, 'soapy' feel to the vessels which are thin walled. The main types are flanged bowls, jars with undercut rims and plain-rimmed dishes. The external surfaces are often rilled. At Towcester, Northamptonshire, this fabric is ware 44d (Brown *et al.* 1983, 75, 79). It appears to be rare in the Lower Nene Valley, but is common to the north, west and south.

2f Roman shell-gritted ware — late fourth-century large bowls

The fabric is comparable to that of general RSG but the forms differ markedly (*e.g.*, Fig. 105, Nos 607–9), suggesting either a different source or sources, or an addition to the vessel range. Similar bowls were made at Harrold, Bedfordshire,[9] but a small programme of fabric analysis has indicated that the Orton Hall Farm vessels were not made in the kilns excavated there.[10] Variations in the basic fabric may represent differing production centres but these need not have been outside the immediate locality.

Shell-gritted ware: hand-made or wheel-thrown?

Most of the coarse Iron Age pottery of the area was hand-made, and there was little attempt at more than the basic smoothing of surfaces. The first indication that local potters began to use a wheel comes from the shape, evenness, symmetry and decoration of certain vessels. The first appearance of shell-gritted pottery that had apparently spent part of its manufacturing process on a wheel is not closely dated but is thought to be at some time between the mid-first century BC and the mid-first century AD. The balance of the available evidence, however, favours an early to mid-first-century date (Jackson and Ambrose 1978, 174–5, Group 3A; Jackson 1977, 31–4, Group 3; Williams 1974, 24–5, Group 2) rather than earlier (Pryor 1984, 157). This pottery co-existed with hand-made shell-gritted vessels for the remainder of the Iron Age. It is important to note the distinction between pre-Roman wheel-made and hand-made shell-gritted ware as the latter comprised most of the pottery used in the late pre-Roman Iron Age, while the former either did not occur at all, or only in very small amounts, on some sites, *e.g.* the Werrington Enclosure (Rollo 1988, 113). It was, in fact, high-class pottery imitating wheel-made 'Roman' types made in other wares and used on the Continent, and in southern Britain, in the pre-Conquest period.

The Roman Conquest probably provided the impetus for the eventual almost wholesale adoption of Roman pottery types and hence the wheel for part or all of the pottery-making process. The shell-gritted 'belgic' types would have been gradually supplemented by other wheel-made vessels, including those in different fabrics, and by forms which were more 'Roman'. In the local area the hand-made varieties probably continued to be produced for ten years or so after the Romans arrived, and vessels obviously remained in use, as survivals, for some years after that. It is difficult to say when the last fully hand-made pottery was made, but it is with these utilitarian pots that the change to the use of the wheel is most interesting and important.

As far as the local area is concerned, it is thought that the Iron Age jar or deep bowl types were first replaced by jars which, while not completely dissimilar in form, had more definite rims and narrower bases, much more even thickness of wall, greater symmetry and, most important, horizontal, linear decoration which would have been out of place on hand-made vessels but was standard on wheel-thrown pottery. These types are termed 'transitional' (see fabric 2b above). They were, in turn, replaced by vessels which were much closer to the standard Roman jar made in other fabrics, and by the end of the first century all traces of the Iron Age precursors had disappeared.

It is not certain how the wheel was used when first introduced. It need not have involved a rapidly spinning plate or kick-wheel and variations such as a 'slow-wheel' or 'turntable' were no doubt used. The problem is emphasised when it comes to recognising the various possible methods on the pottery itself. An examination of the 'transitional' pottery indicates that the body was first hand-made and then smoothed and evened out on a wheel used as a hand-turned 'turntable'. The rim was almost certainly added later, and most probably formed either with the aid of a faster moving wheel, or was separately wheel-made. The grooved decoration is also sufficiently even to suggest that it was formed when the vessel was revolving at a fairly fast speed. In most of the Romano-British pottery industries the use of a proper potting wheel eventually became standard practice, and hand-building ceased.

However, the mixture of techniques obviously had a number of advantages, probably related to economies in materials and manpower, and simplicity of processes. There are a number of types of pottery which continued to combine hand and wheel stages. Foremost of these was BB1 with which pre-Roman potting methods remained in use up to the end of Roman Britain and may, in fact, have been partly responsible, in terms of economy, for the widespread success and distribution that the ware enjoyed. The calcite-gritted wares of East Yorkshire were also used in huge amounts in Northern Britain especially in the later fourth century, and of more regional and local significance were Dales ware and Knapton ware, each of which was both hand- and wheel-made. More locally, the shell-gritted wares of the Bourne/Greetham area (fabric 2d above) also clearly combined the two methods, and it is very possible that many of the Roman shell-gritted wares were also made in a process which involved hand and wheel stages. Unfortunately, the potters had enough skill and dexterity to mask most if not all traces of the possible hand-building element in the later wheel-finishing, and it is not, therefore, usually clear whether a vessel was wholly or partly wheel-made. The nature of the internal rilling, if it survives, can occasionally provide clues. Some of the problems will, hopefully, be answered by the excavation of production sites.

3 Grey wares

Grey ware was one of the most common kinds of Roman pottery. On the whole, preparation of the clay produced very similar fabrics regardless of the source. The principal characteristic of these fabrics was quartz sand temper, which sometimes may have been added or augmented, and variations of the quantity resulted in degrees of overall coarseness. However, the surface feel of the grey ware is not always a guide to the actual coarseness of the fabric and some grey wares had smoothed, perhaps slipped, surfaces which did not bind well with the clay body and were lost while buried in the soil. Occasionally, other natural or unusual inclusions allow particular sources to be noted and, sometimes, identified. For the majority of these wares, however, no amount of analysis will allow allocation beyond a geological zone or even region. The main variations between and within the grey wares of the different areas are related to form and period.

Ordinary grey wares were used in the Lower Nene Valley before the establishment of the major industry around Durobrivae in the first half of the second century, and in the fourth century when the products of other centres again found markets locally. The sources in these two periods were different. In the first and early second century much was perhaps made locally. The fourth-century grey wares were the products of production centres located in the East Midlands/Trent Valley/ Lincolnshire area (Todd 1968a), perhaps including Swanpool (Webster and Booth 1947).

What may have been a 'local' source is not easy to define. It is generally accepted that it was not viable or worthwhile to market basic pottery beyond a radius of some twenty to thirty miles, but there would have been many variations to this 'model', especially if water transport was readily available, or if vessels were traded for what they contained or could be shipped with other articles and produce. As far as the Lower Nene Valley is concerned, a catchment area encompassing east Northamptonshire, south Lincolnshire, west Norfolk, and north Cambridgeshire and Bedfordshire may not have been beyond the bounds of possibility, bearing in mind the available river and road system.

The lack of known kilns producing such wares in this catchment area is a problem along with the gaps in the knowledge of the full production-life and vessel and fabric range of those that have been excavated. The nearest known to Orton Hall Farm are Ecton (Johnston 1969), Weston Favell (Bunch and Corder 1954) and Wakerley (Jackson and Ambrose 1978). Few, if any, of the Orton Hall Farm grey wares, however, are obviously from these centres. There are indications of pottery production at this time in Normangate Field (Swan 1984, 95), and the existence of pottery manufacture at settlements like Rushden (Woods and Hastings 1984), Quinton (Friendship-Taylor 1979), Brixworth (Woods 1970) and Wakerley (Jackson and Ambrose 1978), was perhaps usual before the founding of the major industries, as it had been in the Iron Age, and there may have been pottery-making near Orton Hall Farm, perhaps at Stilton and Yaxley. This would certainly help to explain why most of the Lower Nene Valley late first and early second-century grey wares cannot be paralleled at sites a little further to the west, such as Ashton.[11]

When grey wares were first made and used in the area is unknown. As the colour, hardness, decoration and fabric are significantly different from the local Iron Age pottery, it is possible that initial production was related to the introduction of new methods of manufacture and firing, probably with the arrival of the Roman army and its followers in the area. But it is possible that it and its associated innovations could have preceded the Roman Conquest, but none of the grey wares on Orton Hall Farm suggest pre-Conquest manufacture. In Period 1, grey wares occur together with pre-Flavian and Flavian samian ware, so a date in the earlier part of the third quarter of the first century seems likely. The absence of quantities of grey ware at the Longthorpe fortress, dated *c.* AD 44–61/2 (Frere and St Joseph 1974, 38), may be significant but, as most of the 'finer' pottery was made on site in the prevailing oxidised 'military' style specifically for the garrison, the production may not reflect the prevailing situation in the native rural settlements.

Whether the production of grey wares was at any time contemporary with that of the TSG and the 'belgic' wares is also open to question. Deposits at Orton Hall Farm, Monument 97 (Rollo in Mackreth forthcoming), and to some extent at the Werrington Enclosure (Rollo 1988; Perrin 1988), while showing that they were at least disposed of together, give little indication of any chronological sequence, and the resolution of this particular problem must await other, more well stratified and dated, sites. The use at Longthorpe of both 'fine' military and 'coarse native' pottery indicates that different styles could exist side by side if the need arose, as did 'fine' and 'coarse' wares throughout the Roman period. On sites further up the Nene Valley, for example Brixworth (Woods 1970), grey wares of the same basic forms, if not fabrics, as those at Orton Hall Farm continued in use into the Antonine period. In the Lower Nene Valley, however, the development of the local grey ware industry based around Durobrivae in the early second century cut short the production of other grey wares, though until production centres are excavated the date when manufacture finally ceased cannot be known. The Period 1 contexts at Orton Hall Farm, and those of the Antonine period at Chesterton (Perrin, to be published) show that such grey ware vessels could survive in use for some time.

4 'Belgic' and grog-tempered wares

Though much of the pottery from Period 1 exhibited 'belgic' characteristics, only a few sherds were in fabrics which could be classed as such. The basic fabric of these locally is smooth and laminated, but with very little quartz, and does not always contain grog. The vessels were fired to a lower temperature than Roman pottery, and have a smooth, 'soapy' feel. One vessel, No. 15, was distinguished by black inclusions which were magnetic, and other examples of this distinctive ware have occurred locally at Werrington Enclosure (Perrin 1988, 120, G5) and Monument 97 (Rollo, in Mackreth forthcoming). It is not noted in Thompson's list of fabrics (Thompson 1982), probably because such inclusions are somewhat unusual and unexpected, and would not be allowed or tested for in normal fabric analysis. The source ought to be easy to isolate. Vessels from Piddington and the surrounding area[12] indicate a possible location. A fuller discussion of local 'belgic' and grog-tempered wares is to appear (Monument 97: Rollo, in Mackreth forthcoming).

5 London ware type

The main criteria for identifying pottery of London ware type are an imitation samian form together with decoration which is usually incised or stamped. A number of different production centres have been identified (Rodwell 1978) and it has been suggested that one also existed in the Lower Nene Valley (Perrin 1980). It is important to note that in most areas the fabrics were not solely used for the production of these types, but for a whole range of other, more ordinary, pottery. In the Lower Nene Valley, London ware vessels occur, chronologically, in 'belgic' or grog-tempered ware, then various grey wares, and finally LNVGW with, possibly, LNVCC, for the stamped vessels from Stanground are obviously related.

V. Discussion of the dating and catalogue of pottery from Periods 1 to 5

Period 1

(Figs 80–8)

1 Introduction

That the nucleus of this primary settlement lay north-west of the excavated site is shown by the occurrence of Iron Age material in features in that corner of the excavated area. The main features of the period were mostly 'open' in character, such as ditches and pits, and therefore could have received material until they were full, unless recut or cleaned out. Such features were not conducive to the development of closely-dated, well-stratified groups of pottery, which means that it has not been possible to trace the development of the Period 1 pottery in short stages. The main catalogue (and discussion), therefore, is based on three major ditch systems, F588, F813 and F1048, each consisting of a number of associated ditches of which the specified ditch contained the bulk of the pottery.

There is a chronological progression in the pottery from these systems, with F588 being the earliest, F1048 the latest, and F813 covering the period between. The date spans are broad, however, and there is considerable overlap, but, taken together, they provide a clear and comprehensive indication of the nature and range of the pottery used from the mid-first to the late second centuries. The material from the three systems is supplemented by that from other, less significant ditch systems, and a few additional features of particular note. Some vessels are included because they are of intrinsic interest or represent types which did not occur in the main groups. Period **1** is thought to date from the middle of the first century AD to *c*. AD 175.

The initial date was deliberately left vague because of the difficulty of dating Conquest period pottery closely. None of the pottery made at Longthorpe was found at Orton Hall Farm. Independent pottery dating evidence was provided by Flavian samian ware, and the relative paucity of 'belgic' pottery was thought to be indicative of a date after the Conquest rather than before, despite the occurrence of LIASG vessels. For the remainder of the first century, samian ware together with sherds of Verulamium region manufacture provided independent dating evidence, and the assemblages at sites like Great Casterton were useful.

In the second century, dating was again provided by samian ware, with the addition of mortaria. Once the products of the major local industry appear, there was much more scope for dating by parallel with other local sites. The few colour-coated vessels could be fitted into the dating provided by numerous examples elsewhere.

The end of the period is rather arbitrary and is based on the overall nature and date of the pottery in those deposits belonging to the first stages of Period 2. It is supported in part by the dates of particular samian vessels, some mortaria, and those accepted for certain colour-coated beakers, especially those with 'late cornice' rims, occurring in both the final Period 1 and the first Period 2 deposits.

The six sub-phases, *a–f*, are largely stratigraphical ones, as they could not be easily distinguished ceramically:

- *a+b* mid and later first century AD
- *c* late first–early second century
- *d* early second century
- *e* first half of second century
- *f* mid-second century plus.

These years are extremely important with regard to local pottery. Sequentially they cover the end of the Iron Age pottery traditions, the beginnings of Roman wheel-made pottery, the start and initial growth of the Lower Nene Valley industry based on Durobrivae and the appearance, disappearance and rise and decline of various pottery types and industries. The range of vessels and fabrics and the nature of the deposits do not allow a comprehensive examination of all of these aspects, and the various discussions are designed to cover major points.

The Period 1 layers contained 19A% of the total stratified pottery recovered.

Other fabrics are grogged, various cream or buff, CGCC, Lower Rhineland, Oxfordshire, BB1, BB2, BB-type, London ware type, VR, miscellaneous SG including middle Iron Age, samian ware and amphorae. Additional forms are a colander, two cheese presses, a cup, and vessels which may be either jars or bowls, or dishes or bowls.

	F588	F813	F1048	F814	F1025	F525/6	P1 Total
Fabric							
LIASG	11	<1	-	5	-	-	8
TSG	5	-	7	3	6	-	5
RSG	29	37	17	29	33	21	25
Grey	34	37	29	30	42	35	31
LNVGW	6	8	33	1	11	13	16
LNVCC	-	4	2	5	1	-	2
C/W	8	4	7	21	5	28	7
% P1	15	21	31	2+	9	3	

Table 5 Period 1, main fabrics A%.

	TSG	RSG	Grey	LNVGW	LNVCC	C/W	Other	Total
Jar	19	102	134	61	3	9	22	350
Bowl	-	1	5	10	1	3	5	25
Dish	-	6	14	11	5	-	6	42
Flagon	-	-	2	1	-	11	3	17
Lid	-	6	3	-	-	-	-	9
Beaker	-	-	2	-	20	-	-	22
Mortarium	-	-	-	-	-	4	3	7
Other	1	3	5	3	-	-	2	14
Total	20	118	165	86	29	27	41	486

Table 6 Period 1, main vessel forms/fabric, by entries.

2 Period 1 features

Main Ditches

2.1 F588 System

This comprised F588, F588/592, F592, F626, F627, F636, F688, F775, F776, F797, F805, F805/775, and F807. These were all sections of ditches or gulleys which together formed an enclosure in the north-west corner of the excavated area (Pl. II), dated in the main to phases *a* and *b*, though with elements of phase *c* and perhaps up to phase *e*. Table 5 gives the main fabric A%.

Additional fabrics are London ware type, grogged, buff and uncertain.

Independent dating evidence is provided by one piece of samian ware from F588 of pre-Flavian date and others of Flavian and Flavian or Trajanic date (Samian, below). One or two iron brooch fragments from F776 (Fig. 61, Nos 19–20) suit the pottery from this feature which, together with F775, contained some of the earliest from the site, stylistically of late Iron Age character.

In the following lists, the phase is added at the end, and the context group is also given throughout the report.

	LIASG	TSG	RSG	Grey	LNVGW	C/W	VR	Misc	Total
Jar	6	4	19	22	4	-	-	1	56
Bowl	-	-	-	1	-	1	-	-	2
Dish	-	-	-	3	2	-	-	-	5
Flagon	-	-	-	1	-	2	2	-	5
Lid	-	-	-	1	-	-	-	-	1
Total	6	4	19	28	6	3	2	1	69

Table 7 F588 System, main vessel forms/fabric, by entries.

(Fig. 80)
LIASG
1 CR22+28. F588, L1647 (2)
2 CR22+28. F664, L1572 (6)
3 CR11+21. F588, L1647 (2)
4 CR1+8+21 patches. Large lumps limestone. F775, L1658 (1)
5 CR11/21/8. F797, L1675 (2)

TSG
6 CR5+17+2. F797, L1675 (2)
7 CR3. F588, L1599 (2)
8 CR2+11+22. F588, L1663 (2)
9 CR11/28/12. F588, L1663 (2)
10 CR4+8+12+22. F588, L1646 (2)
11 CR14+15+19+21. Lumps of limestone. F588, L1733 (2)

Bourne/Greetham shell-gritted ware
12 CR8+28. F588, L1733 (2)
13 CR28/9/28. F588, L1733 (2)
14 CR15+17+22. F627, L1619 (2)

'Belgic' ware with magnetic inclusions
15 CR4+21+22. Partly hand-made? F776, L1660 (1)

GW
16 CR19+20/15/19+20. F588, L1663 (2)
17 CR15+19. F588, L1599 (2)
18 CR 20+21//11/19/11//20+21. F588, L1516 (2)
19 CR26/20/6. Slightly underfired? F775, L1690 (2)
20 CR12//19/20/19//14. F588, L1516+L1687 (2). F776, L1689 (2)
21 CR19. F588, L1599 (2)
22 CR16+19. F588, L1598 (2)

(Fig. 81)
23 CR22//15+19/4+18/15+19//22. External surface eroded F592, L1520 (2)
24 CR15+19/20/15+19. Burnished surfaces F592, L1520 (2)
25 CR15+19/19/15+19. F664, L1573 (6)
26 CR19/20. Partly hand-made? F588, L1598 (2)
27 CR21//4/21/4//21. F588, L1687 (2)
28 CR15/21/15. Possibly burnished. F807, L1712 (2)
29 CR16+19+20. F797, L1675 (2)
30 CR11/21/11. F627, L1538 (2)
31 CR22+28. F588, L1733 (2)

LNVGW
32 CR19/18/19. Not LNVGW? F588, L1733 (2)
33 CR20/26/20. Half vessel F588, L1733 (2)
34 CR19+20/19/19+20. F588, L1733 (2)

VR
35 CR13. F775, L1690 (2)
36 CR5+6/4+5/5+6. F592, L1520 (2)
37 CR19/18+26/19. F588, L1733 (2)

C/W
38 CR26/18/26;15 slip? Three-ribbed handle. F588, L1733 (2)
39 CR26. Powdery surfaces. F588, L1599 (2)

2.2 F813 System
This consisted of F13, F40, F657, F675, F737, F766, F813, F873, F901, F912, F953, F969, F970, F971, F973, and F984. These comprised the main east-to-west ditch across the site and the main north-to-south branch which joined it (Pl.II). They are part of the development of the layout assigned to phase *c*, but continuing up to the end of Period 1, and marking the first use of the area south and east of the F588 system and its pre-Roman antecedents. Table 5 gives the main fabric A%.

	RSG	Grey	LNVGW	LNVCC	C/W	VR	Misc	Total
Jar	30	34	12	-	3	-	2	81
Bowl	-	1	2	-	-	-	2	5
Dish	2	3	1	3	-	-	1	10
Flagon	-	-	-	-	2	-	-	2
Lid	4	-	-	-	-	-	-	4
Beaker	-	-	-	10	-	-	-	10
Mortarium	-	-	-	-	-	2	-	2
Other	-	3	1	-	-	-	-	4
Total	36	41	16	13	5	2	5	118

Table 8 F813 System, main vessel forms/fabric, by entries.

The additional fabrics are VR, London ware type and grogged. Other forms are LNVGW and grey ware cheese presses, and grey ware vessels which may be either jars or bowls.

There is more samian ware from this system than from either 2.1 or 2.3, and it ranges in date from the pre-Flavian (a Ritterling 12 or Curle 11) through to the mid-late Antonine periods (Samian, Stamp A, Fig.113, Nos 2 and 5). That from the main feature, F813, is, apparently, late first-century/Trajanic (including No. 2), while F675 had more of Antonine and later second-century material (including Stamp A and No. 5). This distinction is not obvious in the coarse wares. The pottery from the F813 system includes the first mortaria from dated contexts on the site (Mortaria, Fig. 114, M1–M2). The features also contained a number of datable finds including a late first/early second-century glass beaker from F675 (microfiche MF8, Fig. 76, No. 4), and mid first/early second-century copper-alloy pins from F813 (Figs 63–5). Burnt stone and clay, together with daub and some tile from a number of the features suggests that the destruction of at least one building (see Chapter 1) was responsible for some of the deposits containing pottery.

LIASG
40 CR17+28. F972, L1853 (39)

TSG/RSG
41 CR3. F813, L1672, L1707 (19)(51)
42 CR8+10/12/8+10. F813, L1755 (51)
43 CR2+8/17/2+8. Wholly wheel-made? F813, L1707 (51)
44 CR4/15/4;21 patches. Wholly wheel-made? F813, L1672 (19)
45 CR2+4/12+17/2+4. Wholly wheel-made. F813, L1722 (19)

(Fig. 82)
46 CR2/15/2+28. Wheel-made. F813, L1754 (51)
47 CR3+9;28 areas. F813, L1722 (19)
48 CR8+10/21/4+28. F813, L1707 (51)
49 CR3;28 patches. F813, L1707, L1755 (51)
50 CR3. F813, L1672 (19)
51 CR8+28/21/12+28. F675, L1621 (39)
52 CR2+28. F675, L1818 (51)
53 CR3+5/12+17/3+5. F813, L1673 (19)
54 CR10+28/4/10+28. F969, L2242 (31)
55 CR2+28/18/2+28. F969, L2229, L2230 (39)
56 CR12+28. F969, L2270 (39)
57 CR2/4/2. F813, L1672 (19)

RSG
58 CR8/12/8. F813, L1672 (19)
59 CR3/20/3. F675, L1741 (51)

Grogged ware
60 CR4/16/4. F675, L1741 (51)

GW
61 CR11//17+21/4/17+21//11. F969, L1815 (51)
62 CR11+20+21/22/11+20+21. F970, L1852 (39)
63 CR14+18/22/14+18. F969, L2256 (39)
64 CR17//8/21/8//22. F813, L1707 (51)
65 CR15//18/21/18//15. F969, L1815 (51)
66 CR15+19/16/15+19. F813, L1672 (19)

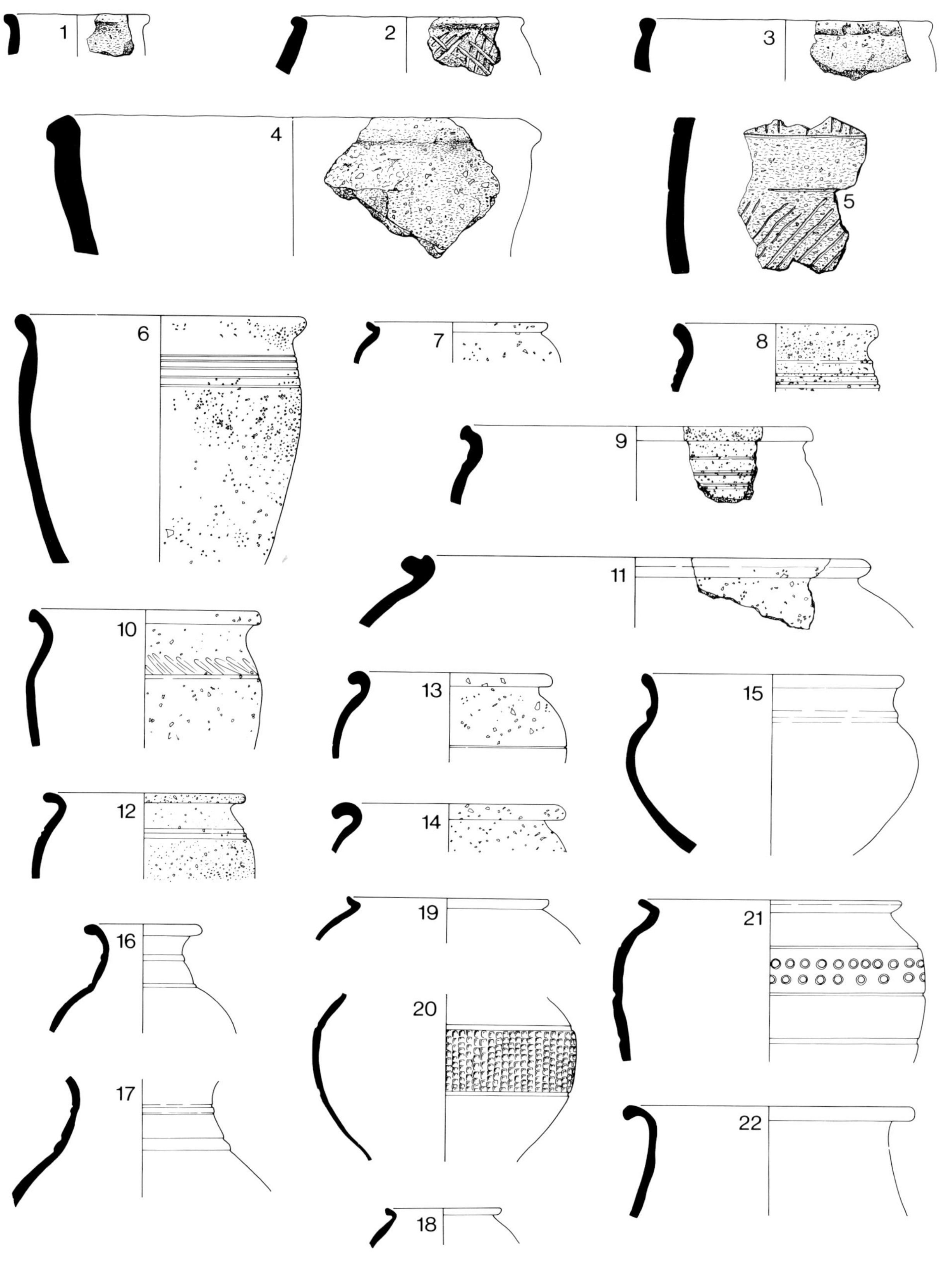

Figure 80 Roman pottery. Period 1: Nos 1–22, F588 system. Scale 1:4.

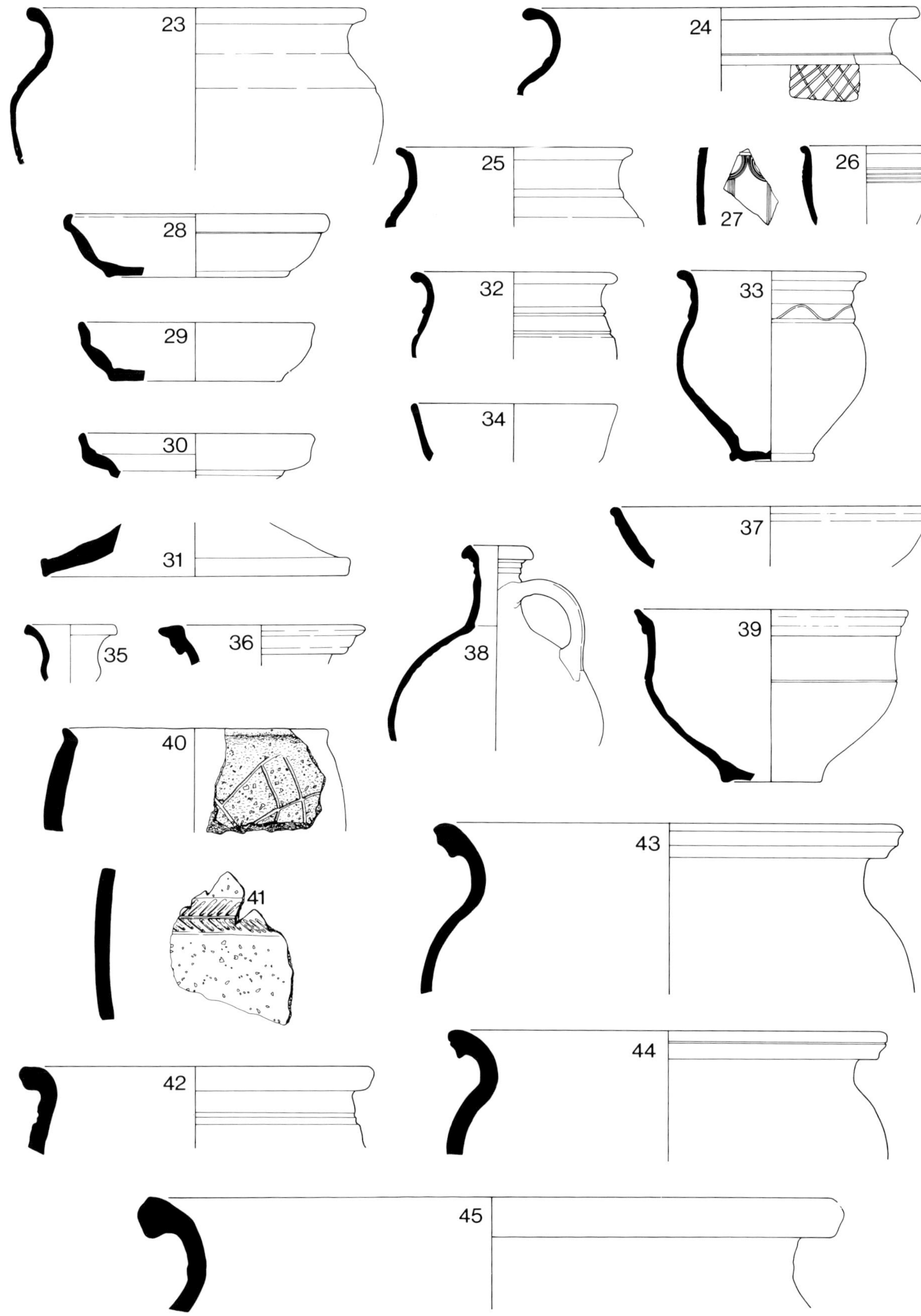

Figure 81 Roman pottery. Period 1: Nos 23–39, F588 system; Nos 40–45, F813 system. Scale 1:4.

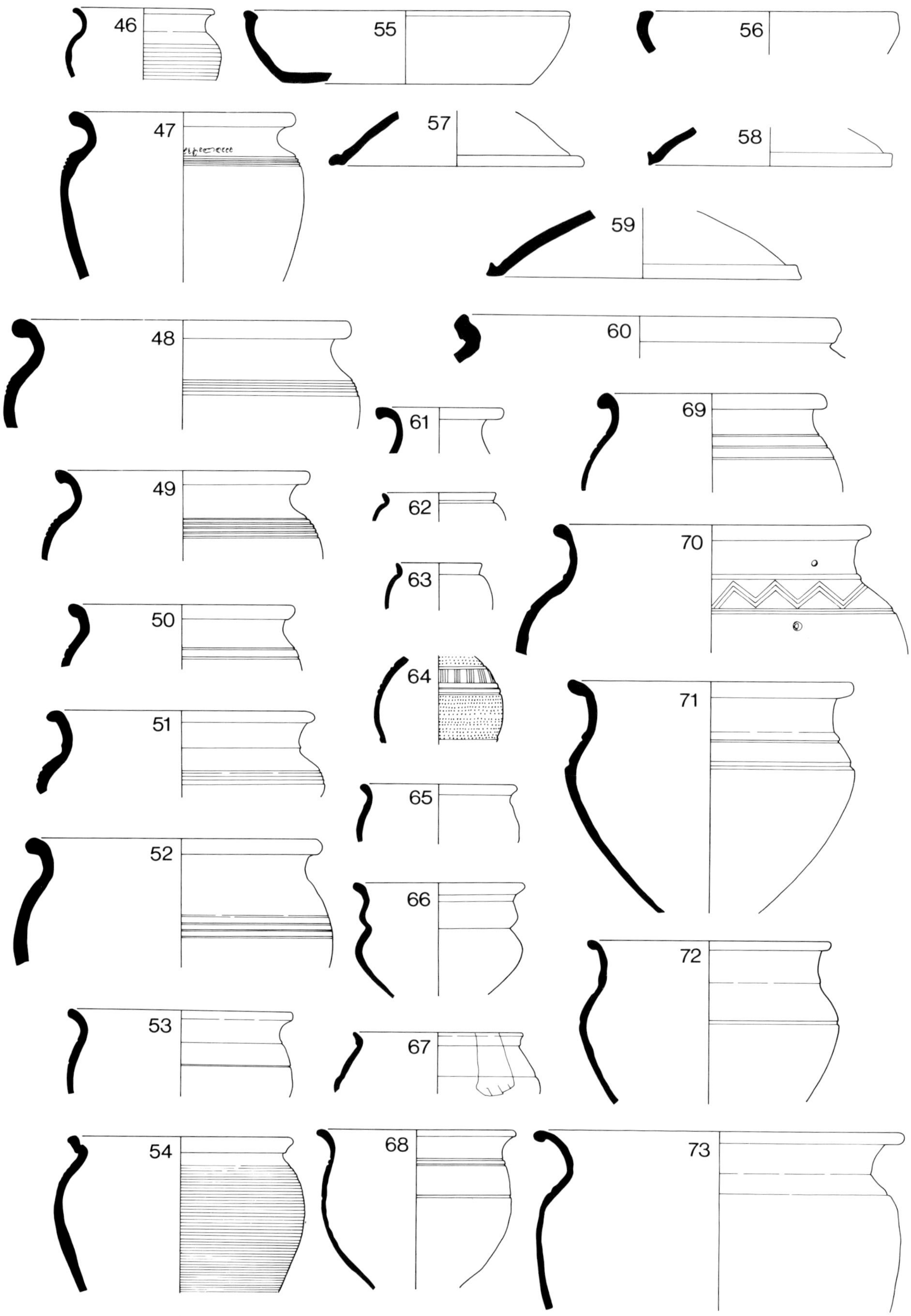

Figure 82 Roman pottery. Period 1: Nos 46–73, F813 system. Scale 1:4.

67 CR19/18/19. F969, L2242 (31)
68 CR21+22//15/22/15//21+22. F675, L1741 (51)
69 CR19/21/21+22. F813, L1672 (19), L1755 (51)
70 CR18/11+17+21/21. F984, L2144 (19)
71 CR4/15/4. F813, L1722 (19)
72 CR15/19/15. F813, L1673 (19)
73 CR11//11+12/19/11+12//16. F813, L1672 (19)

(Fig. 83)
74 CR20//11+15/20/11+15//20. F813, L1672 (19)
75 CR11+16//4/19/4//11+16. F813, L1672 (19)
76 CR 8//8/21/8//11. F813, L1769 (19)
77 CR11/19/11. F675, L1741 (51)
78 CR18+19/19/18+19. Burnished surfaces. F969, L1815 (51) Mica flakes clearly visible in fabric. Joins: Period 1, F907, L1979 (22). F1048, L2220 (HS), L2549, L2584 (42). Period 3, L1783 (134). Unassigned, L2611 (39)
79 CR10//10/20/10//17. F675, L1741 (51)

LNVGW
80 CR19/18/19. 'Speckled' surfaces. F813, L1672 (19)
81 CR20+21/18/20+21. F969, L2257 (39)
82 CR21/26/21. Possibly not LNVGW. F675, L1621 (39)

LNVCC
83 CR11+22/14/11+22. Clay roughcasting. F675, L1818 (51)
84 CR22/5+26/22. F675, L1890 (51)
85 CR22/4/22. F675, L1818 (51)
86 CR4+12/22/4+12. F675, L1593 (51)

VR
87 CR14+26/5/14+26. Verulamium region? F675, L1818 (51). F675 (Period 5), L1772 (316)
88 CR5/21/5. Verulamium region? F813, L1754 (51)
89 CR5+26/22/5+26. F675, L1741 (51)

C/W
90 CR14+26. One handle? F675, L1741 (51)
91 CR5,4 patches (colour-coat?). F813, L1707 (51)

BB2 or imitation
92 CR21//8/21/8//21. F813, L1755 (51)

2.3 F1048 System

Composed of F914, F929, F909, F1032, F1043, F1045, F1048, F1049, F1063, F1083, and F1088, these were all ditches or gulleys associated with the development of enclosures between the western house and the main branch running from the east-to-west ditch line. The largest groups of pottery were from the western end of the southern ditch of this system. These are considered to be derived from the clearance of a house or other area of occupation in phase *f* (see above). The other features had a wider span incorporating phase *e* and possibly *d* as well. Much of the fill probably related to general clearance ahead of the new layout belonging to Period 2. Given the suspected nature of the larger groups, they could perhaps be treated separately, but though they may have resulted from a single action disposal, the pottery represents what had been in recent use and is, therefore, essentially the same as that from the other features in which pottery had perhaps accumulated more gradually. Therefore, all the pottery is treated together. Table 5 gives the main fabric A%.

	TSG	RSG	Grey	LNVGW	LNVCC	C/W	Misc	Total
Jar	8	22	36	32	1	5	-	104
Bowl	-	1	1	5	1	2	2	12
Dish	1	1	2	5	1	-	5	15
Flagon	-	-	-	1	-	4	-	5
Beaker	-	-	-	-	7	-	-	7
Mortarium	-	-	-	-	-	1	-	1
Total	9	26	40	44	10	12	7	148

Table 9 F1048 System, main vessel forms/fabric, by entries.

There are also sherds of London ware type, amphorae and LIASG. The other forms are a lid and a dish or bowl in RSG, a cup in grey ware and a colander in LNVGW.

Samian ware from the system dates from the Hadrianic-early Antonine period through to the end of the second century with an emphasis on the Antonine period (Samian, Stamp B, Fig. 113, No. 4). The coarse pottery includes a mortarium (Mortaria, Fig. 114, M7), but there were no other finds of contemporary date.

TSG/RSG
93 CR8+28/17/8+28. F1048, L2704 (41)
94 CR3. F1048, L2702 (41)
95 CR2. F1048, L2550 (42). F1049, L2756 (53)
96 CR2+8. F1048, L2584 (42)
97 CR2+12. F1048, L2584, L2767 (22)
98 CR4/16+20/4. F1049, L2756 (53)
99 CR9+22. F1048, L2703 (41)
100 CR14/22/14. F1048, L2704 (41)
101 CR14/21/21. F1048, L2585 (42)
Vessels **98–101** may have been entirely wheel-made.

RSG
102 CR8/22. F1088, L2667 (53)
103 CR12+22. F1048, L2550 (42)

(Fig. 84)
Bourne/Greetham shell-gritted ware
104 CR10+15+22. F1048, L2703, L2704 (41)
105 CR21/11/21. F1048, L2550 (42)
106 CR17+21. F1049, L2677 (53). F1088, L2667 (53)
107 CR22. F1048, L2584 (42)
108 CR22//10/21/10//22. F1048, L2703 (41)
109 CR11+22. F1043, L2972 (41)

GW
110 CR14/11/14. 'Speckled' surfaces. F1048, L2704 (41)
111 CR20/4/20. Complete rim. Verulamium region?? F1049, L2756 (53)
112 CR19. Fabric full of tiny shell particles. F1048, L2669 (53)
113 CR14/20/14. Near complete. Slightly underfired? F1048, L2705 (41)
114 CR20. F1045, L2532 (29)
115 CR21/20/21. F1032, L2630 (41)
116 CR12+22/20/12+22. F1049, L2547 (41)
117 CR22/4/22. F1048, L2703, L2705 (41)
118 CR19/21/19. F1048, L2705 (41)
119 CR16+19/21/16+19. F1048, L2703 (41)
120 CR14+16/4/14+16. F1048, L2551 (42)
121 CR14+15/21/14+15. F1048, L2703 (41)
122 CR20+21/21/20+21. F1049, L2547 (41)
123 CR15//4/19/4//15. Third of vessel. Underfired? F1048, L2669 (53)
124 CR10+15/14//20//14/10+15. F1043, L2581 (29)
125 CR15+16+21/20/15+16+21. Two-thirds vessel. F1048, L2669 (53). F1049, L2546, L2547, L2756 (41)(53)

(Fig. 85)
126 CR22/19+23/14+22. Underfired? Micaceous fabric. Possibly once colour-coated, slipped or mica-dusted. F1049, L2677 (53). F1088, L2667 (53)
127 CR11+14+21/21/11+14+21. Underfired? Traces of a CR8 slip or colour-coat. F1048, L2703, L2704, L2705 (41). F1088, L2667 (53)
128 CR14+15/22/14+15. F1043, L2943, L2971, L2972 (41)
129 CR17+21/22/17+21. F1043, L2526 (41)
130 CR16+20/22/16+20. Verulamium region?? F1048, L2703 (41)

LNVGW
131 CR15+19/19+21/15+19. Top third vessel. F1049, L2756 (53)
132 CR19/18/19. Top half vessel. 'Speckled' surfaces. F1088, L2667 (53)
133 CR18//18/20/18//21. LNVGW? F1049, L2587 (41)
134 CR20/26/18. LNVGW? F1048, L2702 (41)
135 CR21/18/21. F1048, L2704 (41)
136 CR21/19/20. Two-thirds vessel. F1048, L2669 (53)
137 CR20/26/20. Two-thirds vessel. F1048, L2704 (41)
138 CR21/20/21. Half vessel. F1049, L2677 (53)
139 CR19/21/19. Third to half vessel. F1049, L2756 (53)
140 CR19/20/19. F1048, L2702; (41). F1049, L2682 (53)
141 CR20/26/20. LNVGW? 'Speckled' surfaces. F1048, L2704 (41)
142 CR18+19/21/18+19. LNVGW? Fabric similar to 112. F1049, L2756 (53)
143 CR19+20/26/19+20. LNVGW? Quarter vessel. F1048, L2679 (41)
144 CR15+19/14+18/15+19. Two-thirds vessel. F1049, L2756 (53)
145 CR20//26/20/26//20. LNVGW? F1048, L2705 (41)

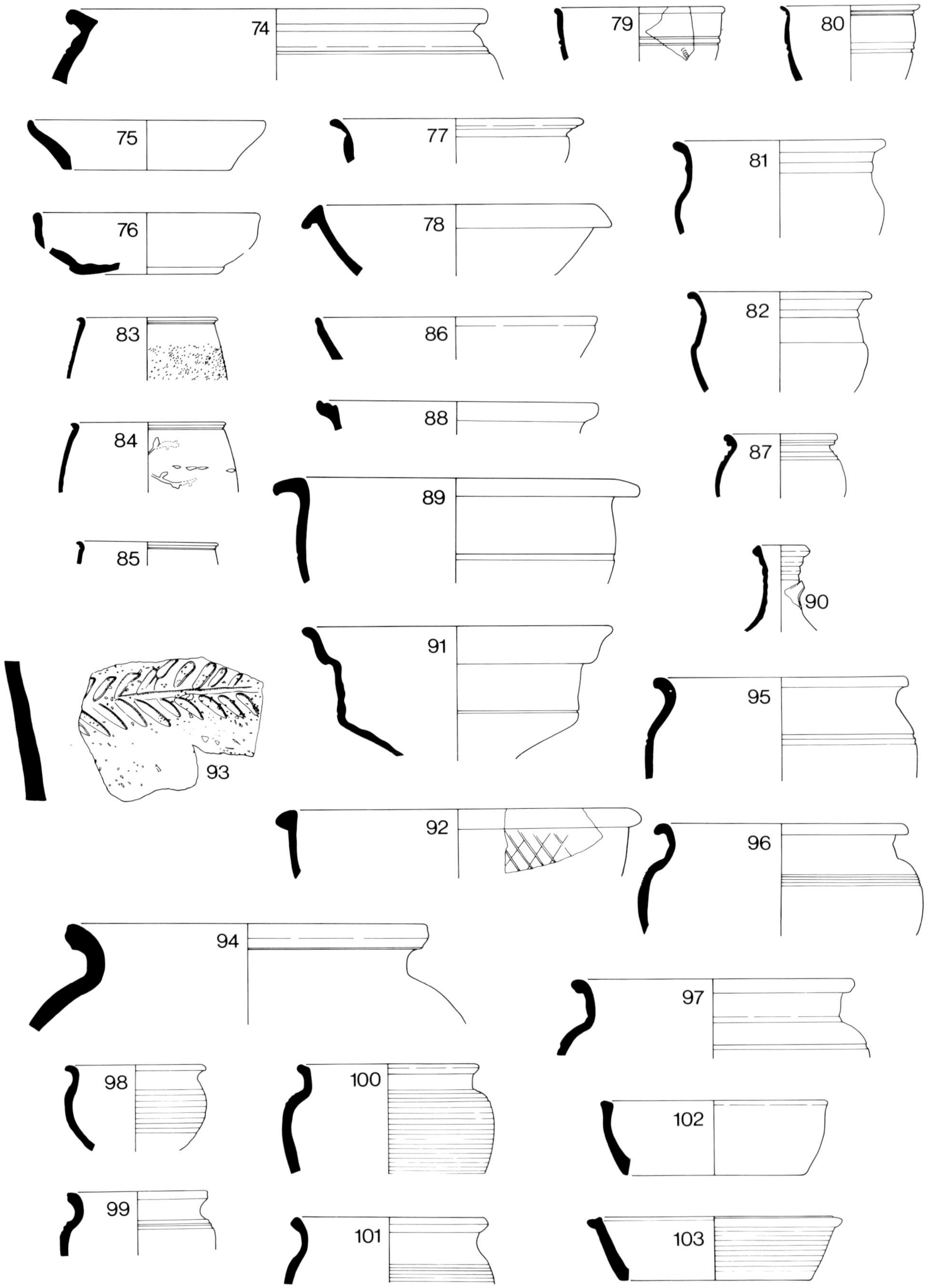

Figure 83 Roman pottery. Period 1: Nos 74–92, F813 system; Nos 93–103, F1048 system. Scale 1:4.

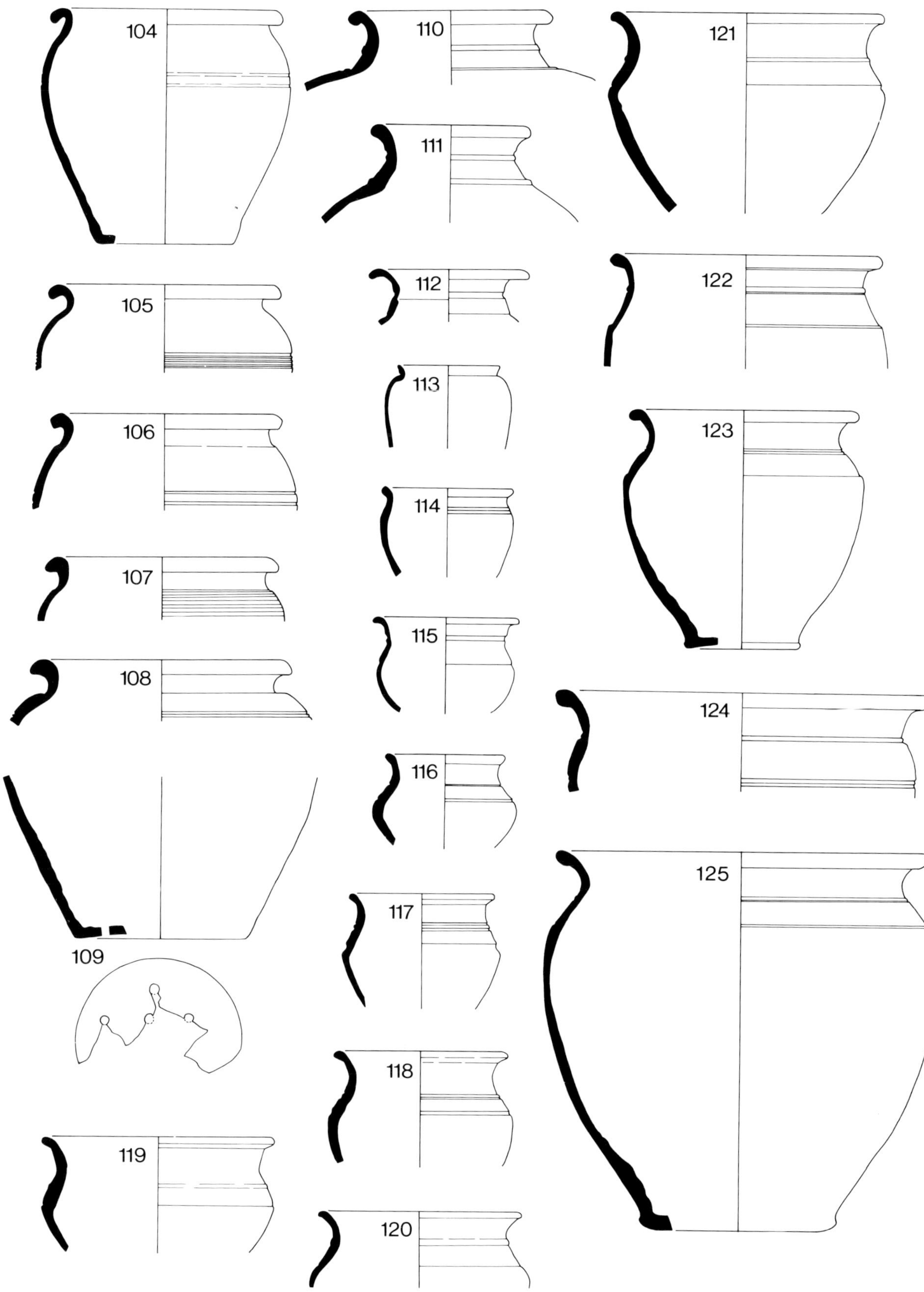

Figure 84 Roman pottery. Period 1: Nos 104–125, F1048 system. Scale 1:4.

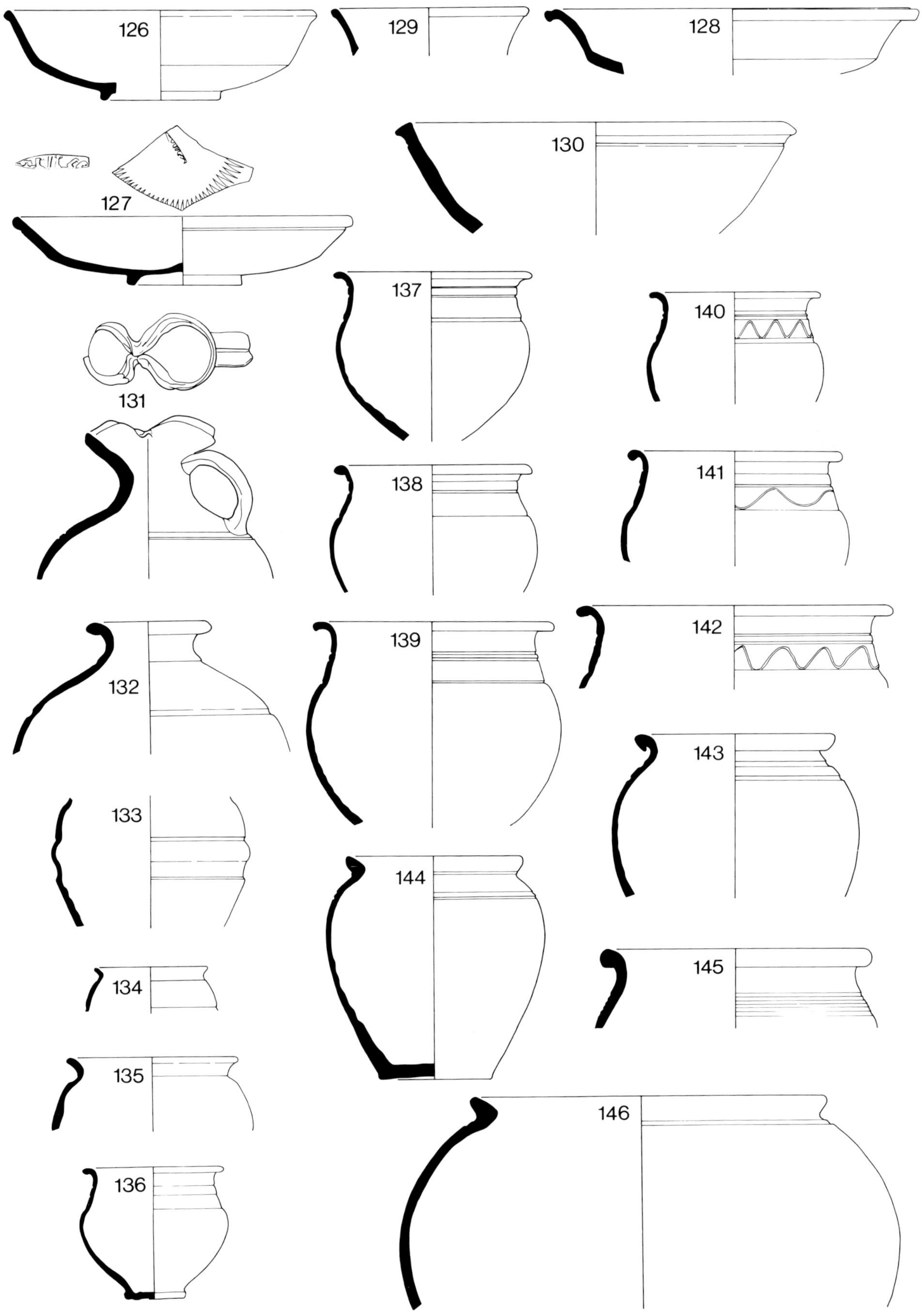

Figure 85 Roman pottery. Period 1: Nos 126–146, F1048 system. Scale 1:4.

146 CR18+19/18/18+19. Third vessel. F1049, L2756 (53)

(Fig. 86)

147 CR16+20/18/16+20,11 patches/slip? Half vessel. 'Speckled' surfaces. F1049, L2756 (53)

148 CR18+26. Burnished. F1049, L2756 (53)

149 CR21/18/21. 'Speckled' surfaces. F1048, L2550 (42)

150 CR20+21/18+19/20+21. Sixth vessel. F1048, L2684 (53)

151 CR20//19/20/19//20. F1083, L2639 (41)

152 CR19/18/19.Third of vessel (base complete). 34 holes pierced before firing; waste not pared off. F1048, L2703 (41)

LNVCC

153 CR11/14+15/20. LNVCC? Clay roughcasting. F1049, L2683 (53)

154 CR8/14+26/22. F1048, L2550 (42)

155 CR4+12/26/4+12. F1048, L2703, L2704, L2705 (41)

156 CR8//5/4/5//8. F1049, L2756 (53)

157 CR4/26/4. F1049, L2677 (53)

C/W

158 CR26. Single handle? F1048, L2669 (53)

159 CR14+26. Single handle? F1048, L2703 (41)

160 CR14+26. Single handle? F1048, L2537 (22)

161 CR14+26. Single handle? F1048, L2584 (42)

162 CR5+26. F1049, L2677 (53)

163 CR26. F1048, L2550 (42)

164 CR26. F1088, L2667 (53)

Miscellaneous

165 CR2+3. Mica-dusted surface. F1049, L2756 (53)

Additional Ditches

2.4 F646 System

This comprised F562, F563, F564, F571, F576, F577, F582, F646, F690, F691, F692, and F773 which together formed a tongue of a ditch possibly associated with the suspected 'kitchen' area of phases *a* and *b*. The pottery was mostly shell-gritted ware, at least half being LIASG. Of the four vessels, three were jars, the other a dish or bowl.

166 LIASG. CR11+12+17+22. F563, L1505 (17)

167 LIASG. CR8+16/20/8. F562, L1504 (17)

168 LIASG. CR10+16/22/10+16. Could be a lid. F582, L1526 (17)

169 TSG. CR3/8/22. F577, L1513 (17)

2.5 F814 System

A probable enclosure ditch, F814, and its subdivisions, F1018 and F1037, were all cut in phase *b*. Table 5 gives the main fabric A%.

Additional fabrics were London ware type, LIASG, buff/cream and VR.

	TSG	RSG	Grey	LNVCC	C/W	Misc	Total
Jar	1	2	5	-	-	-	8
Dish	-	-	1	-	-	-	1
Flagon	-	-	-	-	1	1	2
Beaker	-	-	-	1	-	-	1
Total	1	2	6	1	1	1	12

Table 10 F814 System, main vessel forms/fabric, by entries.

F814 contained pieces from five different samian vessels, four of which are of mid to late first-century date, and the other possibly of second-century date (Samian, one decorated, Fig. 113, No. 1).

170 RSG. CR3/11. F814, L1725 (3)

171 GW. CR11/11+16/17,21 slip? Fabric similar to 112? F1037, L2244 (3)

172 GW. CR14/11+16+21/14,21 slip? F814, L1725 (3)

173 GW. CR20//4/19/4//20. F814, L1725 (3)

174 LNVCC. CR17/4+14/17. F814, L1725 (3)

175 C/W. CR26. Single handle. F814, L1725 (3)

2.6 F1025 System

Consisting of F900, F905, F906, F907, F909, F920, F923, F950 and F1025, these were the various ditches and gulleys delineating the main western enclosure in its initial, pre-expansion, form, laid out in phase *c* and lasting, in part at least, for the rest of the period. Table 5 gives the main fabric A%.

	TSG	RSG	Grey	LNVGW	LNVCC	C/W	BB1	Other	Total
Jar	3	9	16	2	1	-	-	1	31
Bowl	-	-	-	-	-	-	1	-	1
Dish	-	1	1	-	2	-	-	-	4
Flagon	-	-	-	-	-	1	-	-	1
Other	1	-	-	1	-	-	-	-	2
Total	4	10	17	3	3	1	1	1	40

Table 11 F1025 System, main vessel forms/fabric, by entries.

There are also sherds of London ware type bowl, a possible TSG tripod vessel, a jar or bowl in LNVGW and grey ware?

Samian ware ranging in date from the Flavian-Trajanic to Antonine periods (Samian, Fig. 113, No. 3) was recovered.

RSG

176 CR8+11. F907, L1979 (22)

177 CR8+28/21/8+28. Three or four legs? F923, L2019 (22)

178 CR3/19/3,28 patches. F905, L1978 (23)

179 CR2+12/21/2+12,28 patches. F1025, L2767 (22)

(Fig. 87)

180 CR4–5/21/11+16. Two-thirds vessel. F1025, L2910 (22)

181 CR8/12/22. F905, L2018 (23)

Bourne/Greetham shell-gritted ware

182 F906, L2483 (22)

GW

183 CR19+20. F909, L1987 (22)

184 CR21+22//11/17/11//21+22. F1025, L2488 (22)

185 CR16+22/4/16+22. F1025, L2494 (22)

186 CR11+16/21/11+16. F906, L2121 (23)

187 CR16+17. F907, L1979 (22)

188 CR11+16/21/11+16. Near complete. F906, L2141 (23)

189 CR11+16/20/11+16. F906, L2482 (22)

190 CR15+17+21//11/19/11//15+17+21. F906, L2482 (22)

191 CR14+15/21/14+15. Traces of a CR22 slip or colour-coat: accidental? F1025, L2189 (22)

192 CR17+21. F909, L2153 (22)

LNVGW

193 CR19/18/19. Two-thirds vessel. 'Speckled' surfaces. F1025, L2767 (22)

194 CR19/18/19. LNVGW? F1025, L2190 (22)

Other Features

2.7 'Kitchen' area

Lying to the north of F775 was an area of post-holes and pits, some containing pots, thought to have been the focus for some kind of specialised activity. Most of the pottery points to a date in the mid to late first century, but if the nature of the activity remained unchanged some could have survived in use.

Vessels **195–197** are in LIASG and from F684, L1594 (9).

195 CR8+22.

196 CR2+4+21,28 patches.

197 CR4+22.

Vessels **198–199** are from F685, L1601 (9).

198 LIASG. CR8.

199 'Belgic'. CR10/21/10. Underfired.

200 LIASG. CR2+4+8/17/2+4+8. F686, L1603 (9)

201 'Belgic'. CR4/21/4. F687, L1605 (10)

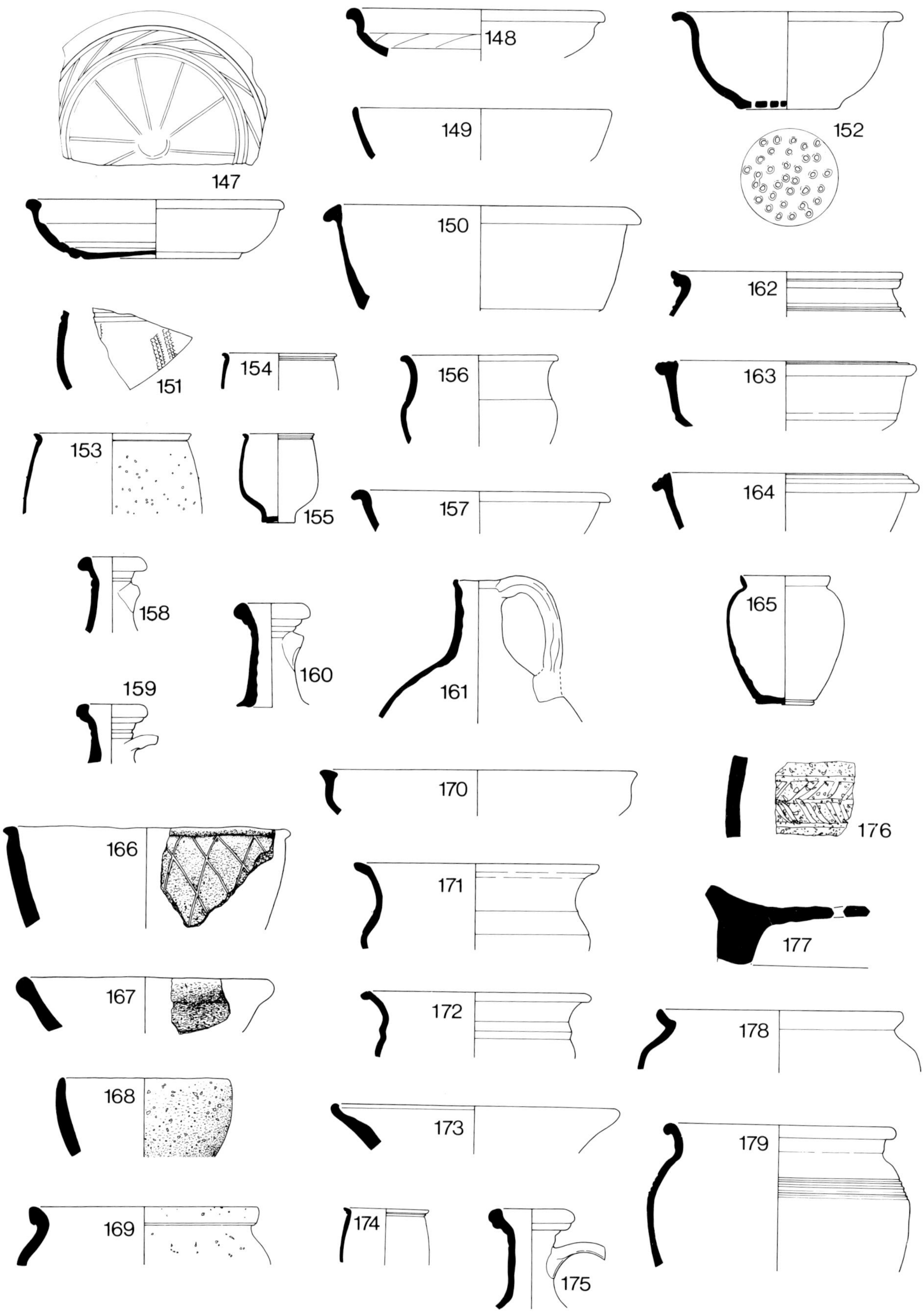

Figure 86 Roman pottery. Period 1: Nos 147–165, F1048 system; Nos 166–169, F646 system; Nos 170–175, F814 system; Nos 176–179, F1025 system. Scale 1:4.

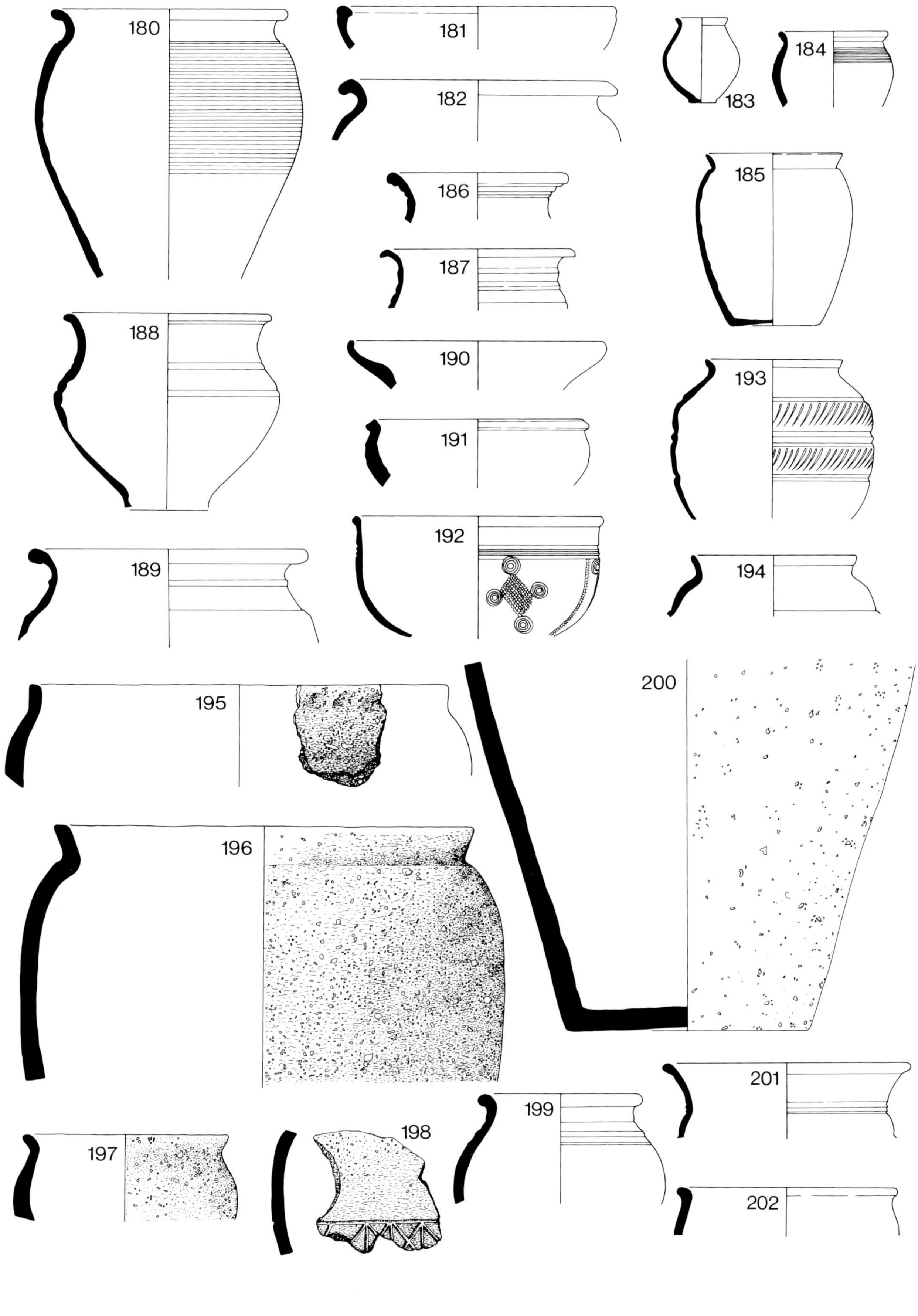

Figure 87 Roman pottery. Period 1: Nos 180–194, F1025 system; Nos 195–202, 'kitchen' area. Scale 1:4.

Vessels **202–204** are from F755, L1340 (15).
202 LIASG. CR11+21+22.

(Fig. 88)
203 LIASG. CR10+12.
204 'Belgic'. CR4+5/21/4+5.
205 LIASG. CR22, 11 patches. F758, L1345 (11)
206 GW. CR15+21. F838, L1460 (15)

Vessels **207–211** are from F858, L1462 (14).
207 TSG. CR2+8.
208 TSG. CR3/21/9.
209 'Belgic'. CR17+22. Also L1464 (14)
210 'Belgic'. CR3+4/21/3+4.
211 'Belgic'. CR4/21/4.

2.8 F875, Gulley, Western House

This was the earliest building found. The pottery from the northern gulley, F875, and various other associated features, suggests that it had been occupied from the mid-first into the second century. In addition to LIASG, F875 contained one sherd of possible LNVGW.

Both **212** and **213** are in LIASG and from L1752 (4).
212 CR4+16+17+22.
213 CR4+22.

2.9 F525 and F526, Gulley, Eastern House

This replaced the other to the west. Pottery from its gulley, F525 and F526, suggests that the changeover had occurred by the early to mid-second century. The vessels which are represented by very large fragments might have been discarded when the dwelling was abandoned, perhaps in the mid-second century. Table 5 gives the main fabric A%. Additional wares are LIASG, TSG and other gritted.

	RSG	Grey	LNVGW	C/W	Total
Jar	2	3	1	-	6
Bowl	-	-	1	-	1
Dish	-	1	1	-	2
Flagon	-	-	-	1	1
Beaker	-	1	-	-	1
Total	2	5	3	1	11

Table 12 F525/6, main vessel forms/fabric, by entries.

All but **218** are from F526, L1052 (32); only **220** is not in GW.
214 CR10+11/4/10+11,17 patches. Near complete.
215 CR15//20/19/20//15. Over half vessel.
216 CR19. LNVGW?
217 CR17+21.
218 CR15+19/11/15+19. F525, L1050 (32)
219 CR21/18/21. 'Speckled' surfaces. VR?
220 C/W. CR14+26. Single handle.

Period 1 Additional Sherds

221 GW. CR22/12/22. Burnished. F1111, L2731, L2775 (45)
222 GW. CR18/4/18. 'Speckled' surfaces. F1130, L2801 (45)
223 GW. CR19/18+19/19. Two-thirds vessel. 'Speckled'. F1130, L2801 (45)
224 B/GSG. CR14+15+16. F1131, L2802 (45)
225 GW. CR11+14/20/11+14. F1131, L2802 (45)

3 The pottery

3.1 Shell-gritted wares

3.1a Iron Age (LIASG)

Table 5 gives the LIASG A% in the main Period 1 features and for Period 1. Most of the vessels are jars, though it is often difficult to be precise. Many of the features containing LIASG were probably parts of the preceding settlement which became incorporated into the area of the later Roman farmstead.

Of the illustrated material from Period 1, parallels occur for Nos 1, 3, 4, 5, 40, 166, 168, 195, 197, 202, 203, 205, 212 and 213 at Werrington Enclosure (Rollo 1988); for Nos 1, 3 and 40 at Wakerley (Jackson and Ambrose 1978); for Nos 2, 40, 166 and 196 at Fengate (Pryor 1984), and for Nos 1, 3, 40, 166, 168, 195, 202, 212 and 213 at Monument 97 (Rollo, in Mackreth forthcoming). All the Werrington Enclosure examples were from its Period 1, dated up to *c*. AD 50/60, whilst those from Wakerley are dated *c*. AD 30–55/60. The Fengate parallels were from mid-late and late Iron Age deposits, and those from Monument 97 occurred in various Period 1 features there, dated up to AD 70/80 at the latest. Though the large pots placed in pits F684, F685 and F686, including Nos 195, 196, 198 and 200, are more late Iron Age in character than Roman, a specialised function may have had an influence on the fabric.

3.1b Iron Age to Roman, 'transitional' (TSG)

Table 5 gives the TSG A% in the main features and Period 1 as a whole. Almost all the vessels were jars (Table 6).

Of these examples, No. 6 is closest to Iron Age vessels, and is similar to vessels from the Werrington Enclosure (Rollo 1988), Longthorpe (Frere and St Joseph 1974, fig. 54, no. 111), and Fengate (Pryor 1984, fig. 227, no. 9, and fig. 231, no. 5). No. 8 is close to ones from the Werrington Enclosure and Longthorpe (Frere and St Joseph 1974, fig. 54, no. 107). The chevron pattern on No. 41 is easily paralleled on later Roman vessels, for example at Wakerley (Jackson and Ambrose 1978, fig. 45, no. 7), and the simple row of burnished short diagonal lines on No. 10 is common from the Flavian period onwards (*e.g.*, Verulamium: Frere 1972, fig. 104, nos 150–5, and fig. 112, nos 431–42).

The local parallels noted above and the contexts in which they occurred at Orton Hall Farm point to a definite Conquest or immediate post-Conquest date for their production and use; in this area perhaps *c*. AD 50–70. Survival in use or even in production could extend this narrow date-range; however, understanding is hampered by the difficulty in identifying definite stages in their development into fully 'Roman' types. It is often hard to distinguish between wheel-thrown or just wheel-finished vessels. Locally, the possible Bourne/Greetham vessels combined both methods, and much of the other Roman shell-gritted ware could have been made in a similar way. At Orton Hall Farm, vessels Nos 46–54, 95–101, 178–80 and 207–8 appear to have a greater 'ratio' of Roman characteristics, and could, therefore, represent the next stage of development. Some of those at Longthorpe, however (Frere and St Joseph 1974, *e.g.*, fig. 54, nos 100–1 and 103–12), are not that dissimilar. The result is that the discovery of obviously gradual rather than dramatic changes must wait for more information, especially a sequence of well-dated groups. These problems in positive identification of supposed TSG ware make it difficult to give accurate figures (Table 5). There are nineteen jars in TSG or similar ware. The possible tripod vessel, No. 177, might also be in TSG.

3.1c Roman shell-gritted (RSG)

Table 5 gives the RSG A% for the main features and the whole of Period 1.

	F588	F813	F1048	F814	F1025	F525/6	Other	P1
Jar	19	30	22	2	9	2	18	102
Bowl	-	-	1	-	-	-	2	1
Dish	-	2	1		1	-	2	6
Lid	-	4	-	-	-	-	1	6
Misc	-	-	2	-	-	-	-	3
Total	19	36	26	2	10	2	23	118

Table 13 Period 1, RSG main vessel forms, by entries.

Vessel 54, and perhaps No. 178, are the only examples of the lid-seated type which is ubiquitous on sites further up the Nene Valley to the west, such as Quinton (Friendship-Taylor 1979, figs 35–7), Stoke Goldington (Field and Mynard 1979, fig. 82), and Brixworth (Woods 1970, figs 25–9). It is also apparently absent from other sites near Orton Hall Farm, including Maxey, Longthorpe, Fengate and Monument 97, suggesting a definite typological boundary.

The only RSG lids from Period 1, including Nos 58–9, are most easily paralleled in second-century contexts, *e.g.*, Chesterton (Perrin, to be published), though No. 59 is close to lids in other wares from Quinton (Friendship-Taylor 1979, fig. 40, nos 136–7) and Godmanchester (Green 1961, fig. 3, no. 5), which are dated *c*. AD 60–80, and late first to early second centuries, respectively.

Vessels 43–5 and 94 are Roman in fabric, hardness, colour, manufacture and decoration. Dated parallels occurred in Hadrianic to early Antonine contexts at Normangate Field (Perrin and Webster 1990,

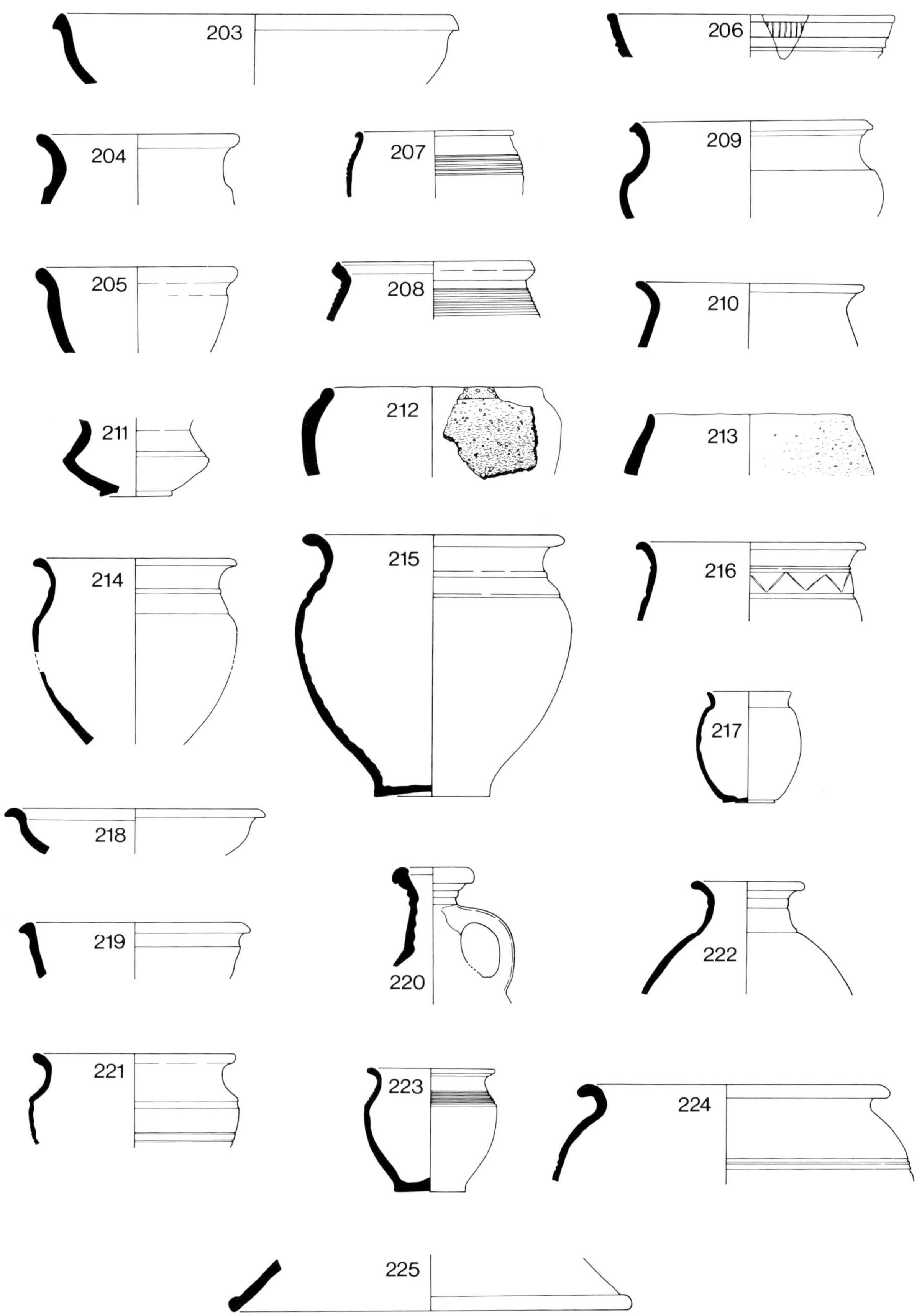

Figure 88 Roman pottery. Period 1: Nos 203–211, 'kitchen' area; Nos 212–213, F875, western round house; Nos 214–220, F525–F526, eastern round house; Nos 221–225, additional Period 1 pots. Scale 1:4.

fig.7, no. 105), and in second half of second-century layers at Chesterton (Perrin, to be published), a date confirmed by their presence at Orton Hall Farm in the F813 and F1048 systems.

3.1d Bourne/Greetham products

Jars 12–14, 104–9, 182 and 224 were most probably manufactured in the Bourne/Greetham area (Bolton 1968). Examples have occurred on other local sites including Fengate (Hayes 1984, fig. 127, no. 18 and fig. 130, no. 68), Normangate Field (Perrin and Webster 1990, 40), Maxey (Gurney 1985, fig. 85, no. 107, and fig. 86, nos 123–4) and Chesterton (Perrin, to be published). The Chesterton and Fengate vessels are dated to the second half of second century, while the Normangate Field provenances are dated *c.* AD 130–150. The Maxey vessels were all from Phase 9 there which contained pottery of second to fourth-century date. A similar jar form occurs in grey ware at Winterton (Stead 1976, fig. 84, no. 104), and is probably from the Antonine kilns at Roxby, though the context is Severan. A vessel at Verulamium (Frere 1984, fig. 91, no. 2222), dated *c.* AD 145–170, is also similar. The Orton Hall Farm dating is somewhat ambiguous, for the vessels occurred in the earlier F588 system and the later F1048 system, but not in the intermediate F813 system, though the F1025 system, which is also dated between the F588 and F1048 systems, did contain the ware. A second-century date seems most likely for their purchase and use on the site. The ware accounts for around 1A% of the Period 1 pottery, and all of the six identifiable vessels are jars.

3.2 'Belgic' wares

The form of No. 15 is very common and can be easily paralleled on local sites, for example Fengate (Pryor 1984, fig. 101, Group 3), and Great Casterton (Todd 1968b, fig. 17, no. 35), and further to the west including Stoke Goldington (Field and Mynard 1979, fig. 83, no. 70) and Moulton Park (Williams 1974, fig. 19, no. 142). Numerous other examples are contained in Thompson's survey of 'belgic' pottery (*e.g.*, Thompson 1982, fig. 69, no. 827; fig. 94, no. 17; fig. 96, no. 42; fig. 104, no. 15). The fabric of No. 15 is unusual (see p. 121), and does not contain any grog.

Grog is not an obvious inclusion in the fabrics of any of the other vessels either, though some of the lumps in Nos 199, 204 and 210 appear to be fired clay of some sort. The surfaces of No. 201 have numerous small thin impressions up to 4mm long, the kind of marks that could have been left by flecks of straw or chaff in the surface of the clay which had been subsequently burnt away during firing. The fabric itself was not organically tempered and as No. 201 is the only vessel with this surface appearance, it must have been the result of some chance occurrence while the pot was being finished or left to dry.

The precise forms of Nos 199, 201, 204 and 209–11 are not easily paralleled but the same general types occurred on most sites with occupation at the appropriate period both locally, for example Fengate (Pryor 1984, Group 3), Longthorpe (Frere and St Joseph 1974), Monument 97 (Rollo, in Mackreth forthcoming), and further afield, for example, Moulton Park (Williams 1974, Group 2). Orton Hall Farm had only a small amount of this material, less than 1 per cent of Period 1 as a whole, and only around 1 per cent of the F588 system pottery, the only one of the three to contain significantly quantifiable amounts.

3.3 Grey wares

Table 5 gives the grey ware A% for the main Period 1 features and Period 1.

The additional forms are a cheese press, a cup, a possible beaker and vessels which might be either jars or bowls.

Though they have been classified together, a number of different sources are involved. Most would undoubtedly have been 'local', but some may have come from further afield. For the purposes of discussion, the vessels are ordered under a number of distinct categories.

3.3a Narrow-mouthed jars (NMJars)

As a type, narrow-mouthed jars are common throughout the Roman period, and the fabric is often the only means by which development and chronology can be traced. The neck cordons and grooves on Nos 16–17, 110–12 and 222 were especially common locally in the later first to early second centuries, for example Wakerley (Jackson and Ambrose 1978, fig. 44, no. 76), Grandford (Potter and Potter 1982, fig. 18, no. 21), and Monument 97 (Rollo, in Mackreth forthcoming), and, occasionally, elsewhere as at Verulamium (Frere 1972, fig. 103, no. 119, fig. 123, no. 830; Frere 1984, fig. 83, nos 1994–5). The smoother profile of No. 61, however, is perhaps indicative of a later date.

3.3b Globular jars

Vessels 18–20 are of the same class as Camulodunum type 108, and possibly type 91 (Hawkes and Hull 1947). The decorated version, No.

	F588	F813	F1048	F814	F1025	F525/6	Other	P1
Jar	20	32	33	5	12	3	16	121
NMJar	2	2	4	-	4	-	1	13
Bowl	1	1	2	-	-	-	1	5
Dish	3	3	-	1	1	-	5	14
Flagon	1	-	-	-	-	-	1	2
Lid	1	-	-	-	-	-	2	3
Other	-	3	1	-	-	-	-	5
Total	28	41	40	6	17	5	26	163

Table 14 Period 1, Grey ware main vessel forms, by entries.

20, is readily paralleled locally at Chesterton (Perrin, to be published), the Werrington Enclosure (Perrin 1988), Maxey (Simpson 1981, fig. 12, no. 7), and further afield at Norwood (Potter 1965, fig. 7, NB4) and Scole (Rogerson 1977, fig. 73, no. 5). The dating is consistently Flavian to early second-century. Nos 18 and 19 may be decorated or plain, and can be additionally paralleled in late first to early second-century contexts at Mileoak (Green and Draper 1978, fig. 9, nos 75 and 83), and Great Casterton (Corder 1961, fig. 15, nos 42 and 43?). The type was also produced in kilns at Little Chester (Brassington 1971, figs 9–10; Dool *et al.* 1985, fig. 77), also in the late first to second century.

No. 21 is possibly of the same general type though it may actually be more akin to a butt-beaker, with Camulodunum type 103 as a possible parallel. A vessel with a similar profile occurred at the Werrington Enclosure. The decoration, impressed, perhaps by means of a straw, plant stalk, or hollow piece of bone, is not easily paralleled, but another vessel from the same site (Perrin 1988), in shell-gritted ware, has a row of circular impressions, while a vessel from Great Casterton (Corder 1961, fig. 15, no. 35) has a herringbone motif. No. 21 is probably of mid to late first-century, possibly Flavian, date.

3.3c Jars with slight carinations, cordons and grooves

The typological variations within this class of jar are illustrated: Nos 23–5, 70–3, 115–25, 171–2, 187–9, 214–16 and 221. The range might have been sub-divided further, but they are considered together here as they have a number of features in common. They also form a separate group in terms of fabric and overall form, when compared with the associated pottery, specifically the grey ware products of the Lower Nene Valley industry.

In many respects, these jars epitomise the standardised, utilitarian product that was so much a feature of the Roman pottery industry as a whole. Apart from occurring on all local and most other sites along the Nene Valley, the same general type can be found throughout south-east England and East Anglia, for example, Scole (Rogerson 1977, fig. 76, no. 67), Braintree (Drury 1976a, fig. 39, no. 39), Nazeingbury (Huggins 1978, fig. 16, no. 143, and fig. 19, no. 274), Southwark (Hammerson and Murray 1978a, fig. 196, nos 1465–70) and Verulamium (Frere 1972, fig. 111, no. 385). Their common antecedent was the 'belgic' cordoned vessel (*e.g.*, Hawkes and Hull 1947, pl. LXXV; Wheeler and Wheeler 1936, pls XLIX and L) adapted to suit Roman methods of production and taste.

At Orton Hall Farm, these types of jar occurred in all of the main ditch systems. Relatively large numbers in the later F1048 system reflect both the nature of the deposit itself and the fact that vessels survived in use beyond the date by which production or trade is thought to have ceased.

Some of the jars have decorated zones, and others may have had decoration which has been eroded away with the surface (Section IV, Greywares). That on No. 70 can be easily paralleled, but the lattice motif on No. 24 is less common locally, though it occurs on vessels made at Brampton (Green 1977, figs 26–30), and in south Lincolnshire (Stead 1976, fig. 82) and on some jars on sites in Cambridgeshire, for example Coldham Clamp (Potter 1965, fig. 2, C4, C127, C138). It is possible that the areas in which this particular motif was more common could have been the sources for some of the grey wares reaching the site in the late first to early second century.

3.3d Small jars or beakers

Vessels 62–3, 113, 185 and 217 are of a type which can be paralleled in BB1 (Gillam 1976, fig. 2, nos 18–22), various grey wares, possibly including BB2 (*e.g.*, Perrin 1981a, fig. 29, no. 360; Hull 1963b, type 124; Dakin 1961, fig. 6, no. 3), and a variety of other wares, including local oxidised ware (Perrin 1981a, fig. 29, no. 358) and mica-dusted ware (No. 165). Most seem to be of Hadrianic to Antonine date.

3.3e Small jars with incised shoulder grooves

A number of small jars broadly similar to Nos 26, 114, 184 and 223 have been found on sites locally and to the west, including Maxey (Simpson 1981, fig. 12, no. 10, and fig. 10, no. 14), Hardwick Park (Foster *et al.* 1977, fig. 15, no. 21), Great Casterton (Corder 1961, fig. 14, no. 10) and Grandford (Potter and Potter 1982, fig. 36, no. 275). The fabric is usually coarse, and in the case of No. 26, exceptionally so. This, together with the nature of the decoration, perhaps suggests a specific function for the vessels. Most dated examples are late first-early second century.

3.3f Miscellaneous jars

No. 22 may be part of a butt-beaker similar to some found at Longthorpe (Frere and St Joseph 1974, fig. 52, nos 39–41, 44). It is the only possible example of this type of vessel from Orton Hall Farm. No. 64 can be directly paralleled at Grandford (Potter and Potter 1982, fig. 20, nos 59–60) in contexts dated *c.* AD 90–150. The type is perhaps a derivative of certain Gallo-Belgic jars (*e.g.*, Hawkes and Hull 1947, types 91a, 92a+b; and Holwerda 1941, pl. 10, nos 477–8, 489–90). **67** is almost certainly from a decorated, 'slashed cordon' jar, and is one of the few examples of this usually common local type from Orton Hall Farm. The fabric is not certainly that of the Lower Nene Valley industry, therefore the vessel may be a product of another kiln source. The rim is also unfamiliar, being reminiscent of the 'cornice' type common on early to mid second-century colour-coated beakers. Nos 68 and 69 are also forms which might usually be expected to occur in Lower Nene Valley grey ware. They may be from the same source as No. 67. No. 183 is another type of small jar which occurs in a number of fabrics (*e.g.*, Gillam 1970, type 165; Holwerda 1941, pl. 10, no. 495).

3.3g Decorated bowls of London ware type

Nos 27, 79, 192 and 206 belong to the range of vessels, usually of imitation samian form, with incised, stamped or rouletted decoration commonly called London ware type. It accounts for approximately 4A% of the Period 1 pottery and, of the four recognisable forms, three were form 37 and one form 30.

The particular motifs on Nos 27 and 79 can be matched at Chesterton (Perrin 1980, fig. 5, nos 6 and 21), while that on No. 27 also occurs on a vessel from Normangate Field and Ashton (*ibid.*, fig. 5, nos 3 and 7). The lozenge and stamp motif on No. 192 can be paralleled at Grandford (*ibid.*, fig. 5, no. 12; Potter and Potter 1982, fig. 30, no. 186) and Chesterton (Perrin 1980, fig. 5, no. 7), and, apparently, further afield (Potter and Potter 1982, 63). The fabrics of Nos 27, 79 and 192 would, on existing evidence, suggest a late first to early second-century date, though most of the noted parallels are from early to mid second-century contexts. The sherds of London ware type in LNVGW are considered below.

3.3h Dishes of imitation samian ware form

It is not certain if vessels 126–7 should correctly be classified as grey ware: each had been altered by burning and partial submersion, and they may equally well be LNVGW or LNVCC, especially since both, and in particular No. 127, show signs of having been slipped. Nos 126 and 127 appear to be imitating samian forms 31 and 18/31 respectively, and copies of the latter form have also occurred locally at Monument 97 (Rollo, in Mackreth forthcoming), Normangate Field (Perrin and Webster 1990), and Chesterton (Perrin, to be published) in Hadrianic and Antonine contexts. No. 128 seems to be closer to a Curle 15 in form and can be paralleled at Grandford (Potter and Potter 1982, fig. 29, no. 171) in colour-coated ware, and at Fengate (Hayes 1984, fig. 132, no. 89) and Chesterton (Perrin, to be published), in deposits of the second half of the second century.

The precise form of No. 129 is difficult to assess, though a devolved imitation form 37, similar to a vessel from Grandford (Potter and Potter 1982, fig. 30, no. 186), is perhaps the most likely.

3.4 LNVGW

Table 5 gives the LNVGW A% in the main Period 1 features and Period 1.

The additional forms are a cheese press, colander and a jar or bowl.

Vessels 32, 82, 133–4 and 194 may not actually have been made locally. The assemblage has few vessels similar to those which may be some of the earliest types of LNVGW produced, except for the 'slashed cordon' jars 133 and 193–4, and vessel 147. The presence of cordons on Nos 32 and 82 (if made locally), and No. 81, however, may suggest that these were also part of the earliest production range. The rest are types which are easily paralleled in the mid to late second century, and are so common that little additional comment is required, except in one or two instances.

	F588	F813	F1048	F1025	F525/6	Other	P1
Jar	4	12	32	2	1	11	62
Bowl	-	2	5	-	1	2	10
Dish	2	1	5	-	1	2	11
Flagon	-	-	1	-	-	-	1
Other	-	1	1	1	-	-	3
Total	6	16	44	3	3	15	87

Table 15 Period 1, LNVGW main vessel forms, by entries.

Most of the known LNVGW flagons or jugs are similar in form to those made at Sibson (Guide 14). However, as the majority of the known vessels are of third-century date, the form of No. 131 was perhaps more normal in the second century and it can be paralleled at this time at a number of sites, including Verulamium (Frere 1972, fig. 116, no. 580).

The colander or similar vessel, No. 152, can be matched at Hardwick Park (Foster *et al.* 1977, fig. 13, no. 52), Grandford (Potter and Potter 1982, fig. 32, no. 212), Wakerley (Jackson and Ambrose 1978, fig. 47, no. 125 and fig. 52, no. 184), and Colchester (Hull 1963b, type 298). Similar vessels were made at the Roxby, Lincolnshire, kilns (Stead 1976, fig. 67, nos 43–4) and one of the Wakerley vessels (Jackson and Ambrose 1978, fig. 52, no. 184) was also a product of a kiln there. That the particular form of this vessel can be so easily paralleled is perhaps indicative of either a popular size, or a specific function within its obvious area of use as a colander or strainer.

Many of the jars, including Nos 33, 80 and 136–45, have incised grooves occurring, occasionally, with burnished wavy line decoration, and both of these would appear to be the chronological successors to cordons and burnished diagonal lines or lattice. The form of dish 150 suggests that the Lower Nene Valley potters, in common with those of most contemporary industries, followed the fashion for certain varieties of dish and bowl forms established by the widespread successful marketing of BB1 and BB2.

3.5 LNVCC

Table 5 gives the LNVCC A% in the main Period 1 features and Period 1.

	F813	F1048	F814	F1025	Other	P1
Jar	-	1	-	1	1	3
Bowl	-	1	-	-	-	1
Dish	3	1	-	2	-	6
Beaker	10	7	1	-	2	20
Total	13	10	1	3	3	30

Table 16 Period 1, LNVCC main vessel forms, by entries.

The absence of LNVCC, especially in the F1048 system, could suggest that the ware was not common on rural sites until the end of the second century. However, a domestic focus may have existed west of the excavated area.

Many of the beakers, including Nos 85, 154–5 and 174, occur only as one or two sherds. Some of these may have come from the same vessels, hence the actual numbers of beakers may have been fewer. Four of the beakers have underslip barbotine decoration, including one 'hunt cup', No. 84, and twelve others are roughcast, including Nos 83 and 153. Their rims are all of cornice type. The other vessels, including Nos 86 and 156–7 are really just colour-coated varieties of forms also produced in LNVGW, and similar vessels are known from other local sites including Chesterton (Perrin, to be published) and Normangate Field (Perrin and Webster 1990, fig. 11, no. 173; fig. 15, no. 276).

The amount of LNVCC with roughcast decoration from Orton Hall Farm was small, none the less it is significant, providing as it does examples from a relatively ordinary settlement. Production of roughcast beakers is attested at Great Casterton (Corder 1961, 50–3), and the technique was probably used at other local kiln sites. Evidence for its use in the area is increasing and it has now been recognised at Ashton, Fengate and Maxey as well as further afield at Stonea, Piddington and Grandford.[13] These sites, with Orton Hall Farm, appear to suggest that production started in the Hadrianic period, though none occurs in any of the Normangate Field or Chesterton groups of that date. The roughcasting on all these examples consists of clay particles and all the rims appear to be of cornice type.

3.6 Verulamium region wares (VR)

Cream and white ware of Verulamium region manufacture accounts for less than 1 per cent of the pottery in the F588 and F813, and none in the F1048 systems. Nos 35–7 and 89 can be paralleled at Verulamium itself (No. 35: Frere 1972, fig. 107, nos 236 and 242; No. 36: *ibid*., fig. 107, nos 243–4, Frere 1984, fig. 82, no. 1946; No. 37: Frere 1972, fig. 129, nos 993–4; No. 89: *ibid*., fig. 106, no. 206, fig. 119, nos 691–3, Frere 1984, fig. 101, no. 2415) and No. 89 is close to some VR products at Southwark (Hammerson and Murray 1978a, fig. 121, nos 679 and 681; Hammerson and Murray 1978b, fig. 194, no. 1397). Nos 87–8 and 219 may also have been made in the Verulamium region. No. 87 is also similar to pottery found at Verulamium itself (Frere 1972, fig. 124, no. 872; Frere 1984, fig. 96, no. 2305), and carinated bowls with reeded rims like No. 88 were a common VR product (Frere 1972, fig. 118, nos 669–70 and 679; Frere 1984, fig. 102, no. 2449), but were probably also made more locally (see Nos 163–4).

Vessels like Nos 37 and 219 are potentially embarrassing as their pale fabric, with grey surfaces, and form can be readily paralleled by locally produced LNVGW vessels and the temptation may be not to look for a non-local source. Therefore, it is possible that a number of VR products of these types may have been misidentified as made in the Nene Valley, though the numbers are unlikely to be large. Fortunately a link with the Verulamium region is proven by other vessel types, and dishes like Nos 37 and 219 were made in the second half of the second century in both the Verulamium region and the Lower Nene Valley.

3.7 Other cream/white wares

Table 5 gives the C/W A% in the main Period 1 features and the whole of Period 1.

	F588	F813	F1048	F814	F1025	F525/6	Other	P1
Jar	-	3	5	-	-	-	1	9
Bowl	1	-	2	-	-	-	-	3
Flagon	2	2	4	1	1	1	-	11
Mortarium	-	-	1	-	-	-	3	4
Total	3	5	12	1	1	1	4	27

Table 17 Period 1, C/W main vessel forms, by entries.

The occurrence of carinated bowls similar to Nos 88 and 163–4, on most local sites, for example Chesterton (Perrin, to be published), Godmanchester (Green 1961, fig. 5, no. 16; Frend 1968, fig. 10, nos 1–8), and Castor (Green and Green 1987, fig. 11, no. 35), are indicative of another, closer, origin than the Verulamium region. The fabrics of Nos 163–4 are not readily identifiable as VR, and the quantities in the Godmanchester area may suggest they were produced nearby. Jar No. 162 has a similar fabric and can also be paralleled at Godmanchester (Frend 1968, fig. 9, no. 1) and at Chesterton (Perrin, to be published). The date for both the jars and the carinated bowls appears to be the second half of the second century.

Bowls 39 and 91 stand out from the bulk of the other vessels. No. 39 is possibly a devolved imitation of a samian ware form and is similar to a vessel from Towcester (Woodfield 1983, fig. 23, no. 76) in orange-red ware, in a layer dated late first to late second century. The fabric is softer than usual and the surfaces are very powdery. No. 91 is a more obvious imitation of a samian ware form 29, and can be paralleled on many sites with late first-century deposits, for example, Wroxeter (Darling 1977b, fig. 6.7, no. 22), Godmanchester (Green 1961, fig. 3, no. 6) and Quinton (Friendship-Taylor 1979, fig. 42, no. 179). There are barely discernible traces of a colour-coat or slip on parts of the vessel, and the fabric is fairly smooth, though not obviously that of the Lower Nene Valley industry. It is not certain, therefore, where bowls 39 and 91 were produced.

The same uncertainty applies to the various white and cream ware flagons, other than those made in the Verulamium region, though a local source is likely. There are at least eleven from Period 1, including Nos 38, 90, 158–61, 175 and 220 which came from the main ditch systems, with both F588 and F813 having two, and F1048 four. They also occur on most local sites including Chesterton (Perrin, to be published), Normangate Field (Perrin and Webster 1990, fig. 7, nos 93–4), Fengate (Hayes 1984, fig. 132, no. 95), Godmanchester (Green 1960, fig. 5, no. 1), Castor (Green and Green 1987, fig. 11, no. 29), Monument 97 (Rollo, in Mackreth forthcoming), and Maxey (Gurney 1985, fig. 9, no. 184).

Most of the deposits in which they occurred were second-century in date, with an emphasis in the second half. The noticeable differences in rim form may be chronological with the internal 'ledge' on Nos 159, 175 and 220 being a more obvious Antonine characteristic (Gillam 1970, type 8), while the triangular or bead top-ring of Nos 38 and 90 perhaps being reminiscent of later first-century or Trajanic types (*ibid*., types 3 and 5?). Nos 158 and 160 could either be intermediate types or potter's variations. Apart from No. 83, the Orton Hall Farm flagons have very few neck rings, and the top one is larger than the rest, suggesting that they are all probably of second-century date (*ibid*., types 5 and 8). The limited chronology provided by the ditch systems does not contradict this suggested development for the potentially earlier Nos 38 and 90 which came from F588 and F813, while F1048 contained the later Nos 158–60. Some of the pottery associated with Nos 175 and 220 also suggests an Antonine date.

3.8 Miscellaneous wares

Apart from amphorae and mortaria, less than 0.5% of the Period 1 pottery comprises wares other than those already noted and discussed. Some are shelly, buff or grey wares of uncertain type or source. There is one sherd of CGCC. The more significant wares are black-burnished and grogged.

3.8a BB1, BB2, and possible local variants

Vessel 92 is reminiscent of BB2 in fabric and form, and is the only example of possible BB2 from Period 1. The main period of marketing of this ware outside its local area was the mid-second to mid-third centuries, but as the principal production centres were in the Thames Estuary region (Williams 1977), the Nene Valley area could just have fallen within the wider limits of its 'local' distribution, and might, therefore, have received BB2 before the Antonine expansion. However, the apparent lack of significant amounts of any Thames Estuary, or even Colchester products, including mortaria and colour-coated ware, on Lower Nene Valley sites suggests that this is unlikely.

As a singleton, however, No. 92 need not have found its way to the site by ordinary trading links, and has limited value for dating and evidence of trade. Its context within the F813 system is ambiguous in that it could equally well fit a date before or around the mid-second century. A more local source than the Thames Estuary cannot be ruled out, moreover, as similar vessels were made at Roxby in Lincolnshire (Stead 1976, fig. 67, no. 53) and Brampton (Green 1977, fig. 31, nos 105 and 107, and fig. 32, no. 113). These kilns were certainly in production in the Antonine period.

The BB1 from Period 1 might not have been of Dorset origin. A similar ware was, apparently, made in the Cambridge/Godmanchester area,[14] and it is possible that this was the source of some of the black-burnished wares from the site, including the few reminiscent of BB2.

3.8b Grogged ware

Jar 60 is the only definite example from Period 1 of a ware incorporating grog. Various grogged wares are more common further up the Nene Valley, and in the Midlands as a whole, at sites such as Towcester (Woodfield 1983, fabric 35c), Ashton and Alcester.[15] Its scarcity in the Lower Nene Valley probably reflects the widespread availability of other local wares, especially those which could be made from clays with naturally-occurring shell inclusions not needing additional temper.

Period 2
(Fig. 89)

1 Introduction

The initial development and use of the reorganised site did not appear to last for very long before modifications to the layout were carried out, and new buildings were erected in Period 3. Period 2 was, therefore, relatively short-lived, running from *c*. AD 175–225/250, and, despite the obvious changes which it encompassed, the amounts and relative quality of the pottery from the various features assigned to it are low. This is perhaps because, initially at least, apart from the period being shorter, it was mainly a time of reorganisation, with a possibly reduced occupation in the excavated area. Some of the rubbish may also have been dumped elsewhere. The layers contained 5A% of the total stratified pottery from Orton Hall Farm.

The occupation residuum was less easy to pinpoint and the dates were based on samian ware, mortaria and typological developments in both the colour-coated and grey wares of the local industry. The date of the final phases of Period 2 was dependent on that assigned to the succeeding Period 3 activities, for example in Barn 1. Thus the occurrence of colour-coated 'funnel-necked' beakers, usually dated from *c.* AD 225, was significant.

	F189	F969	F437	F65/217	P2
Fabric					
RSG	28	29	24	24	32
Grey	12	35	17	-	24
LNVGW	50	22	28	50	29
LNVCC	5	11	19	9	7
C/W	4	1	5	13	5
P2 A%	9	14	7	6	

Table 18 Period 2, main fabric A%.

	RSG	Grey	LNVGW	LNVCC	C/W	Other	P2
Jar	52	23	34	6	2	-	117
Bowl	1	-	5	2	1	1	10
Dish	3	4	5	2	-	-	14
Beaker	-	-	-	10	-	2	12
Flagon	-	1	-	-	3	1	4
Mortarium	-	-	-	-	7	1	9
Other	2	4	4	2	-	-	12
Total	58	32	48	22	13	5	178

Table 19 Period 2, main vessel forms/fabric, by entries.

The additional fabrics are London ware type, CGCC, VR, samian ware, amphorae, miscellaneous SG, BB1 and uncertain. Other forms are a colander and a possible flagon, vessels which could be either jars or bowls, including some wide-mouthed types, and either bowls or dishes.

There were three sub-phases *a*, *b* and *c*:

a *c.* AD 175–*c.* AD 200

b *c.* AD 200–*c.* AD 225?

c *c.* AD 225?–*c.* AD 250?

As with the Period 1 sub-phases, these could not always be clearly separated ceramically.

2 Period 2 features

In contrast with Period 1, there were no large groups of pottery. Only five features or systems, F189 from phase *a*, F437 from phase *b*, and F955, F980 and F218 from phase *c*, had over 5A%, and only F969 of phase *a* over 10A% of the total pottery from the period. Much of the pottery contained in these layers is not obviously any later in date, or different in type, to that previously noted for the feature in Period 1, and residual pottery occurs in all of the Period 2 contexts. Some of the pottery is apparent in levels associated with the early Period 3 reorganisation. In general, there are few features with pottery obviously contemporary with the date of Period 2, the best being that from the Barn 2 development.

There is little dating evidence to supplement that provided by the coarse wares. In addition to that from the main features listed below, phase *a* features F772 and F1010 contained samian ware dated to *c.* AD 120–130 and the Antonine period (Samian, below), and a mortarium from F772 (Mortaria, Fig. 114, M13) is early third-century in date. Phase *c* features F218, F338, F433, F633, F1022 and F1111; all contained samian ware dated to the second century with none later than the second half (Section VIII).

2.1 Main enclosure ditch (Phase *a*)

Consisting of F189, F194, F204, F220, F221?, F231, F580 and F734, only F220, F221 and F231 did not receive material in later periods. Samian ware from F189 is of Hadrianic and Antonine date (below), but two mortaria (Mortaria, Fig. 114, M8–9) are dated *c.* AD 230–400 and after *c.* AD 200. Table 18 gives the main fabric A%.

There are a few sherds of uncertain fabric, and the unlisted forms are a colander in grey ware, and a possible flagon in LNVGW.

	RSG	Grey	LNVGW	LNVCC	C/W	Total
Jar	4	-	4	-	-	8
Bowl	1	-	-	-	-	1
Dish	-	2	1	-	-	3
Beaker	-	-	-	1	-	1
Mortarium	-	-	-	-	2	2
Lid	1	-	-	1	-	2
Other	-	1	1	-	-	2
Total	6	3	6	2	2	19

Table 20 F189 system, main vessel forms/fabric, by entries.

(Fig. 89)

226 LNVGW. CR20/26/20. F189, L364 (107)
227 LNVGW. CR19+26/21/19+26. F189, L396 (58)
228 LNVCC. CR4+11+12/4+26/4+11+12. F189, L396 (58)

2.2 F969. Final fills (Phase *a*)

F969 was part of the F813 Period 1 ditch system, and the layers from it assigned to Period 2 were over its last major fills. The samian ware is all of the second half of the second century (Samian, below). A mortarium (Mortaria, Fig. 114, M16) is dated *c.* AD 230–400. Table 18 gives the main fabric A%.

There are two LNVGW vessels which may be dishes or bowls.

	RSG	Grey	LNVGW	LNVCC	C/W	Other	Total
Jar	8	3	3	-	-	-	14
Bowl	-	-	1	1	-	-	2
Dish	1	1	1	-	-	-	3
Beaker	-	-	-	6	-	-	6
Mortarium		-		-	1	-	1
Other	-	-	2		-	-	2
Total	9	4	7	7	1	-	28

Table 21 F969 system, main vessel forms/fabric, by entries.

229 RSG. CR10/21/10. L1858 (68)
230 GW. CR21//10/26/10//21. Two-thirds vessel. L2204 (68)
231 GW. CR20/26/20. LNVGW? L1858 (68)
232 LNVGW. CR21/26/21. Stanground? L2204 (68)
233 LNVCC. CR12/14+26/12. L2204 (68)
234 LNVCC. CR25/4/10 patches. L1858 (68)
235 LNVCC. CR12/11/8. L1858 (68)

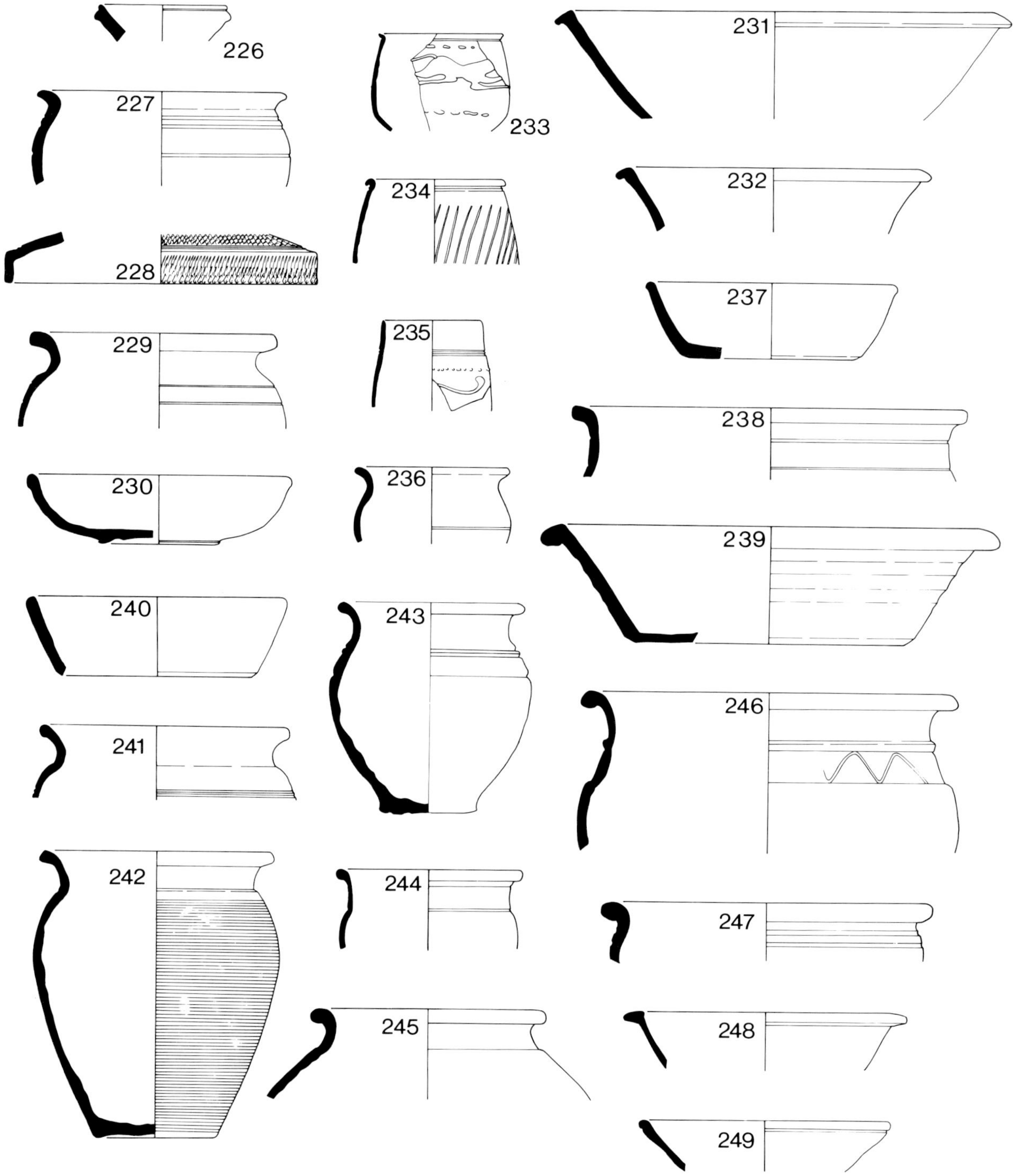

Figure 89 Roman pottery. Period 2: Nos 226–235, phase *a*; Nos 236–237, phase *b*; Nos 238–240, phase *c*; Nos 241–249 additional Period 2 pots. Scale 1:4.

2.3 Enclosure ditches (Phase *b*)

The small rectangular enclosure located in the middle-north of the site attached to the boundary ditch was bounded by ditches F437, F440, F445 and F446. F437 continued to receive material in Period 3. Samian ware from F437 (Samian, Fig. 113, No.7) was of pre-Antonine, Hadrianic-Antonine and Antonine date. Table 18 gives the main fabric A%.

The only additional fabric is BB1.

236 LNVCC. CR2/4/2. F437, L859 (72)
237 LNVCC. CR3/14/3. F437, L824 (72)

	RSG	Grey	LNVGW	LNVCC	C/W	Total
Jar	4	1	2	1	1	9
Bowl	-	-	1	-	-	1
Dish	-	-	-	2	-	2
Total	4	1	3	3	1	12

Table 22 F437 system, main vessel forms/fabric, by entries.

2.4 Barn 2 (Phase *c*)

No floor levels survived and it is the associated drainage ditches, F65 and F217, which provided the dating evidence. The former also received material in Period 3. All the additional dating evidence came from F217. Three mortaria are dated *c*. AD 200–280, *c*. AD 250–350 and after *c*. AD 180 (Mortaria, Fig. 114, M10–M12). Samian ware was of Hadrianic or Antonine date (Samian, below). Table 18 gives the main fabric A%.

There were a few sherds of amphorae.

	RSG	Grey	LNVGW	LNVCC	C/W	Total
Jar	3	1	5	1	-	10
Bowl	-	-	1	-	-	1
Dish	-	-	1	-	-	1
Mortarium	-	-	-	-	3	3
Total	3	1	7	1	3	15

Table 23 Barn 2 ditches, main vessel forms/fabric, by entries.

Vessels **238–240** are in LNVGW, and from F217, L361 (83).
238 CR21/26/21. 'Speckled' surfaces.
239 CR21/18/21.
240 CR20/19/20. Stanground product?

Period 2 Additional Sherds

241 RSG CR3+10,28 patches. F324, L649 (70)
242 RSG. CR4/20/4+8. Two-thirds vessel. F980, L2134 (88)
243 GW. CR20/18/20. Near complete. F964, L2148 (60)
244 LNVGW. CR21/26/21. (F481), L1020 (122)
245 LNVGW. CR20/26/20. F196, L309 (86)
246 LNVGW. CR15+18/26/15+18. F324, L649 (70)
247 LNVGW. CR21//26/21/26//21. F218, L406 (84)
248 LNVCC. CR21+25/26/21+25. Stanground? F1022, L2456 (92)
249 LNVCC. CR3//4/26/4//3. Mica-gilt .F639, L1240 (88)

3 The Pottery

3.1 RSG

Table 18 gives the RSG A% for the main Period 2 features and Period 2.

	F189	F969	F437	F65/217	Other	P2
Jar	4	8	4	3	33	52
Bowl	1	-	-	-	-	1
Dish	-	1	-	-	2	3
Lid	1	-	-	-	1	2
Total	6	9	4	3	36	58

Table 24 Period 2, main RSG vessel forms, by entries.

All the vessels are apparently wheel-made, none having hand-made bodies.

The jars, including Nos 229 and 241–2 have a smoother profile with a less marked shoulder than some of those in Period 1 (Nos 47, 96–7). Although a few retain horizontal shoulder grooves, as on Nos 229 and 241, these do not occur at such an obvious break in the line of the vessel wall. The rilling on No. 242 is reminiscent of that on No. 180 (Period 1 vessels, such as No. 101), and indicates a continuity of practice or function.

In general the overall proportions of the jars are less globular and more elongated, with the width of the rims approaching that of the maximum girth. This may indicate a gradual chronological development akin to that noted for BB1 cooking pots (Gillam 1976, 62–3). The rims of Nos 229 and 241–2 are also of the 'cavetto' type (Webster 1976, 12). A similar vessel from Chesterton (Perrin, to be published) came from a late second to mid-third-century context. Little can be said of the other vessels which are mainly represented by small sherds.

	F189	F969	F437	F65/217	Other	Total
Jar	-	3	1	1	19	24
Dish	2	1	-	-	1	4
Lid	-	-	-	-	2	2
Other	1	-	-	-	1	2
Total	3	4	1	1	23	32

Table 25 Period 2, main grey ware vessel forms, by entries.

3.2 Grey wares

Table 18 gives the grey ware A% for the main Period 2 features and Period 2.

The additional forms are a colander and a vessel which may be a dish or bowl.

Two of the illustrated vessels, Nos 230–1, both from the F969 final fills, are not obviously any later typologically than similar vessels from Period 1. The Gallo-Belgic-derived dish No. 230 is similar to No. 76. However, as it does not have an internal 'step', it probably represents the ultimate devolved version of the type. Its occurrence in a Period 2 context does not necessarily mean that Gallo-Belgic-derived vessels continued to be produced into the late second or third centuries. No. 231 is close to No. 130 in form.

Most of the other vessels, including the jars, can be matched in Period 1. A great deal of the Period 2 grey ware is likely to be residual, with some of the vessels represented by larger sherds, for example Nos 230 and 243, constituting survivals in use, or disposal at the end of Period 1 or early in Period 2.

3.3 LNVGW

Table 18 gives the LNVGW A% for the main Period 2 features and Period 2.

	F189	F969	F437	F65/217	Other	Total
Jar	4	3	2	5	20	34
Bowl	-	1	1	1	2	5
Dish	1	1	-	1	2	5
Other	1	2	-	-	1	4
Total	6	7	3	7	25	48

Table 26 Period 2, main LNVGW vessel forms, by entries.

Had the RSG not included some large heavy storage jars, LNVGW would be the most common ware in Period 2. The additional forms are a possible flagon, and vessels which may be dishes or bowls.

Of the jars, No. 244 is similar to Nos 136–42 from Period 1, though the more upright neck could be a later characteristic as it is most easily matched in the later second and third centuries by jars fired in a number of kilns including Sibson (Guide, fig. 1, no. 4), and occurring on other local sites, such as Chesterton (Perrin, to be published). The absence of burnished decoration, and occasionally grooves may also be later traits. No. 246, however, has a neck cordon and a burnished wavy line and is, therefore, like Nos 81 and 82, most probably residual.

Nos 227, 238, 245 and 247 are all third-century types. Nos 227, 238 and 247 are essentially wide-mouthed jars or bowls, of which No. 238 is perhaps a larger version of No. 142, and vessels like those from the later second-century group at Fengate (Hayes 1984, fig. 126, nos 3–6). The range of pottery made at both Sibson (Guide, fig. 1, no. 10) and Stanground[16] included such jars or bowls. No. 245 is difficult to parallel, though largish narrow-mouthed jars of similar type were also made at Sibson (*ibid*., fig. 1, no. 11), and are known in colour-coated ware.

No. 226 is similar to a VR product (Frere 1984, fig. 82, no. 1964) in a layer dated *c*. AD 170–190. No. 232 is easily paralleled by vessels in LNVGW and LNVCC. At Chesterton (Perrin, to be published), one in LNVGW occurred in a layer of *c*. AD 150–200, while one from the same site in LNVCC came from an early to mid third-century pit. No. 232 could possibly be a Stanground product.

The large bowl or dish, No. 239, and the plain-rimmed dish No. 240 are also third-century types. Vessels like No. 240 are ubiquitous, occurring in large numbers in both LNVGW and LNVCC, and was obviously made at most of the third-century kiln sites. No. 239 is slightly less common, but again occurs in both LNVGW and LNVCC, and was definitely made at Sibson (Guide, fig. 1, no. 17).

3.4 LNVCC

The Period 2 features contained relatively little LNVCC, possibly because much of the pottery was residual from Period 1 which itself did not have a lot of LNVCC, or because LNVCC was still not extensively used on rural sites. It might also reflect the basic character of Period 2.

Table 18 gives the LNVCC A% in the main Period 2 features and Period 2.

	F189	F969	F437	F65/217	Other	Total
Jar	-	-	1	1	4	6
Bowl	-	1	-	-	-	1
Dish	-	-	2	-	1	3
Beaker	1	6	-	-	3	10
Box	1	-	-	-	-	1
Other	-	-	-	-	1	1
Total	2	7	3	1	9	22

Table 27 Period 2, main LNVCC vessel forms, by entries.

The additional form is a vessel which may be a dish or bowl.

The 'hunt cup', No. 233, has a rim form slightly reminiscent of the true 'cornice' type and this, together with its squat proportions, suggests that it was likely to have been a mid to late second-century product and either residual or a survival in use in Period 2. It came from the final fills of ditch F969. The motif is unusual in that it runs from left to right. The 'late-cornice' or simple curved rim of No. 234 occurs on all the other beakers, except No. 235, and is a late second to early third-century type. No. 235 is a common form of small beaker (Gillam 1970, fig. 9, nos 80–1), and the decoration on No. 234, though it does not occur that frequently, is usually indicative of a third-century date.

The forms of Nos 237 and 248 are more common in LNVGW but other LNVCC examples are known, for example at Chesterton (Perrin, to be published). Similar vessels to No. 248 were made at Stanground,[17] probably in the first half of the third century. No. 237 illustrates the caution required when dating some LNVCC, for the form might suggest a fourth-century date, but examples from Orton Hall Farm and elsewhere show clearly that they were produced much earlier. The slight bead on the rim could prove to be a useful diagnostic characteristic. No. 249 is similar in form to 86 from the Period 1 F813 ditch system. Along with Nos 165 and 373 it indicates the variety of vessels with mica-rich slips on the site.

No. 228 is stratigraphically the first 'Castor box' from the farmstead. Comment has been made elsewhere (Guide, p. 24, no. 89), that there is a gradual development in 'boxes'. The earlier examples have angular profiles with well executed rouletting and narrow, beaker-like, bases (Gillam 1970, fig. 32, no. 342), while the later versions have more rounded profiles with poorer rouletting and wide, flat bases (Wild 1974, fig. 8, no. l). This development apparently did not occur in even or regular stages, and there could be considerable variation at any one time, probably as a result of the difference in expertise of the potters producing them. The junction angles and quality of decoration on No. 228 would place it in the first half of the third century.

Finally, the four jars, including No. 236, probably indicate continued small-scale production of a type first apparent at the end of Period 1 (No. 156). It is not certain if the particular form of No. 236 is of chronological or typological significance.

3.5 Other pottery

The Period 2 features contained a few sherds of cream and white wares, including some VR and probably some of LNVC/W. Most are body sherds of uncertain form, but those of probable LNVC/W include three flagons, one ring-necked, and three jars, one lid-seated.

Table 18 gives the C/W A% in the main Period 2 features and Period 2.

In addition there are sherds of BB1, London ware type and CGCC. The latter comprise at least two beakers, and the London ware type include an imitation samian form 30.

	F189	F969	F437	F65/217	Other	Total
Jar	-	-	1	-	1	2
Bowl	-	-	-	-	1	1
Flagon	-	-	-	-	3	3
Mortarium	2	1	-	3	2	8
Total	2	1	1	3	7	14

Table 28 Period 2, main C/W vessel forms, by entries.

Period 3
(Figs 90–6)

1 Introduction

The basic farmstead plan established in Period 2 lasted throughout the rest of the Roman occupation, and influenced that which followed. Periods 3 and 4 were characterised by modifications, additions and alterations to this plan, some of a major nature.

The best dating evidence for this period was provided by the large range of locally made colour-coated vessels, most of which have accepted chronologies based on examples from many sites, both local and national. The developments in beakers involving shape and decoration were particularly significant, especially the use of white paint and the appearance of 'rhenish' types. There is little independent dating evidence available. By Period 3, all the samian ware (Samian, below) is either residual or survivals in use and, therefore, of no use for dating, and the mortaria are mainly types which were prevalent in the fourth century as well as the third. The finds also include first and second-century items, and most of the others cannot be dated closer than the third to fourth centuries. One coin was useful. Once again the re- arrangements that followed, and some that occurred within the period, provided a few *termini ante quem*, and the nature and date of the suspected end-of-period rubbish deposits were important.

The date-range of Period 3 is assessed as AD 225/250 to AD 300/325.

There is considerably more pottery from this period than from Period 2 because it came from the first intensive use of the replanned farmstead in the excavated area.

The Period 3 layers contained 19A% of the total stratified pottery recovered.

	F105	F1094	Yard	F1016	F435	F441	F772	F254	F969	P3
Fabric										
RSG	84	17	33	4	19	24	51	23	28	30
Grey	+	8	13	-	6	8	6	5	11	10
LNVGW	5	40	27	62	20	27	10	48	44	27
LNVCC	9	25	15	32	46	36	29	10	13	23
C/W	+	1	7	2	6	2	4	11	-	5
P3 A%	5	12	29	3	6	5	3	4	3	

Table 29 Period 3, main fabric/feature A%.

The additional fabrics are BB1, BB2-type, Horningsea, miscellaneous SG, grogged, buff/cream, London ware type, Oxfordshire, LR, possible Hadham area, possible CCC, CGCC, other colour-coated wares, amphorae, VR, and uncertain. Most of the other vessels are ones which may be either jars or bowls, or dishes or bowls.

As with Period 1, the illustrations and discussion concentrate on the pottery from a number of significant features, supplemented by examples from additional features and sherds of intrinsic interest.

	RSG	Grey	LNVGW	LNVCC	C/W	Other	P3
Jar	165	44	121	23	5	11	369
WMJar	-	-	14	23	-	1	38
Bowl	2	6	24	31	3	10	76
Dish	3	8	49	33	3	24	120
Flagon	-	1	6	6	3	-	15
Beaker	-	-	1	54	-	8	63
Lid	5	2	1	4	-	-	12
Box	-	-	-	-	-	-	12
Mortarium	-	-	-	-	40	8	48
Other	1	1	3	35	-	11	51
Total	176	62	219	210	54	73	805

Table 30 Period 3, main vessel forms/fabric, by entries.

2 Period 3 features

The primary deposits are the vat base F105, drier F156, and its stoke-hole F201 in Barn 2; an interconnected sump F1094, shallow trench F1113, and pit F1121 associated with Barn 1; a well F1016, also adjacent to Barn 1; and the primary metalling and use of the walled yard between the House and Barn 1, as well as an area of metalling east of it in the Main Yard. The additional features consist mainly of the ditch systems, F435, F441, F772, of various enclosures, and two further wells, one, 'F969', in the walled yard close to F1016 noted above, and the other, F254, in the south end of the central area, probably associated with Barn 3.

2.1 Vat base F105, drier F156 and stoke-hole F201

The building housing these was put up towards the end of Period 2. The features were probably in use from around the middle till the end of the third century, with the fills representing their disuse when Barn 2 was replaced by Barn 4 in Period 4. Table 29 gives the main fabric A%.

The limited range of wares and vessel forms emphasises the specialist function of both features and pottery.

	RSG	LNVGW	LNVCC	Total
Jar	19	2	2	23
WMJar	-	1	-	1
Dish	1	-	1	2
Total	20	3	3	27

Table 31 Barn 2, main vessel forms/fabric, by entries.

(Fig. 90)

Vessels **250–7** are in RSG. **252–3** are from F201, L319 (106). The rest are from F105, L160 (106).

250 CR4+5/21/4+5. Fairly soft fabric.
251 CR5+10/21/5+10.
252 CR4/21/4.
253 CR5+10.
254 CR4. Fairly soft fabric.
255 CR4. Fairly soft fabric.
256 CR4+5. Fairly soft fabric.
257 CR4. Fairly soft fabric.
258 CCC?? CR8, burnt surfaces. F156, L270 (104)

2.2 The sump F1094, narrow trench F1113, and pit F1121

These all relate to the use of Barn 1, also originally constructed in Period 2. The pottery from these features provides one of the largest and most varied assemblages from the site as a whole and is of particular importance. The deposits are thought to relate to the dumping of material at the end of Period 3 or the beginning of Period 4 in features which were not to be part of the reorganised Barn arrangements. Some of the pottery might, however, have been thrown into the features during their 'working' life. A coin of *c.* AD 270–290 (C25) was found in F1094. The mortarium (M48) is dated *c.* AD 230–400. Table 29 gives the main fabric A%.

	RSG	Grey	LNVGW	LNVCC	C/W	Other	Total
Jar	12	4	23	1	-	-	40
Bowl	-	-	4	3	2	2	11
Dish	-	2	9	1	-	3	15
Flagon	-	-	3	-	-	-	3
Beaker	-	-	-	7	-	-	7
Lid	3	-	1	-	-	-	4
Box	-	-	-	5	-	-	5
Mortarium	-	-	-	-	1	-	1
Other	-	-	4	1	-	-	5
Total	15	6	44	18	3	5	90

Table 32 F1094 *etc.*, main vessel forms/fabric, by entries.

Many of the vessels are represented by large fragments. The additional fabrics are samian ware, amphorae, BB1, CGCC, VR and Horningsea, all represented by only a few sherds. There are vessels which may be either jars or bowls, or dishes or bowls.

GW

259 CR19/21/19,18 patches. Fabric contains yellow flecks. Stanground?? F1113, L2339 (129)
260 CR19. F1094, L2359 (125)
261 CR17. F1113, L2385 (129)

LNVGW

262 CR21/26/19. Slipped? F1121, L2347 (129)
263 CR20/18/20. Slipped? Hole apparently made after firing. F1113, L2384 (129)
264 CR19+20+21/26/19+20+21. Slipped? F1121, L2347 (129)
265 CR20/18/19/20. F1094, L2359 (125)
266 CR21/18+26/21. Slipped? F1121, L2347 (129)
267 CR20/18+26/20. Slipped? F1121, L2347 (129)
268 CR20//18/21/18//20. Slipped? F1121, L2344 (129)
269 CR15+19+21/26/15+19+21. Slipped? F1094, L2359 (125)
270 CR20+21/26/20+21. F1094, L2359 (125)
271 CR21/18/21. F1094, L2359 (125)
272 CR19/18/19. F1094, L2359 (125)

(Fig. 91)

LNVCC

273 CR4+11+16/14+26/4+11+16,22 patches. Two-thirds vessel. 'Firing-band'. F1121, L2347 (129)
274 CR3/14+26/12+22. Two-thirds vessel. Slight 'firing-band'. F1121, L2396 (129)
275 CR3/14+26/12+22. Near complete. F1121, L2353 (129)
276 CR2/4+5/12. F1121, L2347 (129)
277 CR2//26/4/26//12,4. Rouletted. F1113, L2385, L2386 (129)
278 CR12+21/14/12+21. LNVCC? Stanground? F1094, L2359 (125)
279 CR8/14/8. F1094, L2359 (125)
280 CR3+12/26/3+12. F1121, L2347 (129)
281 CR2//26/4/26/12. F1113, L2385 (129)
282 CR8+22/5+14/8+22. Rouletting worn smooth. F1121, L2347 (129)

C/W

283 CR14/18/14. VR? Slipped? F1121, L2347 (129)
284 CR14+26. LNVC/W? F1094, L2359 (125)

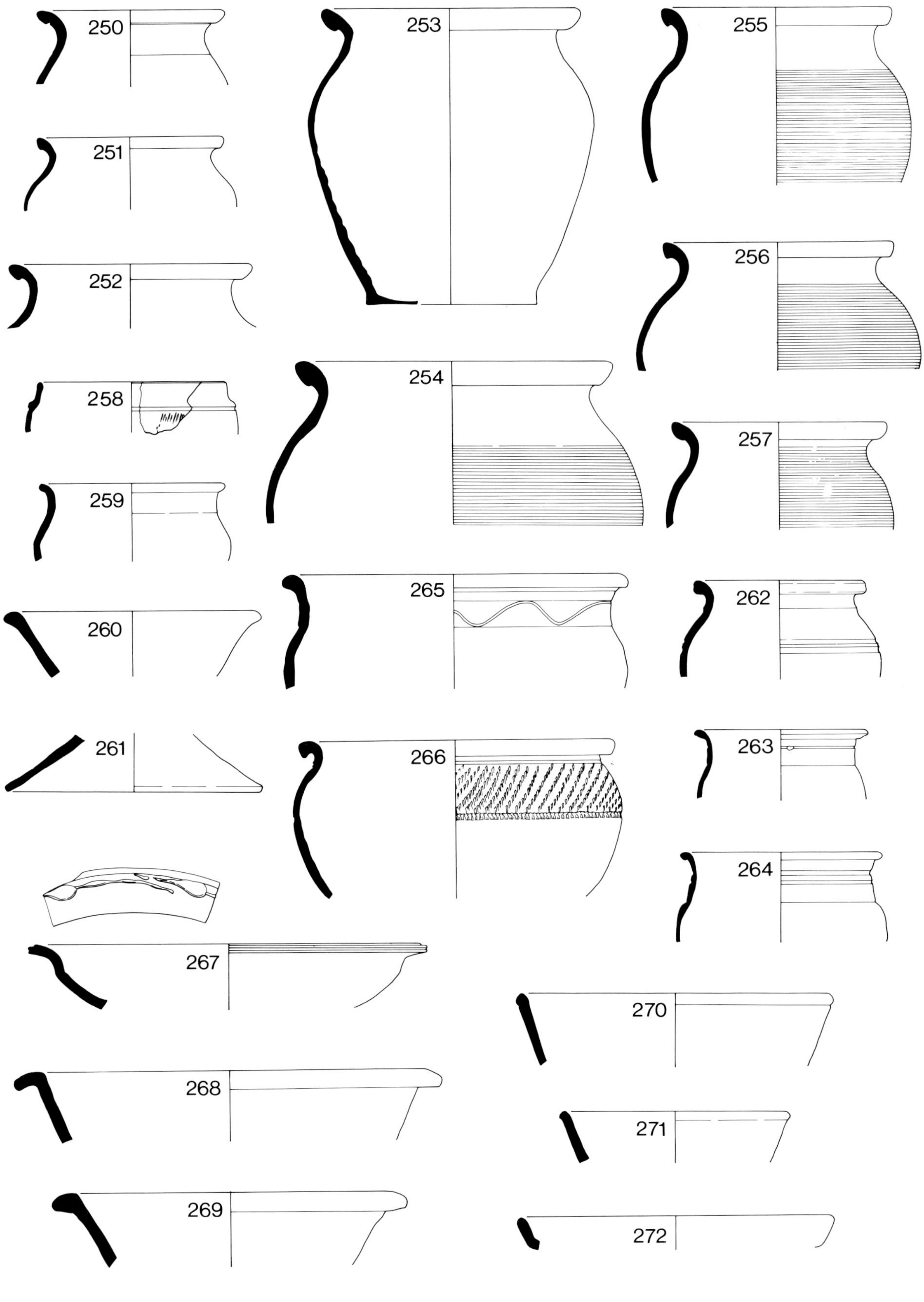

Figure 90 Roman pottery. Period 3: Nos 250–257, F105 'vat base' *etc*.; No.258, F156 drier; Nos 259–272, F1094–F1113–F1121. Scale 1:4.

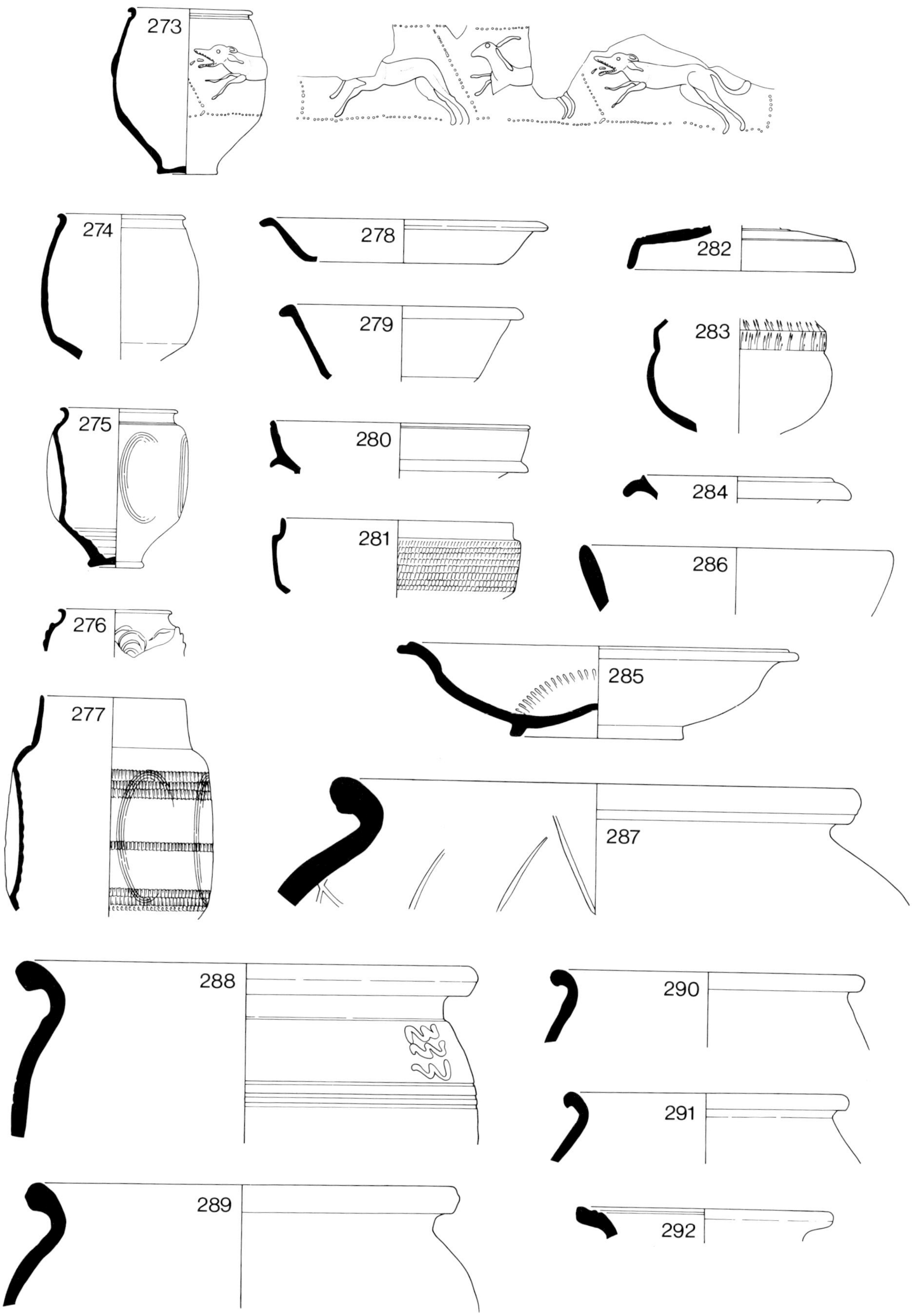

Figure 91 Roman pottery. Period 3: Nos 273–286, F1094–F1113–F1121; Nos 287–292, Main Yard. Scale 1:4.

Miscellaneous
285 CR4+5/15/4+5. LNV? Mica-gilt. Near complete. F1121, L2347 (129)
286 CR17+21+22. BB1 or imitation. F1121, L2347 (129)

2.3 The yards

This group consists of layers which had sagged into hollows formed by the consolidation of earlier features, as well as a few areas designated as features in their own right and individual layers: F554, F656, F675, F802, F906, F964, F968, F969, F995, F1020, F1035, F1047 and F1048.

The yards were laid out in Period 3 as part of the reorganisation encompassing Barn 1 and the new House and were re-metalled at the beginning of Period 4. However, this stratification was largely destroyed by later activity, especially post-Roman ploughing. The pottery from the Period 3 metalling and the layers between it and that of Period 4 is treated as one group relating to the construction and use of the Main Yard and its surrounding area in the third century, but it was not a completely sealed group. The mortaria (Fig. 115, M37, M41, M43, M51–M60) mainly date from around the mid-third century. Table 29 gives the main fabric A%.

	RSG	Grey	LNVGW	LNVCC	C/W	Other	Total
Jar	53	13	43	5	4	4	122
WMJar	-	-	1	4	-	-	5
Bowl	-	2	9	8	1	3	23
Dish	1	2	16	6	-	6	31
Flagon	-	-	1	2	-	-	3
Beaker	-	-	-	15	-	5	20
Lid	1	1	-	-	-	-	2
Box	-	-	-	2	-	-	2
Mortarium	-	-	-	-	8	8	16
Other	-	-	7	1	-	-	8
Total	55	18	77	43	13	26	227

Table 33 The yards, main vessel forms/fabric, by entries.

In addition, there are a few sherds of LIASG, grogged, London ware type, CGCC, LR?, Oxfordshire, amphorae, BB1, BB2-type, possible CCC, Horningsea and uncertain ware. Some other vessels were either jars or bowls, or dishes or bowls.

As most of the pottery is in much smaller fragments than the previous features, it is possible that some may be from the same vessels, thereby reducing the overall total.

RSG
287 CR2+4. Partly hand-made? F1048, L2548 (134)
288 CR7+9+11/17/7+9+11. F1047, L2535 (134)
289 CR2/8+11/2. F1047, L2542 (134)
290 CR2/16/2. F1020, L2541 (134)
291 CR8/17/8. F1047, L2543 (134)
292 CR4/17+21/4. UF, L1031 (134)

(Fig. 92)
GW
293 CR15+18. F1048, L2644 (134)
294 CR19/19+21/19. Fabric contains flecks. Stanground? F656, L1565 (134)
295 CR14//11/19/11//14. F554, L2145 (134)
296 CR18+19. Fabric contains flecks. Stanground? F1047, L2543 (134)
297 CR20//8/21/8//20. F995, L1826 (134)
298 CR21/20/21. F964, L2011 (134)
299 CR19. Fabric contains flecks. Stanground? UF, L1031 (134)

LNVGW
300 CR20+21/18/20+21. F969, L1834 (134)
301 CR19/18/19. F969, L1834 (134)
302 CR20/18/20,11+14 slip? F1020, L2541 (134)
303 CR15/26/15. 'Speckled' surfaces. Slipped? UF, L1139 (134)
304 CR20/11/20. Stanground? F1047, L2512 (134)
305 CR20/18+21/20. F1047, L2561 (134)
306 CR21/7+15/21. F995, L2029 (134)
307 CR20+21/4/20+21. F1048, L2700 (134)
308 CR19/26/19. F1047, L2561 (134)
309 CR21/26/21,15 slip? F995, L2028 (134)
310 CR18+19/19/18+19. Stanground? F1047, L2541, L2561 (134)
311 CR21/26/21. F554, L1536 (134)

LNVCC
312 CR3+8/5+26/12. F1047, L2561 (134)
313 CR4,11 patches/26/4. F1047, L2502, L2504 (134)
314 CR12+22/5+26/12+22. UF, L1031 (134)
315 CR8+22/3+26/8+22. UF, L1267 (134). Topsoil? L1268 (134)
316 CR22/14/22. F1020, L2540 (134). MS, L579

(Fig. 93)
317 CR22/26/22. F1047, L2535 (134)
318 CR12+17/5+26/12+17. UF, L1267 (134). Topsoil? L1268 (134). F801, L1417 (134). Period 3, not yard.

Miscellaneous
319 C/W. CR5/14/5. Slipped? F995, L1826 (134)
320 BB1. F656, L1565 (134)
321 BB2. F995, L1806 (134)
322 BB2. F1020, L2540 (134)

2.4 Well F1016

This was the main well used for the activities in Barn 1, and most of the final fills would have accumulated at the end of Period 3 or the beginning of Period 4, probably, as with F1094, F1113 and F1121, in one concentrated dumping though, again, some of the pottery may have fallen, or been thrown, into the well while it was in use. Table 29 gives the main fabric A%.

Some of the fragments are again large.

	RSG	LNVGW	LNVCC	Total
Jar	1	7	-	8
Bowl	-	2	1	3
Dish	-	3	-	3
Beaker	-	-	4	4
Box	-	-	2	2
Total	1	12	7	20

Table 34 F1016 well, main vessel forms/fabric, by entries.

LNVGW
323 CR28/20/28. L2210 (127)
324 CR18/21/18. Near complete. Warped. Stained. L2209, L2210 (127)
325 CR20/26/20,7+19 slip? Third to half vessel. L2209 (127)
326 CR20/26/20. Over half vessel. L2209 (127)
327 CR19/26/19. 'Speckled' surfaces. Slipped? L2209, L2210 (127)
328 CR21/18/21. L2209 (127)
329 CR21/18/21. L2210 (127)
330 CR20/18+26/18. L2209, L2210 (127)

LNVCC
331 CR11/18/11. L2209 (127)
332 CR8+22/5+26/8+22. Near complete. L2209 (127)
333 CR22/5+14/8. L2209 (127)
334 CR22/4/8. L2209 (127)
335 CR4+22/26/4+22. Two-thirds vessel. L2209 (127)

2.5 Enclosure ditch systems F435, and F441

These were the last remnants of an enclosure in the north-east corner of the Main Yard. They were swept away in the reorganisations of Period 4 and therefore represent essentially third-century occupation and the deposits in the enclosure ditches probably accumulated gradually throughout the Period. The two ditches had remarkably similar amounts of pottery and numbers of vessels.

Some dating evidence is provided by mortaria. There were eight in the F435 system, one, of Lower Rhineland origin (Fig. 114, M35), is dated *c.* AD 150–250+; the other seven (Fig. 114, M21–M27), are mainly of third or third to fourth-century date. The two mortaria from the F441 system (M28, Fig. 114, M30) are third century or later in date.

F435 system — F435, F438 and F540
Table 29 gives the main fabric A%.

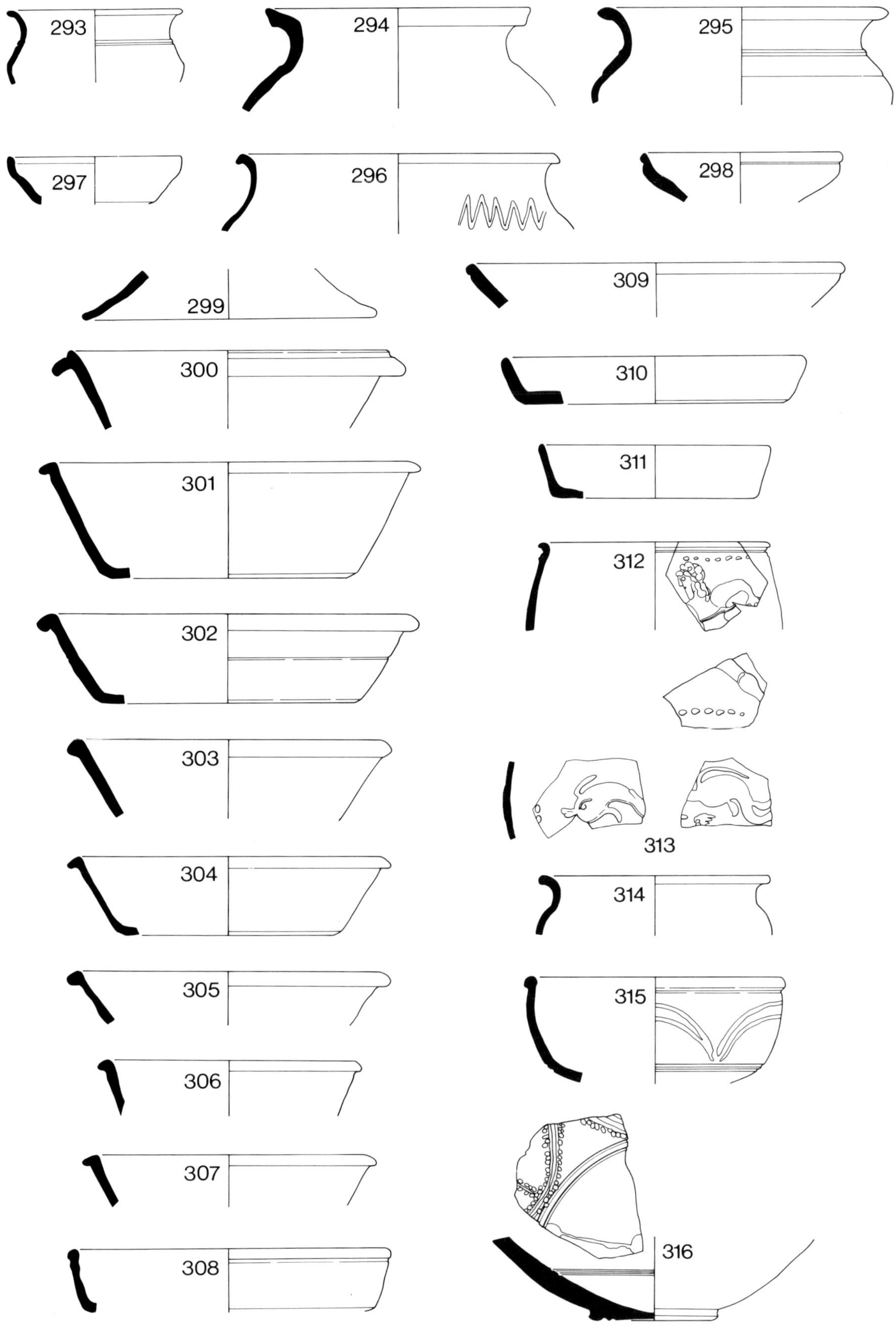

Figure 92 Roman pottery. Period 3: Nos 293–316, Main Yard. Scale 1:4.

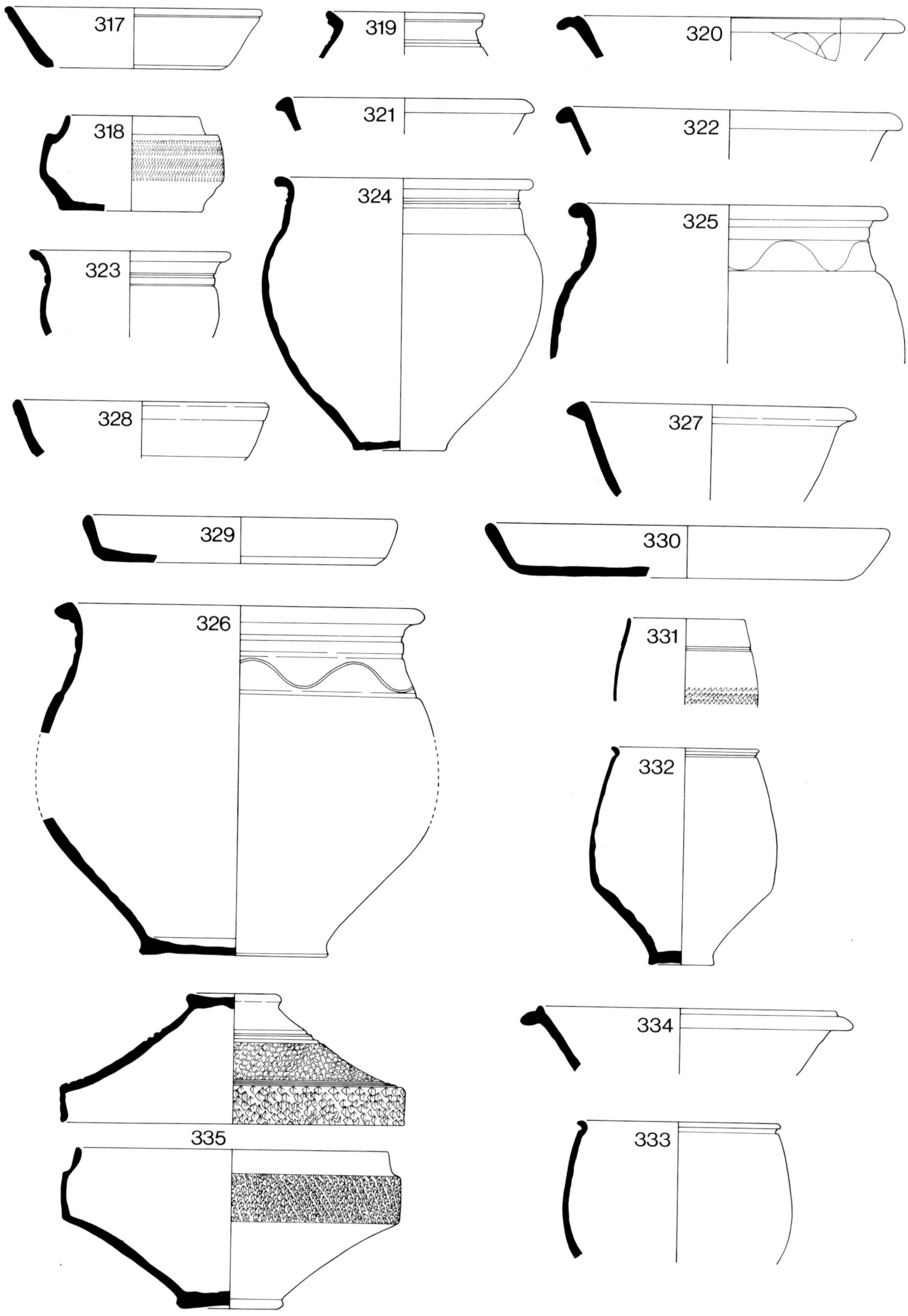

Figure 93 Roman pottery. Period 3: Nos 317–322, Main Yard; Nos 323–335, F1016 well. Scale 1:4.

	RSG	Grey	LNVGW	LNVCC	C/W	Other	Total
Jar	5	-	8	2	-	1	16
WMJar	-	-	1	4	-	-	5
Bowl	-	1	1	1	-	-	3
Dish	-	-	1	5	-	-	6
Flagon	-	-	-	1	-	-	1
Beaker	-	-	-	5	-	-	5
Box	-	-	-	1	-	-	1
Mortaria	-	-	-	-	7	1	8
Other	-	-	1	1	-	-	2
Total	5	1	12	20	7	2	47

Table 35 F435 system, main vessel forms/fabric, by entries.

There are also a few sherds of a BB1 jar, samian ware, amphorae, and a Lower Rhineland mortarium (see above).

(Fig. 94)
RSG
336 CR5/4/5. F435, L822 (120)
337 CR8+22. F438, L838 (122)

LNVGW
338 CR19+20/26/19+20. F438, L827 (120)

LNVCC
339 CR3/14/3. Face pre-moulded and luted on. Vestigial colour-coat. F435, L820 (120)
340 CR22/26/22. F438, L827 (120)
341 CR2/14+27/2. Third vessel. F435, L822 (120)
342 CR2/14+26/2. F438, L827 (120)
343 CR2/5/2. F438, L838 (122)
344 CR22/26/22. Two-thirds vessel. F435, L822 (120)

F441 system — F441, F442, F453, F481, F617, F620 and F621
Table 29 gives the main fabric A%.

	RSG	Grey	LNVGW	LNVCC	C/W	Total
Jar	7	2	8	2	-	19
WMJar	-	-	3	2	-	5
Bowl	1	-	2	7	-	10
Dish	-	-	1	5	-	6
Flagon	-	-	-	1	-	1
Beaker	-	-	-	3	-	3
Mortarium	-	-	-	-	2	2
Total	8	2	14	20	2	46

Table 36 F441 system, main vessel forms/fabric, by entries.

BB1 is also present, together with a few sherds from the Oxfordshire kilns, other colour-coated, uncertain and some possibly from the Hadham area.

LNVGW
345 CR20/19/20. Stanground? F441, L857 (122)
346 CR22/26/22. F441, L857 (122)
347 CR19/26/19. F453, L912 (122)
348 CR19. F481, L1219 (122)

LNVCC
349 CR2/2+3/22. Fabric contains white lumps. F617, L1197 (122)
350 CR12+22/14/12+22. F617, L1197 (122)
351 CR12/4/12. F453, L912 (122)
352 CR12+17+19/4/12+17+19. F441, L857 (122)

2.6 Enclosure ditch system F772
Consisting of ditches F548, F772, F801 and F808, the enclosure was created at the same time as F435 and F441, but was not completely abandoned at the end of Period 3 for its east/west ditch, as an integral part of the House/walled yard/Barn 1 complex, continued in use through Period 4. There was not as much pottery as F435 and F441. Of the two mortaria (Mortaria, Fig. 115, M38–M39), one is of Oxfordshire origin and probably dates after *c.* AD 180, and the other is dated *c.* AD 230–400. Table 29 gives the main fabric A%.

	RSG	LNVGW	LNVCC	C/W	Other	Total
Jar	9	2	2	-	-	13
WMJar	-	-	1	-	-	1
Bowl	-	-	3	-	-	3
Dish	-	1	2	-	-	3
Mortarium	-	-	-	2	1	3
Total	9	3	8	2	1	23

Table 37 F772 system, main vessel forms/fabric, by entries.

Sherds of London ware type and an Oxfordshire mortarium (see above) also occur.

RSG
353 CR4/21/4,12 patches. F801, L1378 (207)
354 CR5/21/5. Sooted area shaded. F801, L1380 (207)

LNVGW
355 CR20/18/20. F801, L1379 (207)

LNVCC
356 CR2+8/4/2+8. F772, L1428 (133)
357 CR12/5+26/12. F772, L1428 (133)
358 CR21/3+9/21. F772, L1428 (133)
359 CR4+17/4/4+17. F801, L1379 (207)

2.7 Well F254
The fills of the well were unequivocally third-century in date. The pottery did not indicate systematic dumping at any particular time, but the material arrived here after the well had gone out of use which may not have been before the end of Period 3. Table 29 gives the main fabric A%.

	RSG	LNVGW	LNVCC	C/W	Other	Total
Jar	2	8	-	-	-	10
Dish	-	1	3	-	1	5
Flagon	-	-	-	1	-	1
Beaker	-	-	1	-	-	1
Total	2	9	4	1	1	17

Table 38 F254 well, main vessel forms/fabric, by entries.

A few sherds of a BB1 dish, samian ware, Hadham? also occur.

(Fig. 95)
RSG
360 CR8. L612 (146)

LNVGW
361 CR18+20. Half vessel. Fabric contains flecks. Stanground? L552 (146)
362 CR18+20. L614 (146)
363 CR18+20. Fabric contains flecks. Stanground? L551 (146)

LNVCC
364 CR22/15+19/22. Stanground? L552, L612 (146)
365 CR21+22/26/21+22. Stanground? L552 (146)

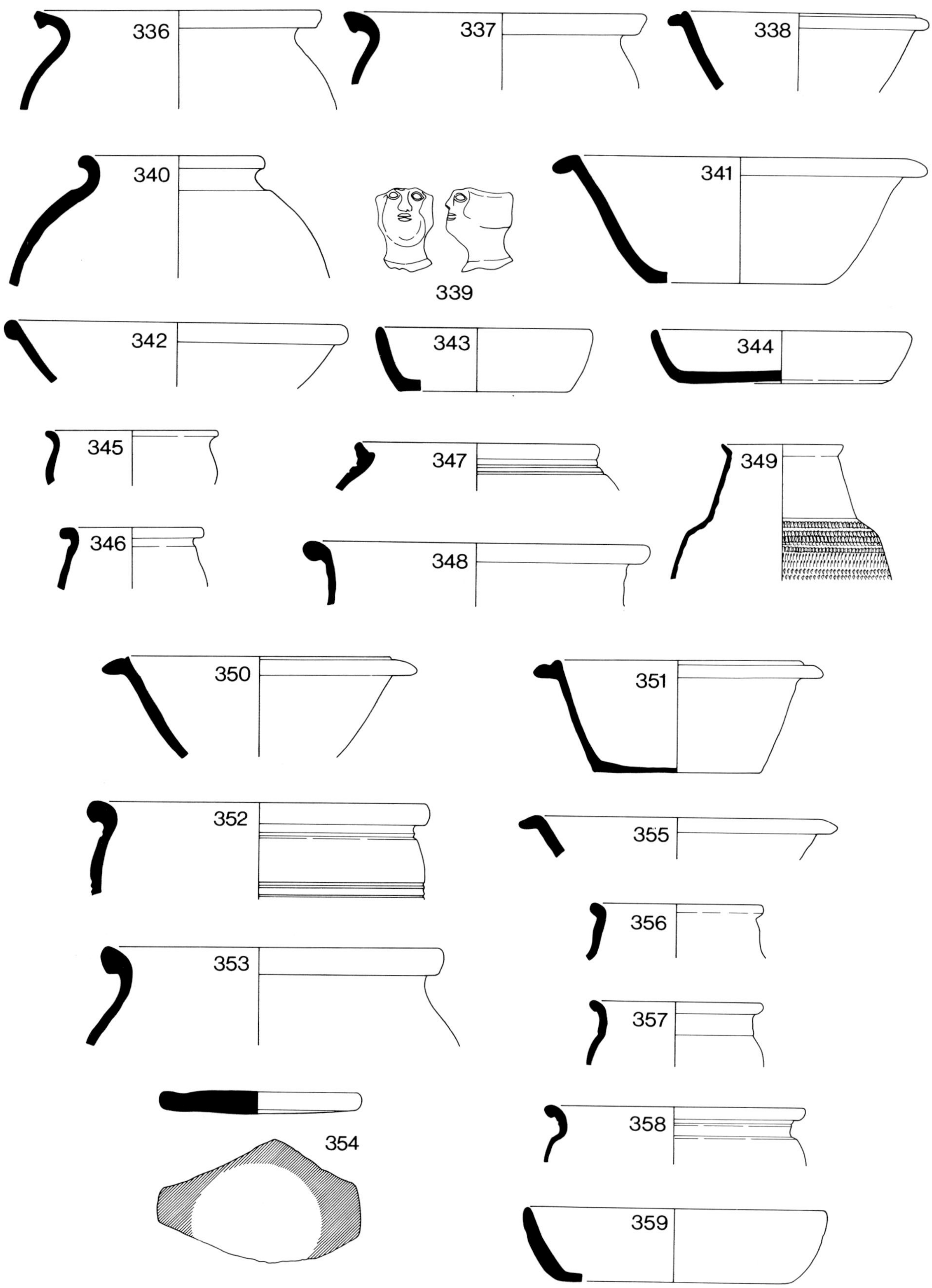

Figure 94 Roman pottery. Period 3: Nos 336–344, F434–F438 sub-enclosures in Main Yard; Nos 345–352, F441–F453 sub-enclosures in Main Yard; Nos 353–359, F772–F801 ditches. Scale 1:4.

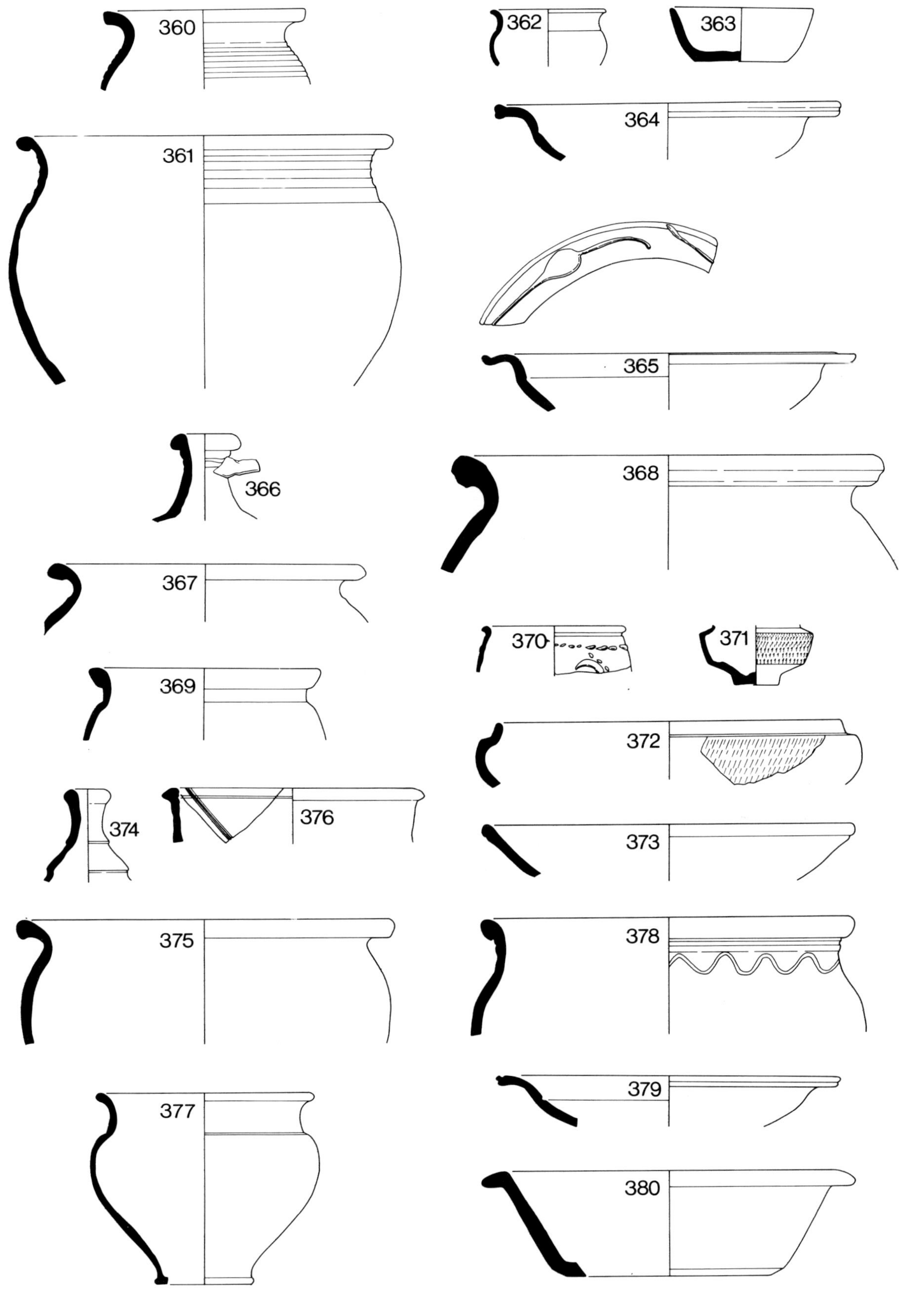

Figure 95 Roman pottery. Period 3: Nos 360–366, F254 well; Nos 367–373, 'F969' well; Nos 374–380, Period 3 additional pots. Scale 1:4.

C/W
366 CR14+26/5+14/14+26. L614 (146)

2.8 Well 'F969'
This was not immediately identified as a well during excavation as it had been sunk into the fills of the Period 1 and 2 ditch F969. It is tentatively allied to well F1016 as having served the needs of the activities in Barn 1 and the walled yard. It was abandoned by Period 4, and its fills are of third-century date, though they contained material derived from F969 itself. The contrast in pottery statistics with those for the adjacent well F1016, which had fewer sherds, weighing more and with a greater number of EVEs, might suggest that it had passed out of use before the end of Period 3 and did not, therefore, receive any of the pottery derived from the clearance of material thought to comprise much of the final fills of F1016. Table 29 gives the main fabric A%.

	RSG	Grey	LNVGW	LNVCC	Other	Total
Jar	9	3	8	-	-	20
Bowl	-	-	2	-	-	2
Flagon	-	-	-	1	-	1
Beaker	-	-	-	3	-	3
Box	-	-	-	2	-	2
Other	-	-	1	-	1	2
Total	9	3	11	6	1	30

Table 39 'F969' well, main vessel forms/fabric, by entries.

The layers also contain a few sherds of samian ware and uncertain fabric.

Vessels **367–373** are from L2207 (126).

RSG
367 CR14/18/14.
368 CR2+11.

LNVGW
369 CR10+20/26/10+20.

LNVCC
370 CR12/14/12. Colour-coat almost worn away.
371 CR11/14/11.
372 CR22/26/22.

Miscellaneous
373 CR14; light grey core in places. Mica-gilt, mostly worn away.

Period 3 Additional Sherds

GW
374 CR16+20. F734, L1370 (107)
375 CR21+22//11/21+22/11//21+22. F1154, L2921 (152) L2922, L2923; (152)(136). F31, L768 (94)
376 CR21/20/21. F425, L799 (118)

LNVGW
377 CR20/26/20. F65, L78 (101)
378 CR19+20/26/19+20. F500, L1958 (116)
379 CR21/26/21. F425, L798 (118)
380 CR20/26/20. VR?? 'Speckled' surfaces. F1142, L2548 (134)

(Fig. 96)
381 CR18+20//4+18/19/4+18//18+20. Stanground? F437, L841 (119)
382 CR19//15/4/19/4/15//19. Stanground? F202, L321 (94)

LNVCC
383 CR8/5+26/8. F28, L314 (96)
384 CR22/4+26/22. F556, L1113 (119). F445, L842 (63) (period 2).
385 CR8//4/18/4//8. F546, L1089 (121). F556, L1113 (119). F439, L1111 (119) (period 2).

C/W
386 CR5+27. LNV? F194, L303 (84)
387 CR27. F437, L841 (119). L867 (period 2) (72)
388 CR7,22 decoration. F65, L78 (101)

Miscellaneous
389 CR26/4+11/11. CCC? Stanground? Misfired. F439, L828, L841 (119). F324, L659 (117)
390 BB1. F681/683, L1988 (114)

3 The Pottery

3.1 RSG
Table 29 gives the RSG A% in the main Period 3 features and Period 3.
The only additional form is a vessel which may be a jar or bowl.

	F105	F1094	Yard	F1016	F435	F441	F772	F254	F969	Other	P3
Jar	19	12	53	1	5	7	9	2	9	48	165
Bowl	-	-	-	-	-	1	-	-	-	1	2
Dish	1	-	1	-	-	-	-	-	-	1	3
Lid	-	3	1	-	-	-	-	-	-	1	5
Others	-	-	-	-	-	-	-	-	-	1	1
Total	20	15	55	1	5	8	9	2	9	52	176

Table 40 Period 3: main RSG vessel forms, by entries.

The vessels from the vat base, F105 in Barn 2, and its stoke-hole, F201, Nos 250–7, show some of the main characteristics of later third-century RSG jars. All the vessels have a curved neck and a rounded shoulder with simple rims, many of which are slightly undercut. Most of the vessels have rilled external surfaces and, as this rilling usually starts at or just below the shoulder, it is likely that some or all of Nos 250–2 were also rilled. The rilling is much finer, less deeply incised and more closely spaced than that on earlier jars like No. 180. The fabric and colour is uniform and may have been the norm, as made, at this time. Of the jars from other contexts, Nos 336 and 337 have similar slightly undercut rims. No. 360 is rilled, though this is more widely spaced and, taken together with the lack of an undercut to the rim, it may be earlier than those from F105 and F201, and could be typical of mid third-century jars.

The larger jars, 288–9, 353, 368, and the very large jar 287 are not dissimilar to those of the same categories from earlier features, for example Nos 43–5 and 94. Some may well either be residual or survivals in use, but they need not be, for these larger vessels underwent less typological change than their smaller, more common, counterparts which were used, broken and replaced more often. Nos 290 and 291 belong to a variety of jar which was not seen in either Periods 1 or 2, and is therefore a third-century type. No. 367 seems to be a mixture of this and some of those from F105 and F201 but, as it is from the well 'F969', it may be residual. No parallel can be cited for No. 292. The base fragment No. 354 is included because the sooting on one side suggests that it had been used, or reused, as a stand or lid.

The small range of vessel classes in RSG of Period 3 is not considered unusual for it probably reflects the ready availability of a wide range of vessels in other fabrics in the third century.

3.2 Grey wares
Table 29 gives the grey ware A% in the main Period 3 features and Period 3.

The additional forms are vessels which may be a jar or bowl, or a dish or bowl.

	F1094	Yard	F435	F441	F969	Other	P3
Jar	4	13	-	2	3	22	44
Bowl	-	2	1	-	-	3	6
Dish	2	2	-	-	-	4	8
Lid	-	1	-	-	-	1	2
Other	-	-	-	-	-	2	2
Total	6	18	1	2	3	32	62

Table 41 Period 3, main grey ware vessel forms, by entries.

A large proportion is probably residual, and it may be significant that the F1016 well group does not contain any grey ware. The high figures for this in well 'F969' is almost certainly due to its disturbance of earlier ditches, including F969, whereas that in the yard deposits might be more indicative of survival in use. The generally low proportions, much less if residuality is taken into account, clearly show the cornering of the market by LNVGW. None of the main and additional groups shows a bias towards one particular class of vessel.

Although the major first and second-century grey wares sources no longer supplied the local area in the third century, others probably did. Their products are not, however, easily distinguishable from the mass of local LNVGW in either form or fabric. Among these could be pottery from the Stanground kilns which is now becoming identifiable.

Of the illustrated grey ware jars, Nos 293, 295 and 352, dishes 297–8 and lids 261 and 299 are probably second-century products occurring residually or as survivals in Period 3 features. Jar 296 and dish 260 may have come from Stanground, while the form of No. 294 is more third than second-century. None of these three can be paralleled. The fabric and finish of No. 375 is reminiscent of pottery found in quantity, and therefore probably manufactured, at or near Godmanchester.[18] The stepped, cordoned profile of No. 374 is a mark of a second or even a late first-century date. As a specialised form, however, it might be expected to have survived in use, but, as no parallels can be quoted, it might be a third-century type. The internal scoring on No. 376 is very unusual but occurs on two similar vessels from Chesterton (Perrin, to be published) in mid to late second and late second to early third-century contexts. The purpose of the scoring is unknown.

3.3 LNVGW

Table 29 gives the LNVGW A% in the main Period 3 features and Period 3.

	F105	F1094	Yard	F1016	F435	F441	F772	F254	F969	Other	P3
Jar	2	23	43	7	8	8	2	8	8	12	121
WMJar	1	4	1	-	1	3	-	-	-	4	14
Bowl	-	4	9	2	1	2	-	-	2	4	24
Dish	-	9	16	3	1	1	1	1	-	11	43
Flagon	-	3	1	-	-	-	-	-	-	2	6
Other	-	1	7	-	1	-	-	-	1	1	11
Total	3	44	77	12	12	14	3	9	11	34	219

Table 42 Period 3, main LNVGW vessel forms, by entries.

Additional forms are a lid, and vessels which may be either jars or bowls, or dishes or bowls.

The illustrated LNVGW pottery from Period 3 provides a reasonably comprehensive view of the range of vessels likely to have been used locally in the third century. Most can be easily paralleled in the products of the known kiln sites at Sibson and Stibbington and would have been made at most or all other local centres.

Jars 263–5, 323–6, 348, 361 and 378 are really size variations in the same basic form (Guide, fig. 1, nos 4 and 10). Some of the vessels approach the boundary between jar and bowl. It is difficult to note definite differences between these and their late second-century predecessors. The upright neck, and scarcity or absence of neck grooves, has already been mentioned, but Nos 265, 325–6 and 378 retain burnished neck decoration, though some of these particular vessels could be survivals from earlier phases. Nos 348 and 362 appear to be Stanground products. The type represented by No. 266 (Guide, fig. 1, no. 9) does not seem to occur before the third century and the shoulder rouletting is a common feature. The other jars, 262, 346–7, 362, 369 and 377, are not, apparently, as common as those noted previously, but occur on most other local sites. No. 362 may also be a Stanground product. No. 377 is slightly unusual, but the highish shoulder and tapering base are reminiscent of LNVCC vessels from a well at Ashton (Guide, fig. 7, no. 77), Chesterton (Perrin, to be published) and graves at Great Casterton.[19] Their contexts appear to be fourth-century, and No. 377, therefore, may be a late type of LNVGW jar.

The standardisation noted in jars is also apparent with bowls and dishes of which there are five basic types in Period 3 all occurring in a variety of sizes. They can be classified by rim form: flanged, Nos 300, 338; flat-topped, Nos 268, 301, 355, 380; triangular or rounded, Nos 269, 302–7, 327; plain and upright with an external groove, Nos 270–1, 308–9; and plain and upright with no external groove, Nos 310–11, 328–30, 363, 381–2. All were among the range of vessels made at Sibson (Guide, fig. 2, nos 17–21). No. 272 could be one of the latter variety but it might also be a lid, possibly even that from a 'box'. Nos 363 and 382 are Stanground products. Some of these types were around in the later second century and it is again difficult to identify what may be purely third-century traits. The more splayed rims of Nos 310 and 330 might be later examples, if LNVGW had a similar typological development to BB1 (Gillam 1976, 77), and the larger vessels have so far only occurred in third-century contexts. The straight-sided flanged bowl is not sufficiently common for its inception to be certain. As a type in other wares, elsewhere in the province, it is dated from the late second century (Gillam 1973, 59–60), but there is no evidence to suggest that it was made at this time in the Lower Nene Valley. It may have first appeared later in the third century when LNVCC flanged bowls were also, seemingly, first produced. As with the latter, size is not a chronological factor, and there is no apparent development in rim form like that again noticeable on BB1 vessels (Gillam 1976, 70). A few of the LNVGW vessels, including Nos 267 and 379, are imitations of samian ware forms, particularly form 36. Such vessels were made at Sibson (Guide, fig. 2, nos 15–16). Most known examples are of third-century date.

3.4 LNVCC

Table 29 gives the LNVCC A% in the main Period 3 features and Period 3.

	F105	F1094	Yard	F1016	F435	F441	F772	F254	F969	Other	P3
Jar	2	1	5	-	2	2	2	-	-	9	23
WMJar	-	-	4	-	4	2	1	-	-	12	23
Bowl	-	3	8	1	1	7	3	-	-	8	31
Dish	1	1	6	-	5	5	2	3	-	10	33
Flagon	-	-	2	-	1	1	-	-	1	1	6
Beaker	-	7	15	4	5	3	-	1	3	16	54
Box	-	5	2	2	1	-	-	-	2	-	12
Other	-	1	1	-	1	-	-	-	-	25	28
Total	3	18	43	7	20	20	8	4	6	81	210

Table 43 Period 3, main LNVCC vessel forms, by entries.

The additional forms are vessels which may be either jars or bowls, or dishes or bowls.

Examples found on sites all over Roman Britain show that there was considerable typological change in, and an increase in the range of, LNVCC vessels produced in the third century. Unfortunately, these are not fully reflected at Orton Hall Farm. It is not possible, therefore, either to discuss any developments in detail or refine current thinking about typology and chronology. This is unfortunate for the date or dates by which certain types of vessel were introduced, notably flanged bowls, 'rhenish-type' beakers, wide-mouthed jars or bowls, bead-rimmed and painted bowls and imitation samian ware bowls and dishes. The Period 3 contexts do, however, show that all these types were in use by the end of the third century, and some of them are discussed more fully in Period 4. Vessels such as Nos 278–9 and 317 are merely colour-coated versions of well-known third-century LNVGW types.

Ignoring the possible ramifications of their particular contexts, the body and rim forms of beakers 273–6, 312–13, 331–3 and 370 are typologically of the later second to early third centuries. There are very few 'hunt cups' from the site as a whole and No. 273 is the most complete example. The barbotine is not particularly well executed. Beaker No. 312 is, similarly, the only one with a motif incorporating a human figure, on which Dr G. Webster has kindly contributed the following note:

> The two sherds, although not joining, clearly belong to the same figure. The stance is that of a *bestiarius*, with a long hunting spear, but the detail of his dress is very perfunctory. The best parallel, also of LNVCC, was found at Bedford Purlieus, and has been published (Roach-Smith 1857, 90, pl. XXII; Perring 1977, fig. 10, no. 6). On this vessel, and probably No. 312, the spear is held behind the body. The dress, which is more like that of a modern circus entertainer, includes a belted tunic and spangled, fringed tights. This last item was known by the Romans as a *subligaria*, a rare word which is mentioned on one of the writing tablets from Vindolanda (Bowman and Thomas 1983, 262). It seems evident that LNVCC potters were copying the dress of gladiators depicted on late samian ware (Déchelette 1904, 93–7).

An overall idea of the style of No. 313 is given by a vessel from Thorplands (Hunter and Mynard 1977, fig. 16, no. 217) though the subject of the main motif is different. No. 277 is the best example of a funnel-necked, indented beaker from Period 3. In all, three of the Period 3 beakers have this variety of rim form, though a further six have funnel-neck and bead rims, including Nos 349 and 356. Five have plain rims, including No. 331. Only eleven are definitely indented, though this figure would probably be larger had more of the profile of some of the other beakers survived. The form of No. 277 places it firmly in the mid-third century, on existing evidence (Guide, fig. 4, no. 42).

Vessels similar to Nos 349 and 356 were made in kilns at Chesterton,[20] dated to the fourth century. A number of painted sherds are of the so-called 'rhenish-type' of beaker (*ibid.*, fig. 5, nos 47–9), examples of which were also made in the kiln at Chesterton. It has been noted that the production of 'rhenish-type' beakers in LNVCC probably began in the mid-third century (*ibid.*, 20), but the Period 3 contexts at Orton Hall Farm only confirm that it was being made then: the first examples could have been contemporary with actual rhenish ware imports starting in the later second century. The fabric of some of these rhenish or rhenish-derived beakers is orange or reddish-yellow in colour, and contains lumps of white material. The beakers made in the Chesterton kiln are also in this fabric. A more iron-rich clay may have been part of the process needed to reproduce the lustrous finish, and perhaps hardness, of the imported ware. The proportion of vessels with this particular fabric is very low, and much of it could, therefore, have been made in the Chesterton kilns. The fabric also occurs in Periods 4 and 5.

The small jars, including Nos 314 and 357–8, do not show obvious changes from those noted in Period 2 (including No. 236) other than a more noticeable shoulder. However, the Period 3 contexts mark the first appearance of wide-mouthed jars or bowls in LNVCC, including No. 352, but there is none known with a rouletted shoulder like No. 266 in LNVGW. The narrow-mouthed jar represented by No. 340 also makes its first appearance in Period 3 contexts.

The Period 3 LNVCC flanged bowls, including Nos 334 and 350–1, emphasise that the type was first made before the end of the third century. The variety of rim form again shows that there is no chronological development in LNVCC flanged bowls as there is in those made in BB1 (see above). The flat-topped bowls, including No. 341, may not have outlasted the third century, and the differences between the plain-rimmed dishes Nos 317, 343–4 and 359 may prove to have chronological significance. No. 278 may be a Stanground product. No. 373 has a mica-rich slip and is another example of this surface treatment on this site. Its 'F969' context could allow it to be of later second or early third-century date, and most of the mica-gilded LNVCC vessels known are in fact of this period.

Three of the Period 3 bowls and dishes, including No. 315, have painted arc decoration, and a number of others are imitations of samian ware forms. In all, there are six of form 38, including No. 280, five of form 31, including No. 342, and ten of form 36, including Nos 364–5, 385 and possibly 316. Nos 364 and 365 are probably Stanground products. The decoration on No. 316 is most easily matched on fourth-century vessels made at Swanpool (Webster and Booth 1947, fig. 4, D24–32), though it is similar to that found on some Stibbington products.[21]

Of the various 'boxes' and 'box' lids from Period 3, the near-complete vessel from well F1016, No. 335, is a particularly fine example. If the typological development of these had been as suggested (Guide, 24, no. 89), No. 335 would be of early to mid third-century manufacture, though a close parallel at Verulamium (Frere 1972, fig. 32, no. 1119) is from a context dated *c.* AD 280–315. Nos 281 and 282 might be a little later in date, perhaps mid or mid to late third century, while the more rounded profile, and the decoration of No. 372, would best suit a late third to early fourth-century date. The small vessel, No. 318, has the wide base which is thought to be characteristic of later 'boxes', but its size probably dictated its shape, and might also imply a slightly specialised use. No. 371 may also have been a variation with a more specific function.

Flagons similar to Nos 339 and 383 were made at Stibbington at the end of the third century (Wild 1974, fig. 8, c and d). A parallel for No. 383 from Chesterton (Perrin, to be published), was dated to the mid-third century, and it is probable that the types of which Nos 339 and 383 are examples were both made from the middle of the third century and possibly earlier. They continued in production into the fourth century.

3.5 Cream and white wares

Table 29 gives the C/W A% in the main features and Period 3 as a whole.

Mortaria are the only vessels for which there is proven local production in the third century, but it is almost certain that the other classes were also locally made.

The flagon, 366, and the jar, 319, can be paralleled in Period 1 (Nos 38, 158–60, 280 and 162 respectively), and may be residual, survivals in use or evidence for continued production of certain types. As a more specialised vessel, the flask 386, also had the potential to survive in use.

	F1094	Yard	F435	F441	F772	F254	Other	P3
Jar	-	4	-	-	-	-	1	5
Bowl	2	1	-	-	-	-	-	3
Dish	-	-	-	-	-	-	3	3
Flagon	-	-	-	-	-	1	2	3
Mortarium	1	8	7	2	1	-	21	40
Total	3	13	7	2	1	1	27	54

Table 44 Period 3, main C/W vessel forms, by entries.

The hemispherical flanged bowl with painted decoration on the flange, including No. 388 and, possibly, No. 284, is a certain third-century type. There are only two of these from Period 3 but they are common locally. At Chesterton (Perrin, to be published) they occur mainly in contexts of the later second to mid-third century, and appear to have been made in the Godmanchester area around the same time.[22] These third-century vessels were perhaps a later development of those made at the turn of the first century in the Derbyshire area (Brassington 1971, fig. 7, nos 37–112; Brassington 1980, fig. 10, nos 326–32 and fig. 15, nos 400–44), on which the decoration consists of painted lines rather than 'blobs'. Current evidence also suggests that their use or production in the Lower Nene Valley did not outlast the third century, and examples from layers of later date, such as at Godmanchester (Frend 1968, fig. 13, no. 19) and Orton Longueville (Dakin 1961, fig. 8, no. 79) are almost certainly residual or survivals in use.

Seven of the mortaria are from the Oxfordshire kilns (Mortaria, M31, M34?, M37–8, M43, M53, M55, and Figs 114–5) and another is of Lower Rhineland origin (M35, and Fig. 114). Some of the other cream wares are residual VR sherds.

3.6 Black-burnished wares

Vessels in BB1, BB2 or similar wares account for around 1A% of the Period 3 pottery. The BB1 forms include cooking pots, flanged bowls, for example No. 320, and plain-rimmed dishes like Nos 286 and 390, but BB2 is only represented by bowls or dishes, including Nos 321–2. On typological grounds, No. 320 ought to be of late second to early third-century date, while Nos 286 and 390 could be of any date between the late second and the fourth centuries. The angle of the vessel walls, however, suggests that No. 286 is closer to the earlier and No. 390 to the later date. Similarly No. 321 is essentially a mid to late second-century variety of BB2, whilst No. 322 is of later second to mid third-century date. Such small total amounts of these extremely well-marketed and popular wares, some twenty vessels at the most, must represent chance purchases, or the desire for a particular vessel type rather than trade as such, and outline the dominance of locally produced wares in the third century. No. 286 is not a product of the major BB1 industries, and is probably of more local manufacture, with the Godmanchester area seeming the most likely source (see above).

3.7 Non-local colour-coated wares

Nos 258 and 389 are vessels which are suspected to have come from a source other than the main Lower Nene Valley kilns, possibly outside the local area. No. 258 is possibly a 'box' and is similar to one from Verulamium (Frere 1972, fig. 116, no. 558) dated *c.* AD 140–150. It may be from the Colchester area, but no parallels can be quoted from sites in that region. The form of No. 389 can be easily matched locally but the fabric and general appearance of the vessel again suggest an origin in the south-east, such as Colchester or Pakenham (Smedley and Owles 1961). However, it might possibly be a poorly-fired Stanground product.

CGCC and LRCC are represented by only a handful of sherds, and the Period 3 pottery includes some sherds of OXCC, one a base apparently from a small vessel.

3.8 Miscellaneous wares

The Period 3 pottery contains fragments of at least one large jar from the Horningsea kilns (Hughes 1902; Walker 1912). Most sites in the Lower Nene Valley have examples of these distinctive vessels. Although the Horningsea kilns are only some thirty miles away, and might therefore be considered as 'local', it is likely that the main basis for the marketing of these huge and weighty jars was a specific content or function. A few sherds of oxidised ware were found and are thought to have been produced in the Cambridgeshire/Hertfordshire/Essex border area, possibly in the Hadham kilns. Contact with this region is more clearly attested in Period 4. A number of locally made ordinary oxidised wares, as well as oxidised ware with fumed surfaces, occurred in the Period 3 features. Some of these probably represent misfired grey ware, and sherds of the fumed variant were noted at Stanground.[23] A vessel similar in form and fabric to No. 285 was found at Towcester in a layer dated to *c.* AD 170–270 (Woodfield 1983, fig. 20, no. 31). No. 285 is the most complete and the best example of a mica-gilded ware from Orton Hall Farm, but its source is unknown.

Period 4

(Figs 96–100)

1 Introduction

Period 4 is characterised by further alterations to the farmstead layout. Major changes involved the replacement of Barn 2 by Barn 4 to the north, the creation of the rectangular building and the Mill-house along the northern side of the site, and the enclosure of the western end of the Main Yard by walls. A new well replaced F1016 of Period 3. The walled yard, the House and the main boundary ditches continued in use during part of the period at least. In the central area the infilling of ditch systems F435 and F441, and well F254, together with the relocation of the pond, indicate a change in the location or type of activity. This was probably associated with the new emphasis given to the site by the Mill-house and its adjacent enclosures.

Period 4 is thought to have lasted from AD 300/325 to *c.* 375, and its contexts contained 16A% of the total stratified pottery from the site.

	F434	F200	F1052	F164	F148	F911	F542	P4
Fabric								
RSG	22	19	29	15	24	21	33	32
Grey	5	6	8	1		4	5	6
LNVGW	11	17	31	30	37	47	15	22
LNVCC	50	42	29	50	34	15	44	32
C/W	8	12	1	3	1	6	1	4
BB1	4	1	1			2		1
P4 A%	10	7	12	5	4	15	9	

Table 45 Period 4, main fabric/feature A%.

	RSG	Grey	LNVGW	LNVCC	C/W	BB1	Other	P4
Jar	136	20	60	24	2	-	8	250
WMJar	-	-	4	32	-	-	-	36
Bowl	4	2	20	44	2	6	1	79
Dish	9	3	29	48	-	6	-	95
Flagon	-	-	1	7	1	-	-	9
Beaker	-	-	-	28	-	-	3	31
Lid	1	1	-	2	-	-	-	4
Box	-	-	-	8	-	-	-	8
Mortarium	-	-	-	3	21	-	9	33
Other	-	2	6	15	-	-	-	23
Total	150	28	120	211	26	12	21	567

Table 46 Period 4, main vessel forms/fabric, by entries.

The other fabrics are LIASG, other shell-gritted, VR, CGCC, OXCC, other colour-coated, Trent Valley type, other BB, Horningsea, samian ware, amphorae, possible H-dham area, and Lower Rhineland and Oxfordshire mortaria. Additional forms are a LNVGW cheese press, a LNVCC colander, and vessels which may be bowls or dishes in LNVGW, LNVCC and grey ware.

Material from sites outside the local area shows that there was continued typological change in Lower Nene Valley products through the fourth century, but many of these developments were not as radical or apparent as those which occurred in the third century. The main changes were related to the demise of certain types of colour-coated ware and the cessation of the production of grey ware, though this may have occurred before the beginning of Period 4. The fact that many of the ubiquitous fourth-century types had first appeared in the third century, and that they remained essentially unaltered in form through the fourth century, often makes it impossible to date deposits closely or to recognise possible sequences. Once again, as with Period 3, the features themselves are not often of use either in refining pottery dates or identifying phases. Residuality is an increasing problem. The pottery is, therefore, considered in broad terms, and, to try to make any conclusions more useful, it has been discussed with reference to existing published suggestions, notably those contained in the Guide and a review of the late pottery from Great Casterton (Perrin 1981b).

2 Period 4 features

2.1 Ditch F434

This could not be easily linked to other features. The deposit, all from one layer, is seen as the result of a clearance at the end of Period 4. Table 45 gives the A% of the main fabrics in F434.

The percentages of LNVCC and table vessels suggest that the deposit derived from domestic occupation. The only additional fabric is samian ware and form is a LNVGW cheese press.

	RSG	Grey	LNVGW	LNVCC	C/W	BB1	Total
Jar	4	3	-	1	-	-	8
WMJar	-	-	-	2	-	-	2
Bowl	-	-	1	4	-	2	7
Dish	-	-	1	7	-	1	9
Flagon	-	-	-	1	-	-	1
Beaker	-	-	-	2	-	-	2
Mortarium	-	-	-	-	3	-	3
Other	-	-	1	-	-	-	1
Total	4	3	3	17	3	3	33

Table 47 F434 ditch, main vessel forms/fabrics, by entries.

(Fig. 96)
Vessels **391–403** are from L819 (179).

RSG
391 CR10,22 patches. Two-thirds vessel.

LNVGW
392 CR19/26/19. Third to half vessel.
393 CR21/18/21. Two-thirds vessel.

LNVCC
394 CR27/26/8. Two-thirds vessel.
395 CR8/3/22,26 paint. Two-thirds vessel.
396 CR8+13/4+5/8+13. Half vessel.
397 CR13/5+14/2+22. Third vessel.
398 CR8/14/8.
399 CR11+22/14/11+22.
400 CR8/26/8. Complete.
401 CR2/4+14/2. Near complete.
402 CR3/4/3.
403 CR3+8/26/3+8.

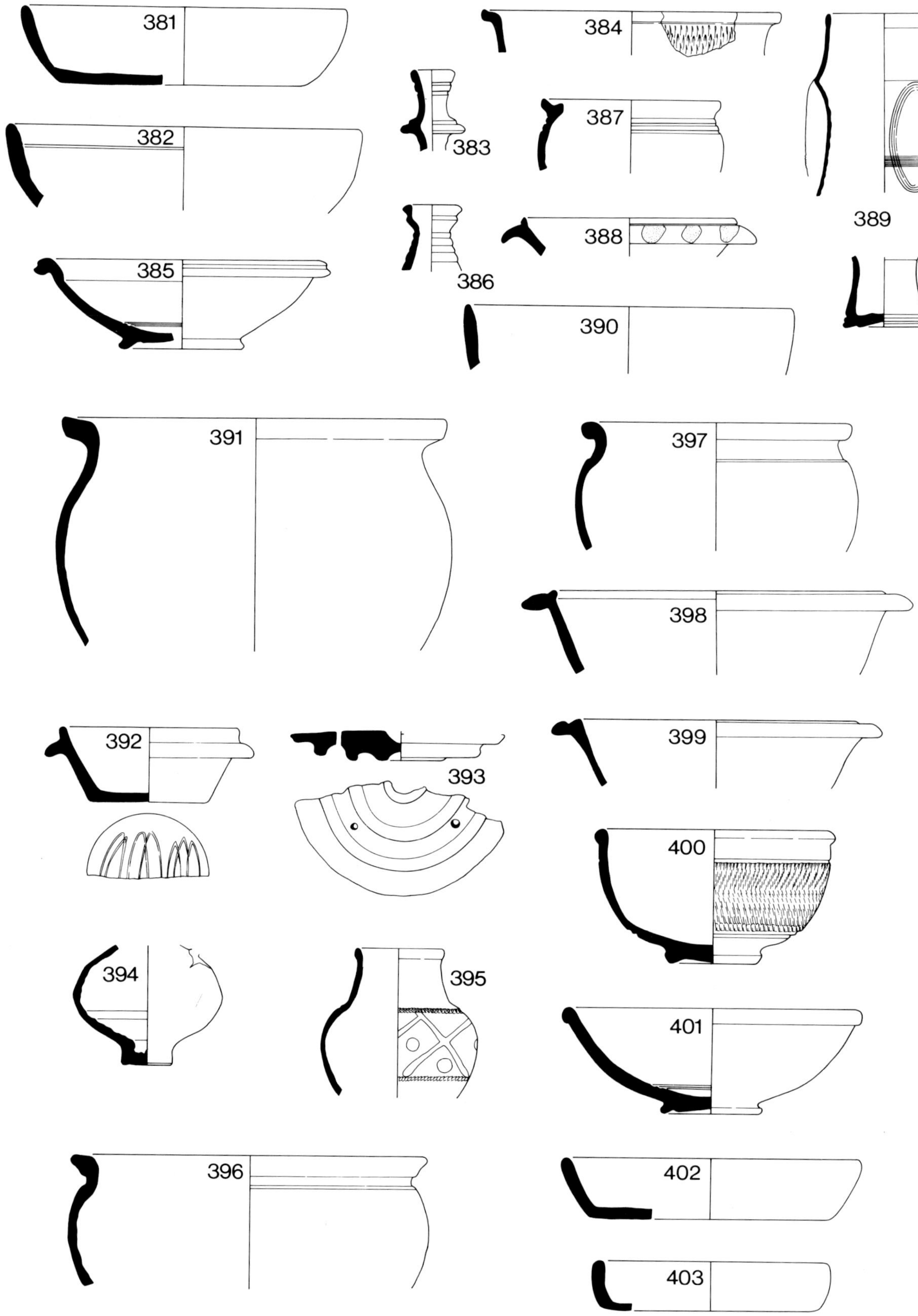

Figure 96 Roman pottery. Period 3: Nos 381–390, additional pots. Period 4: Nos 391–403, F434, ditch outside door to Rectangular Building. Scale 1:4.

2.2 Pond F200

Located in the north-east corner of the Main Yard, this replaced the Period 3 pond, F500. The infilling of the hollow was a gradual process and continued into Period 5. Table 45 gives the main fabric A% for F200.

The only additional fabric is Trent Valley type.

	RSG	Grey	LNVGW	LNVCC	C/W	BB1	Other	Total
Jar	5	1	4	4	-	-	1	15
WMJar	-	-	-	3	-	-	-	3
Bowl	-	-	3	5	-	-	-	8
Dish	-	-	2	7	-	2	-	11
Mortarium	-	-	-	-	7	-	-	7
Total	5	1	9	19	7	2	1	44

Table 48 F200 pond, main vessel forms/fabric, by entries.

(Fig. 97)

GW

404 CR20,22 slip. LNVGW or LNVCC? Stanground? L342 (177)

LNVGW

405 CR20//22/19/22//20. 'Speckled' surfaces. L343 (177)
406 CR20//22/19/22//20. L486 (177)
407 CR19+20. L342 (177)

LNVCC

408 CR4+22/4+14/4+22. L343 (177)
409 CR8+22/14+26/8+22. L341 (177)
410 CR22/14+26/22. L341 (177)
411 CR2+3+8/5+14/2+3+8. L379 (177)
412 CR8+12/5+8+14+26/8+12,26 paint. L379 (177)
413 CR12+17/26/12+17. Third vessel. L343 (177)
414 CR21/26/21. L486 (177)
415 CR17/20/17. L342 (177)

2.3 Well F1052

This replaced the Period 3 well F1016. Its fills represented destruction and are similar in this respect to those in F1016 and 'F969' in Period 3, and its elimination may not have occurred at the end of Period 4. The most likely sources for the material were Barn 1 and the adjacent yard areas. Table 45 gives the main fabric A%.

Additional fabrics are OXCC and samian ware. Two vessels in LNVCC are either a jar or bowl, or a dish or bowl.

	RSG	Grey	LNVGW	LNVCC	C/W	BB1	Total
Jar	14	2	15	3	-	-	34
WMJar	-	-	-	1	-	-	1
Bowl	1	-	2	1	-	1	5
Dish	-	1	7	4	-	1	13
Flagon	-	-	-	3	-	-	3
Beaker	-	-	-	8	-	-	8
Box	-	-	-	1	-	-	1
Mortarium	-	-	-	-	1	-	1
Other	-	-	-	2	-	-	2
Total	15	3	24	23	1	2	68

Table 49 F1052 well, main vessel forms/fabric, by entries.

GW

416 CR15+19. Stanground? Quarter vessel. L2304 (200)
417 CR15+19. Stanground? Third vessel. L2304 (200)
418 CR18+19/18/18+19. Stanground? Third vessel. L2303 (200)

LNVGW

419 CR19/14+18/19. L2306 (200)
420 CR21/18/21. Two-thirds vessel. L2404 (200)
421 CR21/18/21. L2303 (200)
422 CR20/26/20. L2303 (200)

LNVCC

423 CR3+22/26/3+22. Rim complete. L2304 (200)
424 CR22/4+26/22. L2303 (200)
425 CR3/5+20/3. L2303 (200)

2.4 Yard ditch system F164

F130, F157 and F164 were the extension of the eastern boundary ditch across the site of Barn 2 up to Barn 4. As a result of disturbing the site of the former, there was some contamination from earlier levels. Table 45 gives the main fabric A%.

Additional fabrics are amphorae and BB2? The three other forms are LNVCC jars or bowls.

	RSG	LNVGW	LNVCC	C/W	Total
Jar	3	1	-	-	4
WMJar	-	-	2	-	2
Bowl	-	5	3	-	8
Dish	-	-	9	-	9
Flagon	-	1	-	-	1
Beaker	-	-	2	-	2
Box	-	-	1	-	1
Mortarium	-	-	-	2	2
Other	-	-	3	-	3
Total	3	7	20	2	32

Table 50 F164 system, main vessel forms/fabric, by entries.

GW

426 CR19/14/19. F164, L262 (157)
427 CR20/18/20. F164, L251 (157)

LNVGW

428 CR19+22//26/19/26//19+22. F164, L262 (157)
429 CR20//26/19/26//20. F164, L286 (157)
430 CR15+20/26/15+20. F164, L286 (157)

(Fig. 98)

LNVCC

431 CR12/26/12. Quarter vessel. F164, L286 (157)
432 CR2+3/26/2+3. F157, L276 (157)
433 CR12+17/14+26/12+17. F164, L286 (157). F189, L339 (161)
434 CR21/26/21. F164, L286 (157)
435 CR22/26/22. F164, L251 (157)

2.5 Ditch system F148

Made up of F108, F128, F138, F148, F165 and F170, all were drainage ditches round Barn 4 and may have been silted up by the end of Period 4. Table 45 gives the main fabric A%.

The additional fabrics and forms are a LNVGW jar or bowl; a LNVCC lid and three mortaria; one LNVC/W; one Oxfordshire and the other Lower Rhineland or LNVC/W, samian ware and uncertain.

	RSG	LNVGW	LNVCC	Other	Total
Jar	3	4	1	-	8
WMJar	-	1	2	-	3
Bowl	-	2	2	-	4
Dish	1	1	-	-	2
Beaker	-	-	3	-	3
Mortarium	-	-	3	3	3
Other	-	1	1	-	2
Total	4	9	9	3	25

Table 51 F148 system, main vessel forms/fabric, by entries.

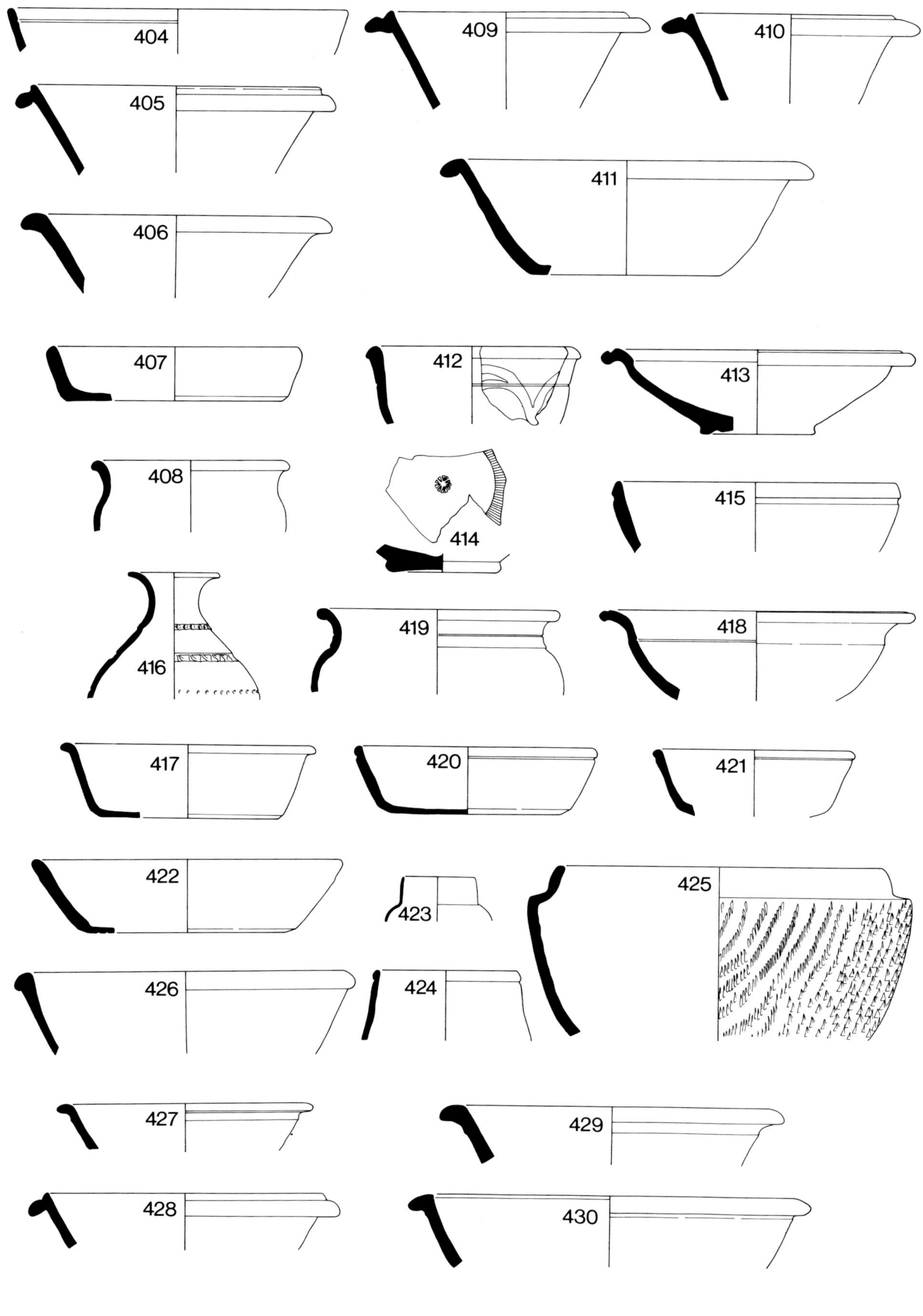

Figure 97 Roman pottery. Period 4: Nos 404–415, F200; Nos 416–425, F1052 well; Nos 426–430, F164 system. Scale 1:4.

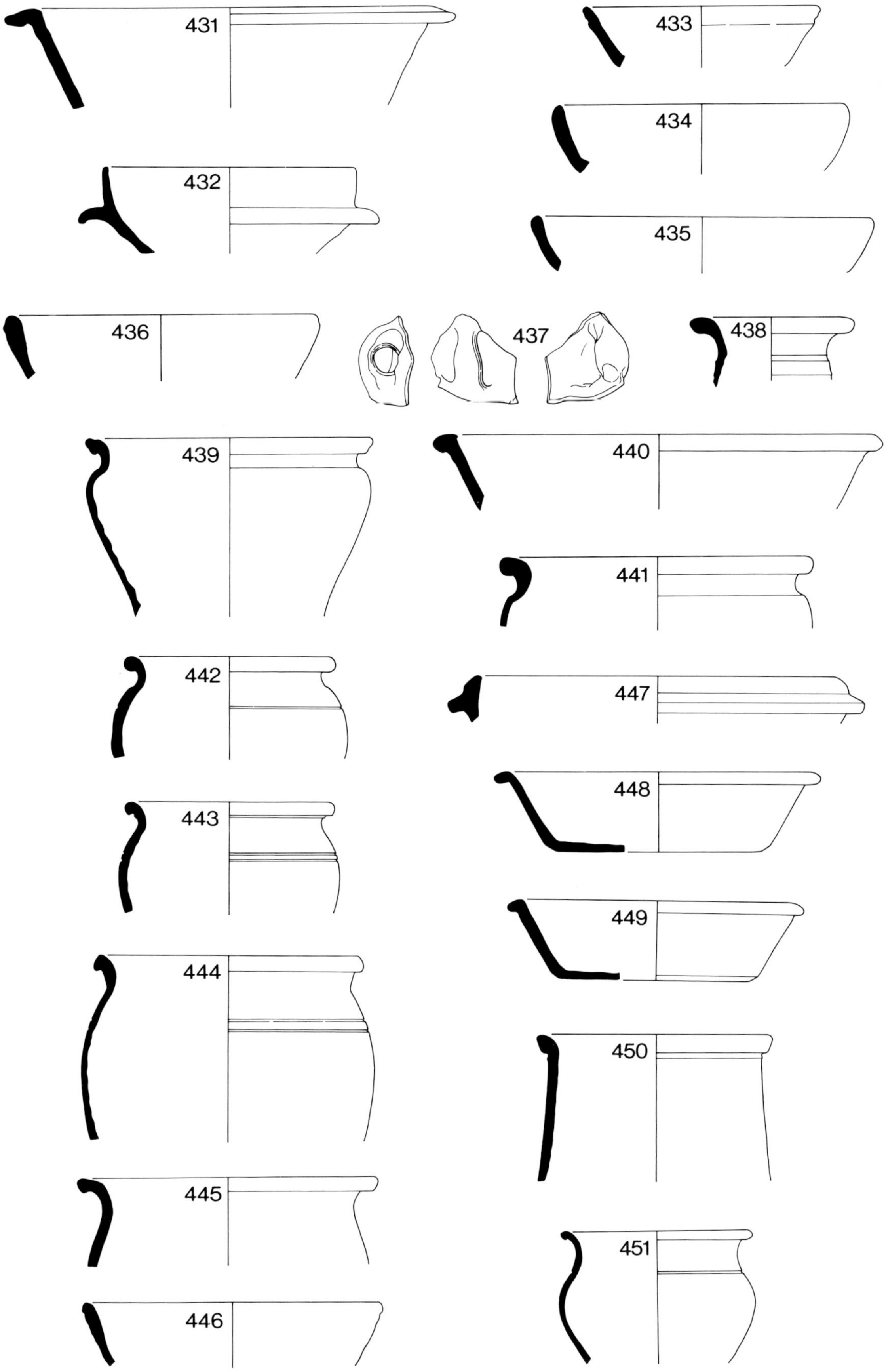

Figure 98 Roman pottery. Period 4: Nos 431–435, F164 system; Nos 436–441, F148 system; Nos 442–451, F911 Small Yard. Scale 1:4.

RSG
436 CR10+15+17. F128, L195 (168)

GW
437 CR19+15/18/19. East Yorkshire? F148, L238 (168)
438 CR18. F165, L268 (168)
439 CR21//26/21/26//21. F148, L337 (168)
440 CR21/20/21. F108, L166 (168)

LNVCC
441 CR2/26/2. F138, L256 (168)

2.6 Yard deposit F911

This was an ambiguous deposit located in the southern part of the Small Yard next to Barn 1. The pottery was all from one layer and its overall character suggests that it was related to work carried out at the beginning of Period 4 rather than later. The stratification around F911 was disturbed and some of the feature may have accumulated or been deposited at the end of Period 3. Table 45 gives the main fabric A%.

The additional fabrics are samian ware and Horningsea, and the unlisted form is a LNVGW dish or bowl.

	RSG	Grey	LNVGW	LNVCC	C/W	BB1	CGCC	Other	Total
Jar	11	-	23	2	-	-	-	1	37
Bowl	-	1	1	1	-	2	-	-	5
Dish	1	1	12	2	-	-	-	-	16
Flagon	-	-	-	1	1	-	-	-	2
Beaker	-	-	-	5	-	-	2	-	7
Box	-	-	-	2	-	-	-	-	2
Mortarium	-	-	-	-	1	-	-	-	1
Other	-	-	1	-	-	-	-	-	1
Total	12	2	37	13	2	2	2	1	71

Table 52 F911 yard, main vessel forms/fabric, by entries.

Vessels **442–469** are from L1887 (203).

RSG
442 CR3–10+22.
443 CR3–12. Joins sherd in L2239, F1052.
444 CR1–22.
445 CR2/17/2. Possibly rilled.
446 CR5+10/21/5+10.

GW
447 CR21/19/21.
448 CR15/14/16+21. Stanground? Three-quarters vessel.
449 CR19. Stanground?
450 CR20/19+21/22.

LNVGW
451 CR20/26/20. 'Speckled' surfaces.

(Fig. 99)
452 CR15+19. 'Speckled' surfaces. Near complete.
453 CR20/26/20.
454 CR19/26/19.
455 CR20/26/20.
456 CR21/14/21.
457 CR21//18/21/18//21.
458 CR20/18+26/20.

LNVCC
459 CR3/5/3.
460 CR11/4/22.
461 CR17+22/4+14/17+22.
462 CR21/26/21.
463 CR17/14/2.

CGCC
464 CR22/3/22.
465 CR22/4+19/22.

C/W
466 CR26.
467 CR26,12 slip? LNV?

BB1
468
469

2.7 Ditch system F542

Consisting of F481, F482, F542, F579, F583 and F584, all were probably associated with the Mill-house from whence the material in them could have derived. Table 45 gives the main fabric A%.

Additional fabrics are VR, samian ware and an Oxfordshire (mortarium). The other form is a LNVCC colander.

	RSG	Grey	LNVGW	LNVCC	C/W	Other	Total
Jar	22	5	3	1	-	1	32
WMJar	-	-	-	8	-	-	8
Bowl	-	-	1	11	-	-	12
Dish	-	-	3	3	-	-	6
Flagon	-	-	-	1	-	-	1
Mortarium	-	-	-	2	2	1	5
Other	-	-	-	1	-	-	1
Total	22	5	7	27	2	2	65

Table 53 F542 system, main vessel forms/fabric, by entries.

GW
470 CR21/18+19/21. Stanground? Slipped? F542, L1094 (194)

LNVGW
471 CR21/26/21. F542, L1069 (194)

LNVCC
472 CR1/14/1. Two-thirds vessel. F542, L1069 (194)
473 CR2+3+8+12/5/2+3+8+12. F482, L991 (194)
474 CR22/5+14/22. F584, L1152 (194)

Period 4 Additional Sherds

RSG
475 CR5/21/5+10. F586, L1155 (187)

(Fig. 100)
476 CR10+21. F1139, L2861 (137) (or L1168 Period 5 (186))
477 CR4/21/4. F1062, L2572 (212)
478 CR4+5/21/4+5. F814, L1797 (206)
479 CR4+9/21+22/17+22,5. Base. F630, L1227 (197). F470, L923 (196)

GW
480 CR20/21/4/20/21. TV? F548, L1494 (207)
481 CR20/19/20. Horningsea. Uncertain diameter. F1122, L2784 (212)
482 CR15+19//16/20/16//15+19. TV? F277, L542 (173)

LNVGW
483 CR21/26/21. F581, L1147 (189)

LNVCC
484 CR1+12/26/21. F813, L1669 (203)
485 CR8/14+27/8. F813, L1669 (203)

3 The Pottery

3.1 RSG

Table 45 gives the RSG A% in the main features and Period 4 as a whole.

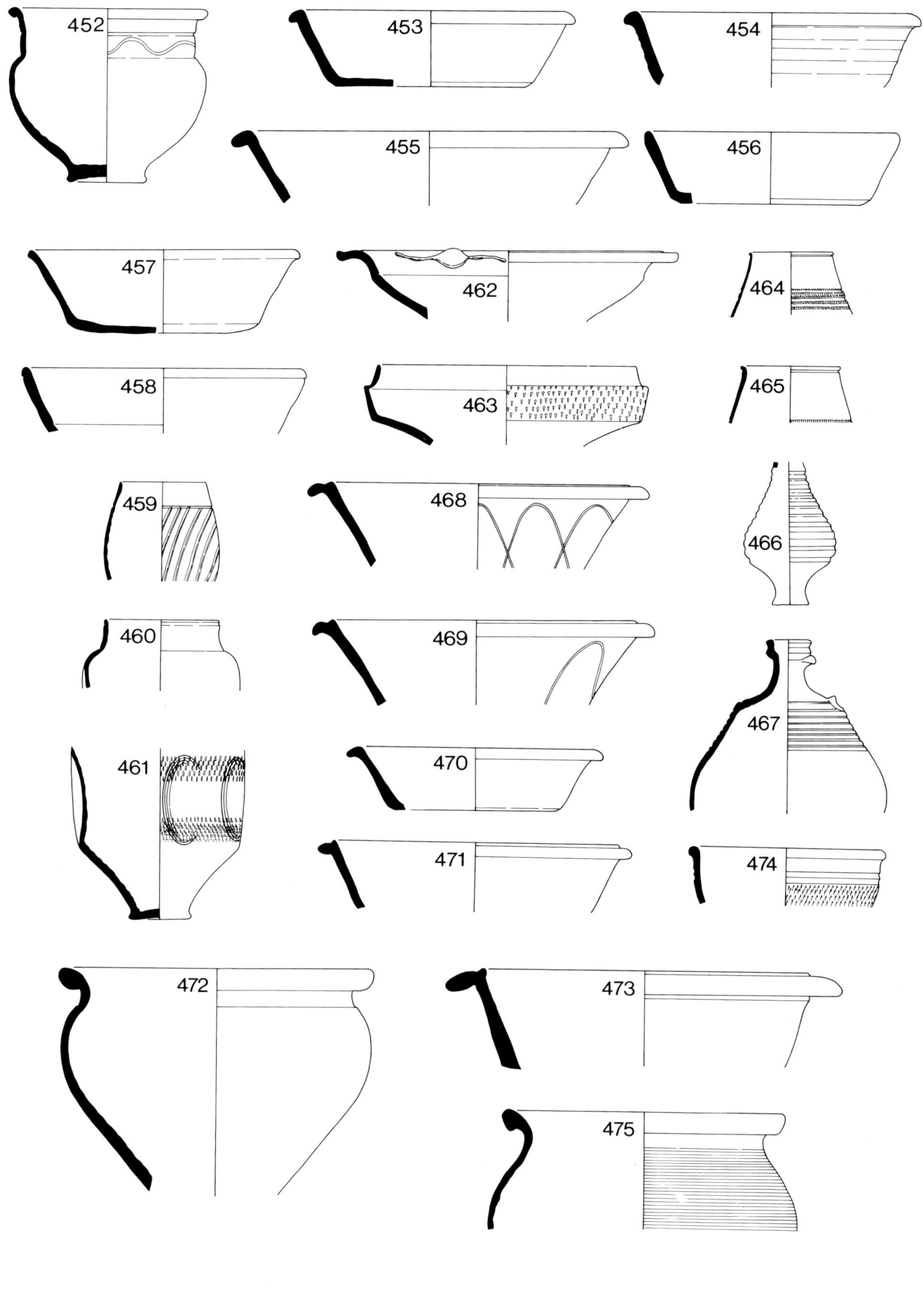

Figure 99 Roman pottery. Period 4: Nos 452–469, F911 Small Yard; Nos 470–474, F542 system; No. 475, Period 4, additional pot. Scale 1:4.

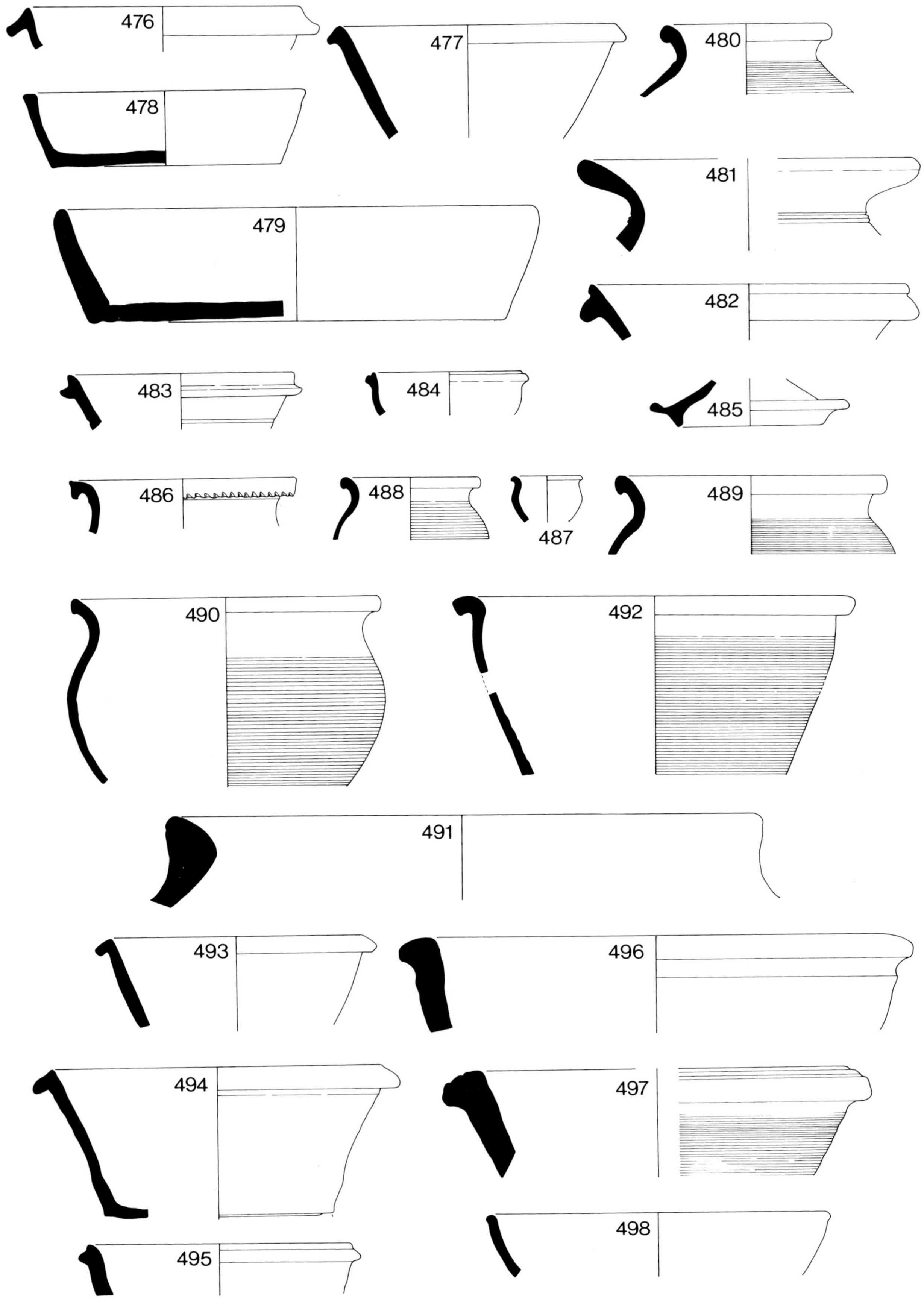

Figure 100 Roman pottery. Period 4: Nos 476–485, Period 4, additional pots. Period 5: Nos 486–498, F14–F31–F203, eastern enclosure ditches. Scale 1:4.

	F434	F200	F1052	F164	F148	F911	F542	Other	P4
Jar	4	5	14	3	3	11	22	74	136
Bowl	-	-	1	-	-	-	-	3	4
Dish	-	-	-	-	1	1	-	7	9
Lid	-	-	-	-	-	-	-	1	1
Total	4	5	15	3	4	12	22	85	150

Table 54 Period 4, main RSG vessel forms, by entries.

It is difficult to pinpoint any obvious typological changes in the jars, probably because many are residual or survivals in use from Period 3. The more globular profiles and definite shoulders of Nos 442–4 from F911, for example, are most reminiscent of third-century types but, given that this feature is thought to belong to the beginning of Period 4, this is perhaps to be expected. Nos 445 and 475 show a gradual development from third-century jars having slacker profiles with almost no shoulder and rims with a more pronounced undercut. Both are probably rilled, showing the continuation of this fashionable or functional trait. No. 391, however, though from the clearly fourth-century F434, is neither rilled, nor has an undercut rim, and aptly demonstrates the drawbacks in applying typological developments indiscriminately.

The other RSG vessels include types which appear for the first time in the fourth century. The most noticeable is the flanged bowl, No. 476, for which existing parallels are all mid-fourth-century and after: *c.* AD 360–410 at Verulamium (Frere 1972, fig. 138, no. 1258); second half of the fourth century at the Park Street villa (O'Neil 1947, fig. 20, nos 5–7); late fourth century at Great Casterton (Gillam 1951, fig. 9, nos 32–3); *c.* AD 330–370+; and mid-fourth to early fifth-century at Towcester (Woodfield 1983, fig. 30, no. 257; Brown and Alexander 1982, fig. 14, no. 161). This is the only example from Orton Hall Farm and the type is not known from sites in the immediate Lower Nene Valley vicinity. Its lack of occurrence in the latest deposits locally perhaps indicates that the type was neither made nor used in the area until the later fourth century. This apparent scarcity could reflect the lack of genuinely late deposits from many local sites. No. 477 may be a related form. Similar vessels from Chesterton (Perrin, to be published) were of fourth-century date.

The dishes, Nos 436, 446 and 478, are essentially RSG equivalents of the ubiquitous plain-rimmed LNVCC dish. The type was also one of those occurring in calcite-gritted ware of East Yorkshire manufacture (Perrin 1981a, 61, fig. 41, no. 566). The size of No. 479 probably indicates that it had a specific function.

3.2 Grey wares

Table 45 gives the grey ware A% for the main features and Period 4 as a whole.

	F434	F200	F1052	F911	F542	Other	P4
Jar	3	1	2	-	5	9	20
Bowl	-	-	-	1	-	1	2
Dish	-	-	1	1	-	1	3
Lid	-	-	-	-	-	1	1
Other	-	-	-	-	-	2	2
Total	3	1	3	2	5	14	28

Table 55 Period 4, main grey ware vessel forms, by entries.

The other vessels are either bowls or dishes.

The reduction in the amount of grey ware noted in Period 3 continued in Period 4. Some will have been residual material from previous periods but little, if any, represents survivals in use, apart from, possibly, some of that in F911. A number of potteries producing grey wares in the fourth century found markets in the Lower Nene Valley area for the first time. The most significant of these is that known collectively as East Midlands burnished ware or Trent Valley ware, from a number of kiln centres in Nottinghamshire and Lincolnshire. Nos 447, 480 and 482, for example, are comparable to vessels made at Swanpool (Webster and Booth 1947, fig. 5, type H) and used in Lincoln (Darling 1977a, figs 2–3, nos 38–48, 56–8) and elsewhere (Todd 1968a, fig. 1, nos 6–8, and fig. 2, no. 6). Flat-topped dishes or bowls, including Nos 417, 427, 440, 448–9 and 470, were also made at Swanpool (Webster and Booth 1947, fig. type E8–9) and used in Lincoln (Darling 1977a, fig. 4, nos 87–8), though those from the essentially third-century group F911, Nos 417 and 448–9, might be products of the Stanground kilns or another Lower Nene Valley source. Vessels similar to Nos 404 and 418 were also made at Stanground.[24]

The narrow-mouthed jars, 416 and 438, are probably also from the East Midlands sources suggested for the previous vessels. The handle 437 is most probably from a two-handled jar type produced in many kilns in Lincolnshire and East Yorkshire in the third and fourth centuries: Dragonby kiln 3 (Stead 1976, fig. 64, no. 6); Roxby (*ibid.*, fig. 67, nos 38–9); Messingham (*ibid.*, fig. 71, nos 13–14); Crambeck (Corder and Birley 1937, fig. 2, types 3 and 3a); Norton (Hayes and Whitley 1950, fig. 10, types 4a and b); and Throlam (Corder 1932, figs 14–15, nos 72–90). The size, neatness and countersunk form of No. 437 is most reminiscent of the Crambeck variety, though it is difficult to suggest how it may have found its way to Orton Hall Farm, except as an individual personal possession (see Nos 500 and 614, Period 5). Nos 426 and 450 might be from as yet unlocated local centres in production at the same time as Stibbington and Sibson, and could well be third-century survivals. The complete form of No. 451 is uncertain and no parallels can be cited with confidence.

The large jar, 481, is a Horningsea product (Hughes 1902; Walker 1912). Fragments of these jars were noted in Period 3 and they would obviously have been robust enough to survive for a considerable time in use. The full date-range of the Horningsea kilns is not known but they appear to have continued into the later third century. The presence of Horningsea pottery in Period 4 features need not, therefore, indicate continued trade in the fourth century, and the small proportion of the ware (less than 1 per cent) also suggests this was unlikely.

3.3 LNVGW

Table 45 gives the LNVGW A% in the main Period 4 features and Period 4.

	F434	F200	F1052	F164	F148	F911	F542	Other	P4
Jar	-	4	15	1	4	23	3	10	60
WMJar	-	-	-	-		-	-	3	4
Bowl	1	3	2	5	2	1	1	5	20
Dish	1	2	7	-	1	12	3	3	29
Flagon	-	-	-	1	-	-	-	-	1
Other	1	-	-	-	1	1	-	3	3
Total	3	9	24	7	9	37	7	24	120

Table 56 Period 4, main LNVGW vessel forms, by entries.

The other forms are a possible beaker, a cheese press, three jars or bowls, and one bowl or dish.

Some of this pottery may be residual or represent survivals in use from previous periods, especially Period 3. That from some features, for instance F911, would have been almost entirely of third-century manufacture, although F911, as has been pointed out, is an ambiguous deposit and should perhaps be discounted. This is emphasised by the proportion of LNVGW being still too high if the ware was not made after the late third century as has been suggested (Perrin 1981b, 455). However, this can be seen by comparing two of the main groups, F911 and F434.

F434 probably belongs to the end of Period 4, while F911 has material in use at the end of Period 3, as well as the beginning of Period 4. A very large proportion, some 39A%, of the F911 pottery was LNVGW compared with 11A% of that in F434 (Table 45). Moreover, F911 contained thirty-seven separate LNVGW vessels compared to only three in F434 (Tables 52 and 47). Apart from date, the very varied figures could represent a difference in the nature of each deposit. Both, however, contained beakers, jars, jugs and imitation samian ware vessels, and the numbers of LNVCC vessels in each were much closer (seventeen in F434, thirteen in F911: Tables 47 and 52) than had been the numbers in LNVGW. A comparison of the two groups could indicate that the production of LNVGW ceased in the fourth century rather than at the end of the third, and *may* have lasted into the second quarter of the fourth century.

The evidence from the other groups is not so clear cut, but the fact that the newly dug pond F200 and ditch F542 had less LNVGW, relatively, than the ditches and well near Barns 1 and 4, in areas with considerable Period 3 activity (F164 system, F1052), does not contradict the suggestion given above, but the conclusion must remain tentative.

There is little to add to the previous typological discussion, for most of the Period 4 vessels are basically third-century forms. There were, however, more flanged vessels, at least nine in all, including Nos 392, 405, 428, 471 and 483. Though there was none from F911, there were four from F434, F200 and F542, which supports the view that these were among the latest types of LNVGW made. The cheese press, No. 393, is a rare item in such a late context as F434, but is precisely the kind of vessel which might be expected to have survived in use. Unillustrated LNVGW from the period includes a flagon and a beaker of possible Stanground manufacture.

3.4 LNVCC

Table 45 gives the LNVCC A% in the main Period 4 features and Period 4.

	F434	F200	F1052	F164	F148	F911	F542	Other	P4
Jar	1	4	3	-	1	2	1	12	24
WMJar	2	3	1	2	2	-	8	14	32
Bowl	4	5	1	3	2	2	11	15	43
Dish	7	7	4	9	-	1	3	16	47
Flagon	1	-	3	-	-	1	1	1	7
Beaker	2	-	8	2	3	5	-	8	28
Box	-	-	1	1	-	2	-	4	8
Mortarium	-	-	-	-	-	-	2	1	3
Other	-	-	2	3	1	-	1	12	27
Total	17	19	23	20	9	13	27	83	211

Table 57 Period 4, main LNVCC vessel forms, by entries.

Other forms are two lids, a colander, ten jars or bowls and four bowls or dishes.

The range of vessels made in LNVCC altered and perhaps contracted during the fourth century. There is, however, little positive evidence for the introduction, or the phasing out of particular types, and the features on the site are not very useful in defining what may have been their start or end dates. Some of the discussion is postponed to Period 5, but a number of points can be noted here.

The presence of a complete imitation samian ware form 37, No. 400, in the late F434 deposit suggests that this type probably continued in production further into the fourth century than had been previously thought (Perrin 1981b, 453 and 455), though it might have continued in use as a survival. The main groups and the Period 4 features as a whole contained a variety of plain-rimmed dishes and flanged bowls confirming the belief that there is little or no chronological significance in the differences. Moreover, the presence of flat-topped and flanged bowls together suggests that the latter was probably not the chronological and typological successor to the former. The pierced lid, No. 485, clearly shows that this particular type was not exclusively late fourth-century in date.

In the discussion of Period 3 LNVCC, the orange or reddish-yellow fabric with white inclusions was mentioned. In Period 4, the total percentage of this variant ware, including No. 395, is low (2%–3%) when compared with that of the more usual white or cream fabric (20%–30%). These figures reinforce the idea that these vessels were made at very few centres, one of which was Chesterton as already suggested. Similar beakers occurred in fourth-century contexts at Chesterton (Perrin, to be published). No. 395 also came from the late deposit F434, though it might also have been a survival in use.

There was quite a wide range of beaker types present. Those from F911 are essentially third-century varieties. No. 461, for example, is very close to No. 277, even showing the same technique of rouletting through the colour-coat to expose the lighter fabric below. The diagonal barbotine of No. 459, though over the slip, is again of third-century date (Gillam 1970, fig. 9, nos 81–3), as are Nos 460 and 423 from well F1052 (*ibid.*, fig. 7, no. 51). Three of the beakers, including No. 395, were painted with geometric or rhenish-style motifs. Of the thirteen with identifiable rims, six had third-century curved or plain rims, and one an even earlier 'cornice' type. The other six all had funnel-necks, four with bead and two with flattened and everted rims.

The range of imitation samian ware vessels, Nos 400–1, 413, 432, 462 and 474, is matched in the production at the New Forest and Oxfordshire kilns (Fulford 1975; Young 1977). The stamped base fragment No. 414 is likely to be residual from Period 3, and is similar in appearance to No. 462 from the F911 deposit. Both are probably Stanground products. The 'box', No. 463, from F911 can be closely paralleled by a mid to late third-century vessel at Chesterton (Perrin, to be published). No. 425, however, has supposedly later characteristics and is probably of early to mid-fourth-century date. The small bowl No. 484 cannot be matched at present.

3.5 Miscellaneous wares

In the main, these consisted of very small amounts of other local wares, other colour-coated wares, including Oxfordshire, possible Hadham area ware, black-burnished wares and residual Iron Age or early Roman pottery including shell-gritted ware and VR ware.

The illustrated material is all from F911 which contained all of the Period 4 CGCC, and over a quarter of the other Lower Nene Valley wares. The beakers, Nos 465–6, are standard CGCC products. The ware is so rare at Orton Hall Farm that these two vessels may have been prized possessions, and could, therefore, have survived in use for a considerable time. The BB1, Nos 468–9, is closer to third-century rather than later types (Gillam 1976, fig. 3, nos 43–4), but other BB1 from later deposits, including F434, shows that the ware was still being purchased and used locally in the fourth century, though in small amounts.

The cream ware flagon No. 467 is unusual and cannot be easily paralleled, but some vessels from London may be related (Richardson 1986, 122–3, I.154– 5). The flask, No. 466, is more easily matched at Colchester (Hull 1963a, fig. 123, no. 389), Verulamium (Frere 1972, fig. 124, no. 885, and Frere 1984, fig. 83, no. 1982) and Southwark (Hammerson and Murray 1978c, fig. 162, no. 1221, and fig. 168, no. 1328). Most of these are from second-century contexts but one from Verulamium (Frere 1984, fig. 83, no. 1982) is dated to *c.* AD 290–310, and the type is one that could easily have survived in use

Period 5

(Figs 100–6)

1 Introduction

The four preceding periods chart the development of a Roman farmstead through three centuries. The changes in each of these periods were part of a discernible growth in the intensity of use of the farmstead, which by the end of Period 3 had extended to cover the whole of the excavated area and possibly beyond. Period 5 is different in that it marks the passing of at least part of the farmstead from Roman into Anglo-Saxon hands.

The Roman farm declined gradually with areas being abandoned in turn rather than all at once, and it is far from certain when it might have ceased to be viable in any form to the Roman population, but this need not have been at the end of Roman control in the area as a whole and it is similarly uncertain when Anglo-Saxon tenure began. It need not have been in the fifth century. The best estimate for the beginning of the period is not before *c.* 375, but not necessarily much later.

The period was the most difficult to assess of all because the character of the site was significantly different from that of previous periods. New, independent, evidence was provided by sherds of Oxfordshire colour-coated and Hadham-area vessels, to which was added the somewhat negative evidence of the demise of certain pottery types, including beakers and 'boxes'. The uncertainty about the Period 5 occupation was epitomised by the Anglo-Saxon pottery, for which there are few parallels in terms of its presence in surviving and functioning elements of a Roman farmstead.

Period 5 had the largest amount of pottery of all the periods, accounting for around 41A% of the total stratified pottery from the site.

	F14	F80	F81	F200	RB	F1061	F55	F171	F12	F172	AF48	F751	P5
Fabric													
RSG	58	49	42	27	48	49	51	57	39	25	58	15	43
Grey	-	3	1	6	5	8	-	4	-	2	-	2	5
LNVGW	6	6	9	15	7	7	5	11	22	7	10	29	11
LNVCC	27	39	40	41	35	32	41	18	40	60	28	48	32
C/W	3	2	6	6	1	1	2	9	-	3	3	3	3
Saxon	1	<1	2	<1	-	-	-	-	-	-	-	-	3
P5 A%	11	2	3	8	17	9	1	1	1	2	1	1	

Table 58 Period 5, main feature/fabric A%.

	RSG	Grey	LNVGW	LNVCC	C/W	Other	P5
Jar	521	28	86	86	5	13	739
WMJar	5	-	7	105	1	2	120
Bowl	39	7	21	149	1	14	231
Dish	24	6	42	144	2	8	226
Flagon	-	-	2	16	-	-	18
Beaker	-	-	1	36	-	3	40
Box	-	-	-	8	-	-	8
M	-	-	-	5	70	16	91
Other	9	3	11	33	1	1	58
Total	598	44	170	582	80	57	1531

Table 59 Period 5, main vessel forms/fabric, by entries.

Additional fabrics are LIASG, grogged, London ware type, LRCC, other CC, CCC, OXCC, cream, VR, OXW, possible Hadham area, Trent Valley type, Horningsea, BB1, other BB, samian ware, amphorae, miscellaneous SG and uncertain. Vessels that may be jars or bowls occur in all the main fabrics, and there are others that are either dishes or bowls in LNVGW, LNVCC and BB1. There is also a C/W colander and lids in RSG, grey, LNVCC and possible VR.

Much of this pottery is likely to be residual from previous periods, including Periods 1 and 2, as some of the features of Period 5 cut into the earliest levels on the site in areas unoccupied in the interim. Some of the vessels will be survivals in use. The Period 5 material included the first significant quantities of Oxfordshire colour-coated ware and Hadham area pottery, as well as new varieties of established wares such as RSG.

The main problem with the pottery from this period is the range of types represented and their respective dates. The only comparable deposits of value for this discussion are those from the later phases of the Great Casterton villa. The pottery of Period 5 is far less complex than in previous periods for there are fewer wares generally and there is a clear split between shell-gritted kitchen and utilitarian wares and colour-coated table wares. Together, these two account for over 75–80 per cent of the pottery. For all the wares discussed, Period 5 represents their final use on the site, and the types present are almost certainly the last acquired by the occupants.

Many of the main groups of pottery in previous periods were the result of definite changes on the site, such as clearances and levelling at the end of one period or the beginning of another, material from recut or newly cut ditches, deposits relating to new structures and those from successive wells. There are few such relatively clear cut groups in Period 5. Much of the pottery was found where it had been left when that particular area or feature became disused, and it is salutary to think that, without the Anglo-Saxon phase of the site, many of the feature sequences suggested for Period 5 might have been thought to belong to Period 4, leading to the belief that most of the site had been abandoned before the end of the fourth century. Six main and six subsidiary groups form the basis of the pottery discussion, and none is as homogeneous as those of previous periods.

In the following listing, * is attached to those layers which contained Anglo-Saxon pottery.

2 Period 5 features

Main Groups

2.1 The Eastern Enclosure ditch F14/31/203

Anglo-Saxon pottery from parts of the ditch indicates that the latter was a distinct feature of the landscape during Anglo-Saxon occupation of the site, and some of the recutting of the ditch is attributed to this period. There was no useful additional dating evidence. Table 58 gives the fabric A%.

	RSG	LNVGW	LNVCC	C/W	Other	Total
Jar	88	9	12	-	3	112
WMJar	-	-	12	1	-	13
Bowl	10	1	9	-	1	21
Dish	-	3	14	-	-	17
Flagon	-	-	4	-	-	4
Beaker	-	-	-	-	1	1
Box	-	-	4	-	-	4
Mortarium	-	-	-	8	6	14
Total	98	13	55	9	11	186

Table 60 F31 system, main vessel forms/fabric, by entries.

Other fabrics represented are samian ware, OXCC, other colour-coated, OXW and LR (mortaria), BB1, Hadham? and uncertain.

(Fig. 100)
RSG

486 CR17/16/7+18. F203, *L475 (213)
487 CR4/17/4. F14, *L87 (213)
488 CR7+18/21/7+18. F203, *L475 (213)
489 CR9+17/21/9+17. F203, *L475 (213)
490 CR12+14/14+21/12+14. F203, *L345 (213)
491 CR10/17/10. F31, L474 (213)
492 CR3+9+10. F31, *L212 (213)

493 CR5/21/11. F14, *L87 (213)
494 CR15/20/15. F31, *L212 (213)
495 CR4+5/21/4+5. F31, ?*L254 (213)
496 CR3/21/3. F14, *L87 (213)
497 CR3+14/21/3+14. Uncertain diameter. F14, L200 (213)
498 CR10/16+20/10. F202, L383 (213)

(Fig. 101)
GW
499 CR18+20. F203, *L345 (213)
500 CR20. East Yorkshire? F203, *L346 (213)
501 CR15/19/15. F31, L474 (213)

LNVGW?
502 CR20. F203, *L345 (213)

LNVCC
503 CR22/27/22. F31, *L212 (213)
504 CR22/27/22. F203, *L345 (213)
505 CR1+3/4+5/1+3. F31, L473 (213)
506 CR8/26/8. F203, *L345 (213)
507 CR8/14/8. F203, *L475 (213)
508 CR8+13/14+26/8+13. F203, L728 (213)
509 CR3+12/14/3+12. F31, *L212 (213)
510 CR11/4/11. Iridescent 'bronze' colour-coat. F31, *L212 (213). F203, *L436 (213)
511 CR8/11+26/8. F203, *L475 (213)
512 CR8+11/11+21+27/8+11. F14, L17 (213)

Miscellaneous
513 CR12+22. Calcite-gritted (limestone and quartz). F203, *L322 (213)
514 CR4+16/3+20/4+16. Hadham? F31,*L212 (213)
515 CR3. Hadham? F14, L200 (213)
516 CR3/3+20/28, burnt. Hadham? F203, *L346,*L475 (213)

2.2 Southern ditch F80
Table 58 gives the main fabric A%.

	RSG	Grey	LNVGW	LNVCC	C/W	Other	Total
Jar	13	1	2	6	-	-	22
WMJar	1	-	-	2	-	-	3
Bowl	2	-	-	4	-	-	6
Dish	-	-	1	4	-	-	5
Beaker	-	-	-	2	-	-	2
Box	-	-	-	1	-	-	1
Mortarium	-	-	-	-	3	1	4
Other	-	-	-	1	-	-	1
Total	16	1	3	20	3	1	44

Table 61 F80 ditch, main vessel forms/fabric, by entries.

There are a few sherds of London ware type and an OXW mortarium.

RSG
517 CR10. Vesicular. *L120 (240)
518 CR3/22/3. Part vesicular, fairly soft. L472 (240)
519 CR4/17/4. Vesicular. Soft and 'corky'. *L120 (240)
520 CR15+20. *L120 (240)

GW
521 CR17+21//8/20/8//17+21. Fairly soft. L472 (240)

LNVCC
522 CR4+11/4/4+11. *L120 (240)
523 CR4+8/26/4+8. *L120 (240)
524 CR4/26/4. *L120 (240)

2.3 Southern ditch F81
Table 58 gives the main fabric A%.

The ditch is thought to be the successor to F80; the pottery from it is very similar though greater in quantity. The only additional fabric is a type of BB1. One colour-coated beaker is of uncertain source, but may just be a local variant. The other vessels are all either jars or bowls occurring in both LNVGW and LNVCC.

	RSG	LNVGW	LNVCC	C/W	Other	Total
Jar	19	1	4	-	-	24
Bowl	1	1	8	-	-	10
Dish	-	1	2	-	-	3
Flagon	-	-	1	-	-	1
Beaker	-	-	2	-	1	3
Mortarium	-	-	1	4	-	5
Other	-	2	5	-	-	7
Total	20	5	23	4	1	53

Table 62 F81 ditch, main vessel forms/fabric, by entries.

RSG
525 CR4+10. L148 (243)

LNVCC
526 CR11/4+14/11. *L121 (243)
527 CR12+17/14/12+17. *L121 (243)
528 CR8/14/8. *L121 (243)
529 CR8/26/8. Complete. *L121, L148 (243)
530 12+17/14/12+17. *L159 (243)
531 CR4+11/4/4+11. *L159 (243)

BB1 or imitation
532 *L151 (243)

2.4 Pond F200
F200 ceased to be usable as a pond in Period 5, having largely and gradually silted up during Period 4. The final fills probably represent continued, perhaps occasionally deliberate, infilling of the remaining hollow and some of these contained Anglo-Saxon pottery. Table 58 gives the main fabric A%.

	RSG	Grey	LNVGW	LNVCC	C/W	Other	Total
Jar	18	3	15	5	-	1	42
WMJar	-	-	1	15	-	1	17
Bowl	4	1	3	18	-	2	28
Dish	3	-	3	17	-	1	24
Flagon	-	-	-	2	-	-	2
Beaker	-	-	-	2	-	-	2
Mortarium	-	-	-	1	8	-	9
Other	3	-	-	-	1	-	4
Total	28	4	22	60	9	5	128

Table 63 F200 pond, main vessel forms/fabric, by entries.

Additional fabrics are samian ware, LR, amphorae, OXCC, BB1, Trent Valley type, London ware type and possible OXW. There are three RSG jars or bowls and a C/W colander.

(Fig. 102)
RSG
533 CR4/18+21/4. Probably the lid for 534. *L589 (277)
534 CR4/18+21/4. *L1363 (277)
535 CR7+10/21/7+10. L670 (277)

GW
536 CR19+20. *L589 (277)
537 CR19+20. L670 (277)

LNVGW
538 CR20/18/20. *L1363 (277)
539 CR22//18/20/18//22. *L313 (277)
540 CR20+21//26/21/26//20+21. L590 (277)
541 CR20/18/20. *L313 (277)

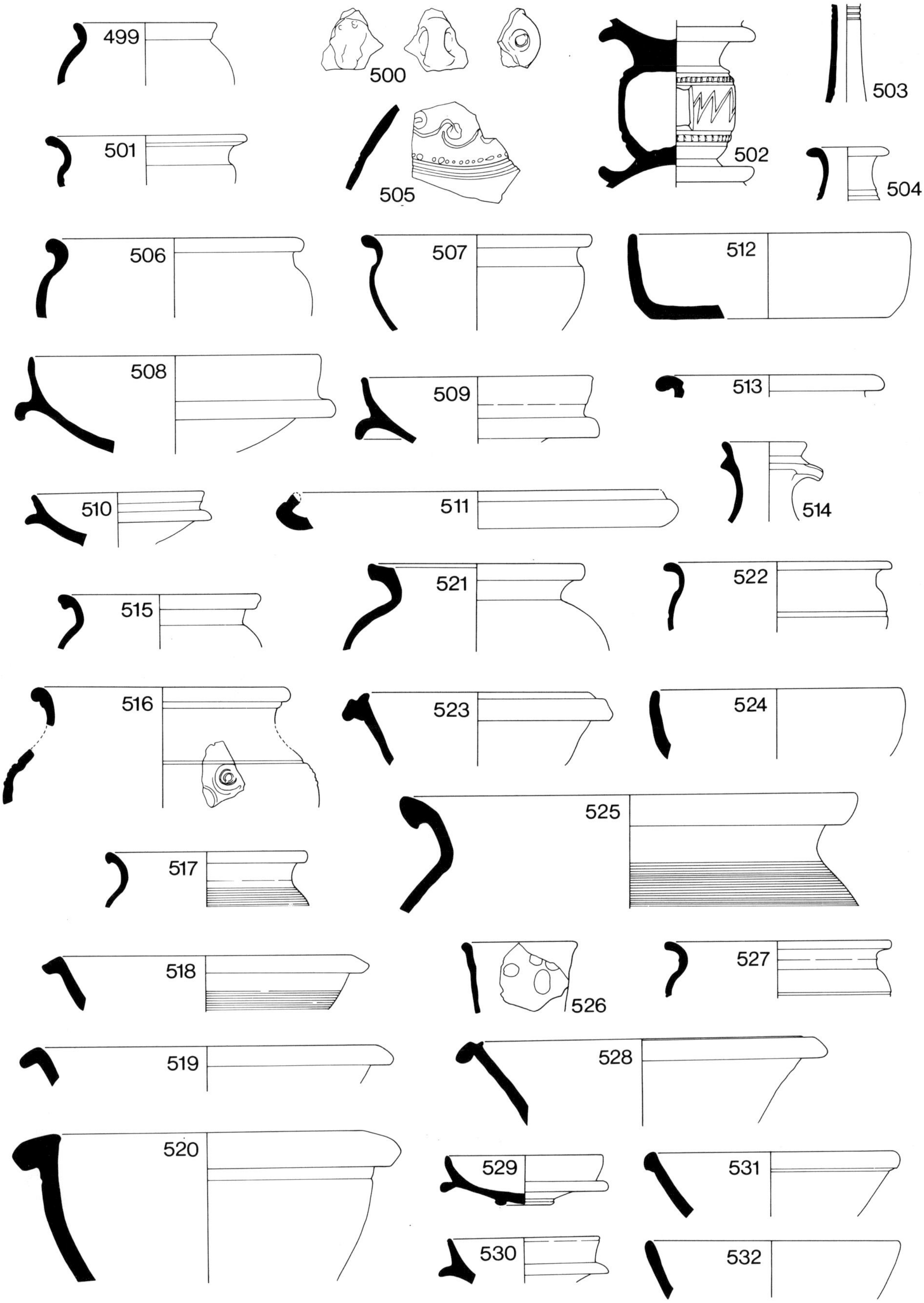

Figure 101 Roman pottery. Period 5: Nos 499–516, F14–F31–203, eastern enclosure ditches; Nos 517–524, F80 south-east enclosure ditch 1; Nos 525–532, F81 south-east enclosure ditch 2. Scale 1:4.

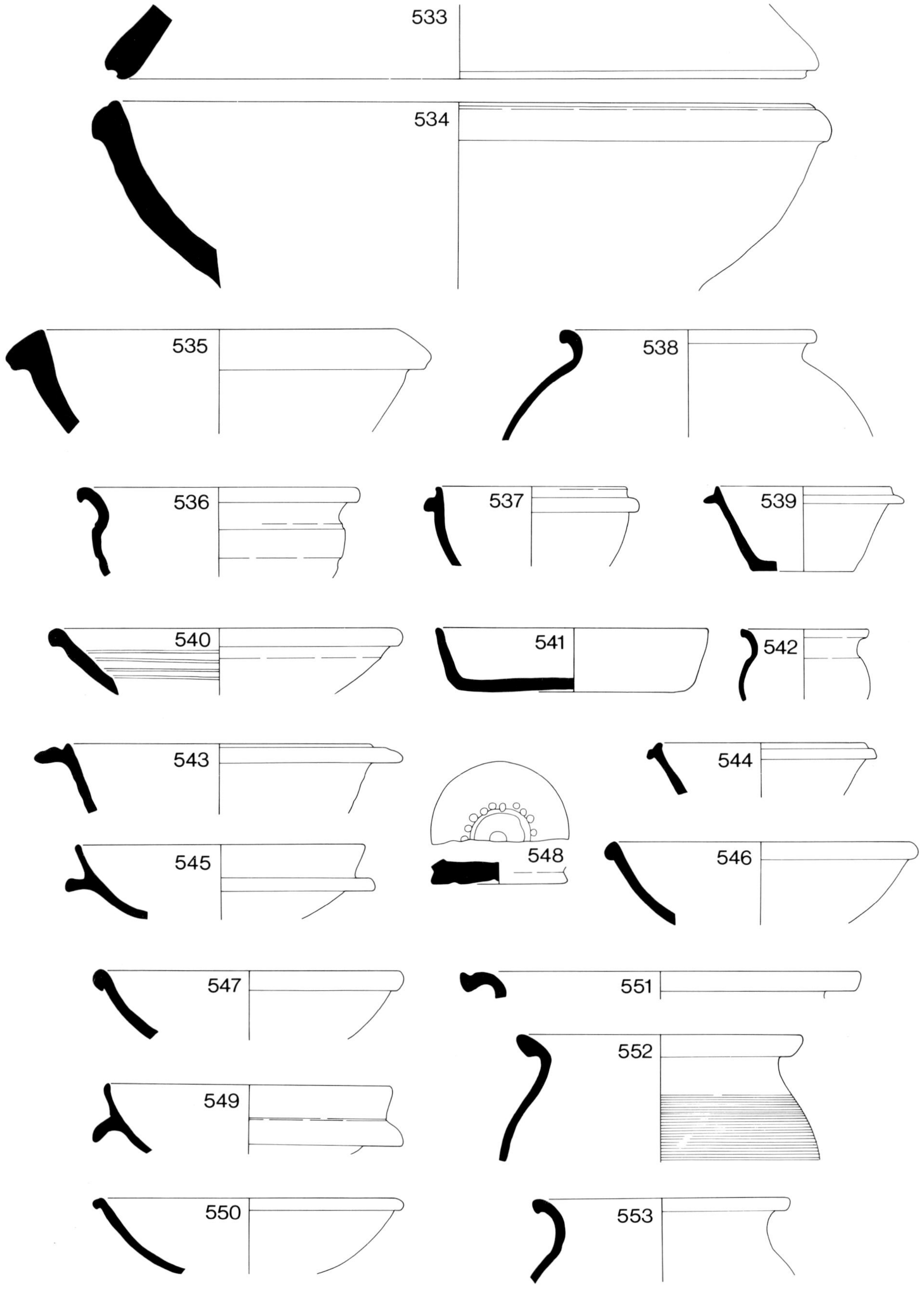

Figure 102 Roman pottery. Period 5: Nos 533–551, F200 pond; Nos 552–553, floor in Rectangular Building. Scale 1:4.

LNVCC
542 CR8+21/26/8+21. *L313 (277)
543 CR1+8//14/18/14//1+8. *L313 (277)
544 CR20/4/20. L669 (277)
545 CR2/5+19/2. *L1363 (277)
546 CR2/10/2. *L313 (277)
547 CR2/4+5/2. *L1363 (277)
548 CR8/26/3. *L1363 (277)

OXCC
549 CR2/3+20/2. Micaceous surfaces. *L1363 (277)
550 CR2/5+20/2. Micaceous surfaces. L590 (277)

Other
551 CR4. Oxfordshire? Hadham?? *L1363 (277)

2.5 Floor levels in the rectangular building
The ill-defined layers representing occupation floors in the rectangular building to the west of Barn 4 were cut by furrows. Table 58 gives the main fabric A%.

	RSG	Grey	LNVGW	LNVCC	C/W	Other	Total
Jar	131	5	8	18	-	2	164
WMJar	-	-	3	26	-	-	29
Bowl	5	1	7	21	-	4	38
Dish	3	1	7	30	1	3	45
Flagon	-	-	2	4	-	-	6
Beaker	-	-	-	10	-	-	10
Mortarium	-	-	-	-	9	2	11
Other	-	-	-	1	-	-	1
Total	139	7	27	110	10	11	304

Table 64 Floors, main vessel forms/fabric, by entries.

The pottery again includes BB1, Trent Valley type, OXCC, OXW and possible Hadham area wares, plus amphorae and uncertain. The one extra form is a LNVCC jar or bowl.

RSG
552 CR4+14. *L845 (292)
553 CR4+8/11/4+8. *L574 (292)

(Fig. 103)
554 CR4+8/21/4+8. *L845 (292)
555 CR4/21/4. *L575 (292)
556 CR10+11/21/10+11. *L575 (292)
557 CR4+5/22/4+5. *L574 (292)
558 CR8+12+20. *L574 (292)

GW
559 CR21+22/17/21+22. *L575 (292)

LNVGW
560 CR20/26/20. L586 (293)
561 CR19+20/26/19+20. *L574 (292)

LNVCC
562 CR3+12/14+27/3+12. L918 (292)
563 CR12/26/12. *L575 (292)

BB1
564 Oval. 3-ribbed handle. *L574,*L575 (292)

2.6 Ditch system F1061/1065
This comprised F1061, F1065, F1132 and F1133, and was located at the south-west corner of the central area. It contained far more pottery than either F80 or F81. The fact that the ditches cut a feature which also contained Anglo-Saxon pottery suggests that the system should have been entirely Anglo-Saxon. Two coins from F1065 are dated *c.* AD 350–360 and *c.* AD 393–395 (C42, C51). Table 58 gives the main fabric A%.

The layers contain samian ware, amphorae, BB1, OXCC, Trent Valley type, Horningsea, buff ware of uncertain source, Hadham? and uncertain. Unlisted forms are an RSG lid, LNVCC jars or bowls, and LNVGW and LNVCC dishes or bowls.

	RSG	Grey	LNVGW	LNVCC	C/W	Other	Total
Jar	46	4	6	7	-	3	66
WMJar	2	-	-	8	-	-	10
Bowl	2	-	-	25	-	-	27
Dish	4	1	5	14	-	-	24
Box	-	-	-	1	-	-	1
Mortarium	-	-	-	-	9	-	9
Other	5	-	1	5	-	1	12
Total	59	5	12	60	9	4	149

Table 65 F1061/F1065 system, main vessel forms/fabric, by entries.

(Fig. 103)
RSG
565 CR4+10/20/4+10. F1065, L2565 (320)
566 CR3+4/20/3+4. F1065, L2778 (320)
567 CR11+14/21/11+14. F1065, L2818 (320)
568 CR4/21/4. F1065, *L2597 (320)
569 CR10+11/20/10+11. F1065, L2565 (320)
570 CR4/20/10. F1065, L2777 (320)
571 CR9/17/22. F1061, L2576 (319)

LNVCC
572 CR12/14/12. F1061, L2576 (319)
573 CR2//5/26/5//2. F1065, L2777 (320)
574 CR2//5/26/5//2. F1065, L2564 (320)
575 CR1+22/7+26/1+2. F1065, L2565 (320)
576 CR8/26/8. F1065, L2754 (320)
577 CR8/26/8,26 paint. F1061, L2576 (319)
578 CR8/26/8. F1065, L2754 (320)

Subsidiary Groups

2.7 Ditch F55
This lay to the west of the southern ditches F80 and F81 and may have been contemporary with them. Table 58 gives the main fabric A%.

There are a few sherds of samian ware. The one other vessel is a LNVCC jar or bowl.

	RSG	LNVCC	C/W	Total
Jar	5	-	-	5
WMJar	-	3	-	3
Dish	1	1	-	2
Mortarium	-	1	1	2
Other	-	1	-	1
Total	6	6	1	13

Table 66 F55 ditch, main vessel forms/fabric, by entries.

(Fig. 104)
Vessels **579–582** are from *L109 (242).

RSG
579 CR9+12/21/9+12.

LNVCC
580 CR3/14/3.
581 CR17+22/14/17+22.
582 CR17+22/26/17+22.

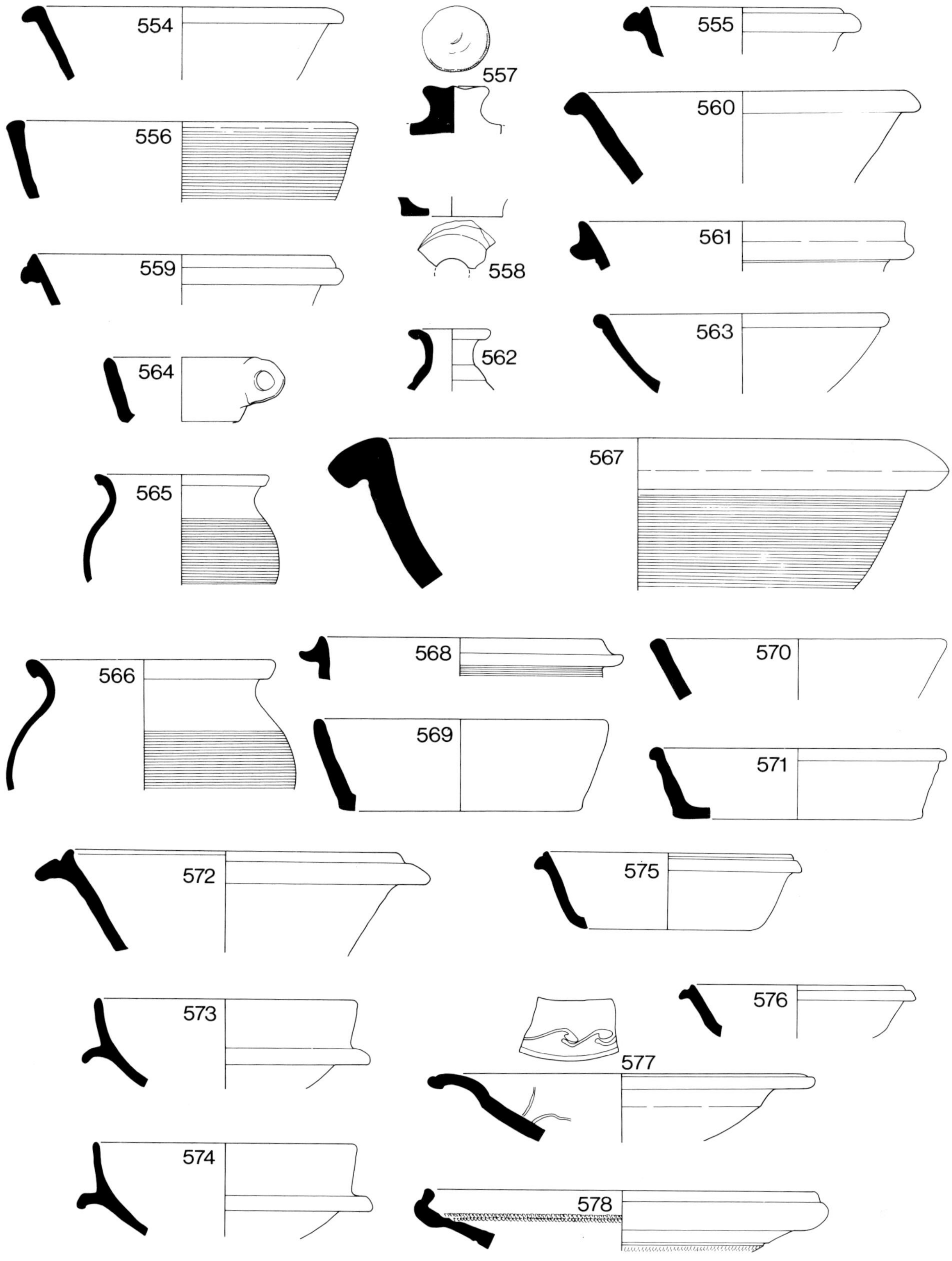

Figure 103 Roman pottery. Period 5: Nos 554–564, floor in Rectangular Building; Nos 565–578, F1016–F1065, late ditch in south-west corner of Main Yard. Scale 1:4.

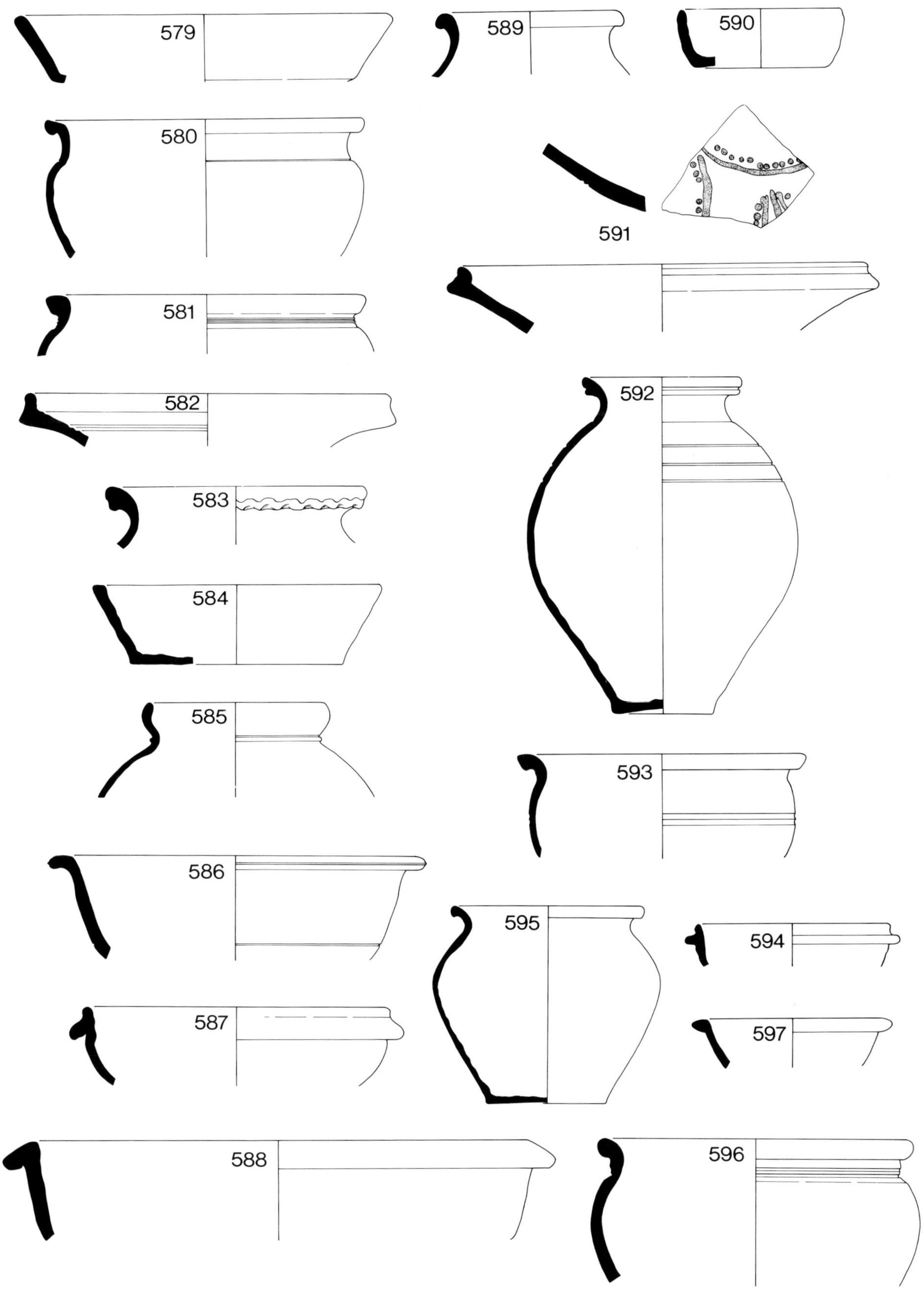

Figure 104 Roman pottery. Period 5: Nos 579–582, F55, south-east enclosure; Nos 583–587, F171 system; Nos 588–591, F12 gulley east of Barn 4; Nos 592–594, F172 well; Nos 595–597, AF48. Scale 1:4.

2.8 Ditch system F171

F149, F162 and F171 were drainage ditches for the reduced Barn 4. The lack of non-local and Anglo-Saxon pottery suggests that F171 may be earlier in general date than most of the other main and subsidiary groups. Table 58 gives the main fabric A%.

	RSG	Grey	LNVGW	LNVCC	C/W	Total
Jar	10	-	1	-	-	11
WMJar	-	-	-	2	-	2
Bowl	-	2	1	-	-	3
Dish	1	-	-	2	-	3
Mortarium	-	-	-	-	3	3
Total	11	2	3	4	3	23

Table 67 F171 system, main vessel forms/fabric, by entries.

There are a few sherds of samian ware and the one unlisted form is a LNVGW jar or bowl.

RSG

583 CR4/21/4. F149, L239 (255)
584 CR4+10+12/17+21/4+10+12. Near complete. F171, L374 (255)

GW

585 CR20/15+18/20. F171, L273 (255)
586 CR8//18/10/18//8. Stanground? F171, L374 (255)
587 CR19+20. F162, L363 (255)

2.9 Gulley F12

The feature may have been a drainage ditch round one end of a building, but most of the ditch had been removed in the 1971 excavations without record. The remaining levels contained a very small amount of pottery. The feature could, however, have been somewhat earlier than many of the noted groups. Table 58 gives the main fabric A%.

	RSG	LNVGW	LNVCC	Total
Jar	3	2	-	5
WMJar	-	-	1	1
Bowl	2	-	-	2
Dish	-	-	3	3
Total	5	2	4	11

Table 68 F12 gulley, main vessel forms/fabric, by entries.

There are no Anglo-Saxon or non-local wares present, but the lost upper levels may have had them.

Vessels **588–591** are from L15 (268).
588 CR4/21/4. RSG.
589 CR18//9/18/9//18. GW.
590 CR21/26/21. LNVCC.
591 CR2/26/2. LNVCC.

2.10 Well F172

This was a feature surviving in use from Period 4, to be finally disused and infilled during Period 5. The well may not have continued in use far into the period. Table 58 gives the main fabric A%.

The LNVCC percentage is boosted by an almost complete narrow-mouthed jar. There is no Anglo-Saxon pottery but there are a few sherds of OXCC and, probably, BB1. The one unlisted form is a probable BB1 dish or bowl.

592 CR8+13/26/8+13. LNVCC. L283 (259)
593 CR8/14+26/8. LNVCC. L282 (259)
594 BB1? L282 (259)

	RSG	Grey	LNVGW	LNVCC	C/W	Other	Total
Jar	6	-	1	1	-	-	8
WMJar	-	-	-	3	-	-	3
Bowl	-	-	-	-	-	2	2
Dish	-	1	-	3	-	-	4
Flagon	-	-	-	1	-	-	1
Mortarium	-	-	-	-	3	1	3
Other	-	-	-	-	-	2	2
Total	6	1	1	8	3	5	22

Table 69 F172 well, main vessel forms/fabric, by entries.

2.11 Pit AF48

Partly excavated in 1971, this pit was located midway along the eastern enclosure ditch. It contained a coin dated *c.* AD 330–335 (C28), and may have been dug by Anglo-Saxons, perhaps initially for storage, though the fills represent rubbish disposal. Table 58 gives the main fabric A%.

Additional fabrics are London ware type and possible Hadham area.

	RSG	LNVGW	LNVCC	C/W	Total
Jar	8	1	2	-	11
WMJar	-	-	2	-	2
Bowl	-	-	1	-	1
Dish	1	1	-	-	2
Beaker	-	-	1	-	1
Mortarium	-	-	-	2	2
Total	9	2	6	2	19

Table 70 Pit AF48, main vessel forms/fabric, by entries.

595 CR4/21/4. RSG. L90, L91 (219)
596 CR4/21/4. LNVCC. L91 (219)
597 CR1+11/14/1+11. LNVCC. L88 (219)

2.12 Slot F751

This belongs to what may have been an annex to the Anglo-Saxon building which overlay and superseded the Periods 3 and 4 House. The assemblage probably includes material which had been in use during the final occupation in and around the House. Table 58 gives the main fabric A%.

	RSG	LNVGW	LNVCC	C/W	Other	Total
Jar	2	2	1	-	-	5
WMJar	-	-	2	-	-	2
Bowl	-	-	3	-	-	3
Dish	-	1	2	-	1	4
Mortarium	-	-	-	1	-	1
Total	2	3	8	1	1	15

Table 71 F751 slot, main vessel forms/fabric, by entries.

There are a few sherds of samian ware and BB1, but no Anglo-Saxon or any other non-local wares.

(Fig. 105)

Vessels **598–600** are from L1349 (308).
598 CR18/26/18. LNVGW. Half vessel.
599 CR1+8/14+27/1+8. LNVCC. Third vessel.
600 CR2+3//4/14/4//2+3. LNVCC. Third vessel.

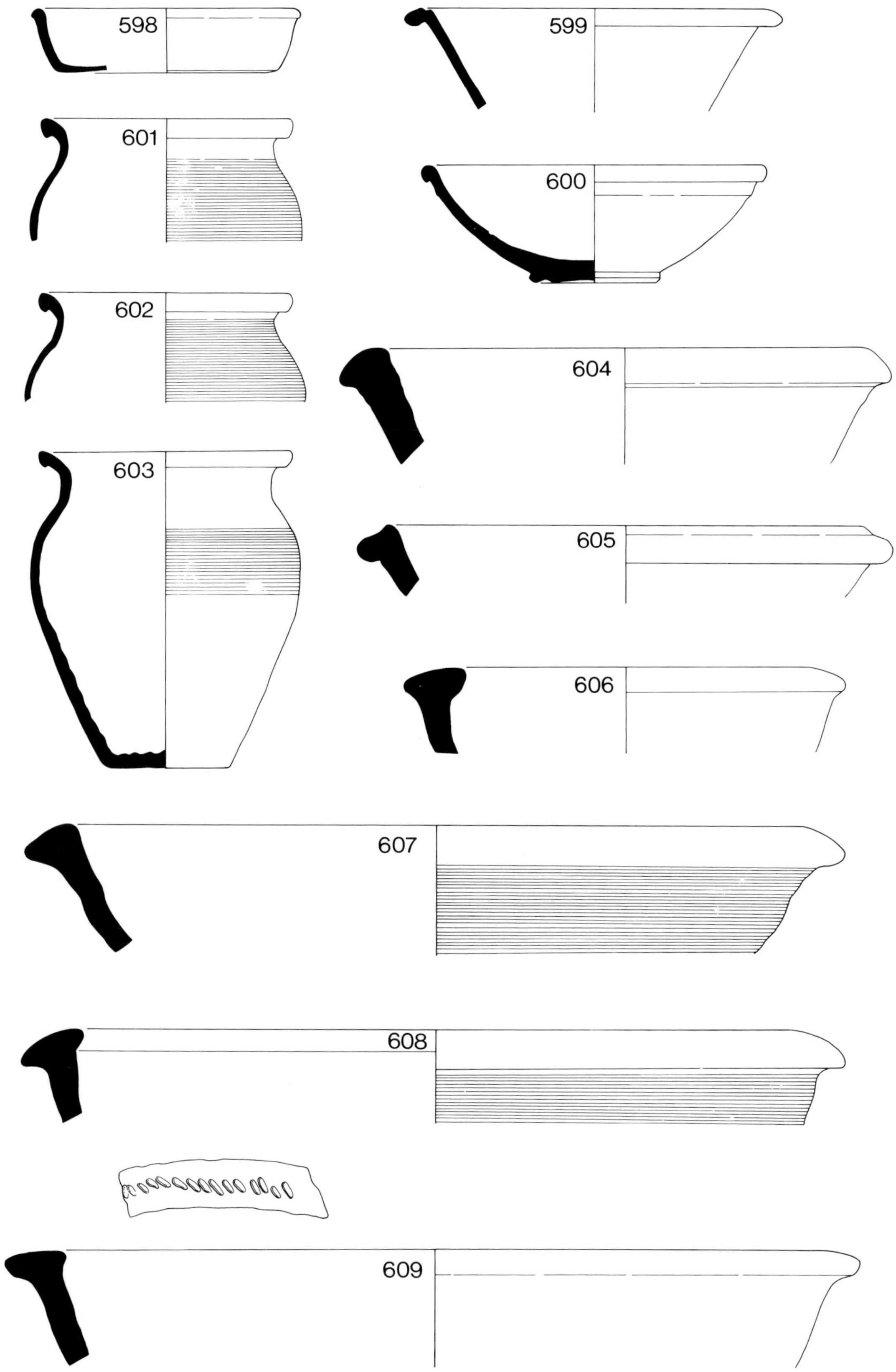

Figure 105 Roman pottery. Period 5: Nos 598–600, F751 annex to Anglo-Saxon Hall; Nos 601–609, additional pots. Scale 1:4.

Period 5 Additional Sherds

RSG

601 CR4/20/4. F533, L1093 (298)
602 CR11+12/20/11+12. F28, L41 (216)
603 CR4+8+12/17+21/4+8+12. Complete. F897, L1823 (314)
604 CR4/16+17/4. Yard? *L1241 (316)
605 CR5+14/20/5+14. F734, L1333 (297)
606 CR3/16+20/3. F20, L81 (260)
607 CR8+9+10/16/8+9+10. F1114, *L2762 (328)
608 CR4+14/20+21/4+14. F126, L193 (261)
609 CR4/16/4. F161, L260 (258)

(Fig. 106)
610 CR4+11+16/16+17/4+11+16. F345, L679 (278)
611 CR4+11/21/4+11. F675, L1771 (316)
612 CR5+14/20/5+14. F20, L28 (262)

GW

613 CR15+20/20/15+20. F192, L266 (261), L290 (267)
614 CR19+20/19/19+20. East Yorkshire? F5, *L8 (262)
615 CR16/20/16. F192, L290 (267)

LNVCC

616 CR4/4/4. F204, *L483 (286)
617 CR11+12/26/11+12,4 patches. F1149, L2956 (322)
618 CR2+12/14/2+12. F531, L1063 (298)
619 CR22/26/22. F252, L480 (279)
620 CR2/3+4/2,26 paint. F11, *L14 (274)

Other

621 CR3. Hadham? Yard ,*L2131 (316)

Additional Sherds fromUnassigned Layers

622 CR4/14/4. LNVCC. 1971, Area IV, L38.
623 CR8/14+26/22. LNVCC. L1.

3 The Pottery

3.1 RSG

Table 58 gives the RSG A% in the main features and Period 5 as a whole.The additional vessels are two lids and seven jars or bowls.

There are large quantities of RSG from Period 5, of which an unknown amount was probably residual or survivals from Period 4 and possibly earlier. The predominance of RSG can be largely explained by the lack of alternative utilitarian pottery as LNVGW had ceased to be produced many years earlier, the gap in the market having been only partly filled by imported grey wares now no longer traded to the area. RSG also predominates at this period in other areas for the same sort of reason. One of the largest and best known centres for the production of RSG was Harrold, Bedfordshire,[25] but this was probably not the source for most of the ware used here. As in previous periods there must have been a number of local production centres, all of which await discovery and excavation.

Most of the jars have the accepted 'late' characteristics of undercut rims, slack shoulders and zones of rilling, and most appear to be wheel-made. No. 490 appears to be more bowl-like in form and may hint at a possible merging of Late Roman and Anglo-Saxon traits. It is similar to a vessel from the destruction deposit at the Great Casterton villa (Gillam 1951, fig. 8, no. 20), and the layer in which No. 490 occurred also contained Anglo-Saxon pottery. The frilled rim on Nos 486 and 583 is hard to parallel in RSG but was common on grey ware especially in the East Midlands and Lincolnshire, for example at Swanpool (Webster and Booth 1947, figs 3–4, types C42, C47, C48, D14), and was therefore possibly a decorative trait copied by potters making RSG. The large storage jar No. 491 is sufficiently fragmentary to suggest that it may be residual and there is little evidence for the continued production of such large jars into the later fourth century. Very small vessels like No. 487 are uncommon in RSG, and it probably had a specific use.

Despite their low overall numbers compared to jars, it is the bowls and dishes that exhibit the most noticeable changes, as they did in Period 4. The Period 5 range of bowls includes some, Nos 494 and 568, similar to No. 476 of Period 4, and there are other flanged varieties, including Nos 495 and 555, of which the latter is so close in form to those occurring in LNVCC, that direct imitation appears likely. The flat-topped type, including Nos 492–3, 518–19, 554 and 611, some of which may be dishes, did not occur in previous periods, but may have been made before Period 5. The type is not easy to parallel, but is similar to vessels from Verulamium (Frere 1972, fig. 138, no. 1259; Frere 1984, fig. 101, no. 2424) dated *c*. AD 280–360 and *c*. AD 365–375, respectively. It is also close in form to a grey ware vessel from the Mileoak villa, Northamptonshire (Green and Draper 1978, fig. 10, no. 127) but, as this is thought to be of second-century date, the similarity is perhaps fortuitous, though it could point to an area of origin. A third-century shell-gritted ware vessel from Baldock (Stead and Rigby 1986, fig. 153, no. 703) is also similar. The rilling on No. 492 links the type to certain of the late jars and the large bowls. Rilled bowls were part of the repertoire of the Harrold potters,[26] as was incised decoration on the rim like that on No. 611. The fabric of No. 519 is significantly different to that of most of the vessels being rather soft and 'corky' to the touch, and vesicular. There was another bowl of the same type and fabric from the layer.

Visually, it is the large bowl which is the most noticeable new RSG type in the period. The illustrated vessels Nos 496–7, 520, 534–5, 567, 588 and 604–9 give an indication of the range in form and rim type. Parallels for a number of these occur at Harrold[27] and similar vessels were found more locally at Chesterton (Perrin, to be published). Others occur further afield at Verulamium (Frere 1972, fig. 136, no. 1212; Frere 1984, fig. 101, nos 2422 and 2425); Baldock (Stead and Rigby 1986, fig. 156, no. 779); Brixworth (Woods 1970, fig. 37, no. 261) and Shakenoak (Brodribb *et al.* 1971, fig. 39, nos 377–81; Brodribb *et al.* 1973, fig. 35, nos 628–39). Large bowls were produced at Harrold throughout the fourth century,[28] but at Verulamium and Shakenoak, as at Orton Hall Farm, their use appears limited to the late fourth century. No. 533 is probably the lid for No. 534. It is not clear if all the large bowls were accompanied by lids and no others have been recognised as such. Such a rather large casserole-type vessel may have had a specific function or, as with the large bowls in general, reflect changes in the availability of certain vessels, perhaps mortaria, or changing diet and methods of food preparation.

The dishes, including Nos 498, 556, 569–71, 579 and 584, show little, if any, variation from those prevalent in Period 4. The rilling on No. 556 can be paralleled on dishes made at Harrold. Of the remaining RSG vessels, Nos 558 and 612 are bases with cut holes, probably from jars, which will have had a specific use. They may be residual. No. 557 is obviously the knob from a lid, which would probably have been too small to have fitted the large bowls.

No. 610 has some curious characteristics. It appears to have been made by luting a thick, solid hand-fashioned base on to a wheel-made vessel. The area around which the base was attached has been knife-trimmed or pared down to give a smooth wall. The rilling is thinly incised. The wear on the interior suggests that it had been used as a lid. It weighs an incredible 3.25 kilos. No parallel has been found.

	F14	F80	F81	F200	RB	F1061	F55	F171	F12	F172	AF48	F751	Other	P5
Jar	88	13	19	18	131	46	5	10	3	6	8	2	170	519
WMJar	-	1	-	-	-	2	-	-	-	-	-	-	2	5
Bowl	10	2	1	4	5	2	-	-	2	-	-	-	13	39
Dish	-	-	-	3	3	4	1	1	-	-	1	-	11	24
Other	-	-	-	3	-	5	-	-	-	-	-	-	1	9
Total	98	16	20	28	139	59	6	11	5	6	9	2	197	596

Table 72 Period 5, main RSG vessel forms, by entries.

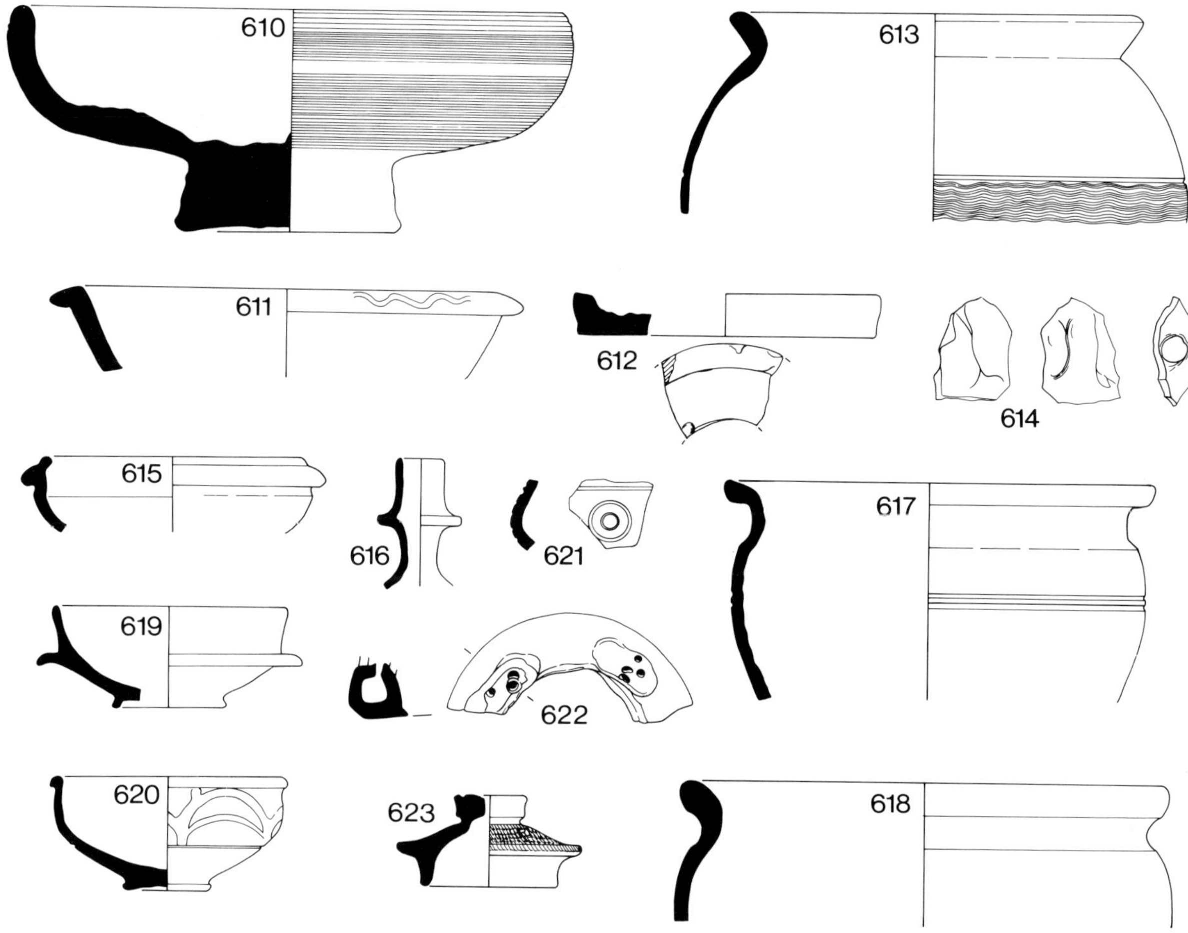

Figure 106 Roman pottery. Period 5: Nos 610–621, additional pots; Nos 622–623, unstratified additional pots. Scale 1:4

3.2 Grey wares

Table 58 gives the grey ware A% in the main Period 5 features and Period 5. The other forms are two lids and a jar or bowl.

	F80	F200	RB	F1061	F171	F172	Other	P5
Jar	1	3	5	4	-	-	15	28
Bowl	-	1	1	-	2	-	3	7
Dish	-	-	1	1	-	1	3	6
Other	-	-	-	-	-	-	3	3
Total	1	4	7	5	2	1	24	44

Table 73 Period 5, main grey ware vessel forms, by entries.

Only a small proportion could be identified as late grey ware of East Midlands (Trent Valley) type. Much was of late first to early second-century date deriving from early levels disturbed during activities in Periods 4 and 5. The illustrated material is probably all late, though the form of No. 589 is not very diagnostic and could be earlier.

Some of the jars and bowls can be easily paralleled. No. 585 is similar to vessels from Burgh Castle (Johnson 1983, fig. 42, nos 147–8) and made at Swanpool (Webster and Booth 1947, fig. 3, types C40 and C43). No. 613 is almost identical to others found in the Great Casterton villa destruction deposit (Gillam 1951, fig. 9, no. 23) and at Normangate Field (*cf* Perrin and Webster 1990, fig. 15, no. 272), while a similar rim occurs on a bowl in a late-fourth century deposit at Piddington.[29] Jars of this type were also made at Swanpool (Webster and Booth 1947, fig. 3, type C22). Jar 499 is less easy to parallel but the fabric is similar to that of the other late grey wares. Lid-seated jars with basic similarities to No. 521 are common in Lincolnshire (Darling 1977a, fig. 6, nos 105–22) and were made in the Swanpool kilns (Webster and Booth 1947, fig. 5, type H).

The hemispherical flanged bowl, including Nos 537, 587 and 615, is a common East Midlands (Trent Valley) and Lincolnshire type (Todd 1968, fig. 1, no. 6; Darling 1977, fig. 3, nos 43–8), and was also made at Swanpool (Webster and Booth 1947, fig. 4, type D19–23). It occurred locally at Great Casterton in the late fourth to early fifth-century drier deposit (Corder 1961, fig. 24, no. 3), and at Chesterton (Perrin, to be published). The biconical bowl, some examples of which are flanged, including No. 559, does not occur in large numbers on Lower Nene Valley sites, probably because it would have been in direct competition with the locally made colour-coated varieties. It is well known elsewhere, for example Towcester (Woodfield 1983, fig. 25, no. 123 and fig. 28, nos 185–8), as are bowls similar to Nos 501 and 536 (*ibid*., fig. 30, no. 237). No. 586 is not really a late fourth-century type and could in fact be a residual third-century Stanground product.

The handles Nos 500 and 614 are further examples of the type of handled jar first noted in Period 4 (No. 437). They are similar in also having the pinched rather than luted variety of handle, for which the Crambeck kilns of East Yorkshire are the only known source. Together, Nos 500, 614 and 437 obviously represent a minimum of two vessels, but if they are indeed from East Yorkshire, they must have been brought to the site by means other than trade in the accepted sense (see No. 513 in 'Miscellaneous wares' below).

3.3 LNVGW

Table 58 gives the LNVGW A% in the main Period 5 features and Period 5. Additional forms are a possible beaker, ten jars or bowls and a dish or bowl.

The amount of LNVGW in the period was high, considering it is not thought to have been produced far into the fourth century. It is best explained as a combination of various factors. As with the grey ware, some probably derives from disturbed earlier features (*e.g.*, F224 cut into a second-century pit containing nearly complete pots); vessels Nos 538–41 are all from pond F200 and may be residual from its use in Period 4. Nos 560–1 and 598 are from occupation or possible clearance levels and could be long-term survivals. Finally, No. 502 must have had some specialist use and this would also have helped it to survive. None of the

	F14	F80	F81	F200	RB	F1061	F171	F12	F172	AF48	F751	Other	P5
Jar	9	2	1	15	8	6	1	2	1	1	2	38	86
WMJar	-	-	-	1	3	-	-	-	-	-	-	3	7
Bowl	1	-	1	3	7	-	1	-	-	-	-	8	21
Dish	3	1	1	3	7	5	-	-	-	1	-	21	42
Flagon	-	-	-	-	2	-	-	-	-	-	-	-	2
Other	-	-	2	-	-	1	1	-	-	-	-	8	12
Total	13	3	5	22	27	12	3	2	1	2	1	78	170

Table 74 Period 5, main LNVGW vessel forms, by entries.

Period 5 LNVGW, therefore, can be taken as evidence for its continued production further into the fourth century than previously suggested, and all the vessels can be easily paralleled in earlier features.

3.4 LNVCC

Table 58 gives the LNVCC A% in the main Period 5 features and Period 5.

Additional forms are twenty-eight jars or bowls, two dishes or bowls and a lid.

Some of the LNVCC is again likely to be residual or survivals from previous periods, a factor which does not aid the consideration of the later LNVCC industry. The discussion of LNVCC in Period 4 touched on the uncertainties concerning the date when each type ceased to be produced, and what might have been the final range of LNVCC vessels. As has been mentioned, the Great Casterton deposits are the only useful late groups of LNVCC published, and they are of paramount importance to the consideration of LNVCC in Period 5. In a previous article, an attempt was made to isolate what may have been the latest LNVCC vessels in those deposits, and the Period 5 material is best examined with reference to the conclusions put forward there (Perrin 1981b, 447–63): 'three basic vessel types — the straight-sided flanged bowl, the plain dish and the wide-mouthed jar or bowl — were made ... right up to the end of pottery production '.

Each occurred in considerable numbers in Period 5 features which lends support to the suggestion. None of these types, as represented by Nos 523, 528, 543–4, 572, 575–6, 599, and 506–7, 580–1, 593, 596, 617–18, and 512, 524 and 590 has any characteristics which could be considered as diagnostic of late products. Certainly, the quality of manufacture and thickness of wall are not diagnostic, as the Period 5 LNVCC exhibited considerable variation in both. Moreover, as deep and shallow flanged bowls both occur, it is doubtful that the former succeeded to the latter.

Of the imitation samian ware vessels, the conclusion was: 'Only those of form 38 lasted into the later fourth century, with those of forms 37 and 45 probably not lasting beyond the first quarter, and those of forms 31 and 36 declining after the middle of the fourth century' (Perrin 1981b, 455).

The complete form 37 in Period 4 (No. 400) led to the tentative conclusion that this type continued in production for longer than was thought at the time of the article. There are twenty-three examples of form 38, including Nos 508–10, 529–30, 545, 573–4 and 619; twenty-four of form 31, including Nos 531, 546–7, 563 and 600; and thirteen of form 36, including No. 577, from Period 5. These figures suggest that form 31, as well as form 38, continued into the later fourth century, but that the earlier end date given for form 36 was probably correct. The complete absence of form 45 indicates that this particular type was not produced in large numbers, and confirms the view that it did not outlast the early fourth century. The place of form 36 may have been taken by the wide platter, including Nos 511, 578, 582 and 591, which also occurred in the Great Casterton deposit (*ibid*., 452).

None of the flagons from Period 5, which included Nos 503–4, 562 and 616, is of the possible late types noted in the article (*ibid*., 451), and there were no jugs of beaker form. Many of the Period 5 beakers only occurred as small sherds and, of the sixteen with classifiable rims, four were of residual type. Most of the others had funnel-neck with bead or flat-topped everted rims. A few had the orange or reddish-yellow fabric with white inclusions previously noted in Periods 3 and 4. There is not enough Period 5 material to allow discussion of the conclusions regarding these three vessel classes given in the article.

The Period 5 LNVCC also includes at least four small bowls with white painted-arc decoration, like No. 620, and four 'boxes' and 'box' lids. The article suggested that both probably lasted into the third quarter of the fourth century (*ibid*., 453 and 455) and this view is not refuted by the Orton Hall Farm evidence. The remaining vessels occur in too few numbers (some were singletons), to allow any conclusions to be drawn. The presence of five narrow-mouthed jars, including No. 592, would appear to agree with the suggestion that the type lasted into the late fourth century (*ibid*., 451–2).

3.5 Miscellaneous wares

There are approximately seventeen other wares represented excluding samian ware, Anglo-Saxon pottery, mortaria and amphorae. Most occur in small amounts, and a large proportion is probably residual. Over half are other Lower Nene Valley wares, especially the LNVC/W varieties. Of the others, the most relevant to Period 5 features are the OXCC, the black-burnished wares and the Hadham area ware.

The OXCC includes types C44 or 45 – No. 550, C49 – No. 551, and C51 – No. 549 (Young 1977). These probably represent a continuation of the trading links already established for mortaria. The overall amounts are too small to suggest that there had been an increase in this trade due to a decline in the local industry at this time. Of the samian ware

	F14	F80	F81	F200	RB	F1061	F55	F171	F12	F172	AF48	F751	Other	P5
Jar	12	6	4	5	18	7	3	-	-	1	2	1	27	86
WMJar	12	2	-	15	26	8	-	2	1	3	2	2	32	105
Bowl	9	4	8	18	21	25	-	-	-	-	1	3	60	149
Dish	14	4	2	17	30	14	1	2	3	3	-	2	52	144
Flagon	4	-	1	2	4	-	-	-	-	1	-	-	6	18
Beaker	-	2	2	2	10	-	-	-	-	-	1	-	19	36
Box	4	1	-	-	-	1	-	-	-	-	-	-	2	8
Mortarium	1	-	1	1	-	-	1	-	-	-	-	-	1	5
Other	-	1	5	-	1	5	1	-	-	-	-	-	3	18
Total	56	20	23	60	110	60	6	4	4	8	6	8	217	582

Table 75 Period 5, main LNVCC vessel forms, by entries.

imitations, form 38 was the most common at Orton Hall Farm and at Great Casterton (Gillam 1951, fig. 9, no. 30; Perrin 1981b, 455).

Oxidised ware of probable Hadham area origin comprised a slightly larger proportion of the pottery of the period than that of Period 4. Flagon No. 514 is reminiscent of a vessel from Burgh Castle (Johnson 1983, fig. 39, no. 52), and additional sherds from No. 516, which are too fragmentary to draw, show that the decoration also included a moulded animal. This can be closely paralleled at Chesterton (Perrin, to be published), and at other sites in Essex and Suffolk (Roberts 1982, 118–20, Class D38; Drury 1976b, fig. 5, no. 62). The bossed decoration on No. 621 occurs on many of the Hadham ware bowl types (Roberts 1982, Classes A16, A22, A21, B37, C15, C22, C38 and D22; A22 seems the most likely parallel). It is thought that most Hadham area ware was not substantially marketed outside its immediate locality until after *c.* AD 360–370 (Going 1987, 116).

BB1 is represented by all the main types, like Nos 532 and 594, as well as the less common, handled, oval dish No. 564. It was probably two-handled. BB1 is not thought to have been widely exported after *c.* AD 350–375 (Gillam 1976, 59), and the small amounts at Orton Hall Farm are indicative of the incidental use of a ware not necessarily acquired through actual trade. It may all be residual from Period 4.

One other vessel is of especial interest. The form and fabric of No. 513 are very close to vessels which are extremely common, and were almost certainly made, in east Yorkshire. Unfortunately, there is not quite enough surviving here to indicate whether it is of a general fourth-century type, similar to vessels found at Crossgates (Rutter and Duke 1958, type 1C), or the later fourth-century 'Huntcliff' variety (Corder and Birley 1937, type 16). Either would appear to confirm the suggestion, made in connection with the handles Nos 437, 500 and 614, that there was some contact with the east Yorkshire area in the fourth century, but at present the precise nature of this can only be conjectured.

VI. Wares and forms: function and status

Introduction

While the wares provide the clearest indication of the local, regional and provincial marketing of pottery, it is the vessel forms which allow an assessment of the actual nature of the occupation on any given site. A study of forms can lead to an appraisal of a site's status and, in some cases, it is possible to either predict what might be expected on certain sites, or to establish models for future reference. For valid suggestions and conclusions to be made and drawn, however, data from many sites are required and, unfortunately, at the time of writing, such a body of material is not available. The discussion of the nature of the occupation at Orton Hall Farm, therefore, has to be general and site-specific; much is also tentative.

One of the advantages of this site is that the complete core of the Roman farmstead, excluding its field systems, was recovered, for, although not the case with the late Iron Age and Anglo-Saxon farms, no traces of additional Roman buildings were found beyond any side of the excavated area (p.xi). The character, number and arrangement of the various buildings are discussed in Chapters 1 and 2. There are justifiable reasons, however, for considering the pottery and assemblage as a whole to be representative of the type of mixed occupation current on an average Roman farmstead, though it should be noted that aspects of Periods 4 and 5 were, most probably, not 'normal' (Chapter 10).

Wares

The pottery could be divided into about twenty-six different wares. Some of these were merely sub-divisions of one industry or centre, such as the local production, or Oxfordshire colour-coated ware and mortaria. If it is accepted that most of the shell-gritted wares (RSG, TSG and LIASG) and the various grey wares up to the end of the second century were not made that far away, then over 90 per cent of the pottery used on the site was 'local'. The remaining wares were either specialised or reached the site at times when marketing regimes were perhaps different from before. Some may have arrived because of what they contained, and others may have been left by visitors, friends or relations. None of those made on the continent can be cited as evidence for direct trade as such; it is most likely that they were available at the main local market, Durobrivae, and were bought for their appearance or some special purpose, as all could be matched by locally made equivalents. That they occur at all attests to the general mobility of pottery in rural areas in the Roman period.

The 'local' wares which predominate are LNVGW, LNVCC, RSG and grey ware. Their Period average percentage (A%) totals are:

	LNVGW	LNVCC	RSG	Grey	Total
Period 1	16	2	25	31	74
Period 2	28	7	32	24	91
Period 3	27	23	30	10	90
Period 4	21	32	32	6	91
Period 5	10	32	43	5	90
Periods 1–5	18	23	33	12	86

A few general conclusions can be gleaned. The decline in grey ware after Period 1 mirrors the growth of the local industry, especially, it would seem, the colour-coated wares. The large drop in grey wares in Period 3 suggests that most were residual in Period 2. LNVGW comprises around a quarter of the pottery until Period 5 and, as with the grey wares, the large drop in Period 5 similarly suggests that most was residual in Period 4. The RSG is fairly constant, with the lower figure of Period 1 being due to the other shell-gritted wares around at the time. The increase in Period 5 reflects the drop in amounts of the other utilitarian wares during the fourth century. The increase of LNVCC in Period 4 is probably related to the cessation of LNVGW production.

Forms

The main drawbacks to a proper understanding of the form assemblage found on the site, apart from the noted lack of comparable material, are the uncertainty about how many people lived and worked there, and to what extent other materials such as horn, wood, skins, glass and metal were also used.

Table 76 gives the numbers and percentages of the main forms per period and in total, including samian, but excluding amphorae. The miscellaneous category comprised colanders, jars or bowls, and various uncertain forms. The abbreviations are as follows:

J – Jar
- S: small
- M: medium
- L: large
- VL: very large
- NM: narrow-mouthed
- WM: wide-mouthed

B – Bowl (FL – conical flanged)
D – Dish
BKR – Beaker
FL/J – Flagon/Jug
L – Lid
M – Mortarium
TS – Samian ware
IMTS – Imitation samian ware

	Period 1		Period 2		Period 3		Period 4		Period 5		Total	
	no.	%	no.	%	no.	%	no.	%	no.	%	no.	%
SJ	141	25	46	23	132	16	51	9	151	10	521	14
MJ	107	19	29	14	132	16	115	19	342	21	725	19
LJ	22	4	10	5	42	5	38	6	147	9	259	7
VLJ	22	4	15	7	24	3	15	2	48	3	124	3
NMJ	16	3	1	-	15	2	10	2	12	<1	57	>1
WMJ	2	-	6	3	38	5	36	6	120	<7	202	5
All Jars	350	>61	117	58	407	49	286	48	859	54	2019	53
B	18	3	8	4	48	6	37	6	122	8	233	6
FLB	-	-	-	-	22	<3	36	6	82	5	140	4
IMTSB	7	1	2	1	6	<1	6	1	27	2	48	1
TSB	15	3	7	3	27	3	4	1	5	-	58	>1
All Bowls	40	7	17	8	103	12	83	14	236	15	479	13
D	35	6	13	>6	99	12	72	12	184	11	403	11
IMTSD	7	1	1	-	18	2	23	4	42	3	91	>2
TSD	47	8	13	>6	15	2	11	2	31	2	117	3
All Dishes	89	15	27	13	132	16	106	18	257	16	611	16
All Bowl/dishes	132	23	50	25	249	30	195	33	497	31	1123	30
BKR	22	4	12	6	63	8	31	5	40	>2	168	<4
TSBKR	4	<1	-	-	1	-	-	-	-	-	5	-
All Beakers	26	<5	12	6	64	8	31	5	40	>2	173	5
M	7	1	9	4	48	<6	33	<6	91	6	188	5
TSM	-	-	-	-	3	-	3	-	1	-	7	-
All Mortaria	7	1	9	4	51	6	36	6	92	6	195	5
FL/J	17	3	4	2	15	2	9	2	18	1	63	2
L	9	1	4	2	12	>1	4	1	6	-	35	1
BOX	-	-	1	-	12	>1	8	2	8	-	29	1
Misc	11	2	1	-	5	1	16	3	48	3	81	2
Misc TS	18	3	4	2	13	<2	4	1	20	1	59	<2
All	486		178		769		567		1531		3531	
All samian	84		24		59		22		57		246	
TOTAL	570		202		828		589		1588		3777	

Table 76 Numbers of main vessel forms.

Certain trends are noticeable. The ratio of bowls and dishes to jars appears to decrease with time. There is evidence for a change in the size of the most common jar with an increased preference for medium rather than small types. In Period 1 for example, small and medium jars accounted for approximately 41 per cent and 31 per cent of the total jars respectively, while by Period 5 the proportions were reversed with 20 per cent being small and 46 per cent medium. The figures are equal in Period 3 at 36 per cent each. The very large and the narrow-mouthed jars appear to have been used at a fairly constant level.

The changes in the numbers and percentages of beakers would appear to match the various periods of actual beaker production with the high figure in Period 3 corresponding to the main *floruit*. The Period 4 tailing off is followed by a relative absence in Period 5 reflecting the gradual decline in the importance of the type and its eventual demise around the mid to late fourth century.

The figures for imitation samian ware forms other than mortaria are interesting. The high figure of 21 per cent of the bowls and dishes in Period 1 obviously reflects the greater numbers of actual samian ware vessels around at the time, but the lasting influence that samian ware forms had on the pottery market is clearly shown by the fact that the figure remains at around 15 per cent for later periods.

Other changes almost certainly reflect variations in food types, preparation, storage and consumption, and the changing nature of the site, its activities, and the size and character of its population. Firm evidence for these is lacking.

It is worth noting that the assemblage does not include any of the supposed latest products of the Lower Nene Valley industry, especially the 'beaker-jug' and handled bowl or jar (Guide 63, 74, 78). This could mean that inhabitants contemporary with their manufacture did not need, or could not afford, such pottery, or that the site was not occupied at the time that it was available. The first alternative could apply to Roman occupants but may be additionally significant in view of the presence of Anglo-Saxons on the site, and could indicate that they were in residence before the end of the Roman period. Certainly in view of the continued occupation by Anglo-Saxons it is perhaps unlikely that the site was unused for any length of time, if at all.

Of the vessels, 28 per cent were in LNVCC, 17 per cent in LNVGW, 29 per cent in RSG and 9 per cent grey wares. The main forms occur in the following percentages:

	Jars	Bowls/dishes	Beakers	Flagons	Total
LNVCC	13	46	14	3	76
LNVGW	56	33	-	1	90
RSG	89	8	-	-	97
Grey	75	17	-	-	92

The figures show that RSG was primarily used for utilitarian vessels, and the percentage of jars would have been higher but for the incidence of the late large bowls. There is an indication that the grey wares were also used mainly for utilitarian purposes. The LNVGW figures fit a ware which was probably used both in the kitchen and on the table and it is clear that LNVCC vessels were mainly reserved for eating and drinking.

While the percentages of the main vessel types in RSG and grey ware remained relatively constant through the five periods, there were a number of changes in LNVCC and LNVGW. The figures for LNVCC are:

	Jar	Bowls/dishes	Beakers	Flagons	Total
Period 1	10	21	69	-	100
Period 2	18	23	45	-	96
Period 3	11	42	26	3	82
Period 4	11	45	13	3	96
Period 5	15	50	6	3	74

The figures suggest that beakers were the main vessel type in colour-coated ware purchased by the occupants of the site in Period 1, perhaps because they were specialised pots not made in any other ware. The numbers of colour-coated bowls and dishes are quite high, however, showing that they were more than a minor part of colour-coated production at the time. It is not certain if these were also bought for specialist purposes. The most noticeable trend in the figures is the increase in the relative percentage of bowls and dishes, seemingly at the expense of beakers, with the numbers of jars and flagons remaining fairly static. The low figure for beakers in Period 5 reflects the decline of this vessel form by the later fourth century. In most periods, the additional percentage comprises wide-mouthed jars or bowls.

The figures for LNVGW are:

	Jars	Bowls/dishes	Beakers	Flagons	Total
Period 1	71	24	-	1	96
Period 2	66	28	-	-	94
Period 3	55	33	-	3	91
Period 4	50	42	1	1	94
Period 5	50	37	-	-	87

In all the periods, including Period 5 when the figures are perhaps irrelevant, jars, bowls and dishes account for most of the LNVGW vessels on the site. The main trend is for the relative percentage of bowls and dishes to increase, becoming almost equal with jars in Period 4, that is, at the end of LNVGW production.

Specific and surmised function

Within the general areas of utilitarian, kitchen, table or 'industrial' pottery, the number of vessel classes for which a definite function can be cited is small. Few would argue that vessels such as candlesticks, lamps, strainers or colanders, inkwells, crucibles, lids and mortaria did not have specific functions for which they would have been used throughout their lives, and there is also a high level of certainty with some other types. Jugs and flagons must have been used for liquids, though the varieties of the latter with especially narrow necks would hardly have poured that easily, and may, therefore, have been used for sprinkling vinegar or oil onto food, or to contain perfumes, scented water and oils.

A large proportion of the beakers, and most of the cups (nearly all in samian ware), were undoubtedly used for drinking. The size of some of the latest LNVCC beakers (*cf.* Guide 43) indicates other uses, however, and other vessels must have been used for drinking in the later fourth century when beakers were, apparently, no longer made. The people occupying the Werrington Enclosure almost certainly used horns (Mackreth 1988, 148) and it is possible that, in addition, glass, metal and possibly wood vessels or skins were used at Orton Hall Farm, although only glass and metal survived. It should be noted, however, that no metal drinking vessels *per se* were recognised, and that there was little glassware that could be dated to contexts in or after Period 3. On the other hand, the relative paucity of pottery beakers and cups suggests that the occupants did use other containers for drinking.

The constricted mouths of the narrow-mouthed jars would obviously be easily sealed, and they may therefore have been used for the storage of particular goods or materials. The use of some of these vessels for liquids might help to explain the apparent lack of flagons and jugs, which may have been reserved for use at the table.

The soot on many of the jars, bowls, dishes and lids indicates that they were heated though not necessarily over a fire, which was a place probably largely reserved for metal vessels. Pottery vessels must have been used to heat food in ovens or on tripods, 'hot-plates', and griddles, as well as for keeping it warm in the ashes of the fire in a manner similar to that depicted in paintings of the post-medieval kitchen (Moorhouse 1981). The casserole-type vessels formed by BB1 dishes and bowls (Gillam 1976, fig. 6, nos 89–91) and, possibly, some of the large RSG bowls with lids at Orton Hall Farm, Nos 533–4, could be termed Roman 'oven-ware'. It is conceivable that the occurrence of BB1 on rural sites in areas outside its main distribution zone, like the Lower Nene Valley, might be the result of a desire for such a specialised vessel type. The 'Castor box' is also a type of casserole, though perhaps more appropriately termed a tureen. The LNVCC flanged bowls and plain rimmed dishes would form a similar vessel when put together. These could obviously have been used for containing food at table or elsewhere.

Much, however, of the pottery could have been put to any use depending on everyday circumstances, the availability of better, alternative vessels, and the user. Most of these many uses can only be surmised. Moreover the use of pottery vessels could continue after they were broken, and the evidence from amphorae shows that the possibilities of reuse for functions or contents far removed from the originals were almost endless. These factors all inhibit the recognition, but increase the potential categories, of actual usage.

The Period 3 vat base F105 and its stoke-hole F201 in Barn 2 were obviously indications of a particular process, perhaps brewing (Chapter 9), but the RSG pots associated with this activity, Nos 250–7, are unexceptional. The large vessels buried in Period 1 pits F684–86, Nos 195–6 and 200, could have been associated with food preparation (Chapter 1, p.3), and buried pots which may have had a similar purpose have been noted elsewhere, for example, Ashton (Hadman and Upex 1979). A large vessel was found set into the ground at the Northchurch villa (Neal 1976, fig. 27,158) and a number of possible uses were postulated including storage, as a latrine or for pouring libations into the earth.

The Northchurch vessel had a hole cut through the base before firing. Two vessels from Orton Hall Farm, Nos 558 and 612, had single cut holes and there are additional examples from other sites in the Nene Valley area, for example Quinton (Friendship-Taylor 1979, fig. 35, no. 45 and fig. 40, nos 133–4). Vessels with similar cut holes were used in medieval times for various recipes (Moorhouse

1981, 117–18, fig. 90b, and Appendix 1b) and an equivalent or identical use in the Roman period cannot be ruled out. Other vessels from Orton Hall Farm had pre- or post-firing pierced holes of varying number and character, and a quick glance through any published Roman pottery report would also reveal a number of pots with pierced holes.

Cheese presses are an obviously specialised form, and vessels like No. 152 seem to have been used as colanders or strainers, but the differences suggest that actual requirements varied considerably depending on the precise materials being separated. Some holes were clearly to enable the vessel to be repaired, or to help to keep it intact, as at Verulamium (Wheeler and Wheeler 1936, 166, fig. 18, no. 55), but none of those from Orton Hall Farm appear to have been pierced for this purpose.

The holes in No. 70 are matched at Quinton (Friendship-Taylor 1979, 108, fig. 53, no. 327) by a vessel with holes that appear to have had lead plugs. It is possible that some of the holes on other pots were to take a wooden handle or had luted handles which have not survived, or were even to allow the vessel to be hung or lids to be attached by string. The holes in the bases of Nos 109 and 177 are also of uncertain purpose but it is possible that they were used, as with Nos 558 and 612, for certain recipes which may have had medieval equivalents. As Moorhouse says: 'earthenware pots were frequently used in medical and craftsmen's recipes ... including pots with pierced bases, some forms of which have not yet been recognised amongst pottery collections' (Moorhouse 1981, 116).

In common with many other sites' pottery, lids in any fabric are quite rare at Orton Hall Farm and it is obvious that vessels were sealed or covered with skins, cloth, wooden lids, perhaps bungs, and convenient flat stones or reused pottery bases, for example No. 354. The latter can be paralleled at Quinton (Friendship-Taylor 1979, fig. 43, nos 185–8 and fig. 51, no. 292), but the diameter of the unsooted area perhaps points to it being a stand. There are also many examples of pottery vessels other than purpose-made lids being used, especially bowls and dishes, for example at Northchurch (Neal 1976, fig. 27, no. 159). The nature of the sooting on many of the Orton Hall Farm RSG bowls and dishes clearly shows that they were used as lids which either fitted inside or over the rim of the jar. Those which have denser sooting around the rim might have been additionally sealed or luted-on with some organic substance or have become stained and burnt as the liquid inside the jar boiled over or was poured out.

The association of No. 479 with the Mill-house may be significant. Perhaps it was a measure of some kind, or an outsized baking dish. The 'casserole' formed by Nos 533–4 has already been noted but the occurrence of such large RSG vessels primarily in Period 5 contexts might additionally imply a change in methods of food preparation, or even the scarcity of certain others, for example, mortaria, toward the end of the fourth century.

No. 481 is a Horningsea jar of which a few were used at Orton Hall Farm. As it is likely that the cost of these would have been greater than that of an equivalent locally made vessel, a specialised purpose is likely. The bulk storage of a particular produce or liquid is implied and perhaps these jars were also connected with the brewing process. Of course they might have served as water butts if for some reason equivalent vessels in local clays were not suitable, and it is possible that they found their way on to the site because of their contents. The region in which they were produced had deposits of fossil coprolite (Hughes 1902, 186); perhaps there was a trade in this substance locally. The other large jars would have been used for bulk storage of dry goods or liquids and in industrial processes.

Vessels Nos 26, 114, 184 and 223 are made from clay containing a great deal of sand inclusions or temper, and it is just possible that these were used for a purpose that necessitated more than average heat. The deep grooves could have had a related purpose, perhaps to aid the grip of tools or tongs. It has been suggested (Wheeler and Wheeler 1936, 166, 61a–f) that grooving or rilling in general could have assisted in gripping wet vessels. No use can be postulated for the incised internal grooves on No. 376. The two small vessels, Nos 183 and 487, could have been used as measures or to contain particular oils or ointments. Nos 371 and 487 are perhaps small enough to have been part of a child's 'tea-set', though the former might have equally well been a lamp.

The double-ended vessel, No. 502, is strange to say the least. In shape, it resembles a pulley-wheel and the bottom of each of the end bowls is heavily worn. The fabric would suggest that it was residual in Period 5. The reason for the cut hole is also unclear, but a hole of some description would have been essential for the pot to fire successfully. However, the size and regularity of this one points to the use of a handle or the insertion of some other object.

The hollow clay ring with seating for, most probably, luted-on vessels, No. 622, is similar to some multiple or triple vases. The pots, however, would have been better used as lamps if they had been filled with oil and had wicks dipped into them. The hollow ring would then ensure that the level of oil in each remained constant. An iron object in a hoard at Silchester (Evans 1894, 153–4, fig. 20) had a similar common reservoir arrangement.

The occurrence of a particular form in a number of different fabrics is worthy of comment. This is usually attributed to the desire of potters in different centres to reproduce a well known or popular vessel for the local market. With the similarly sized Nos 62–3, 113 and 165, however, it is suggested that collectively they could have served as a kind of condiment set, being used together at the same time, but with different contents which were readily distinguished by, and equated with, vessel colour. It was also noticed during the cataloguing of the pottery that occasionally two or more vessels from the same contexts had a similar appearance in terms of fabric and surface colour and treatment. Examples are Nos 274–5, 277 and 281, 340 and 343, 341–2, and it is thought that these each represented pots made by the same potter within a short period, and subsequently purchased together. They were obviously bought either to replace broken vessels or to increase the pottery available to the occupants of the farmstead. In the latter sense they could provide a slight insight into the purchase and sale of pots in 'sets', which is fairly obvious with regard to casserole-type vessels such as 'boxes', but less clear within the more basic pottery.

Status

Archaeological reports nearly always attempt to define the nature of an excavated site, but the conclusions are rarely related to the character of the finds and the pottery. There is an increasing tendency, however, for researchers to consider the interaction of site-type and finds in order to better understand the overall assemblages that occur. Loughlin (1977) based his distributional survey of Dales Ware on 'classes' of sites ranging from legionary fortresses (Class 1) to 'native settlements' (Class 7), with an additional 'other rural settlements' (Class 8) to cover such things as surface scatters. He was able to show variations in marketing and distribution relating to the 'class' of a particular site.

In a more local context, Pryor noted that the pottery and other artefactual material from the Cat's Water site was 'the domestic refuse of a small farmstead or a rural hamlet' and recognised that: 'the pottery ... cannot readily be compared with that from the major urban sites of Hertfordshire and Essex; it even seems restrained alongside assemblages from proto-urban sites such as Irchester, Old Sleaford and Dragonby' (Pryor 1984, 161).

It was noted in the description of Period 1 (Chapter 1, p.1) that the main Iron Age site focus lay to the north-west of the excavated area, and the pottery of this period found represented minor activity at this time on the fringes of the Iron Age complex. The lack of 'belgic' pottery might suggest that the site was of fairly low status at the Conquest period, but there is, of course, no way of knowing if the extant assemblage reflects the full range of pottery in use on the site at this time. The nearby settlement at Monument 97 (Rollo in Mackreth forthcoming) had a fair range of 'belgic' pottery, and it is therefore deemed unwise to draw any firm conclusions as to what the actual status of the Orton Hall Farm site may have been at the same time. It is possible that it was essentially sites along the Fen margin which were of the lowest status locally at the time.

It is difficult to assess the development of the farmstead through Periods 2 to 5 in terms of status. Of the various buildings, the Mill-house and its immediate neighbour are anomalous, and an explanation is given elsewhere (Chapters 2 and 9). The other structures would not be out of place on any rural settlement of the Roman period. The House of Periods 3 and 4 has no mosaics or *opus signinum*. There was no trace of decorated plaster, and tile was rare. There was no bath-house. The suggestion is, therefore, that Orton Hall Farm was essentially a fairly large, unpretentious, self-contained, self-sufficient farmstead.

The pottery reflects this well. As noted before, over 90 per cent is of probable local manufacture. A full range of pottery types was used from the most simple and utilitarian jars through to finely made table wares, but there were few if any 'exotics'. There is little evidence for extravagance. The beakers, for example, tended to be rather plain and the more highly decorated, and more expensive, varieties were few in number. Moreover, there is a considerable body of evidence to suggest that some of the finer vessels (*e.g.*, No. 273) survived in use for a considerable time and that, by implication, these were prized objects. Together with the general lack of glass and metal vessels after Period 2, this confirms the view that the farmstead's inhabitants were neither rich nor had a particularly refined standard of living. Although within its lifetime the site underwent considerable and noticeable changes, these are not matched by obvious changes in the pottery assemblage, other than an increase in numbers. The overall picture is of a site which did not require anything special or out of the ordinary and did not, therefore, look or ever really need to look beyond the local area for its pottery requirements. The only occasions when it did so were probably as a result of a combination of chance, or even fashion, together with perhaps some adventurous marketing.

VII. Types of pottery related to areas of the site

Figures 108–12 show the distribution on the site of samian ware, beakers, and mortaria by period, and the large shell-gritted bowls and Anglo-Saxon pottery in Period 5. All except the last, which is plotted by weight, are by number of entries.

a. Samian ware

In Period 1 there are two main clusters corresponding to the F813 and F1048 systems. The numbers of vessels in Period 2 are much fewer but, in the main, are associated with Barn 2, together with a few around Barn 1. The Period 3 distribution covers much of the site, but there are concentrations around Barn 1 and along the northern side of the site. In Period 4 the only possible cluster occurs in Barn 1 and the Small Yard, and in Period 5 the vessels are dispersed except for a group close to the F1048 system and a spread along the eastern side of the site.

b. Beakers

Although these were not 'on the market' until the latter part of Period 1, enough were purchased and discarded to make the distribution meaningful. The main groups are in the F1048 and F813 systems. In Period 2, numbers were few and occurred either around Barn 2 or in the area to the north of Barn 1. Barn 1 also figured prominently in Period 3 together with the walled yard, Barn 3, and the F435 and F441 ditch systems. The numbers were fewer in Period 4 and occur as a group in the south of the Small Yard or as a spread along the north side of the site. The distribution in Period 5 is more dispersed but tends to favour the north and east sides of the site.

c. Mortaria

These were not common in Period 1 and most occur near both 'houses', while the area around Barn 2 has the bulk of the few belonging to Period 2. In Period 3 the area including and between Barns 1 and 3 is well represented together with the north side of the site especially the F435 and F441 ditch systems. The latter two areas feature strongly in Period 4, but in Period 5 there appears to have been a shift towards the north-east and eastern sides of the site.

d. Large RSG bowls

Most of these belong to Period 5. The distribution is definitely concentrated in the east of the site, with a few 'outliers' to the west of Barn 3 and along the northern side.

e. Anglo-Saxon

All the sherds belong to Period 5. The find spots are scattered but there are dispersed 'groups' in the yard area to the west, and in the south-west corner (including Groups 1 and 2 in the Anglo-Saxon Pottery report). There is also a definite spread along the eastern side.

Samian vessel	●
Mortarium	■
Beaker	▼
Large shell-gritted bowl	♦
Saxon: weight gms	
< 10	○
10 – 15	□
50 – 100	Δ
100 – 150	○
150 – 200	□
200 – 250	Δ
> 2000	○
> 2500	□
Features	991

Figure 107 Key to Figures 108–12.

Interpretation

The purpose of the exercise was to see if the results in any way matched the activities postulated for various parts of the site in the five periods. It was hoped that the samian ware and the beakers would represent the domestic 'table', the mortaria the food preparation, while the large bowls could indicate new areas of late occupation. The distribution of the Anglo-Saxon pottery should suggest which areas of the site might have been utilised by the immigrants. The results were favourable, with a great deal of correlation between the various wares and vessels. However, no ware or pottery type could be identified which would accurately represent solely non-domestic activity.

In Period 1, both samian and mortaria occurred in association with the eastern 'house', and the absence of beakers confirms the belief that this was abandoned in favour of a new focus, not identified, in the west. The evidence for this area is a little less clear because an earlier 'house' was also sited here, and the samian of Flavian and earlier date could be related to this. The mass of samian, and beakers, in the F813 system clearly shows, however, that a domestic dwelling existed here at the end of Period 1. A similar concentration in the F1048 system also hints at another, probably located just outside the limits of the excavation. Of these two main concentrations, the former has the greater number of vessels. This could indicate that the population of the site had increased around the middle of the second century, and then declined by the end, although the general increase in the amounts of samian in the Antonine period generally might be more relevant (see below).

Though there were fewer vessels to plot for Period 2, all the categories show two possible domestic centres, one around Barn 1 and the other close to Barn 2. The former could relate to the suggested timber first phase of this structure. The Period 3 distributions clearly show a more intensive use of the site as a whole and a concentration of domestic activity in certain areas. The most obvious is the south-western corner and incorporates both Barns 1 and 3 and the area between them. The Barn 3 distribution consists mainly of beakers with a few mortaria, but it is not certain if this represents a specialised occupation. The Small Yard has a number of find spots, and there is a small group north of the House, suggesting that much of the western end of the site had a domestic bias in Period 3. The other noticeable spread is in the F438 and F435 ditch systems. This material might have come from the House, or represent a separate domestic-type dwelling within or near the pound itself. The scatter of beakers along the eastern side is not matched by samian or mortaria, and it is possible that these vessels ought really to be counted as Period 4 or 5.

On the evidence of the distributions, there is a shift in the domestic emphasis in Period 4. The south-western corner looks to have been vacated and, but for the vessels in well F1052, it could be surmised that Barn 1 was no longer domestic. The other main deposit is from F911 in the Small Yard which may not be contemporary Period 4 material at all. The main focus now seems to be the north-eastern corner of the Main Yard and perhaps the pond area. The former features strongly in Period 5 as well, but as much of the pottery occurs in pond fills, a precise location for the occupation is less obvious. The concentration in the rectangular building, however, makes this the prime candidate. It is encouraging to note the absence of 'domestic' material from the adjacent Mill-house. The large RSG bowls are a new element in Period 5 and they match in very well with the distributions of the beakers and mortaria. There are two areas in Period 5, the south-western corner and the east side, where beakers do not, however, occur with mortaria or bowls. This could mean that either the domestic occupation was of a different type, or that beakers were not still available or used when these parts of the site were utilised.

The remaining distribution to consider is that of the Anglo-Saxon pottery. There is a scatter over most of the site but, apart from the two large concentrations, the greatest density is along the eastern side of the site and in the rectangular building area. The large groups in F254 and F1114 must have derived from dwellings, but the rest might have come from Anglo-Saxon activity of any kind. As some of this could date as late as the early sixth century, there are obviously problems in considering associated pottery. The occurrence of Roman wares with the Anglo-Saxon is not considered entirely fortuitous (see above), but it is not clear if it represents survival of Roman wares into the Anglo-Saxon period or if the Anglo-Saxons were on the site before the end of the Roman period and Roman pottery production.

Pottery and population size

The simple premise that the amounts of excavated pottery will reflect the numbers of people using a site was tested and the results are shown in Table 77. The vessel numbers include samian ware (Table 78). In order to make it easier to compare the quantities one period with another, the material was divided into twenty-five-year blocks in the same way as the bones in Chapter 9 (Table 83): columns A in Table 77. Thus the Period 1 totals were divided by five, those of Periods 2 and 3 by three, Period 4 by two and Period 5 by six. No attempt was made to account for residuality or survival except, as before, to reduce Period 5 by 20 per cent in the final column.

The most obvious feature is the drop in amounts during Period 2 which would seem to confirm the belief that the site was not used as intensively, or there were fewer people living and working on it at that time. This is most striking with beakers, for it is probable that all of those from Period 1 relate to the last twenty-five-year block, giving a drop from twenty-six to just four for the whole of Period 2. The mortaria totals, however, are closer, but it is again possible that the use of these specialised vessels was confined to

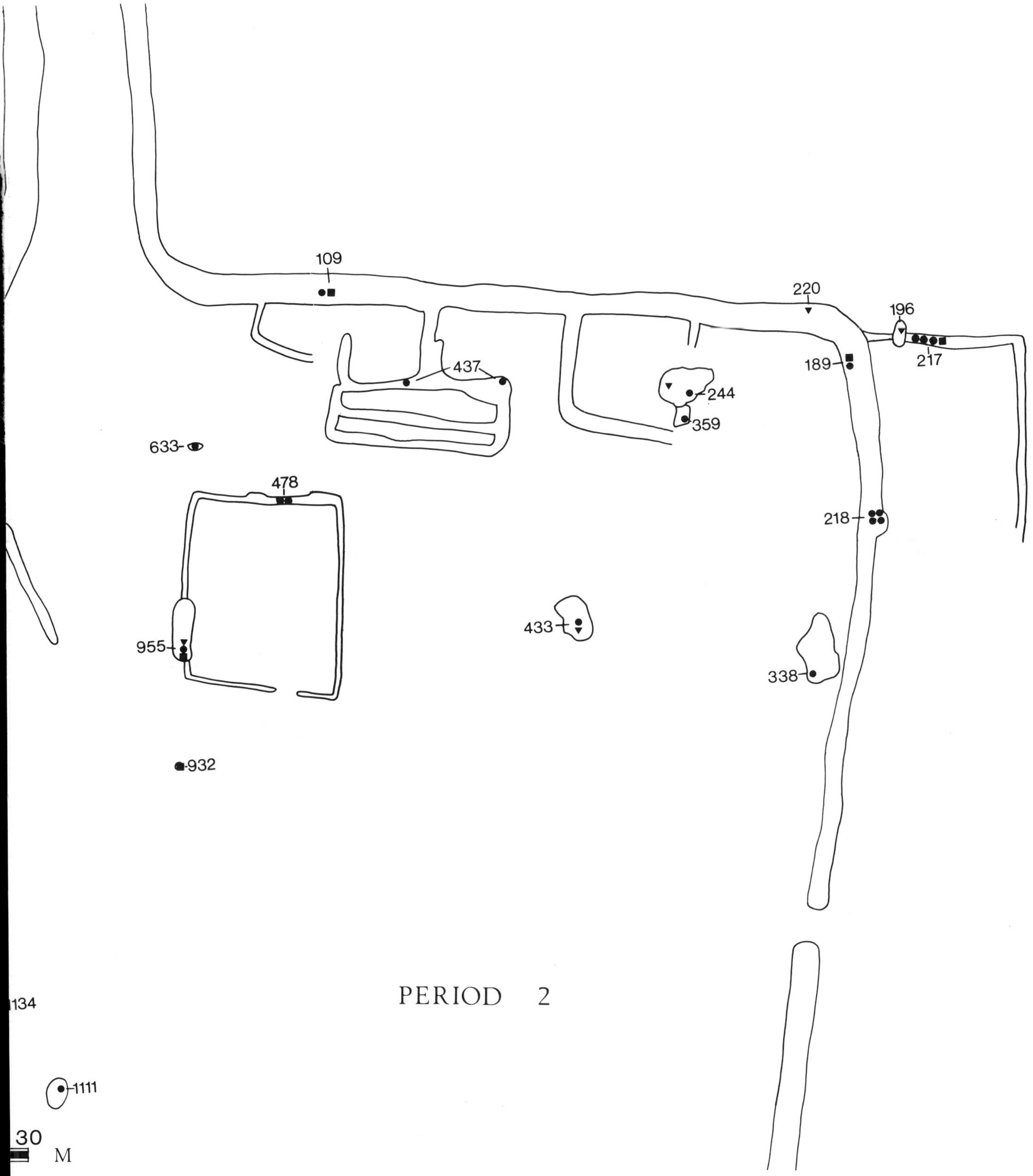

Figure 109 Period 2, distribution of types of Roman pottery.

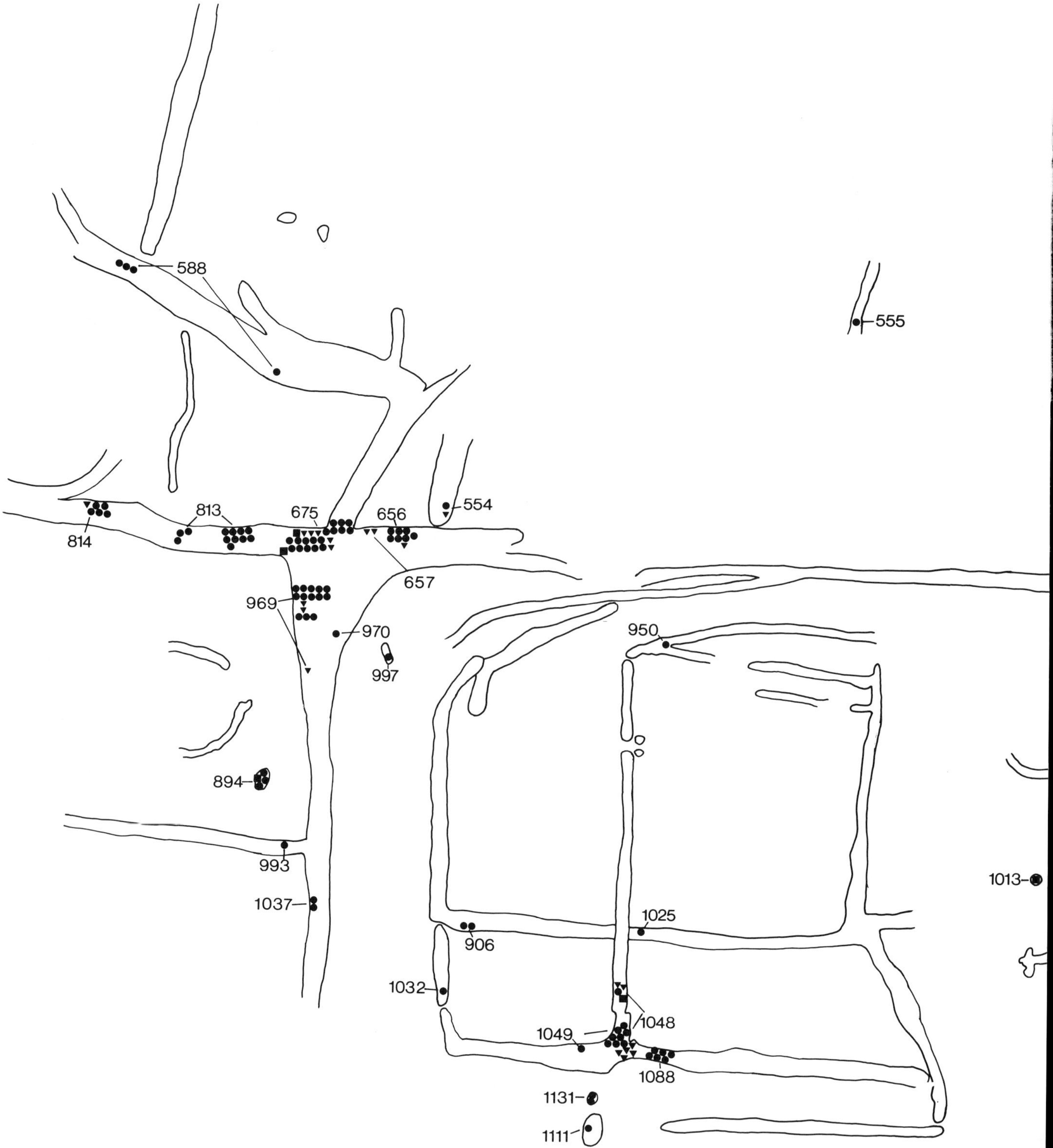

Figure 108 Period 1, distribution of types of Roman pottery.

PERIOD 1

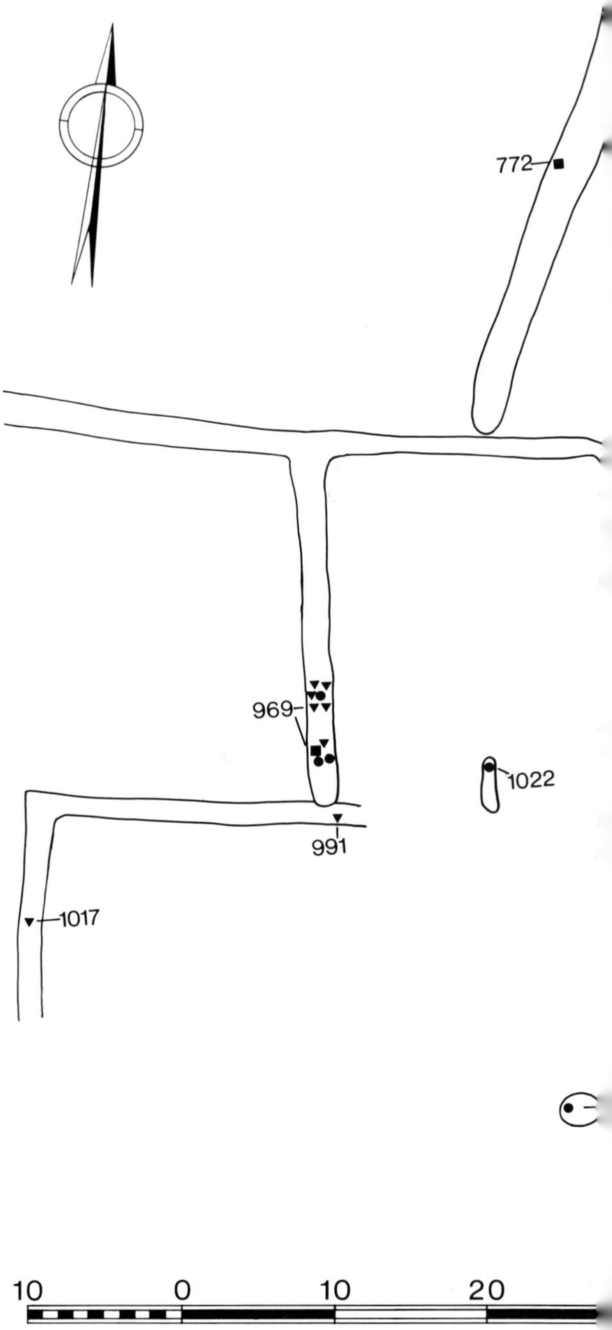
772
969
991
1022
1017
10 0 10 20

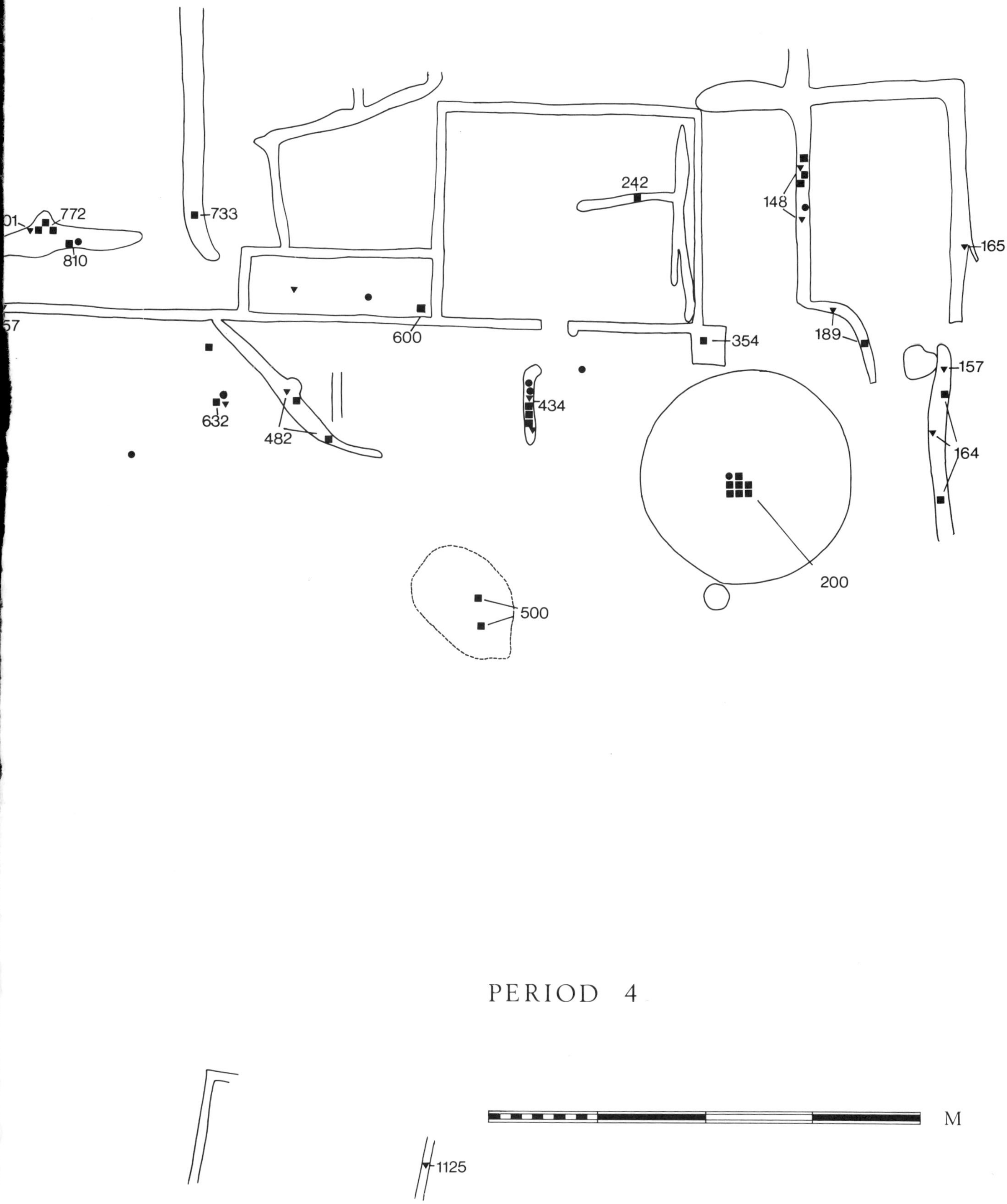

Figure 111 Period 4, distribution of types of Roman pottery.

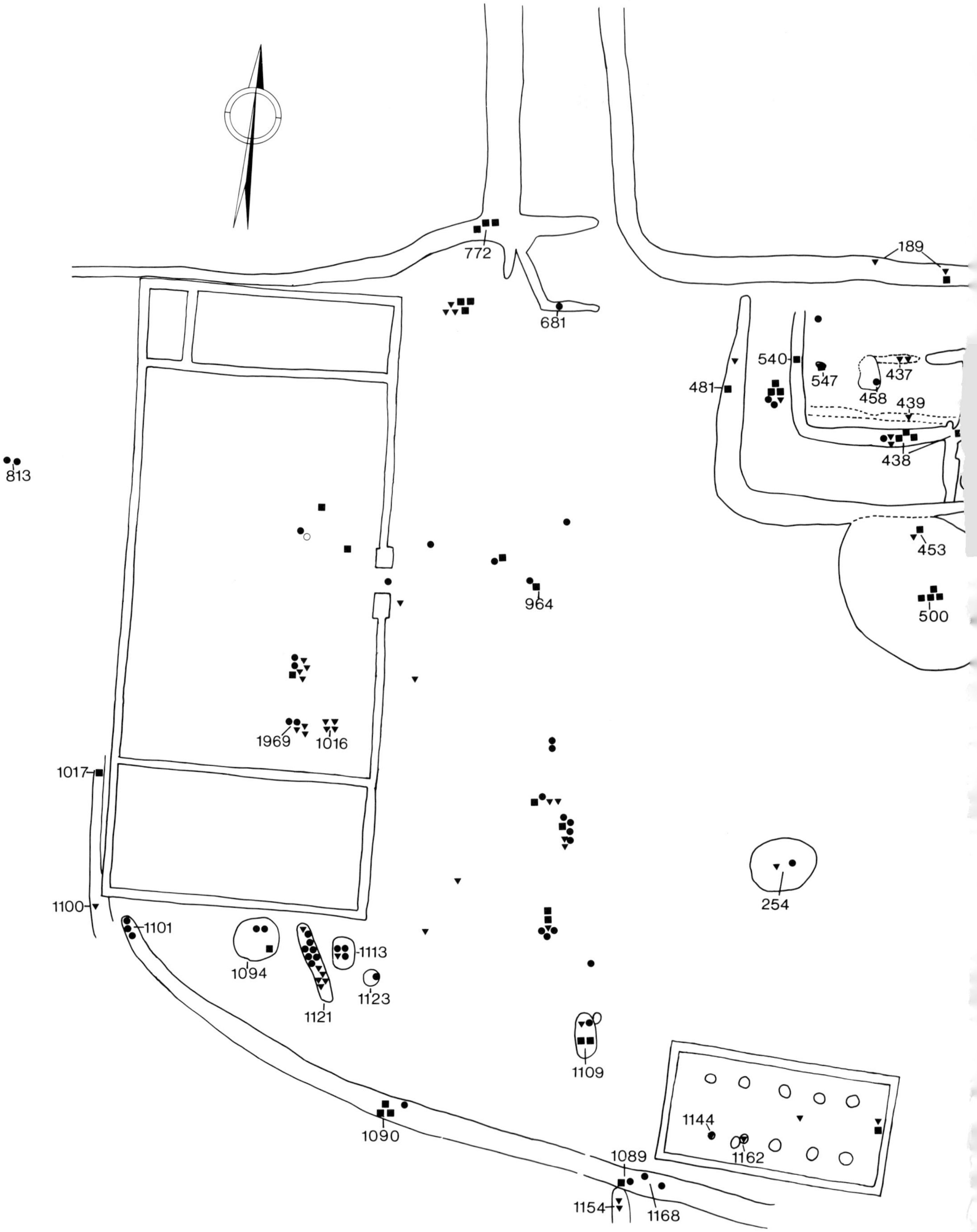

Figure 110 Period 3, distribution of types of Roman pottery.

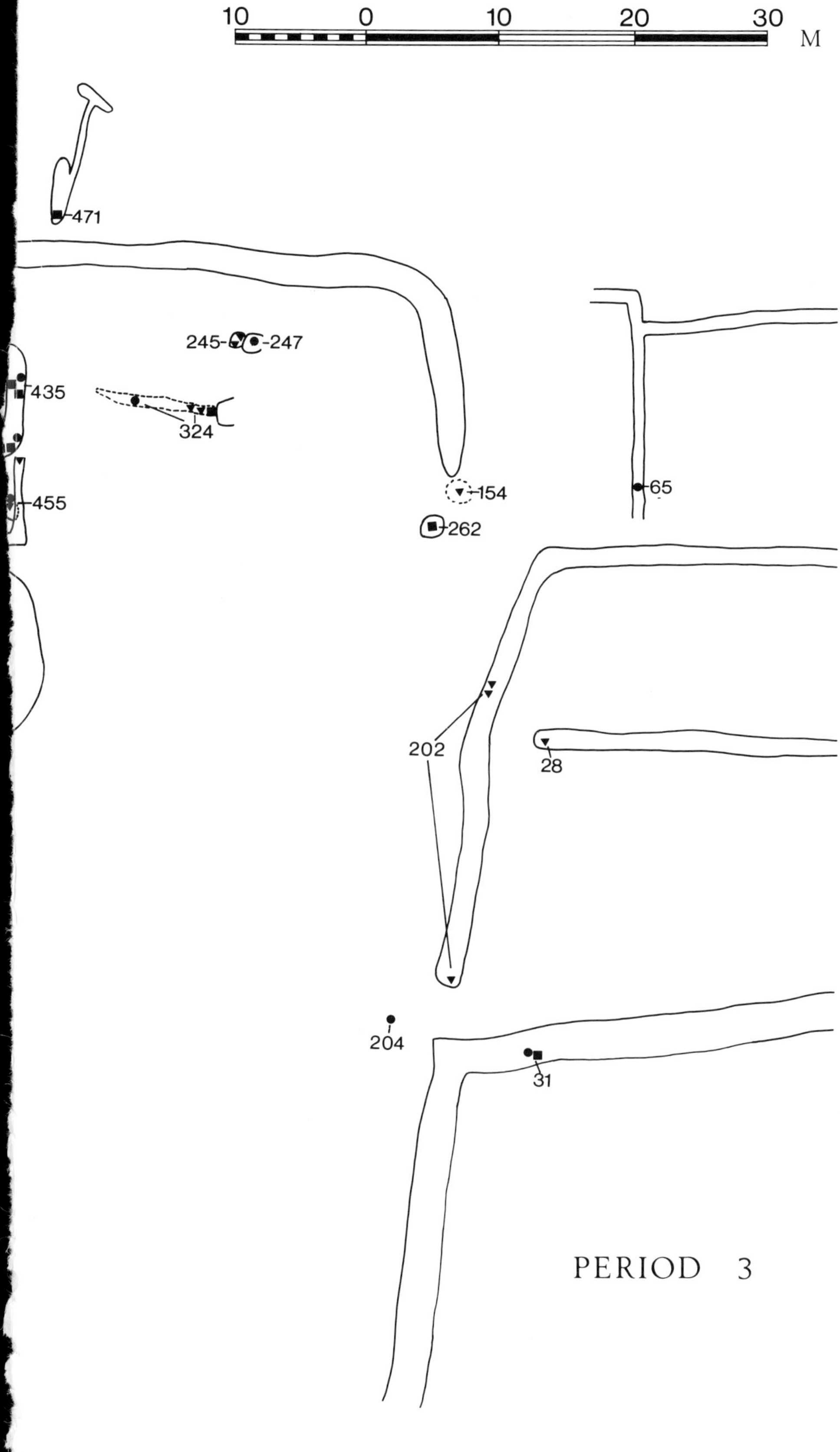

PERIOD 3

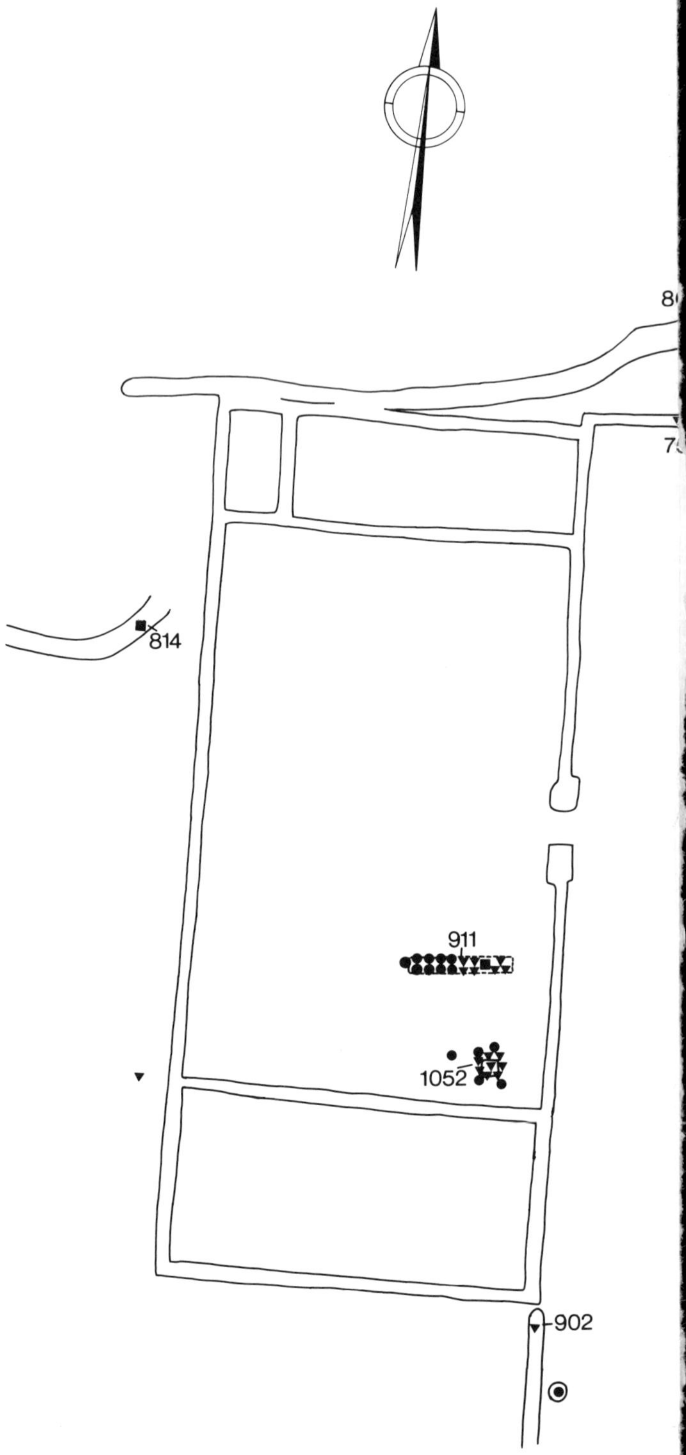

814
911
1052
902

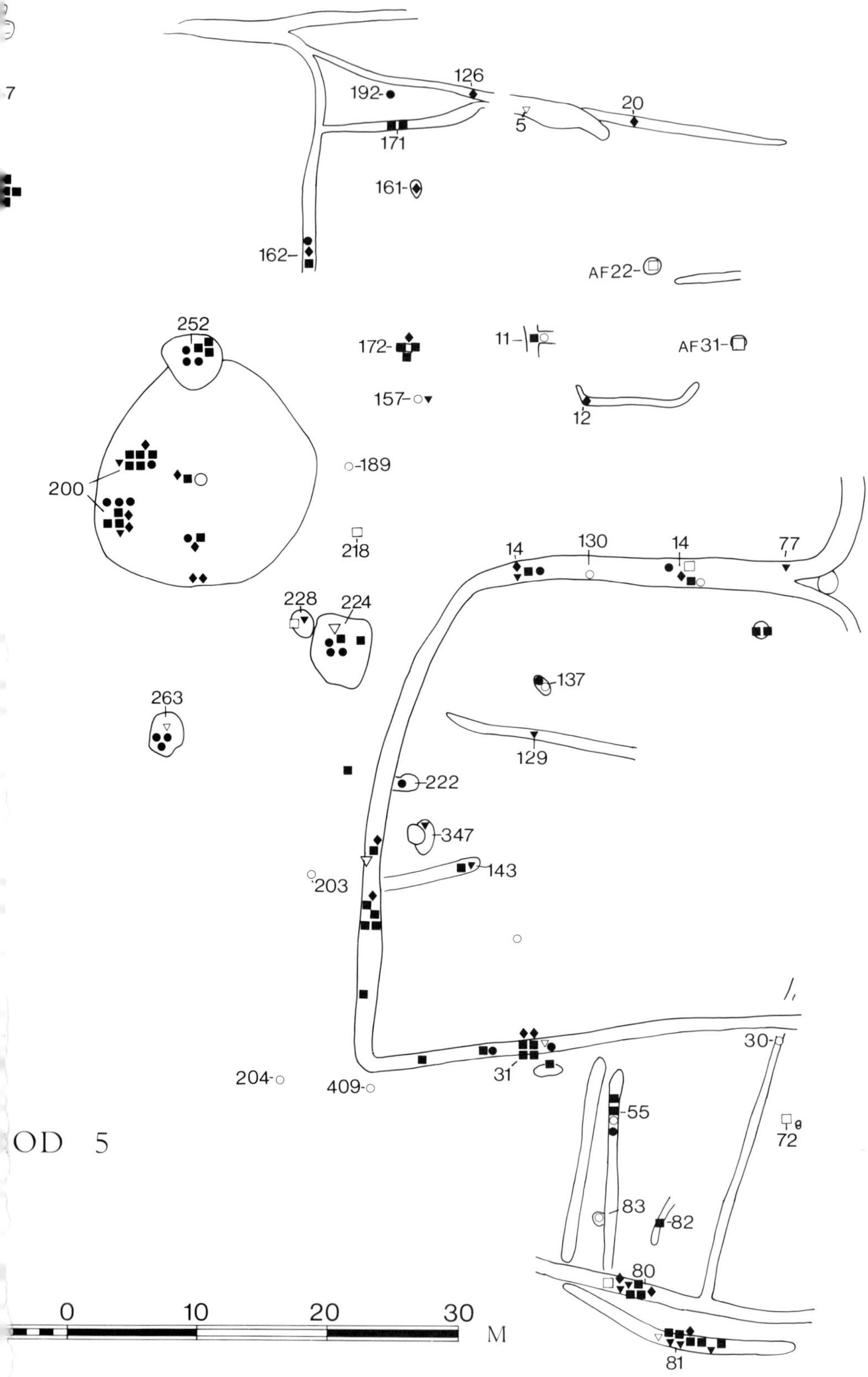

126
192
171
5
20
161
162
AF22
252
172
11
AF31
157
12
189
200
218
14
130
14
77
228
224
137
263
129
222
347
143
203
204
409
31
30
55
72
83
82
80
81
0
10
20
30
M

OD 5

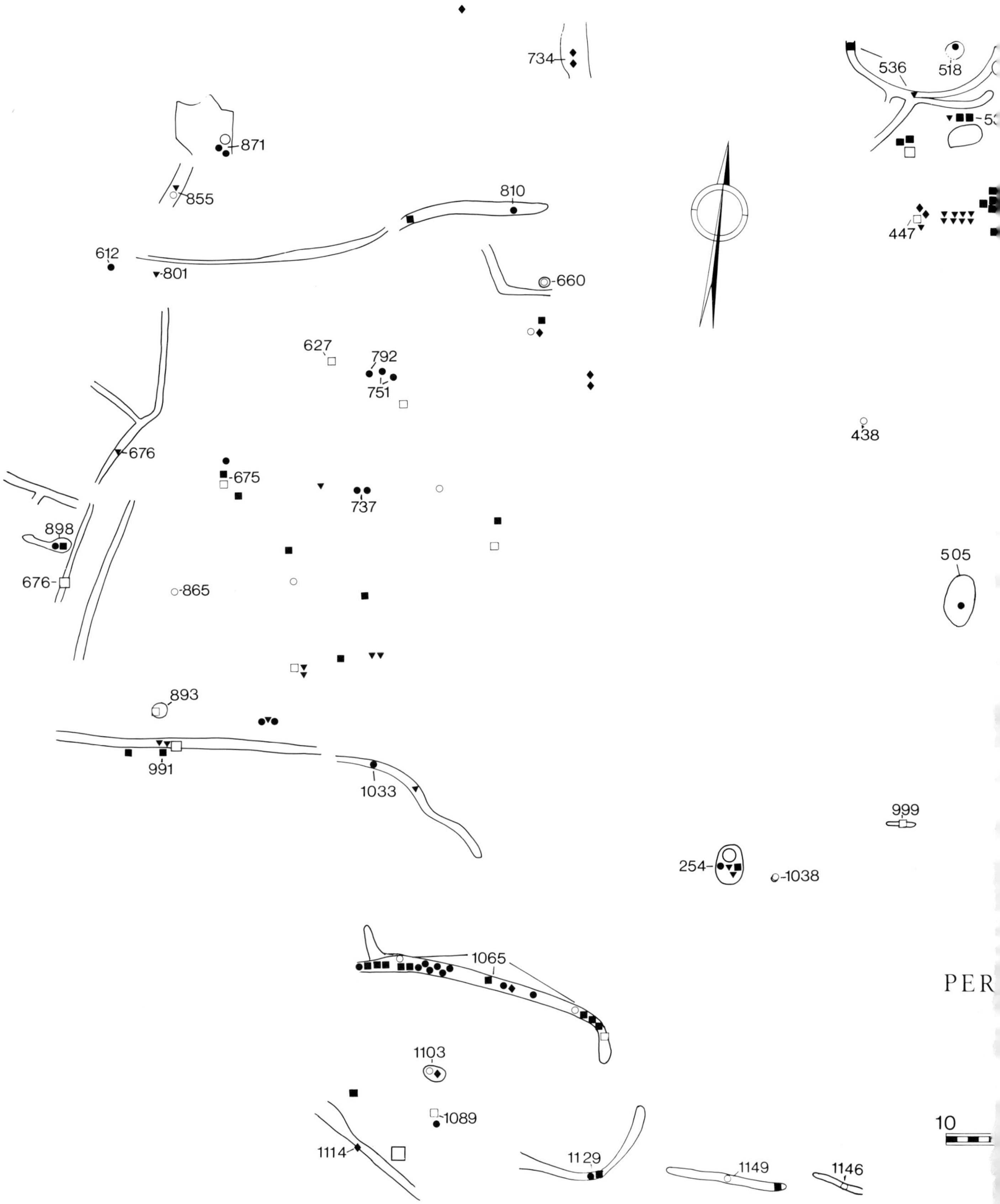

Figure 112 Period 5, distribution of types of Roman pottery.

	Period 1		Period 2		Period 3		Period 4		Period 5		P520% residual
	T	/25yrs	T	/25yrs	T	/25yrs	T	/25yrs	T	/25yrs	
	570	114	202	67	828	276	589	294	1588	264	212
Jars	350	70	117	39	407	136	206	143	859	143	114
Bowls/dishes	132	27	50	17	249	83	195	98	497	83	66
Beakers	26	5	12	4	64	21	31	16	40	6+	5
Mortaria	7	1+	9	3	51	17	36	18	92	15	12

Table 77 Numbers of vessels and size of population.

	P1	P2	P3	P4	P5	Total
Beaker	4	-	1	-	-	5
Bowl/dish	62	20	42	16	36	176
Cup	15	1	5	2	5	28
Mortarium	-	-	3	3	1	7
Inkwell	-	-	-	1	-	1
Other	3	3	8	-	15	29
Total	84	24	59	22	57	246

Table 78 Samian vessel types in Periods 1–5.

just the latter part of Period 1. Even so, the difference is nowhere near that of the beakers. A substantial drop in numbers of people could still be possible, if one accepts that mortaria, as prized items, were taken away by the occupants who left the site at the end of Period 1.

The Period 2 reduction is even more significant when compared with the three to fourfold increase in Period 3. This clearly shows that the site had burgeoned into a sizeable operation by the mid to late third century. Arguing from numbers of vessels, there would appear to have been twice as many people as in Period 1, a figure confirmed by the mortaria even if these are considered as belonging to just one twenty-five-year block. The differences in the other vessel classes perhaps reflect changes in fashion and preference for certain types at various times and, with beakers, possibly the status of the expanded occupation.

The Period 4 totals are rather varied with an increase in overall amounts and in specific classes. The static mortaria figure could be the most representative, however, and suggests that the population figure remained fairly constant at its Period 3 level during Period 4. The number of beakers is the only one which shows a significant decrease; this might, again, reflect the declining use of this type.

Period 5 is the most difficult to interpret. The residual material cannot be ignored in the last phases of a site which had been in operation for 300 years, and which for at least 100 was only occupied by Anglo-Saxons. When dealing with the bones, 20 per cent residuality was allowed for. If the same is done here (last column of Table 77), the result gives a more even picture with all showing a fall in possible population. The largest reduction is again with the beakers, but this is almost certainly because they were not being produced at this time. It is impossible to estimate the Anglo-Saxon population on the same basis as before, but it is just conceivable that these new settlers might have swelled the population of the site in the late Roman period.

Finally, to use the vessel figures in an even more simple way, if it is assumed that each person would have had the sole use of four pots — dish, bowl, jar and beaker — then the possible population of the site would have been 28 in Period 1, 17 in Period 2, 69 in Period 3, 73 in Period 4 and 66 (53) in Period 5.

VIII. Pottery and continuity, residuality or survival

All the pottery found during excavation had been brought to the site for a particular purpose. Once this was completed or abandoned, or if the pot was broken or became superfluous, it was discarded. The occupants had nothing approaching an organised rubbish disposal system other than removing the more noxious waste from current immediate areas of occupation, perhaps to middens or fields. Most of the debris would probably have been scattered around and about, thrown into nearby ditches, or areas not in use at the time: abandoned buildings and features, filling up unwanted hollows and the like. This dispersal of rubbish has been classified in three ways (Schiffer 1976). *Primary* rubbish is that which collects around and under activity or occupation; *secondary* rubbish is material which is accidentally or deliberately deposited away from the area of activity or occupation, and *de facto* rubbish is material still useful but deliberately rejected when an activity or occupation is completed or abandoned.

Each of these categories could provide groups of pottery which were both homogeneous and contemporary. Orton Hall Farm was, however, occupied continuously for 500 years. The buildings and other elements of the farmstead were refurbished many times. This is best shown by the enclosure ditches which were frequently replaced, realigned and recut. Of the site's features, 10 per cent lasted for more than one period, 30 per cent of these went into a third period, and 3 per cent of the last were still prominent enough to receive pottery in a fourth period, which means that they had lasted for around two hundred years. With all this reorganisation within a limited area, added to the later disturbance by medieval and modern ploughing, there is a tendency for the rubbish of separate periods or phases to become intermixed.

The main problem in considering the groups which do survive, and the rest of the pottery as well, is the identification of residual material. The three commonly accepted criteria for determining residuality are sherd size, degree of abrasion and date, but they need to be applied with care.

Unquestionably residual sherds will normally be small with worn surfaces and edges while the more obvious survivals are complete or represented by large sherds with sharp 'breaks'. The dividing line is not always clear cut, however, as soil type and the nature of the deposit can either enhance decay or protect discarded pottery from damage. Moreover, the pottery itself, especially its hardness, the presence of soluble inclusions and sometimes even the type of vessel, can affect the way in which it survives in the ground. In reality, therefore, some residual pottery can occur as large, unabraded sherds, or even complete vessels, while a true survival might only be represented by small pieces.

As far as date is concerned, it is obvious that sherds of first-century manufacture occurring in fourth-century contexts are residual but there are problems when the time difference is not so large. Here the need is to be able to determine whether any of the pottery could have been survivals-in-use. The accepted definition of the term 'survival' (Webster 1976, 4) contains the phrase 'out of fashion when last used' which is a theme echoed by Gillam in his discussion of a large group of pottery from Great Casterton: 'vessels that had survived in use long enough to have become noticeably old-fashioned' (Gillam 1951, 24).

The term 'fashion' is currently defined as: 'prevalent custom or taste' but though there is some evidence for its existence in Romano-British pottery, primarily from the imitation at different times of particular vessel forms or wares, it remains a rather imprecise term, somewhat difficult to relate to a fixed number of years. It will obviously vary according to the type of pot and the use for which it was designed or has been put. Cooking-pots would generally have had a short life expectancy, but storage jars could have had a very long life, especially if used for dry goods. Lamps, candlesticks and *tazze* are others which, because of their irregular or restricted function, could survive for long periods. The life of some classes of pottery might also have been affected by the availability of similar vessels in glass, metal, wood, horn or skin.

The bulk of the pottery would last as long as it remained unbroken, anything from a few hours to many years, or as long as it was wanted. Herein lies the most difficult factor to quantify, but one which has a great influence on the pottery that is found in any particular deposit, and its date. It is extremely difficult to assess and measure the potential for any single specimen or class of pots to survive because they were of greater relative value to the owner. That such vessels existed is not in question; it is rather the level at which the phenomenon operated that is sought. It is thought that the overall wealth of the individual or the site itself was the major influence. On the simple Romano-British settlements in the border areas, for example, proper Roman pottery of any sort, and fine ware in particular, is sufficiently rare to suppose that any such vessel or vessels would have been prized. Conversely, it is likely that the wealthy occupants of the grander villas or town houses would have been less likely to 'treasure' any pottery when they had fine glass and metal vessels at their disposal.

Another associated factor is age when purchased. Some evidence from 'pottery shops' at Castleford, Corbridge and Wroxeter, and the Roman wharfs in London (Richardson 1986, 96–8) shows that stock could be at least twenty years old and still saleable when discarded or overtaken by destruction. Once supplies of any vessel petered out, the price of the existing stock, for which there was no alternative in quality or type, might increase if it was still sought after, thereby extending the time that it might remain unsold, its relative value to the customer and how long it might survive in use. It is almost impossible to quantify or allow for such market forces, or to tell whether they operated differently in the towns as opposed to the country or even from site to site. They could have had a considerable affect on certain sites and with particular vessels.

At Orton Hall Farm the amounts of glass and metal vessels, at least those which remained for the archaeologist to find, are small, especially after Period 2. Amongst the pottery assemblages the best candidates for potential prolonged survival are samian ware, beakers, flagons, boxes and other types of finer table ware. A number of the larger *de facto* groups contain possible survivals.

Given the late second-century date assigned to the deposits from the F1048 system, many of the grey ware and RSG vessels, for example Nos 116–23, 96–7 and 102–3, respectively, appear to be well 'out of fashion'. Similarly Nos 259, 273–6, 281, 312 and 332–3 from the F1094/F1113/F1121 deposits of the end of Period 3, and with a coin of *c.* AD 270–290, are typologically much earlier. The F911 yard deposit belonging to early Period 4, or even to late Period 3, has many vessels which could be survivals, though they are of long-lived production types; the assemblage does, however, include most of the imported CGCC from Orton Hall Farm as a whole (Nos 464–5) as well as an odd flagon or bottle (No. 467) and the only definite unguent flask (No. 466) from the site. Some of the F1052 and F434 material also looks out of fashion.

The bulk of the pottery from Orton Hall Farm is not as easy to assess, but a lot, even some occurring as small sherds which would usually be dismissed as residual, could have in fact come from survivals-in-use. The simple distribution maps for beakers, samian ware and mortaria (Figs 108–12) might provide some evidence. The distributions seem to echo the spread of occupation around and across the site and mirror the main areas of each period's activity.

In terms of date, the samian ware, for example, would be residual after the early part of Period 2, yet a significant amount occurs later in the period in association with Barn 2 which is sited in an area devoid of coeval Period 1 occupation. Similarly, the cluster of samian ware in the Period 3 Barn 1 and north central area enclosures, some of which were *de facto* groups, could be too coincidental to be completely explained away by residuality. The key group in Period 4 is in F911, for this shows that some of the material used on the site for levelling up and so on could have been stored for some time. Moreover, irrespective of the time of its deposition, the group itself exhibits a fairly wide typological date-range with some of the pottery, including the samian ware, seeming to be survivals. Similarly in Period 5, beakers occur regularly with mortaria and large bowls, even though current evidence suggests they were not made in large numbers, if at all, towards the end of the fourth century.

In addition, mortaria occur in some secondary Anglo-Saxon deposits, for example in F1065, in areas where the latest previous Roman occupation was earlier than the date attributed to the mortaria, and the distribution of the large RSG bowls also reflects Anglo-Saxon activity. These vessels, and the mortarium in an Anglo-Saxon fabric, could suggest that Anglo-Saxons used some Roman pottery. The more robust nature of Roman pottery, when compared with ordinary Anglo-Saxon wares, might have influenced its treatment and potential to survive; in other words, the Anglo-Saxons valued it and thereby themselves caused some pottery to survive. The use of Roman pottery, however, need not have been contemporary with just the later Roman phases of the site. Examples elsewhere, for example at York, indicate that Roman pottery was used by much later inhabitants, presumably because it was of better quality than that otherwise available at the time. Some of the Roman pottery in Anglo-Saxon contexts at Orton Hall Farm conceivably represents similar reuse, one of the best candidates being the complete LNVCC imitation form 38, No. 529, from F81.

In conclusion, the mechanics of the occupation of the site clearly led to the creation of mixed deposits often containing residual pottery. On the whole, it would be wrong to apply the criteria for identifying residual pottery too rigidly, for this would lead to the underestimation of lengthy survival in use or store, especially on rural sites: at Orton Hall Farm, for example, there is some evidence to show that certain vessels probably survived for fifty years or more.

IX. The Samian

by Felicity Wild[30]
(Fig. 113)

In all, the site produced sherds from approximately 290 vessels, of which about 12 per cent were South Gaulish. Most of the material was of Antonine date, and the late Antonine forms were well represented in all periods.

Period 1

South Gaulish form 29, 1; 37, 1; 27, 3; 15/17, 1; 18, 4; 18/31, 1; 36, 1; Curle 11, 1; Curle 11 or Ritt.12, 3; uncertain, 3.

Central and East Gaulish form 29, 1; 37, 6; 27, 1; 33, 11; 18/31, 8; 31, 11; 18/31 or 31, 14; 18/31R, 1; 31R, 1; 18/31R or 31R, 1; 36, 1; Curle 11 or 38, 1; 38, 1; 79, 3; 79 or 80, 1; beaker scraps, 4.

Of the South Gaulish material, the form 29 (**1** below) is certainly of pre-Flavian date, the 15/17 and the three pieces of Ritterling 12 or Curle 11 may be. It seems likely, therefore, that there was activity on the site from at least the Flavian period, if not before. The proportion of first-century wares is, however, very small, amounting to not more than about eighteen vessels out of a total of about eighty-six.

The latest material from the group, on the other hand, is of late second-century date. The stamp of Marcus (**B** below) dates to the period *c.* AD 170–200, and the later Antonine forms 31R and 79 are present, which were not produced before AD 150 and probably not before *c.* AD 160. The samian therefore confirms the suggested dating of *c.* AD 175–200 for the end of the period.

The decorated ware and stamps are listed in detail below. The potter and die numbers are to appear in Hartley and Dickinson (forthcoming). According to their notation, recorded there, **a** placed after the name of a pottery denotes that the stamp in question has been recorded there, **b** that other stamps of the same potter have been recorded there, and **c** that the stamp has been assigned to that pottery on form and fabric.

A [C]ΛMVLIN[I] Form 33, Central Gaulish, showing Die 2a of Camulinus of Lezoux[a]. The stamp has been noted from Chesters and South Shields. Its use on form 15/31 and the context in which it appears at Lezoux suggests a mid-Antonine date at the earliest.
L1593 (51), L634 (215)

B MARCIM Form 31, Central Gaulish, showing Die 5c of Marcus v of Lezoux[b]. This is one of Marcus v's less common stamps, not yet noted from a dated context. His other stamps occur on Hadrian's Wall, at Pennine forts evacuated when the Wall was rebuilt and at Pudding Pan Rock. His forms include 31R, 79, 79R and Ludowici TgR.
c. AD 170–200, L2677 (53), L1049 (P1)

1 Form 29, South Gaulish, showing scroll decoration in the upper zone, and festoons with the bird (0.2290) and Nile goose (O.2286) in the lower. The scroll is closely similar to that illustrated by Hermet (Hermet 1934, pl. 36, no.34). A scroll with similar rosettes and circle was used by Aquitanus (Knorr 1919, *Taf.* 8A), and both these and the tendril bindings show similarities to a bowl by Crestio (Knorr 1952, *Taf.* 18D) who also used similar arrangements to that in the lower zone (Knorr 1952, *Taf.* 17C). A bowl from Camulodunum shows the bird and a similar festoon (Hawkes and Hull 1947, pl. XXIX, no. 3). The festoon was used by Murranus on a bowl in the Museum of London, and by Calvus (Knorr 1919, *Taf.* 17B). The connections suggest a date *c.* AD 50–70.
L1725 (153)

2 Three fragments, two joining, the third not illustrated, probably from the upper zone of form 29, Central Gaulish, in the fabric of Les Martres-de-Veyre. The decoration shows an inverted acanthus and part of an animal, possibly the stag (O.1791) which was used by the Potter of the Rosette (Stanfield and Simpson 1958, pl. 26, no. 322). The acanthus also appears on bowls in this style, sometimes inverted (*ibid.*, pl. 24, nos 297–9, *etc.*). To the right of the stag is a tree motif, probably similar to that illustrated by Stanfield and Simpson (*ibid.*, pl. 26, no. 323, *etc.*). Form 29 was made at Les Martres-de-Veyre during the first quarter of the second century AD, but is not one of the forms recorded for the Potter of the Rosette.
c. AD 100–120, L1672 (19), L1755 (51)

3 Form 37, Central Gaulish, in the fabric of Les Martres-de-Veyre. The decoration shows a zone of double festoons containing the stork (O.2197). The style is probably that of Stanfield's potter X.2, who used the festoons (Stanfield and Simpson 1958, pl. 4, nos 33–4) and the stork (*ibid.*, pl. 4, no.36).
c. AD 100–120, L2186 (22)

4 Form 37, Central Gaulish, showing scroll decoration with leaf (Rogers H23) used by Sacer and Attianus. Attianus also used the circle and a ridge below the decoration. A close parallel, also showing the leaf overlapping the basal ridge, is the signed bowl from Colchester (Stanfield and Simpson 1958, pl. 86, no. 18).
c. AD 130–160, L2631 (41)

5 Form 37, Central Gaulish, burnt. Twelve fragments in the style of Casurius, showing his ovolo (Rogers B223) over panel decoration with large-beaded borders. The fragments do not join and the arrangement of the panels is uncertain, but they contain his festoon (Stanfield and Simpson 1958, pl. 134, no. 31), male figure (O.638), upright leaf (Rogers J40) and probably the prisoner (O.1146, *cf.* Stanfield and Simpson 1958, pl. 137, no. 57).
c. AD 160–190, L1566 (51), L1890 (51)

Period 2

Form 37, 4; 33, 1; 36, 1; 31, 3; 18/31 or 31, 3; 79, 3; Lud. Tg, 2; Curle 15 or 23, 1; bowls, 3; scraps, 2; uncertain, 1.

There was only a maximum of twenty vessels from this period, all of second-century date apart from one scrap of South Gaulish form 37. Again, the later Antonine forms, 31, 79 and Lud. Tg were represented. The decorated pieces consisted of a sherd probably in the style of Rogers' X13, datable to *c.* AD 120–130, and the following:

6 Form 37, Central Gaulish, showing the ovolo (Rogers B105) and astragaloid border used by Albucius, over free-style decoration with his leaf-tip space-filler (*cf.* Stanfield and Simpson 1958, pl.123) and probably the hound (O.1984).
c. AD 150–180, L934 (59)

7 Form 37, Central Gaulish, showing seated Venus (O.334) and an astragalus column (Rogers P6) which Rogers attributes to his potter P21. At the edge of the sherd is a corded column, possibly Rogers P18, also attributed to potter P21. The general style suggests a mid- to late Antonine date.
L824 (72)

Period 3

The material from Period 3, all second century apart from two scraps of South Gaulish ware, was clearly residual in context. It consisted of: form 37, 4; 33, 5; 18, 1; 18/31, 4; 31, 4; 18/31R, 3; 31R, 1; 18/31R or 31R, 2; 36, 2; Curle 11, 1?; 38, 5; 79, 4; Curle 23, 1; Curle 15 or 23, 1; 45, 3; beaker scrap, 1; uncertain, 5.

The decorated pieces were Trajanic or Hadrianic in date.

Period 4 and later

The material from the later periods was clearly residual and has not been listed in detail. Again the later Antonine forms 31R, 79, 80, 79R and 45 were present. Also present were two small scraps of the samian inkwell, Ritterling 13. The stamps and only significant decorated bowl fragments are listed below.

C CELSIANI.OF Form 31, Central Gaulish, showing Die 1a of Celsianus of Lezoux[a]. The stamp was used on forms 31R, 79 and 80 and has been noted from Bainbridge, Binchester and Catterick. Stamps from other dies occur at Chester-le-Street and on form 79R.
c. AD 160–200, L2450 (HS)

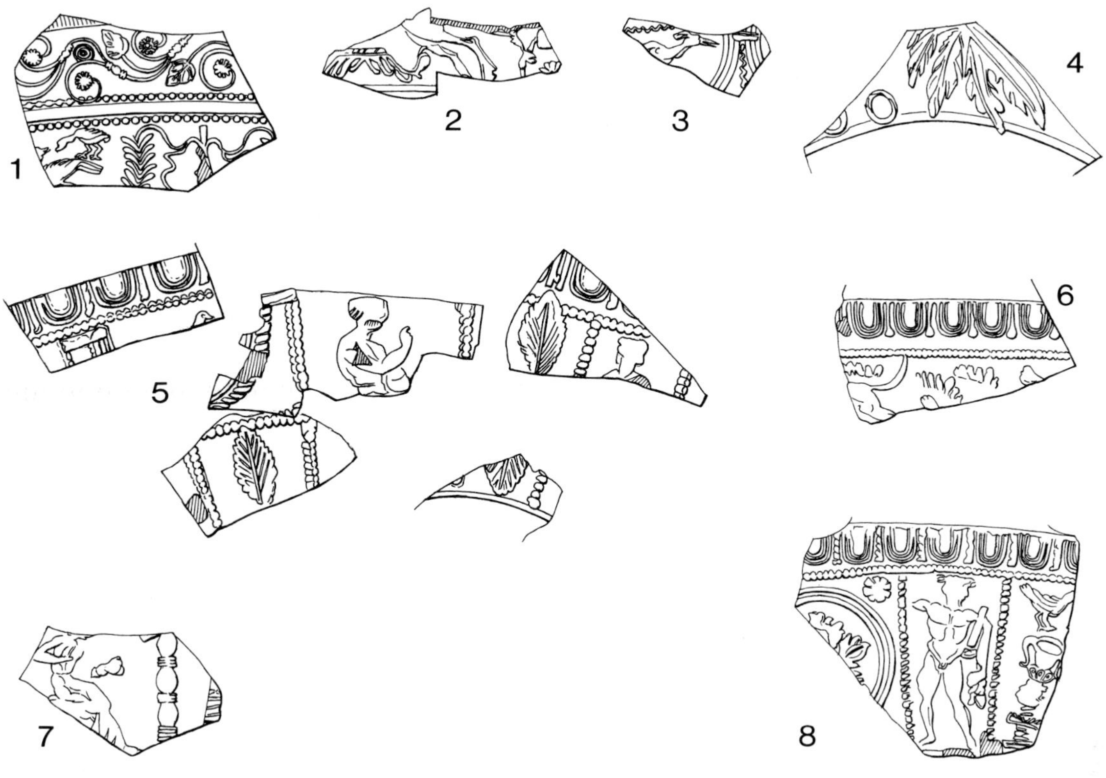

Figure 113 Samian pottery, Nos 1–8. Scale 1:2.

D C O.S.Λ.X[.T.I.S.Γ] Form 79R, Central Gaulish, showing Die 2a of Cosaxtis of Lezoux[c] and Vichy (Terre-Franche)[b]. There is no site dating evidence for this stamp, which also occurs on forms 31R and 79. The forms suggest a late Antonine date.
L1887 (203), L925 (70)

E MAINΛCИ Form 33, Central Gaulish, showing Die 5a of Mainacnus of Lezoux[b]. The stamp has been noted from Catterick and on form 31R. One of his other stamps is in the group of late Antonine samian from Pudding Pan Rock.
c. AD160–200 L1887 (203), L924 (196)

F [P]ATERNI Form 18/31, Central Gaulish, showing Die 2a of Paternus iii of Lezoux[a]. This potter worked in the mid-second century and was associated with Ianuaris ii in the making of decorated ware. His plain forms include 18/31R, 27 and 81. This particular stamp has been noted at Carrawburgh.
c. AD 140–170 L1857 (MS), L899 (—)

G [PIIP.P]OFIIC Flat base fragment, possibly form 79 or 32, East Gaulish, showing Die 2a of Peppo of Rheinzabern[a] (Ludowici 1927, Die e). The stamp was used on forms 31R and 32 and occurs in a group of late second or third-century date from the London Docks.
c. AD 180–260 L37

H SΛC[EROM] Form 31, Central Gaulish, showing Die 1a of Sacero of Lezoux[c]. The stamp has been noted on forms 18/31R, 27, 31R and 37 (on the rim of a bowl in the style of Albucius ii or Paternus v).
c. AD 150–180, L37 (HS), L331 (160)

8 Form 37, Central Gaulish. Two fragments in panel style showing medallion containing leaf; figure, perhaps Perseus or Mercury, apparently a reduced version of O.147A, which was used on bowls in Cinnamus' style; bird (O.2239B) also used by Cinnamus, over cup (Rogers T16) and pedestal. The style incorporated elements of decoration used by Iullinus, Doeccus and Cinnamus, though none of these potters used them all, and Cinnamus is less closely associated with the style than the other two. The ovolo (Rogers B164) and border of rhomboidal beads are Iullinus', and he also used the cup. The bird is also on a bowl in his style from Lezoux (Roanne Museum). The ovolo, pedestal and rosette are on another bowl from Lezoux in Iullinus/Doeccus style (Coll. Sauvaget, formerly Coll. Chabrol-Janelle). In view of the connections, a date *c*. AD 160–180 is likely.
L1359 (MS), L2503 (HS)

X. The Mortaria

by J.R. Perrin with K.F. Hartley
(Figs 114–17)

Over 80 per cent of the mortaria used by the inhabitants of the farmstead were locally produced. Most date to the third and fourth centuries, and there is considerable variety of rim form.

The earliest use of mortaria, of whatever origin, on the site appears to have been in the Hadrianic-Antonine period. This matches the evidence from other rural sites in the area, such as Fengate (Hayes 1984, 180), Castor (Green and Green 1987), Normangate Field (Perrin and Webster 1990, 40) and Chesterton (Perrin, to be published). It is not clear why mortaria were not used before this date, but it is possible that they were not 'adopted' until local potteries began to make them or they first appeared at local markets; the date certainly seems to coincide with that suspected for both the beginning of the local industry and the growth of Durobrivae.

It should be noted, however, that there is little evidence from other than rural sites in the late first to early second centuries, and mortaria might well have been used on higher status sites at this time. Although the occupation at the Longthorpe fortress was of a specialised nature and lasted for a short time, it is possible that it may have influenced some of the local population to adopt Roman methods of food preparation. It is unwise, however, to link the use of mortaria on rural sites with something as intangible as the 'final' adoption of Roman culture.

The low numbers in Period 1, though all really belong to the last fifty years of that, and Period 2, might reflect the time it took for the vessels to be fully accepted, but that there may have been other dwellings outside the excavated area at this time. However, M15, M57 and M64 should be added to Period 1, if their general dating is to be relied upon. The increase in Periods 3 and 4 clearly shows the growth of the farmstead and, probably, the fact that domestic occupation was located in the excavated area.

The period sequences on the site provide useful preliminary dating evidence for many of the types and subtypes used in the third and fourth centuries. It is interesting to note that the first use of ironstone trituration grit may have been a little earlier than suspected. A lot of this particular grit resembles slag, and it is possible that the potters availed themselves of the waste from local ironworking.[31]

Some of the mortaria (*e.g.* M111, M12, M125, M128) which are usually given a mid-fourth-century end date, appear to have lasted much later. It is not certain if this represents survival in use or continued production. Possible use by Anglo-Saxons is considered elsewhere, especially as there had been at least one mortarium in an Anglo-Saxon fabric (Anglo-Saxon Pottery, Group 1, No. 15).

Mortarium M11 (Fig. 114) might have been made in the Stilton area, for it resembles vessels found associated with a possible kiln.[32] Similar mortaria were noted at Chesterton (Perrin, to be published). M17 (Fig. 114) and possibly M45 (Fig. 115), and M150 (Fig. 117) are, respectively, Mancetter-Hartshill and Oxfordshire forms occurring in Lower Nene Valley fabric. Other such mortaria were noted at Chesterton (*ibid.*), and these indicate the migration or movement of potters.

The numbers of non-local mortaria show strong links with the Oxfordshire region (OXW, OXCC), but the numbers of vessels from Mancetter-Hartshill (MH) at Orton Hall Farm are fewer than at other sites like Chesterton (*ibid.*) and Normangate Field (Perrin and Webster 1990). The difference is due to date, for imports from Mancetter-Hartshill are mainly of the second century, when there were few mortaria at Orton Hall Farm, and these were quickly eclipsed by local production. Most of the Oxfordshire vessels are later third or fourth-century types, and represent a different phase of trade altogether.

The two Verulamium region (VR) products, M1–2 (Fig.114), are not unexpected, for the kilns of that area were highly productive and had a wide distribution in the late first and early second centuries. The Lower Rhineland (LR) product, M35 (Fig. 114), is a more unusual find, and was perhaps bought for a particular purpose. It might, however, have been just a chance purchase, or even a vessel brought by a visitor. The only type of mortarium from the site and not represented among the stratified pottery, was a white fabric colour-coated vessel from the Oxfordshire kilns (Young 1977, type WC7): it occurred in a machine-stripping layer (L1362).

The mortaria identifications and form notes were provided by K.F. Hartley.

Catalogue

The fabric is local, and entries are body or base sherds unless specified. Those illustrated are marked with *. Fabric abbreviations as follows: LR, Lower Rhineland; MH, Mancetter-Hartshill; OXCC, Oxfordshire colour-coated; OXW, Oxfordshire white; OXWCC, Oxfordshire white colour-coated; VR, Verulamium region.

Period	P1	P2	P3	P4	P5	Total
Origin						
NV	4	8	38	26	77	153
MH	-	1	1	-	-	2
LR	-	-	1	-	-	1
OX	1	-	7	7	14	29
VR	2	-	-	-	-	2
?	-	-	1	-	-	1
Total	7	9	48	33	91	188

Table 79 Mortaria in Periods 1–5, numbers and origin.

Period 1
(Fig. 114)

M1 F675, L1593 (51), VR. Second century.

M2* F675, L1741 (51), VR. Full curved flange and high, split bead. Reminiscent of Oxfordshire types Young M10–12. Present in Antonine deposits in Scotland. Typological and other site dating: AD 140–200 (Frere 1972, 341, fig. 131, nos 1053–4). Probably same vessel as L2303 (200), F1052, Period 4. **Type 80**[33]

M3* F894, L1811 (5), OXW. Drooping, curved flange turned up towards body. High, backward-facing bead. Type Young M21.1; less probably M11. Typologically AD 170–240; other site dating: 180–300. **Type 83**

M4 F1000, L2169 (34). Late second century and after.

M5 F1000, L2179 (34). AD 130?–170?

M6 F1013, L2174 (37). Indeterminate. **Untyped**

M7* F1048, L2550 (42). Shallow, flattish flange ending in a distal bead. High bead. Only five surviving trituration grits, all iron slag. Typological date AD 130–180; other site dating: 130–175. **Type 5**

Period 2

M8 F189, L350 (58). More likely to be later than *c.* AD 200 than earlier because the trituration grit is fairly abundant and consists entirely of iron slag. **Untyped**

M9 F189, L1118 (58). AD 230–400.

M10* F217, L361 (83). Wide shallow flange, curled sharply inward and coming to a point at the distal end. High bead pointing inward. Typologically later than AD 180. **Type 19**

M11* F217, L361 (83). Slightly sloping flange with a rectangular section. Horizontal shelf on top of flange. Bead broken and turned out over flange to form spout. Typologically *c.* AD 200–280. Possibly a Stilton product? **Type 72**

M12* F217, L361 (83). True hammerhead with bead and flange in alignment. Distal end of the flange is turned under and up into a knob. Finger depression spout. Typologically AD 250–350. Similar to M123. Brown-buff slip. **Type 35**

M13 F772, L1374 (69). Early third century.

M14 F932, L2043 (—). AD 230–400.

M15* F955, L2060 (89), L2061 (23). Probably MH. Flange rising slightly above the bead. Distal end of flange turned slightly inwards. Typological and other site dating: AD 120–160. Unidentifiable stamp. **Type 6**

M16 F969, L2204 (68). AD 230–400.

Period 3

M17* F31, L768 (94). Reeded hammerhead mortarium indistinguishable in form from those made in the Mancetter-Hartshill potteries. Finger depression spout. Typologically AD 250–350+. **Type 50**

M18 F189, L854 (107). AD 200–400.

M19* F262, L514 (117). Flange tapers at distal end, and the underside is welded in. Shallow reeding. Bead upright or sloping slightly inwards. Finger depression spout. Typologically AD 250–400; other site date late third to fourth century. Joins L379 (177), F200, Period 4. **Type 38**

M20 F324, L659 (117). AD 230–400.

M21 F435, L820 (120), L822 (120). After AD 200. **Untyped**

M22 F435, L820 (120). AD 230–400.

M23* F438, L827 (120). Well-rounded flange, reeded throughout and beginning to slope downward and outward. High bead. Typologically probably third century. Same vessel: L2303 (200), F1052. Period 4. **Type 30**

M24 F438, L827 (120). AD 230–400.

M25 F438, L827 (120). AD 230–400.

M26* F438, L833 (120). Thick, stubby reeded flange, triangular in section. High bead. Thick wall. Typologically third to early fourth century. **Type 60**

M27 F438, L833 (120). AD 230–400.

M28 F453, L912 (122). AD 230–400.

M29* F471, L927 (108). Similar to M123, but the whole rim is more nearly horizontal. The end of the flange appears to have been neatly turned under to join the body at the point where it begins. Finger depression spout. Typologically late third or fourth century. Possibly the same vessel as M135. Spout not drawn. **Type 37**

M30 F481, L940 (122). Large part of the body of an unusually small mortarium. Iron slag trituration grit finely fragmented, though not closely packed, and dispersed evenly over the whole interior right up to the bottom of the bead. The solely iron slag grit indicates a date later, rather than earlier, than AD 200. Joins L924 (196), F470, and L939 (194), F481. Both Period 4. **Untyped**

M31* F500, L901 (116). OXW. Small, squat flange somewhat rectangular in section, probably formed by folding flange underneath into itself. Very high bead, often grooved, fairly upright, leaning outward. No distal bead. Finger depression spout. Type Young M22. Typological and other site dating: AD 240–400. **Type 85/87**

M32* F500, L1958 (116). Typologically AD 250–350. **Type 35**

M33 F500, L1958 (116). Body and bead fragment. AD 230–400. **Untyped**

M34* F500, L1958 (116). OXW? Downward and outward pointing flange with grooves. High bead. Type Young M21.3. Typologically third century; other site dating: AD 240–300. **Type 81**

M35* F540, L964 (120). LR. Wall-sided with faintly differentiated bead and very deep collar. Typologically *c.* AD 150–300; (MacDonald and Curle 1929, 526, fig. 92, nos 33 and 34; Haupt 1984, *Taf.* 172, 183 *etc.*). Joins Yard L1126 (134), Period 3; F502, L963 (193), Period 4, and hand-stripping L1116 (—). **Type 79**

M36 F547, L1107 (121). AD 230–400.

M37 F675, L1889 (134). OXW. AD 100–400, but unlikely to be earlier than AD 180. **Untyped**

M38 F772, L1428 (133). As the last. **Untyped**

M39 F772, L1428 (133). AD 230–400.

M40* F801, L1380 (207). Wall-sided with a deep collar divided into 'reeds' separated by unusually wide grooves. Finger depressed spout. Typical of the Lower Nene Valley, but would be unusual in the Mancetter-Hartshill potteries. Typologically AD 250–400. The context is also assigned to Period 4. **Type 53**

(Fig. 115)

M41* F964, L2114 (134). Small flange with slightly concave ledge where it meets a very high bead. Spout formed by depressing the bead with a finger. Typologically third century, and possibly later. **Type 12**

M42 F1017, L1944 (128). AD 230–400.

M43 F1047, L2543 (134). AD 100–400, but unlikely to be earlier than AD 180. **Untyped**

M44 F1089, L2924 (136). AD 230–400.

M45* F1090, L2871 (136). Typologically AD 250–400. Mancetter-Hartshill origin is unlikely but not impossible. **Type 50**

M46* F1090, L2873 (136). Rounded flange, with a distal end somewhat flattened and turned sharply inward. Very high bead, facing inward. The bead is cut away on the inside of the spout as in second-century examples, and almost certainly indicates a date before AD 250. The same basic form, with a lower bead, was made in the Antonine period. Typologically AD 170–240. **Type 17**

M47* F1090, L2873 (136). Narrow, almost horizontal, reeded flange. High bead. Typologically AD 250–400. **Type 45**

M48 F1094, L2359 (125). AD 230–400.

M49 F1109, L2726 (148). Rim fragment. Probably of third-century manufacture. **Untyped**

M50 F1109, L2726 (148). AD 230–400.

M51* Yard, L1031 (134). Typologically AD 250–350. **Type 35**

M52 Yard, L1031 (134). AD 230–400.

M53* Yard, L1031 (136). OXW. Similar to M31, but flange less thick and narrow. Typological and other site dating: AD 240–400. **Type 85**

M54 Yard, L1267 (134). Topsoil?, L1268 (134). Plain, wall-sided, and colour-coated, similar to samian form Dr.45 and Young type C97. Spout probably as with these forms, sometimes with a mock lion or bat head. There may be vessels without a spout or with an undrilled spout. Typologically AD 250–400; other site dating: (Oxfordshire) 240–400. **Type 56**

***** Yard, L1126 (134). See M35. **Type 79**

M55 Yard, L1834 (134). AD 100–400, but unlikely to be earlier than AD 180. **Untyped**

M56 Yard, L1859 (134). After AD 200. **Untyped**

M57* Yard, L2011 (134). Greyish cream fabric with moderate inclusions of sub-rounded quartz and sparse brownish-black iron compound which often stains the surrounding clay. No certain trituration grit survives. Traces of a pinkish-brown slip. The fragmentary impression could be from a die of GERMANVS 2 (with ER and probably MA ligatured). The only other stamp from this die is from Lancaster. The brownish slip which is associated with both these mortaria would best fit an origin in the East Midlands. The single extant rim profile would best fit a date in the period *c.* AD 90–140. **Untyped**

M58 Yard, L2501 (134). AD 230–400.

M59 Yard, L2540 (134). AD 230–400. Possibly joins M60. **Type 68**

M60* Yard, L2541 (134), L2561 (134). Nearly horizontal, reeded flange, ending in a thickened, and often tilted up, distal 'bead'. High bead. Typologically late third and fourth century. Possibly joins M59. **Type 47**

M61 Building 3, L2847 (138). AD 200–400.

M62 Topsoil?, L1268 (134). Stubby, thick horizontal flange. Bead above flange. Form of spout variable. Typologically third century? **Type 68**

M63 Topsoil?, L1268 (134). Several fragments from perhaps more than one mortarium. Rim too fragmentary for close dating. AD 200–400. **Untyped**

M64* L1352 (134). Unassigned. Similar to M7, but flange less flat and bead less high. Bead not prominent. Distinctive and rather unusual. Style of Conrilus (Potter and Potter 1982, 78–9, fig. 39, no. 300). Typological date AD 120–170; other site dating: AD 130–160. **Type 4**

Period 4

M65* F148, L237 (168). OXW. Type Young M21.1. Related to M2 and M34. Deep groove at distal end. Typologically third century? Other site dating AD 240–300. Probably same vessel as one from L1591 (—), unassigned. **Type 82**

M66 F148, L238 (166), L337 (168). Two flakes from a large, thick flange. Insufficient survives to identify the rim profile, but it is unlikely to be earlier than AD 150. The fabric is not readily identifiable, but is either Lower Nene Valley or the Rhineland. **Untyped**

M67* F148, L238 (166). OXW. AD 240–400. **Type 85**

M68 F164, L262 (157). AD 230–400.

M69 F164, L286 (157). AD 230–400.

M70 F189, L339 (161). Reeded flange fragment. AD 250–350.

M71* F200, L341 (177). Typologically AD 250–350. **Type 35**

M72* F200, L341 (177). Thin, delicate version of this type. Trituration grit finely fragmented. Joins F200, L588 (277), Period 5. Typologically AD 250– 350. **Type 35**

M73* F200, L341 (177). Typologically late third or fourth century. **Type 37**

M74 F200, L342 (177). AD 230–400.

M75* F200, L343 (177). Thin, high and wide, almost horizontal, reeded flange; distal end folded under and flattened. High bead. Typologically late third or fourth century. Same vessel L918 (292), Rectangular Building, Period 5. **Type 42**

M76 F200, L343 (177). AD 230–400.

***** F200, L379 (177). Joins M19. **Type 38**

M77* F200, L486 (177). As M29, but with upright bead and plain flange instead of reeding. Joins F200, L1363 (277), Period 5. **Type 65**

M78* F242, L457 (173). Typologically late third or fourth century. **Type 37**

M79 F354, L688 (181). Second century.

M80 F434, L819 179). Wide sloping reeded rim. The angle of rim to body is very unusual. Typologically late second to third century. **Type 63**

(Fig.116)

M81* F434, L819 (179). Bead and flange virtually horizontal. The flange might still be regarded as reeded, but the Lower Nene Valley practice of fluting rather than reeding is indicated. Typologically late third to fourth century. Same vessel Yard, L1782 (316). **Type 41**

M82 F434, L819 (179). AD 250–400. **Untyped**

F470, L924 (196). See M30. **Untyped**

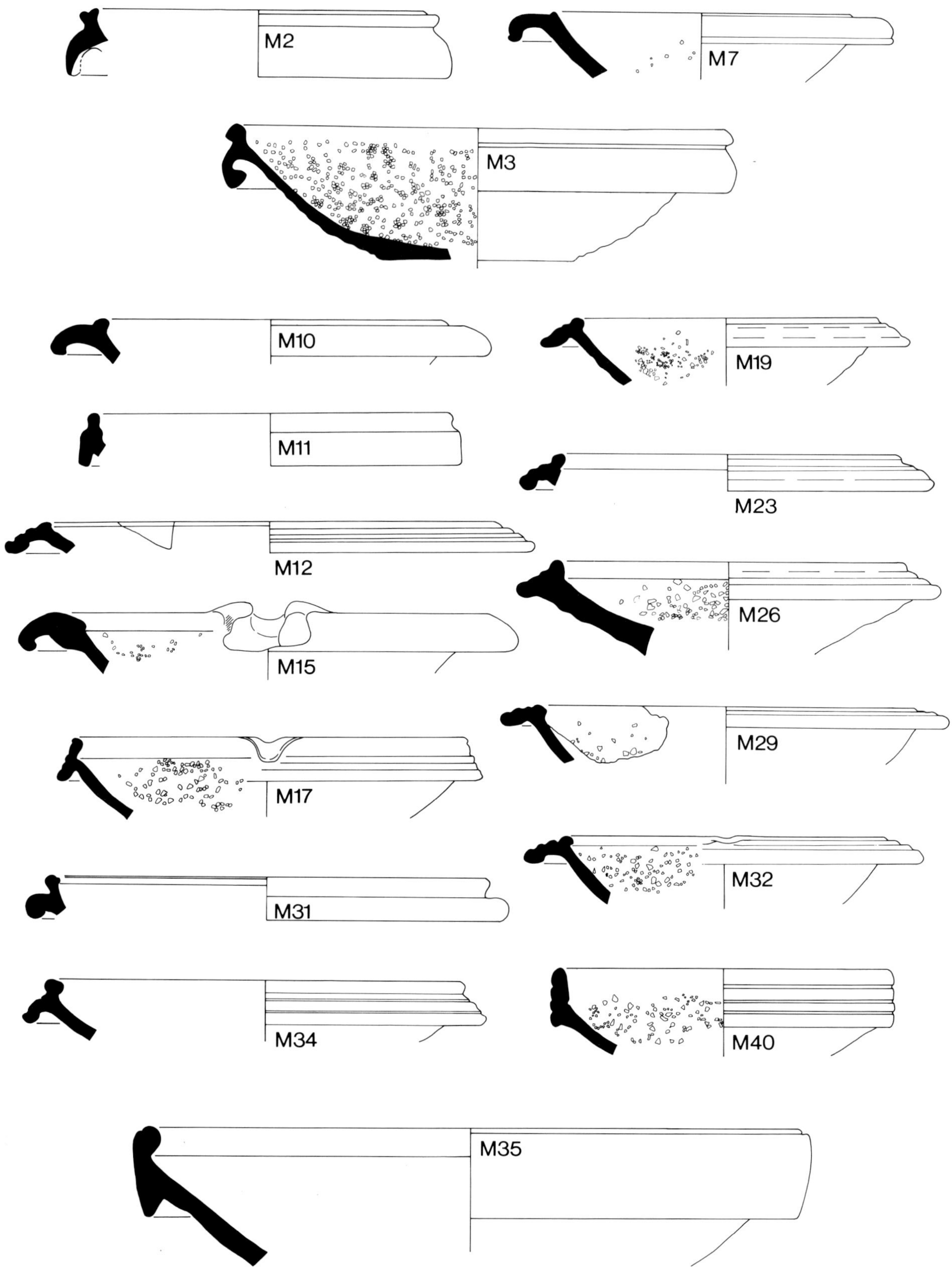

Figure 114 Roman mortaria. Periods 1, 2 and 3. Scale 1:4.

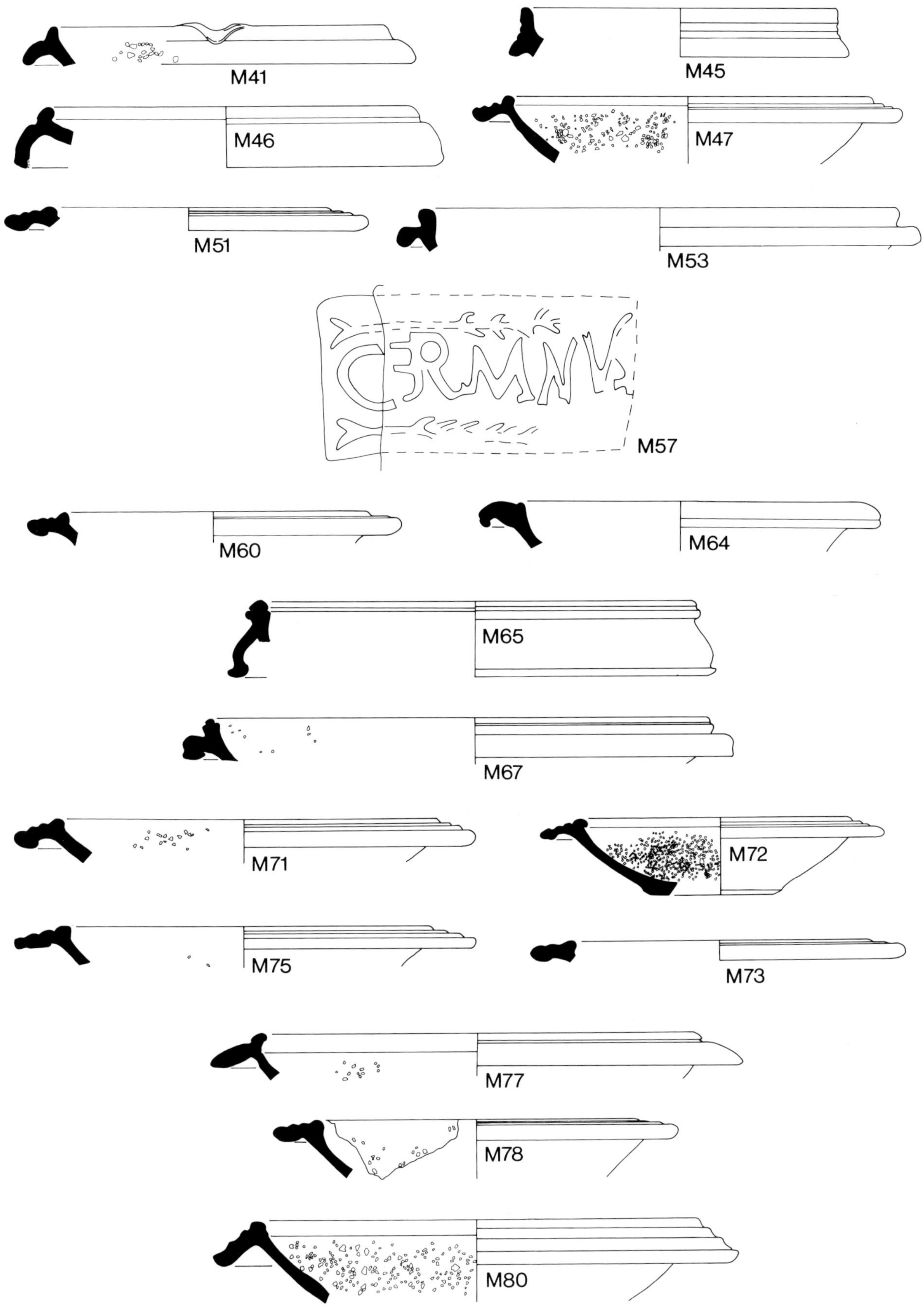

Figure 115 Roman mortaria. Periods 3 and 4. Scale 1:4.

***** F481, L939 (196). See M30. **Untyped**
M83* F482, L984 (194). OXW. Similar to M53, but smaller flange, squarish in section. Type Young M22.14. Typological and other site dating: AD 240–400. **Type 88**
M84 F482, L991 (194). AD 230–400.
M85 F500, L911 (178). OXCC. Typologically AD 250–400. **Type 56**
M86* F500, L1043 (178). Narrow and shallow flange with shallow grooving. Large, inward facing bead. Typologically third century. **Type 26**
***** F502, L963 (193). LR. See M35. **Type 79**
M87* F542, L1002 (194). Typologically late third and fourth century. **Type 47**
M88* F'542', L1132 (194). Reeded flange turned under, and folded rather flatly up to or near the point where the body begins. Bead usually fairly upright but sloping slightly inward. Typologically AD 230–350+. The consistency in the context dates of this form at Orton Hall Farm (see M142, M168, M183) suggests that the type is of late third and fourth-century date. **Type 33**
M89 F600, L1168 (186). Third century?
M90 F632, L1233 (198). AD 100–400, but unlikely to be earlier than AD 180. **Untyped**
M91 F733, L1369 (189). Typologically AD 250–350. **Type 53**
M92* F772, L1427 (207). OXW. Typological and other site dating: AD 240–400. **Type 85**
M93 F801, L1376 (207). AD 230–400.
M94 F801, L1377 (207). AD 100–400, but unlikely to be earlier than AD 180. **Untyped**
***** F801, L1380 (207). See M40. **Type 53**
M95 F810, L1450 (191). AD 100–400, but unlikely to be earlier than AD 180. **Untyped**
M96* F814, L1797 (206). Wide, thin, almost horizontal, reeded flange. High bead. Typologically AD 230–350. **Type 43**
M97 F911 (Yard), L1887 (203). AD 230–400. **Type 60**
***** F1052, L2303 (200). Probably same vessel as M2. **Type 80**
***** F1052, L2303 (200). See M26. **Type 60**

Period 5
M98* F11, L14 (274). Typologically AD 250–400; other site dating: (Oxfordshire) AD 240–400. **Type 56**
M99* F14, L87 (213). OXCC. Typologically AD 250–400; other site dating: AD 240–400. Type Young C97 with rouletting. **Type 56**
M100 F14, L200 (213). AD 230–400.
M101* F30, L43 (241). Typologically AD 250–400. **Type 50**
M102* F31, L212 (213). Typologically AD 250–400. **Type 50**
M103 F31, L212 (213). OXW. Typological and other site dating: AD 240–400. **Type 87**
M104 F31, L254 (213). AD 100–400, but unlikely to be earlier than AD 180. **Untyped**
M105 F31, L255 (213). AD 230–400.
M106 F31, L473 (213). AD 230–400.
M107* F31, L474 (213). Horizontal, reeded flange ending in a thickened bead. Finger depression spout. Typologically probably fourth century. **Type 48**
M108* F55, L109 (242). Typologically AD 250–400; other site dating: (Oxfordshire) AD 240–400. **Type 56**
M109 F55, L109 (242). AD 230–400.
M110 F79, L114 (245). Typologically probably AD 250–400. **Type 56**
M111* F80, L120 (240). Typologically third century, and possibly later. **Type 53**
M112 F80, L120 (240). AD 100–400, but unlikely to be earlier than AD 180. **Untyped**
M113* F80, L120 (240). Wide, shallow and convexly curved flange with shallow grooving. Bead above flange, facing inwards. Typologically AD 200– 250. **Type 24**
M114 F80, L120 (240). AD 230–400.
M115* F81, L121 (243). Typologically third century. **Type 26**
M116 F81, L121 (243). Typologically AD 250–400; other site dating: (Oxfordshire) AD 240–400. Fine textured fabric. **Type 56**
M117 F81, L121 (243). AD 230–400.
M118* F81, L148 (243). Typologically AD 200–250. **Type 24**
M119* F81, L163 (243). Typologically third century. **Type 26**
M120 F82, L122 (252). AD 100–400, but unlikely to be earlier than AD 180. **Untyped**
M121* F143, L664 (223). Typologically AD 250–400. **Type 50**
M122* F162, L363 (255). Typologically AD 250–350. **Type 53**
M123* F171, L273 (255). Similar to M12, but with a broader distal end of flange. Typologically AD 250–350. **Type 34**
M124 F171, L273 (255). AD 230–400.
M125* F172, L282 (259). Reeded flange which tends to be convex. Underside of flange usually rounded but can be straight. Bead high and more prominent than on other reeded mortaria, and in the same plane as the flange. Typologically AD 230–350+. **Type 36**
M126 F172, L282 (259). AD 230–400.
M127 F172, L283 (259). AD 230–400.
M128* F200, L313 (277). Similar to M17, but without undercutting beneath collar; almost wall-sided. Finger depression spout. A type also made in the Mancetter-Hartshill potteries. Typologically AD 250–350. **Type 51**
M129 F200, L313 (259). AD 230–400.
***** F200, L588 (277). See M72. **Type 35**
M130 F200, L589 (277). AD 230–400.
M131* F200, L590 (277). Typologically AD 250–350. **Type 35**
M132 F200, L590 (277). AD 230–400.
M133 F200, L669 (277). AD 230–400.
M134* F200, L1363 (277). Typologically AD 230–350+ Possibly two vessels. Chocolate-brown colour-coat. **Type 36**

(Fig.117)
M135* F200, L1363 (277). Possibly same vessel as M29? Joins: Building 4, L920, Period 5. **Type 37**
M136* F200, L1363 (277). Similar to M29, but with a much wider rim, tapering at the distal end. Finger depression spout. Typologically late third and fourth century. **Type 39**
***** F200, L1363 (277). Joins M77. **Type 65**
M137 F203, L345 (213). AD 100–400, but unlikely to be earlier than AD 180. **Untyped**
M138* F203, L475 (282). Typologically AD 250–400. **Type 50**
M139* F203, L475 (282). OXW. Horizontal flange turned under, but not fused to underside of flange. High bead. Bead broken and turned out to form spout. Type Young M17.5. Typologically third century; other site dating: 240–300. Joins F224, L409 (282). **Type 84**
M140* F203, L475 (282). OXW. Narrow, horizontal flange with distal groove. Underfold of rim is completely welded in, unlike M53 where it can still be clearly distinguished. High, grooved bead. Finger depression spout. Type Young M22.11. Typological and other site dating: AD 240–400. **Type 86**
M141 F203, L475 (282). AD 100–400, but unlikely to be earlier than AD 180. **Untyped**
M142* F203, L728 (213). Typologically AD 250–350+ Slightly unusual. **Type 33**
M143* F224, L409 (282). OXW. Typological and other site dating: AD 240–400. **Type 85**
M144 F224, L409 (282). OXW. Second century? See M139. **Type 84**
M145 F252, L481 (279). Typologically late third or fourth century. Trituration grit unusually tiny for a Lower Nene Valley product. **Type 37**
M146* F252, L481 (279). Trituration grit very tiny and closely packed. AD ?300–400. **Untyped**
M147 F252, L494 (279). AD 230–400.
M148 F254, L487 (280). Wall-sided fragment. Probably AD 250+.
M149 F536, L1154 (300). Typologically AD 250–350. **Type 51**
M150* F675, L1735 (316). Narrow, neatly rounded flange. High bead. Unusual form for Lower Nene Valley, approximating to Young type M22. Most likely to be the product of an Oxfordshire potter working in the Nene Valley, rather than a local imitation. Trituration grit entirely iron slag. Typologically AD 240–400. **Type 74**
M151 F675, L1772 (316). AD 230–400.
M152* F751, L1349 (308). Typologically late third and fourth century. **Type 39**
M153 F801, L1455 (296). AD 230–400.
M154 F898, L1830 (309). AD 100–400, but unlikely to be earlier than AD 180. **Untyped**
M155 F991, L2275 (318). AD 230–400.
M156 F991, L2285 (318). AD 230–400.
M157 F1061, L2574 (319). AD 230–400.
M158 F1061, L2576 (319). Typologically AD 250–400. **Type 50**
M159 F1061, L2576 (319). Typologically AD 250–350. **Type 51**
M160 F1065, L2564 (320). Typologically AD 250–350. **Type 51**
M161 F1065, L2597 (320). AD 230–400.
M162 F1065, L2597 (320). AD 230–400.
M163* F1065, L2738 (320). Wide, plain flange, slightly uptilted at the distal end. High bead. Typologically fourth century, perhaps AD 350–400+; other site dating: late fourth century (Corder 1951, fig. 9, no. 29; Frere 1972, fig. 109, no. 2631). **Type 69**
M164 F1065, L2777 (320). AD 230–400.

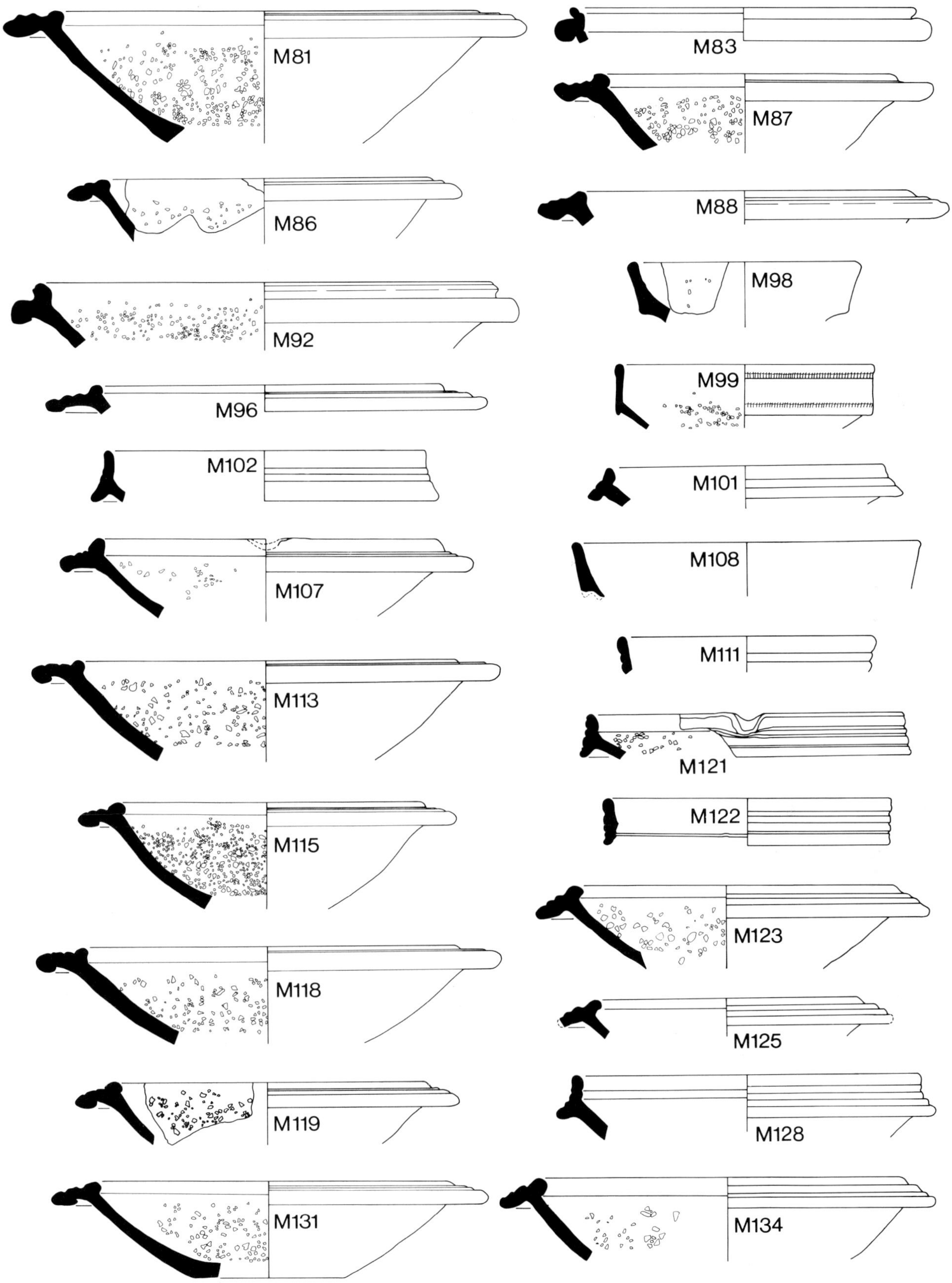

Figure 116 Roman mortaria. Periods 4 and 5. Scale 1:4.

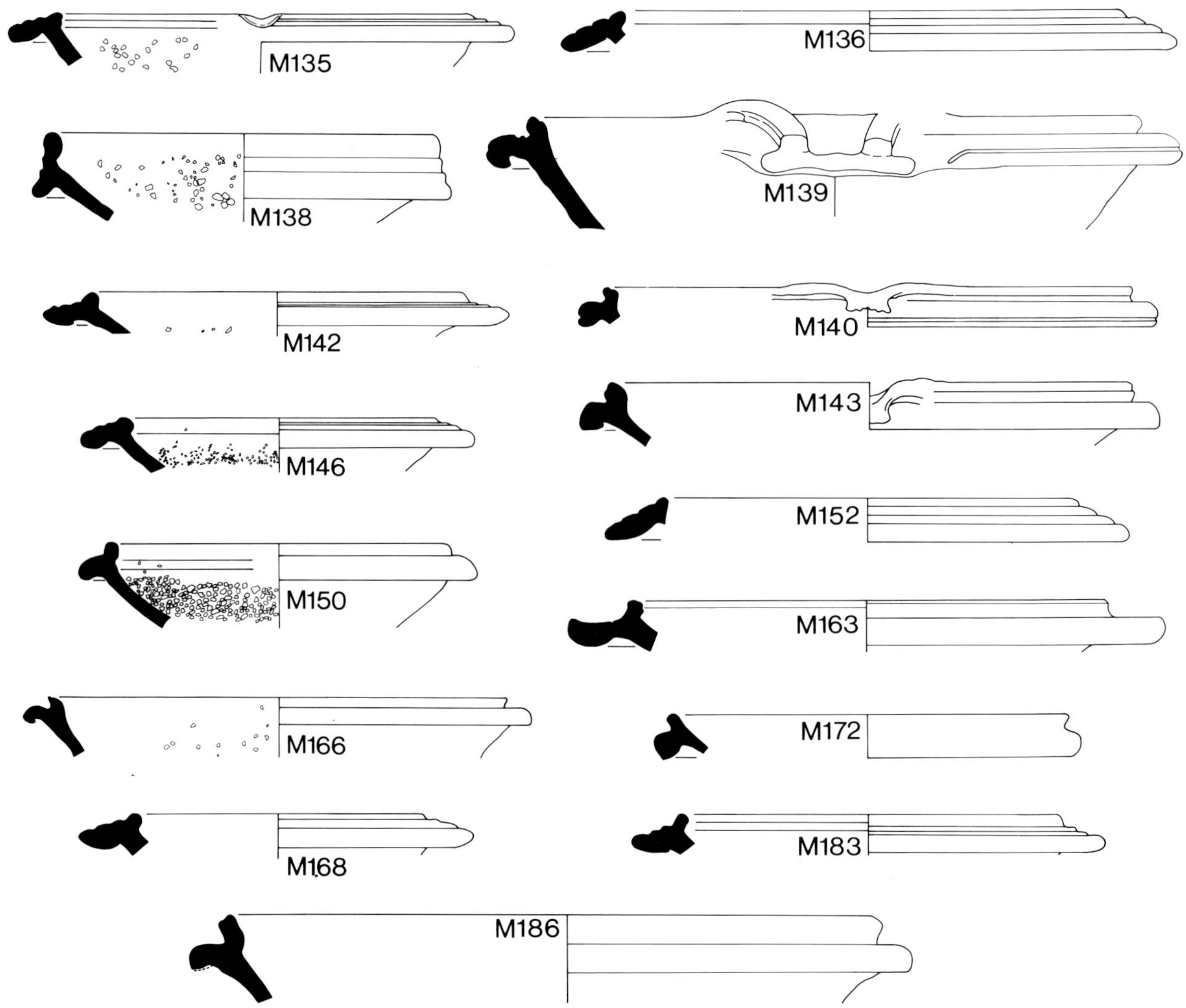

Figure 117 Roman mortaria. Period 5. Scale 1:4.

M165 F1065, L2777 (320). AD 230–400.
M166* F1129, L2915 (324). OXW. Small, delicate flange turned under, and raised above horizontal. High, grooved bead. Type Young M22. Typological and other site dating: AD 240–400. **Type 89**
M167 F1149, L2907 (324). AD 230–400.
M168* Rectangular building, L574 (292). Typologically AD 250–350+. **Type 33**
M169 Rectangular building, L574 (292). Typologically AD 250–350. Joins L918 (292). **Type 35**
M170 Rectangular building, L574 (292). Early third century. **Type 34**?
M171 Rectangular building, L574/575 (292). AD 230–400.
M172* Rectangular building, L585 (293). OXW. Typological and other site dating: AD 240–400. **Type 86**
M173 Rectangular building, L585 (293). AD 230–400.
M174 Rectangular building, L918 (292). AD 230–400.
Rectangular building, L918 (292). See M169. **Type 35**
* Rectangular building, L918 (292). See M75. **Type 42**
M175 Rectangular building, L920 (291). AD 230–400.
M176 Rectangular building, L920 (291). AD 230–400. **Type 37**
* Rectangular building, L920 (291). Joins M135. **Type 37**
M177 Rectangular building, L1056 (293). OXCC. AD 240–400. Same vessel as M85? **Untyped**
M178 Rectangular building, L1056 (293). AD 230–400.
M179 Yard, L1244 (316). AD 230–400.
M180 Yard, L1782 (316). AD 230–400.
* Yard, L1782 (316). See M81. **Type 41**
M181 Yard, L1824 (316). AD 230–400.
M182 Yard, L2008 (316). AD 230–400.
M183* AF48, L89 (219). Typologically AD 230–350+. **Type 33**
M184 AF48, L91 (219). AD 230–400.
M185 Hand-stripping, L587 (277). AD 230–400.
M186* Hand-stripping, L1980 (316). OXW. Typological and other site dating: AD 240–400. **Type 85**
M187 Unassigned L512 (286). AD 230–400.
M188 Unassigned L2866 (328). AD 230–400.

Hand-stripping
M189 L1027. As M29, but bead higher than on other examples. **Type 37**
M190 L1027. AD 100–400, but unlikely to be earlier than AD 180. **Untyped**
* L1116. See M35.
M191 L2757. AD 230–400.

Machine-stripping
M192 L1362. OXW. Typologically third century. **Type 84**
M193 L1362. OXWCC. Young type WC7. AD 240–400+. **Type 86**

Dakin backfill
* L1591. OXW. See M65. **Type 82**
M194 L1873. AD 230–400.

Unassigned
M195 AF35, 1. AD 230–400.
M196 AF35, 2. AD 230–400. **Type 51**
M197 Area III, I. See M40. **Type 53**
M198 Area IV, 1. AD 230–400.
M199 Area IV, 14. Typologically AD 250–400. **Type 50**
M200 Area IV, 38. First half of third century?

Clearance
M201 L2910. Area of F1025, Period 1. Typologically AD 250–350. **Type 53**

A type series for mortaria found in the Lower Nene Valley

by K.F. Hartley

The Fabrics (see Section IV)

Fabric 1

Castor-Stibbington area. Greyish white to pale brownish cream, sometimes with a darker or drab grey core. Variable hardness and inclusions including sparse to moderate quartz, red-brown fragments and occasional white or yellowish chalk. Trituration grit: entirely composed of blackish iron slag. An orange or darker red-brown slip may be present.

Fabric 2

Lower Nene Valley. Greyish cream, but may be yellowish or brownish cream, sometimes with a darker or drab grey core. Hard. Texture varies according to amount and size of inclusions which are usually of quartz, together with more occasional black, red-brown and white fragments. Trituration grit: quartz, black iron slag and red-brown fragments. Vessels can have an orange-brown slip.

The off-white Fabric 1 is most typical of third and fourth-century mortaria made in the Lower Nene Valley. The more extreme variations of this are included under Fabric 2 which is generally rather coarser in texture with larger and more ill-sorted inclusions. The trituration grit is also not always entirely iron slag. Most of the rim-profiles associated with Fabric 2 either pre-date the *floruit* of the mortaria industry in the Lower Nene Valley or are atypical. It is probable, therefore, that Fabric 2 encompasses the work of several minor workshops operating at various points within the second to fourth centuries.

Fabric 3

Lower Nene Valley. Orange-brown, sometimes with a drab grey core. Hard. Few inclusions, mainly quartz and occasional black and red-brown fragments. Trituration grit: entirely black iron slag.

Fabric 3 represents the second most commonly distinguishable mortarium fabric used in the Lower Nene Valley in the third and fourth centuries. It is, however, possible that the variation between it and Fabric 1 is more apparent than real and could be the result of differences in firing. Some mortaria exhibit what appears to be a combination of Fabrics 1 and 3 (here placed under Fabric 1), and one fragment from Stibbington has Fabric 1 and Fabric 3 in different parts of the same piece (Type 41, example (b)).

Fabric 4

Lower Nene Valley and perhaps Mancetter-Hartshill. Cream to brownish cream. Hard. Fine textured with sparse to moderate inclusions of quartz and red-brown, black, and occasional white fragments. Trituration grit: black iron slag, quartz and red-brown fragments. Traces of an orange-brown or cream slip.

Some of the second-century Lower Nene Valley fabrics can be either difficult or impossible to distinguish from some of those of the Mancetter-Hartshill potteries, especially where no trituration grit survives. As a result, some mortaria made by accredited Mancetter potters like Iunius are included in Fabric 4. There is a possibility, based on distributions and changes in rim-profiles, that Similis opened a workshop in the Lower Nene Valley. There is, however, no evidence that Iunius, or any other Mancetter-Hartshill potters, definitely worked in the Lower Nene Valley.

Fabric 5

Lower Nene Valley. Pinkish brown. Hard. Inclusions of translucent quartz with occasional black and red-brown fragments. No trituration grit survives. Dark greyish brown slip.

Fabric 6

Cambridgeshire, Norfolk or Lower Nene Valley. Orange-brown with drab core. Frequent inclusions of quartz, brownish black and opaque white fragments. Trituration grit: quartz and red-brown fragments. Thick cream slip.

Fabric 7

Mancetter-Hartshill. Creamy white with pinkish brown core. Softish. Moderate inclusions of quartz and red-brown fragments. Brownish iron stains are common. Trituration grit: quartz and red-brown fragments.

Fabric 8

Oxfordshire. Cream to brownish cream, sometimes pinkish or with a pinkish core. Sparse to moderate inclusions of quartz and red-brown fragments. Trituration grit: entirely quartz, transparent, translucent and opaque, often brownish.

Fabric 9

Oxfordshire. Red-brown and micaceous. Fine-textured. Few inclusions of tiny quartz and black fragments, with occasional white chalk fragments. Trituration grit: as Fabric 8. Red-brown slip.

Fabric 10

Verulamium region. Cream with a thick pinkish core. Granular texture. Frequent inclusions of quartz with occasional black and flint fragments. Trituration grit: flint and quartz.

Fabric 11

Rhineland. Very pale brownish-pink. Abundant temper. Moderate inclusions of tiny quartz with larger red-brown fragments, and occasional white chalk(?). No trituration grit survives.

The Types

Where a type is represented by a single example, the entry order is fabric, type description and comments, provenance, context date, and finally typological and other site dates, if these are available. If there is more than one example, each vessel with its provenance, context date and any specific comments is given below the general description and comments. Multiple entries are headed (a), (b) and so on to allow for correct identification of the particular example. Occasionally a type occurs in more than one fabric; in these cases, the fabric is omitted from the opening description and is noted in the individual entry. The bulk of the entries are vessels from Orton Hall Farm, and to avoid repetition, therefore, the period dates of this site are given below. The overlap of dates at the start and end of most periods is deliberate and emphasises the fact that the dates given are not fixed. It should also be noted that most layers contained earlier material, the proportion of which increases the later the period. Much of the Period 5 material, for example, is not unequivocally late fourth century. To allow easy identification of the provenance of the illustrated mortaria, all those from sites other than Orton Hall Farm are annotated with the appropriate site code. These are also given below.

Orton Hall Farm Period dates

Period 1 *c.* mid first century – AD 175
Period 2 *c.* AD 175 – 225/250
Period 3 *c.* AD 225/250 – 300/325
Period 4 *c.* AD 300/325 – 350/375
Period 5 post *c.* AD 375

Site codes

BB Durobrivae, Billing Brook
CH Durobrivae, Chesterton
WN Durobrivae, Water Newton
LF Lynch Farm
MH Mill Hill, March
ST Stibbington

Catalogue

1 Fabric 10. Hooked rim with flange rising above small bead. Shallow spout. Stamp attributed to MELVS (Frere 1972, 376, fig. 145, no. 28). Chesterton CH 7975-6 (M129-30). Grid O, L2, dated second to fourth century. Typological and other site dating: AD 80-130. Restored.

2 Fabric 5. Flange rises above bead, distal end flat and turned inwards. Form derived from Verulamium region (Frere 1972, fig. 115, nos 551–2). Stamp of otherwise unknown potter SABINVS. Chesterton CH 7977 (M 131), Grid O, L2, dated second to fourth century. Typological date *c.* AD 110–140/150.

3 Fabric 2. Fairly wide, shallow hook, turned slightly inward at distal end. Lowish bead. Mortarium of VIATOR. Water Newton WN 1181, Trench A, F13. Uncertain date. Typological and other site dating: AD 100–140/150.

4 Fabric 4 (Lower Nene Valley). Shallow flange ending in a distal bead. Bead not prominent. Distinctive and rather unusual. Style of Conrilus (Potter and Potter 1982, 78–9, fig. 39, no. 300). Orton Hall Farm **M64**, L1352, Period 3. Typological date AD 120–170. Other site dating: AD 130–160.

5 Fabric 2. As type 4 but bead higher and flange much flatter. Only five surviving trituration grits, all iron slag. Orton Hall Farm **M7**, L2550, F1048, Period 1 second-century deposit. Typological date AD 130–180. Other site dating: AD 130–175.

6 Flange rising slightly above the bead. Distal end of flange turned sharply inwards. Typological and other site dating: AD 100–145.
(a) Fabric 1. Chesterton CH 2774 (M44), Building 3, L3, dated: mid/late second into third century.
(b) Fabric 7 (probably). Orton Hall Farm **M15**, L2060–61, F955, Period 2. Unidentifiable stamp.

7 Fabric 2. Thick, wide shallow flange rising above a well-defined bead. Stamp of VARINNA. Chesterton CH 7406 (M120). Test Hole 80, F164 dated second to third century. Typological date: AD 130–180.

8 Fabric 4 (Lower Nene Valley?). Wide, well curved, and somewhat splayed flange rising above the bead. Probably the same potter who started work at Mancetter, and may have had a second workshop in the Lower Nene Valley *c.* AD 150–180. Typological date: AD 140–180. Other site dating: AD 130– 180.
(a) Water Newton (Oakham Museum). Undated. Stamp of SIMILIS.
(b) Chesterton CH 7239 (M113), Test Hole 50, F155 dated: second to third century.

9 Fabric 4 (Mancetter-Hartshill or Lower Nene Valley). Short, thick, curved flange, often stubby with a large bead. A typical Mancetter-Hartshill form. Typological and other site dating: AD 140–180.
(a) Chesterton CH 1350 (M23). Building 1, L4, dated: second to third century. Stamp of IVNIVS.
(b) Chesterton CH 4189 (M59). Building 3, F20, dated: second to fourth century.
(c) Chesterton CH 7463 (M123). Building 4, F184, L17, dated: mid/late second century. Stamp of a semi-literate potter.

10 Fabric 4 (probably Lower Nene Valley). Short, thickish sloping flange with a high, large bead. Water Newton WN 751, Trench A, L18, dated second century. Unidentified stamp. Typological date: AD 150–200.

11 Fabric 4 (Lower Nene Valley). Short, sharply angled flange with a high bead. Unusual form. Chesterton CH 1977 (M35). Building 4, L2, dated: second-fourth century. Typological date: AD 150–200.

12 Fabric 1. Small flange with slightly concave ledge where it meets a very high bead. Spout formed, unusually, by depressing the bead with a finger. Orton Hall Farm **M41**, L2114, F964, Period 3. Typologically third century, and possibly later.

13 Fabric 2. Unusual beadless form, normally associated with raetian-type mortaria. Chesterton CH 919 (M18), Building 1, Pit F8, dated: late second to mid-third century, and CH 1142 (M21), Building 1, south-west extension, L2, dated: second to fourth century. Typologically second century.

14 Fabric 4 (Lower Nene Valley). Well curved, shallow flange with a high bead. Thick shoulder. Chesterton CH 8899 (M160), Grid Q, Pit F265, dated: second quarter second century. Typologically AD 150–200.

15 Fabric 4 (Mancetter-Hartshill or Lower Nene Valley). Well curved flange turned inward at distal end. Moderate bead higher than flange. Water Newton 1187, Trench A, L18, dated: second century. Typologically AD 140–200.

16 Flange slopes quite sharply down from a horizontal ledge next to the bead, and turns sharply inwards at the distal end. High upright or inward-facing bead. This example never stamped.
(a) Fabric 1? Chesterton CH 2966 (M48) and 2968 (M50), Building 4, Pit F44, dated: second to fourth century. Typologically AD 150–220.
(b) Fabric 4 (Lower Nene Valley). Chesterton CH 3136 (M52), Building 4, F36, dated: second to third century. Typologically AD 120–170.

17 Rounded flange, with a distal end somewhat flattened and turned sharply inward. Very high bead, probably facing inward. The bead is cut away on the inside of the spout as in second-century examples, and almost certainly indicates a date before AD 250. The same basic form, with a lower bead, was made in the Antonine period.
(a) Fabric 1. Chesterton CH 1809 (M32), Building 2, F19, dated: second to fourth century. Typologically AD 170–240/250.
(b) Fabric 4 (Lower Nene Valley). Orton Hall Farm **M46**, L2873, F1090, Period 3. This vessel is thinner and more delicately moulded than (a). Typologically AD 170–240.

18 Fabric 1. Similar to type 17, but flange wider, and sloping not rounded. Chesterton CH 3664 (M8), Building 1, L3, dated: second to fourth century. Typologically AD 180–250+.

19 Wide shallow flange, curled sharply inward and coming to a point at the distal end. High bead pointing inward. Typologically later than AD 180.
(a) Fabric 4 (Lower Nene Valley). Orton Hall Farm **M10**, L361, F217, Period 2.
(b) Fabric 1? (Stibbington Kiln). Stibbington ST 57.2. Date: fourth century. Thicker behind bead than (a).

20 Wide shallow flange sloping rather weakly and curled slightly inward at distal end. Bead above flange. Typologically after AD 180.
(a) Fabric 1? (Stibbington Kiln). Stibbington ST 57.12. Date: fourth century.
(b) Fabric 4 (Castor-Stibbington). Chesterton CH 2965 (M49), Building 4, Pit F44, dated: second-fourth century.

21 Fabric 1? Wide, almost horizontal, flange hooked under at the end. High bead. Stibbington ST. Typologically after AD 200.

22 Fabric 1? (Stibbington Kiln). A later version of Type 8. Wider, more horizontal, and splayed flange; larger and bolder, inward facing bead. Stibbington ST 57.12. Typologically later than AD 180.

23 Fabric 4 (Castor-Stibbington). Curved convex flange with deep grooves just beyond the change in angle. High bead. Chesterton CH 2658 (M38), Building 3, L3, dated: mid/late second-third century. Typologically AD 180–250.

24 Fabric 4 (Castor-Stibbington). Wide, shallow and convexly curved flange with shallow grooving. Bead above flange, facing inwards.
(a) Orton Hall Farm **M113**, L120, F80, Period 5.
(b) Orton Hall Farm **M118**, L148, F81, Period 5. Typologically AD 200–250.

25 Fabric 1? (Stibbington Kiln). Wide, convexly curved flange with wide, deeply cut grooves. Stibbington ST 57.2. Date: fourth century. Other site dating: *c.* AD 275–325.

26 Fabric 1. Narrow and shallow flange with shallow grooving. Large, inward facing bead. Typologically third century.
(a) Orton Hall Farm **M86**, L1043, F500, Period 4.
(b) Orton Hall Farm **M115**, L121, F81, Period 5.
(c) Orton Hall Farm **M119**, L163, F81, Period 5.

27 Fabric 1. Wide, shallow, well-rounded flange, with end turned sharply inward. Moderately deeply reeded overall. High, large bead. Billing Brook BB 143 (M7). Area 2. Test Hole 30, L2, dated: third-fourth century. Typologically third century, AD 200–260?

28 Fabric 1. Well curved, reeded flange, with horizontal ledge next to bead, and incised wavy line on flange. High, inward-facing bead. Billing Brook BB 144 (M8), Area 2, Test Hole 30, L2, dated: third to fourth century. Typologically AD 250–350.

29 Fabric 1. Well-rounded flange, reeded throughout. High bead. Water Newton WN 2139, Trench A, L26, dated: probably third century. Typologically third century.

30 Fabric 1. As Type 29, but flange not so convex. Flange beginning to slope downward and outward. Typologically probably third century.
(a) Orton Hall Farm **M23**, L827, F438, Period 3.
(b) Chesterton CH 6056 (M97), Cutting O. Uncertain date. Bead broken and turned out over flange to form spout.
(c) Billing Brook BB 44 (M2), Area 1, Test Hole 9, L2, dated: second to fourth century.

31 Fabric 1. Flange straighter than Type 28, coming almost to a point; distal end flat, wide and rounded at the inside edge. Typologically AD 230–350.
(a) Water Newton WN 33 Trench A, L1. Ploughsoil.
(b) Chesterton CH 3745 (M10), Building 1, L7, dated: second to third century.

32 Fabric 1. Exactly as No.31 but with incised wavy line on flange. Chesterton CH 8820 (M158), Grid U, L2, dated: second to fourth century. Typologically after AD 275 and likely to be fourth century. Brown, metallic slip.

33 Fabric 1. Reeded flange turned under, and folded rather flatly up to or near the point where the body begins. Bead usually fairly upright but sloping slightly inward. Typologically AD 230–350+. The consistency in the context dates of these mortaria at Orton Hall Farm suggests that the type is of late third and fourth century date.
(a) Orton Hall Farm **M88**, L1132, F'542', Period 4.
(b) Orton Hall Farm **M142**, L728, F203, Period 5. Slightly unusual.
(c) Orton Hall Farm **M168**, L574, Building 4, Period 5.
(d) Orton Hall Farm **M183**, L89, AF48, Period 5. Coin C27, AD 330–335.

34 Fabric 1. True hammerhead with bead and flange in exact alignment; distal end of flange turned under and up. Bead broken and turned out over flange to form spout. Orton Hall Farm **M123**, L273, F171, Period 5. Typologically AD 250–350.

35 Fabric 1. Slightly convex, reeded rim ending in a knob; bead in same place as flange. Finger depression spout. Typologically AD 250–350.
(a) Orton Hall Farm **M12**, L361, F217, Period 2. Brown-buff slip.
(b) Orton Hall Farm **M32**, L1958, F500, Period 3.
(c) Orton Hall Farm **M51**, L1031, Yard, Period 3.
(d) Orton Hall Farm **M71**, L341, F200, Period 4.
(e) Orton Hall Farm **M72**, L341, F200, Period 4, and L588, F200, Period 5. Thin, delicate version. Trituration grit finely fragmented.
(f) Orton Hall Farm **M131**, L590, F200, Period 5.
(g) Stibbington (not in kiln).

36 Fabric 1. Reeded flange which tends to be convex. Underside of flange usually rounded but can be straight. Bead high and more prominent than the other reeds; and on the same plane as the flange. Typologically AD 230–350+.
(a) Orton Hall Farm **M125**, L282, F172, Period 5. Underside of flange straight.
(b) Orton Hall Farm **M134**, L1363, F200, Period 5. Possibly two vessels. Chocolate-brown colour-coat.
(c) Chesterton CH 154 (M3), Building 1, L2, dated: second to fourth century. Bead much more upright.

37 The whole rim is more nearly horizontal than in Type 34. The end of the flange appears to have been neatly turned under to join the body at the point where it begins. Finger depression spout. Typologically late third or fourth century.
(a) Possibly **M29**, L927, F471, Period 3.
(b) Fabric 1. Orton Hall Farm **M73**, L341, F200, Period 4.
(c) Fabric 1. Orton Hall Farm **M78**, L457, F242, Period 4.
(d) Fabric 1. Orton Hall Farm **M135**, L1363, F200, Period 5.
(e) Fabric 1. Orton Hall Farm **M145**, L481, F252, Period 5. Trituration grit unusually tiny for a Lower Nene Valley product.
(f) Fabric 1. Orton Hall Farm **M176**, L920, Building 4, Period 5.
(g) Fabric 3. Orton Hall Farm **M189**, L1027, hand-stripping layer, not dated. Bead higher than on other examples.
(h) Fabric 1. Water Newton WN 3, Trench A, L1, ploughsoil. Buff to brown slip.
(i) Fabric 3. Chesterton CH 152 (M1), Building 1, L2, dated: second to fourth century.

38 Flange tapers at distal end, and the underside is welded in to give an elegant line. Shallow reeding. Bead upright or sloping slightly inwards. Finger depression spout. Typologically AD 250–350+. Other site dating: late third to fourth century.
(a) Fabric 1. Orton Hall Farm **M19**, L514, F262, Period 3.
(b) Fabric 1. Orton Hall Farm **M76**, L379, F200, Period 4.
(c) Fabric 1? (Stibbington Kiln), Stibbington ST 57.2, dated: fourth century.

39 Fabric 1. Similar to Type 37 but with a much wider rim, tapering at the distal end. Finger depression spout. Typologically late third and fourth century.
(a) Orton Hall Farm **M136**, L1363, F200, Period 5.
(b) Orton Hall Farm **M152**, L1349, F751, Period 5.

40 Fabric 1. Identical to Type 39, except that the underside of the flange is welded in as with Type 38. Typologically late third or fourth century.
(a) Stibbington (not in kiln).
(b) Chesterton CH 8450 (M147), Grid T, L2, dated: second-fourth century.
(c) Billing Brook BB 61 (M3), Area 1, Test Hole 10, L3, dated: third-fourth century.

41 Bead and flange virtually horizontal. The flange might still be regarded as reeded, but the Lower Nene Valley practice of fluting rather than reeding is indicated in (a). (b) is definitely fluted. Typologically late third or fourth century.
(a) Fabric 3. Orton Hall Farm **M81**, L819, F434, Period 4.
(b) Fabric 1/3. Stibbington ST 57.8, date: fourth century. Both fabrics occur in this vessel.

42 Fabric 3? Thin, high and wide, almost horizontal, reeded flange; distal end folded under and flattened. High bead. Orton Hall Farm **M75**, L343, F200, Period 4. Typologically late third or fourth century.

43 Fabric 3. Wide, thin, almost horizontal, reeded flange. High bead. Orton Hall Farm **M96**, L1797, F814, Period 4. Typologically AD 230–350.

44 Fabric 1. Narrow, somewhat convex, reeded flange. High bead. Chesterton CH 5745 (M81), Building 3 F90, L2, dated: fourth century. Typologically AD 230–350.

45 Fabric 1. Narrow, almost horizontal, reeded flange. High bead. Orton Hall Farm **M47**, L2873, F1090, Period 3. Typologically late third or fourth century.

46 Fabric 1. Narrow, horizontal, reeded rim. Stibbington (not from kiln) ST 57.9. Typologically probably fourth century.

47 Fabric 1. Nearly horizontal, reeded flange, ending in a thickened, and often tilted up, distal 'bead'. High bead. Typologically late third and fourth century, probably later than AD 250.
(a) Orton Hall Farm **M60**, L2561, Yard, Period 3.
(b) Orton Hall Farm **M87**, L1002, F542, Period 4.
(c) Chesterton CH 6006 (M95), Building 3, F90, dated: fourth century.

48 Fabric 1. Horizontal, reeded flange ending in a thickened bead. High bead. Finger depression spout. Orton Hall Farm **M107**, L474, F31, Period 5. Typologically probably fourth century.

49 Fabric 1. Similar to Type 48, but wider and thinner flange. Sharply indented reeding. Higher bead. Chesterton CH 155 (M4), Building 1, L2, dated: second to fourth century. Typologically probably fourth century.

Types 50–6 Wall-sided and hammerhead mortaria

50 Reeded hammerhead mortarium indistinguishable in form from those made in the Mancetter-Hartshill potteries. Finger depression spout. Typologically AD 250–400.
(a) Fabric 1. Orton Hall Farm **M17**, L768, F31, Period 3.
(b) Fabric 1/4. Orton Hall Farm **M45**, L2871, F1090, Period 3. Mancetter-Hartshill origin not impossible.
(c) Fabric 1/4. Orton Hall Farm **M101**, L43, F30, Period 5.
(d) Fabric 1. Orton Hall Farm **M102**, L212, F31, Period 5.
(e) Fabric 1. Orton Hall Farm **M121**, L664, F143, Period 5.
(f) Fabric 1. Orton Hall Farm **M138**, L475, F203, Period 5.
(g) Fabric 1. Orton Hall Farm **M158**, L2576, F1061, Period 5.
(h) Fabric 1. Orton Hall Farm **M199**, Area IV 14. Unassigned.

51 Similar to Type 50, but without undercutting beneath collar; almost wall-sided. Finger depression spout. A type also made in the Mancetter-Hartshill potteries. Typologically AD 250– 350.
(a) Fabric 1. Orton Hall Farm **M128**, L313, F200, Period 5.
(b) Fabric 1. Orton Hall Farm **M149**, L1154, F536, Period 5.
(c) Fabric 1/4. Orton Hall Farm **M159**, L2576, F1061, Period 5.
(d) Fabric 1. Orton Hall Farm **M160**, L2564, F1065, Period 5.
(e) Fabric 1. Orton Hall Farm **M196**, AF35 2. Unassigned.
(f) Fabric 1. Chesterton CH 5634 (M77), Building 3, L1, dated: second to fourth century.

52 Fabric 1. Wall-sided. Reeded rim with all 'reeds' rounded in section. Bead broken and turned out to form spout. Type also made in the Mancetter-Hartshill potteries. Typologically probably third century.
(a) Chesterton CH 5545 (M76), Building 3, L4, dated: second to fourth century.
(b) Chesterton CH 7194 (M111), Test Hole 43, F129, dated: third century? Fabric 1?

53 Wall-sided with a deep collar divided into 'reeds' separated by unusually wide grooves. Finger depressed spout. Typical of the Lower Nene Valley, but would be unusual in the Mancetter-Hartshill potteries. Typologically AD 250–350.
(a) Fabric 1. Orton Hall Farm **M40**, L1380, F801, Periods 3 and 4.
(b) Fabric 1. Orton Hall Farm **M91**, L1369, F733, Period 4.
(c) Fabric 1. Orton Hall Farm **M111**, L120, F80, Period 5.
(d) Fabric 1. Orton Hall Farm **M121**, L363, F162, Period 5.
(e) Fabric 1/4. Orton Hall Farm **M197**, Trench III 1.
(f) Fabric 1. Orton Hall Farm **M201**, L2910, clearance layer.

Unassigned

54 Wall-sided with a deep collar, fluted, not reeded. Probably a finger depression spout. Typologically after AD 250, and probably late third to fourth century. Other site dating: late third to early fourth century.
(a) Fabric 1? (Stibbington Kiln). Fourth century.
(b) Fabric 1. Chesterton CH 5875 (M87), Building 3, L2, dated: second to fourth century.

55 Fabric 1. Plain, wall-sided, with inward sloping rim. Stibbington. Colour-coat. Fine textured fabric. Typologically late third to fourth century. Other site dating: late third to fourth century.

56 Plain, wall-sided, and colour-coated, similar to samian form Dr.45 and Oxfordshire form C97 (Young 1977). Spout probably as with these forms, sometimes with a mock lion or bat head. There may be vessels without a spout or with an undrilled spout. Typologically AD 250–400. Other site dating: (Oxfordshire) AD 240–400.
(a) Fabric 1. Orton Hall Farm **M54**, L1268, topsoil?, Period 3.

(b) Fabric 9. Orton Hall Farm **M85**, L911, F500, Period 4. Young form C97.
(c) Fabric 1. Orton Hall Farm **M98**, L14, F11, Period 5.
(d) Fabric 9. Orton Hall Farm **M99**, L87, F14, Period 5. Young form C97, with rouletting.
(e) Fabric 1. Orton Hall Farm **M108**, L109, F55, Period 5.
(f) Fabric 1. Orton Hall Farm **M116**, L121, F81, Period 5. Fine textured fabric.

Types 57–78: unusual forms

57 Fabric 1. Collared form. Convex collar divided into three zones; central one very wide. Bead broken and turned out over flange to form spout. Based on a form produced in the Mancetter-Hartshill potteries by such potters as Iunius, Bruscius and so on. First made *c.*AD 150, but uncommon before *c.* AD 180 at the earliest. Chesterton CH 7144 (M108), Test Hole 35, F182, dated: second to third century. Typologically *c.* AD 180–250.

58 Fabric 1 (Stibbington Kiln). Collared form. Sloping, straight-sided collar. Wide central zone with bead above and below as in Type 57. Scored wavy line decoration on central zone. Stibbington ST 57.12. Typologically after *c.* AD 250. Other site dating: late third to early fourth century.

59 Fabric 1 (Stibbington Kiln). Upright bead with sloping beaded collar. Stibbington ST 57.8. Typologically AD 200–350. Other site dating: late third to early fourth century.

60 Fabric 1. Thick, stubby reeded flange, triangular in section. High bead. Thick wall. Typologically third to early fourth century.
(a) Orton Hall Farm **M26**, L833, F438, Period 3.
(b) Orton Hall Farm **M97**, L2303, F1052, Period 4.

61 Fabric 1. Similar to Type 60, but bead higher and flange shorter. Bead broken and turned out over flange to form spout. Chesterton CH 4371 (M59 or 62), Building 4, L2, dated: second to fourth century. Typologically probably third century.

62 Fabric 1. Stubby reeded-hammerhead rim. Solid at base of rim giving a triangular section. Chesterton CH 3118 (M51), Building 4, F53, dated: second to third century. Typologically probably third century.

63 Fabric 3. Wide sloping reeded rim. The angle of rim to body is very unusual. Orton Hall Farm **M80**, L819, F434, Period 4. Typologically late second to third century.

64 Fabric 1 (Stibbington Kiln). Similar to Type 40, but with plain instead of reeded flange. Stibbington ST 57.2. Typologically AD 250–400. Other site dating: late third to early fourth century.

65 Fabric 1. Similar to Type 64 but with more upright bead. Orton Hall Farm **M77**, L486, F200, Period 4, joins L1363, F200, Period 5. Typologically AD 250–400.

66 Fabric 1. Very weakly reeded, drooping flange. High bead. Very unusual. Mill Hill, March. Stamp of SIMILIS (see Type 8). Typologically AD 160–200.

67 Fabric 4 (probably Lower Nene Valley). Similar to Type 66, but with plainer, thick and clumsy flange. Lynch Farm, Orton Longueville. Stamp of SIMILIS (see Type 8). Typologically AD 160–200.

68 Fabric 2. Stubby, thick horizontal flange. Bead above flange. Form of spout variable. Typologically third century?
(a) Orton Hall Farm **M62**, L1268, topsoil?, Period 3.
(b) Chesterton CH 4883 (M66), Building 4, L2, dated: second-fourth century.

69 Fabric 1. Wide, plain flange, slightly uptilted at the distal end. High bead. Typologically fourth century, perhaps AD 350-400+. Other site dating: late fourth century (Corder 1951, fig. 9, no. 29; Frere 1972, fig. 109, no. 2631).
(a) Orton Hall Farm **M163**, L2738, F1065, Period 5.
(b) Chesterton CH 4452 (M64), Building 3, Wall Trench B, dated: third to fourth century.
(c) Chesterton CH 6005 (M94), Building 3, F90, dated: fourth century.

70 Fabric 2. Slightly sloping flange with a rectangular section. Groove on top of flange at distal end. High bead. Chesterton CH 1662 (M27), Building 2, L4, dated: second to third century. Typologically AD 160–230.

71 Fabric 2. Similar to Type 70 but flange narrower. Bead broken and turned out over flange to form spout. Chesterton CH 2775 (M45), Building 3, L3, dated: late second to mid-third century. Typologically AD 170–230.

72 Fabric 2. Similar to Type 71 but with deeper wall on the flange. Retains horizontal shelf on top of flange. Bead broken and turned out over flange to form spout. Typologically *c.* AD 200–280. Possibly a Stilton product?
(a) Orton Hall Farm **M11**, L361, F217, Period 2.
(b) Water Newton WN 2107, Trench A, L27, dated: second to third? century.

Types 70–72 are likely to be new forms being tried out in the mid to late Antonine period. Possibly Stilton products?

73 Fabric 1/2. As Type 72 but flange extremely narrow and bead extremely high. Water Newton WN 2084, Trench A, L27, dated: second to third? century. Typologically third century.

74 Fabric 1. Narrow, neatly rounded flange. High bead. Unusual form for Lower Nene Valley, approximating to Oxfordshire form M22 (Young 1977). Trituration grit entirely iron slag. Orton Hall Farm **M150**, L1735, F675, Period 5. Typologically AD 240–400.

75 Fabric 2. Deep, narrow collar, curving back toward the body. Neatly made. High bead. Bead broken and turned out over flange to form spout. Trituration grit includes iron slag and quartz. Chesterton CH 1352 (M25), Building 1, L4, and CH 3176 (M54), Building 3, L2, both dated: second to fourth century. Clearly an Oxfordshire Type M10–12 (Young 1977), dated there *c.* AD 180–240. Typologically late second to third century.

Types 74 and 75 are more likely to be the work of potters moving from the Oxfordshire area rather than imitation by some of those working in the Lower Nene Valley.

76 Fabric 2. Deep and narrow flange. High bead. Very thick. Chesterton CH 6422 (M104), Building 4, L4, dated: mid to late second century. Typologically AD 170–200.

77 Fabric 4. Deep, outswept flange. High bead. Chesterton CH 543 (M15), Building 1, under F10. Uncertain date. Typologically AD 170–250.

78 Fabric 6. Deep, narrow drooping collar, turned toward body at distal end. Near horizontal ledge at top. Form derived from Colchester products (Hull 1963b, fig. 87, no. 1). Chesterton CH 542 (M14), Building 1, under F10. Uncertain date. Typologically AD 160–200+. Style of Virapius (White, D.A. 1967, fig. 7, no. 20). The fabric is similar to one used by Virapius and it could be one of his products. He had a workshop at Snettisham in Norfolk, but his use of at least two different fabrics and a wide variety of rims suggests that he was active in more than one area. This vessel is unlikely to have been made at Snettisham.

Types 79–89 are of definite non Lower Nene Valley manufacture.

79 Fabric 11. Wall-sided with faintly differentiated bead and very deep collar. Orton Hall Farm **M35**, L964, F540, Period 4; also L1116, handstripping layer, undated, and L1126, yard, Period 3. Typologically *c.* AD 150–300. Other site dating: *c.* AD 150–250 (MacDonald and Curle 1929, 526, fig. 92, nos 33 and 34; Haupt 1984, *Taf.*172, 183, *etc.*).

80 Fabric 10. Full curved flange and high, split bead. Reminiscent of Oxfordshire types M10–12 (Young 1977). Present in Antonine deposits in Scotland. Typologically and other site dating: AD 140–200 (Frere 1972, 341, 1053–4).
(a) Orton Hall Farm **M2**, L1741, F675, Period 1 (second-century deposit).
(b) Orton Hall Farm **M97**, L2303, F1052, Period 4.

81 Fabric 8? Downward and outward pointing flange, with grooves. High bead. Orton Hall Farm **M34**, L1958, F500, Period 3. Young form M21.3. Typologically third century. Other site dating: AD 240–300.

82 Fabric 8. As Type 81 but bead grooved. Deep groove at distal end. Orton Hall Farm **M65**, L237, F148, Period 4 and, probably, L1591, unassigned. Young form M21.1. Closely related to Types 80 and 81. Typologically third century? Other site dating: AD 240–300.

83 Fabric 8. Drooping, curved flange turned in towards body. High, backward-facing bead. Orton Hall Farm **M7**, L1811, F894, Period 1. Young M11.11. Closely related to Type 75. Typologically AD 170–240. Other site dating: AD 180–300.

84 Fabric 8. Horizontal flange turned under, but not fused to underside of flange. High bead. Bead broken and turned out to form spout. Orton Hall Farm **M139**, L475, F203, Period 5. Young form M17.5. Typologically third century. Other site dating: AD 240–300.

85 Fabric 8. Small, squat flange somewhat rectangular in section, probably formed by folding flange underneath into itself. Very high bead, often grooved, and usually leaning outward. Finger depression spout. Young form M22. Typological and other site dating: AD 240–400.
(a) Orton Hall Farm **M53**, L1031, yard, Period 3.
(b) Orton Hall Farm **M67**, L238, F148, Period 4.
(c) Orton Hall Farm **M92**, L1427, F772, Period 4.
(d) Orton Hall Farm **M143**, L409, F224, Period 5.

(e) Orton Hall Farm **M186**, L1980, hand-stripping layer, Period 5, with coin C50, dated: AD 367–375.

86 Fabric 8. Narrow, horizontal flange with distal groove. Underfold of rim is completely welded in, unlike Type 85 where it can still be clearly distinguished. High, grooved bead. Finger depression spout. Young form M22.11. Typological and other site dating: AD 240–400.
(a) Orton Hall Farm **M140**, L475, F203, Period 5.
(b) Orton Hall Farm **M172**, L585, Building 4, Period 5.
(c) Orton Hall Farm **M193**, L1362, machine-stripping layer.

87 Fabric 8. Similar to Type 85 but flange thicker and narrower, no distal bead, and bead more upright. Typological and other site dating: AD 240–400.
(a) Orton Hall Farm **M31**, L901, 500, Period 3.
(b) Orton Hall Farm **M103**, L212, F31, Period 5.

88 Fabric 8. Similar to Type 85 but smaller flange, squarish in section. Orton Hall Farm **M83**, L984, F482, Period 4. Young form M22.14. Typological and other site dating: AD 240–400.

89 Fabric 8. Small, delicate flange turned under, and raised above horizontal. High, grooved bead. Orton Hall Farm **M166**, L2915, F1129, Period 5. Young form M22. Typological and other site dating: AD 240–400.

Untyped fragments from Orton Hall Farm by Period

Period 1
L2174, F1013. Indeterminate.

Period 2
L350, F189. **M8** Fabric 1. This body fragment is more likely to be later than *c.* AD 200 than earlier because the trituration grit is fairly abundant and consists entirely of iron slag.

Period 3
L822, F435. **M21** Fabric 1. After AD 200.
L940, F481, also L924 and L939. **M30** Fabric 1. Large part of the body of an unusually small mortarium. Iron slag trituration grit finely fragmented, though not closely packed, and dispersed evenly over the whole interior right up to the bottom of the bead (below). The solely iron slag grit indicates a date later, rather than earlier than AD 200.
L1268, topsoil? **M63** Fabric 1. Several fragments from at least one mortarium. Rim too fragmentary for close dating. AD 200–400.
L1428, F772 and L1834, Yard. **M55** Fabric 8. AD 100–400, but unlikely to be earlier than AD 180.
L1859, Yard. **M56** Fabric 1. After AD 200.
L1889, F675. **M37** Fabric 8. AD 100–400, but unlikely to be earlier than AD 180.
L1958, F500. **M33** Fabric 1. Body and bead fragment. AD 230–400.
L2011, Yard. **M57** Greyish cream fabric with moderate inclusions of sub-rounded quartz and sparse brownish-black iron ore? which often stains the surrounding clay (approximates to Fabric 4). No certain trituration grit survives. Traces of a pinkish-brown slip.

The fragmentary stamp gives the G of a stamp which reads GERMANVS (ER and probably MA ligatured), presumably for Germanus. The only other stamp from the same die is from Lancaster. The brownish slip which is associated with both these mortaria would best fit an origin in the East Midlands. The single extant rim profile would best fit a date in the period *c.* AD 90–140.
L2543, F1047. **M43** Fabric 8. AD 100–400, but unlikely to be earlier than AD 180.
L2726, F1109. **M49** Fabric 1. Rim fragment approximating very roughly to Type 59. Probably of third-century manufacture.

Period 4
L238, L337, F148. **M66** Two flakes from a large, thick flange. Insufficient survives to identify the rim profile, but it is unlikely to be earlier than AD 150. The fabric is not readily identifiable, but is either Lower Nene Valley (Fabric 2) or the Rhineland.
L819, F434. **M82** Fabric 1. Two body sherds. AD 250–400.
L924, F470 and L939, F481: see **M30**, L940, Period 3.
L1377, F801+L1450, F810 . **M94** Fabric 8. AD 100–400, but unlikely to be earlier than AD 180.

Period 5
L120, F80. **M112**
L122, F82. **M120**
L254, F31. **M104**
L345, F203. **M137**
L475, F203. **M141**
These five are Fabric 8. AD 100–400, but are unlikely to be earlier than AD 180.
L481, F252. **M146** Fabric 1. Body sherd. Trituration grit very tiny and closely packed. AD ?300–400.
L1056. **M177** Fabric 9. Body sherd. AD 240–400.
L1830, F898. **M154** Fabric 8. AD 100–400, but unlikely to be earlier than AD 180.

Post-period 5
L891. **M188** Fabric 8. AD 100–400, but unlikely to be earlier than AD 180.

Unassigned
L1027. **M190** Fabric 8. AD 100–400, but unlikely to be earlier than AD 180.

Endnotes

1. Results of chemical analysis and thin-sectioning of clays by Mr J. Storey and Mr N. Cooper, to whom I am indebted for this information prior to their separate publications.
2. See note 1.
3. See note 1.
4. Kiln material from Stilton was brought to the Nene valley Research Committee's Field Centre for examination in 1984.
5. My thanks go to Mrs G. Johnston who re-fired slipped grey ware in oxidising conditions.
6. Information from Dr J.P. Wild and Mr G.B. Dannell who carried out the excavations whose report is shortly to be published.
7. On kiln material from Utrecht. Information was kindly provided by Mr T. Hoekstra.
8. I am grateful to Dr R. Pollard, who provided information from his preliminary survey.
9. My thanks go to Mr B. Dix and Mr A.E. Brown for providing details.
10. Mr A.E. Brown kindly sent information of the analysis carried out by Dr F.A. Hart.
11. Details from Miss P. Aird and Miss E. MacRobert, to whom I return my thanks.
12. From excavations and field survey by Mr R. Friendship-Taylor, who sent information and to whom I am grateful.
13. Material from all these sites has been examined by the author.
14. Evidence from pottery recovered during a watching brief at Godmanchester in 1980 by the author and confirmed by Mr H.M.J. Green and Mrs J. Pullinger.
15. My thanks go to Miss P. Aird, Miss E. MacRobert, Mrs C. Woodfield and Mr P. Booth for sending details.
16. See note 6.
17. See note 6.
18. See note 14.
19. Oakham Museum: SK 26 and Grave 6.
20. Excavation carried out by Dr G. Webster in 1958 and to be published.
21. I am grateful to Dr J.P. Wild for the initial information. The publication of the site is underway.
22. Mr H.M.J. Green very kindly sent details.
23. See note 6.
24. See note 6.
25. See notes 9 and 10.
26. See note 9.
27. See note 9.
28. See note 9.
29. See note 12.

30. I should like to thank Miss Brenda Dickinson of Leeds University for suggesting the parallels to item No.8, and providing the details of the potters' stamps, and Mr J.R. Perrin for preparing the table.
31. A suggestion made by Mrs K.F. Hartley when examining the pottery in Peterborough.
32. Similar vessels were amongst material from a suggested kiln at Stilton: see note 4.
33. The references to a type series are to one which will appear in the publication of Ernest Greenfield's excavations at Durobrivae.

Chapter 7. The Anglo-Saxon Pottery[1]

I. Introduction

The collection was not large, weighing barely 7.5kg. The quantity recovered, while influenced by the limited excavation policy, could also have been affected by the possibility that important areas of Anglo-Saxon occupation lay outside the area of formal excavation. Although the pottery found is crucial for the chronology of the main part of Period 5, its quantity does not justify an exhaustive discussion of large numbers of near parallels for both form and decoration. Only the Corpus brought together by J.N.L. Myres (1977), the catalogues of the cremation urns at Spong Hill (Hills 1977; Hills and Penn 1981), and the collection recovered from West Stow (West 1985) have been used.

Similarly, the fabrics have not been subjected to minute study. The work carried out by John Walker (1978) demonstrates the pitfalls of basing arguments on visual examination. The fabrics here are described from an examination of fractures using a x10 hand lens, the object being to give a rough guide to the tempering density and sizing, with some idea of the different materials used. The terms employed are explained below.

Every sherd displays signs of having been coil-made, although in the case of the biconical, No. 1, the indications are less evident. The potting technique was good enough to produce well-shaped pots, competently finished, even finely so, and fired to a hard state. Unless there is a specific comment about the firing of the pot, every item in the catalogue should be assumed to have been hard-fired. Where the term 'very hard-fired' is used, this means that the sherd gave out a noticeably different tone when lightly dropped.

The overwhelming impression given by the surviving sherds is that the intention was to produce a reduced, near black, colour. Where the surfaces deviate markedly from this, there are sufficient signs that the sherd had received secondary burning for it to be suspected that, in the rare instances of almost complete oxidisation, this is adventitious. Some pots show that firing had varied, ending with oxidising conditions which needed further reduced firing to achieve the desired result. Colours are described subjectively but the terms used should be readily appreciated. Frequently there is a change of shade across the surface and it is the overall impression which is recorded, unless the changes are marked.

The surface finish of the pot has been looked at in every case. When the pot has distinct gloss and there is no sign of the technique used, it is described as 'burnished', otherwise a prepared surface is said to be smoothed. No comment is made if the surface is either missing or too damaged for assessment. Three specific styles of finish are of note: one relatively common, one not entirely assured and possibly related to the first, and one very rare. The first two used organic matter. Many pots are said to be grass-wiped. This conclusion is based on the observation of narrow smear marks with minor reeding. These bands are consistent with blades of grass having been used and some pots bear the impressions of chopped grass displaying the same character. Chopped grass may have been used in manufacture, but not as a temper and it is not certain that it had been used deliberately in finishing. The grass marks tend to be seen best on the interior face of the pot, an area not usually fine-finished, unless the vessel was more or less an open form.

The third kind of surface finish is highly distinctive in being a deliberately roughened surface. Only two sherds, from L1 and L37, show unequivocally how the effect was achieved: when the pot was still damp, handfuls of sharp grit, medium to large in size, were pressed into the surface of the clay to be brushed off when the pot had dried. On one sherd not all the grit had been detached. The surviving surface grit is clearly not part of the fabric and, under a hand lens, the rest of the hollows in the surface can be seen to be the impressions made by similar grains. The intention was probably to improve the grip of the user and the technique did not necessarily run over the whole surface; one sherd shows a definite junction with a well-smoothed surface, showing that the grits were applied after the smoothing had been done.

The description of the tempering is intended to convey an impression only, its degree being given first, divided into sparse, moderate or dense. For some fabrics, terms such as 'sparse to moderate' have been used to indicate an intermediate condition. The grain size is given in four categories: fine, small, medium and large. The presence of fine material is usually not detectable without a hand lens. In other instances, 'small/medium' means that, subjectively, the grain size is neither one nor the other, but 'small to medium' means that there is a mixture, 'small to large' covering a wider range; 'medium with large' indicates that the bulk is of one size with a noticeable content of the other. The grain sizes conform roughly with the following actual sizes:

fine – less than 0.25mm
small – less than 0.5mm
medium – less than 1.0mm
large – more than 1.25mm

The discussion below makes it clear that relatively close dating of the pottery is not easy. The parallel material is diffuse and does not lead to a convincing conclusion because none of the usually accepted diagnostic early styles — chevron and dot, faceted carinations, plain horizontal corrugation — is present. The only exceptions are No. 71 with its possible corrugation and a sherd in OHF71 IV (+). It is the basic coherence of all the material from the site which is the most convincing demonstration that the pottery has a limited date-range.

There are only fourteen stamps. Only in two instances are stamps associated, two different ones: Nos 12 and 29. Of the rest, two are incomplete, those on Nos 56 and 73. Edges of stamps occur two or three times and are not included in the total. There were no repeats from one pot to another. A view of the stamp drawings in both J.N.L. Myres' Corpus (1977) and Spong Hill (Hills 1977; Hills and Penn 1981), coupled with a visit to Peterborough

Museum to check pots from Woodston and Longthorpe, showed that comments on the dangers of using drawings (Hills 1977, 13; West 1985, 130–2) are justified. The stamps are, therefore, not discussed. The cruciform stamp used on No. 29 has no parallel amongst the reference material used here.

Save for the bowl, No. 13, and the Great Pot, No. 21, there are no surprises amongst the forms represented — even the mortarium, No. 15, can be paralleled — but for one. This consists of a small number of rims of consistent form: Nos 35, 36?, 52, 66 and 76. The vessels would presumably have been globular and the effect would have been the more marked as none has what could be described as a proper rim. Each pot would have had a plain hole in its top and in the case of No. 52, with a diameter of only 60mm, the question arises as to the function of the pots themselves.

A characteristic of many of the ordinary pots is the way in which the rims were left deliberately unlevelled: many, like Nos 22, 24 and 27, are competently potted and fired and there is no particular reason why the rims could not have been better finished, unless the effect was intentional, perhaps to allow their contents to 'breathe' when a cover was put over the vessel.

II. Group 1, (328), F1114

(Figs 118–19)

All the sherds were together in a single ditch. With them was found the Anglo-Saxon comb, Cat. No. 251. The arrangement of the catalogue conforms with general practice: the plain urns come first, Nos 1–4; decorated urns, Nos 5–12; other forms, Nos 13–15; and fragments of rims, Nos 16–20.

The initial date of the deposit is given by the comb and the fragment of mortarium, No. 15. Discussion of the comb dealt with its probable date and it is unlikely that the mortarium dates far into the fifth century. Therefore, the beginning of the deposition could have been as early as *c*.400 or as late as *c*.425. The assessment of which is the more likely should depend on the rest of the pottery in the group. However, there is little real guide and much depends on the plain forms Nos 1–3 and 13–14 and what signs there are of the full decorative elements in the group as a whole.

Plain forms seldom offer much help and it is tempting to argue that No. 1 is an early biconical, but it seems that it is not paralleled in other dated contexts and it is its fabric, when compared with the rest of the Anglo-Saxon pottery from the site, which is taken to show this: before the rim and base had been found body sherds had been accepted as a standard Roman ware. A good, but undated, parallel for its form comes from Rothwell (Myres 1977, fig. 8, no. 2888). Resort to decorated versions of the form (Myres 1977, figs 201, 203–9 *passim*) shows that it is ornamental style rather than form which dictates the argument and, although this is a proper approach, it clouds the issue when it comes to cases like the present one. As an instance, it may be doubted if Myres 3589 and Myres 1148 (both fig. 201) are actually related, the former being a devolution of the latter. One is an open form, the other closed. The plain biconicals are assigned by Myres essentially to the fifth century with a hint that decorated ones may be early rather than late (Myres 1977, 3). The abiding impression, however, is that small elegant pots like No. 1 are more likely to be not only fifth century, but not particularly close to the end: there is no reason to suppose that, because they are plain, they should post-date the decorated ones (Myres 1977, 3).

As for Nos 13 and 14, bowls were perhaps more common than finds coming from cemeteries would suggest. The balance is probably better shown by the pottery from West Stow. Choosing only those whose profile more or less match the two here, there are eleven (West 1985, fig. 31, no. 4; fig. 37, no. 2; fig. 46, nos 2 and 3; fig. 58, no. 3; fig. 98, no. 2; fig. 119, no. 2; fig. 136, no. 3; fig. 156, no. 1; fig. 211, no. 7; fig. 223, no. 7). The ostensible date-range runs from the early fifth century to the early seventh, but it is impossible to assess the effect of residuality. Of the ten Sunken-featured Buildings involved, one was undated, two were indeterminate, three belonged to the last phase of the site, and four were fifth century at least in origin. Not a single foot like that of No. 13 was recovered and amongst all the pottery there were only three roughly comparable bases (West 1985, fig. 50, no. 8; fig. 57, no. 6; fig. 208, no. 2).

The mortarium, No. 15, has a parallel in one, also in an Anglo-Saxon fabric, at West Stow (West 1985, 27, fig. 92, no. 7). Apart from being correctly identified in the catalogue, it is ignored in the discussion on the Anglo-Saxon pottery (*ibid*., 128). The date of its context begins in the early fifth century (*ibid*., table 63, SFB 21). The extreme rarity of such vessels would, on the face of things, suggest that there was little call for them. However, it is worth pointing out that the main distribution of Roman mortaria in Period 5 at Orton Hall Farm (p. 181, Fig. 117) is overwhelmingly across the main area of excavated Anglo-Saxon settlement and this cannot be fortuitous. In other words, in fifth-century contexts on apparently purely Anglo-Saxon sites, the presence of Roman mortaria may be too important to be lightly dismissed as material scavenged from abandoned Roman sites (West 1985, 85): a tough-walled bowl in itself is useful, but would hardly lead on to imitation of both form and gritting.

The globular urns, Nos 2 and 3, are much more easily paralleled, but the conclusion must be that adequate dating is largely lacking. However, both vessels would not be out of place in the fifth century (Myres 1977, 6) even if there is little sign of whether they are either early or late.

The ornamental traits from Orton Hall Farm may hold out more hope for an agreed overall date than fine argument over either No. 1 or No. 15. The decorative tricks are mainly grooves, horizontal or diagonal, Nos 5–10, stamps, Nos 5, 10–12, or bosses and here only Nos 9 and 12 provide prominent examples although, had more of No. 11 survived, that might have been included. The stamps used are few and limited in range: a simple circle containing a small raised dot, a cross in a circle, both on No. 12, and a double annulus, No. 11. The cross stamp does not show a containing circle as such, but a deeper impression may well have done. There are no extensive stamped schemes or elaborate grooved patterns, the most complex pot being No. 12.

No pot can be shown to have had only horizontal grooving. Equally, none can be demonstrated to have had only horizontal grooves and plain bosses, or chevrons, or *stehende bogen* designs. Only Nos 5 and 10 could have come from pots with mixed horizontal lines of grooves and stamps, and in neither case can it be shown that this was their actual design. Little can be made of individual style when the relevant pots are so fragmentary. In general, the generous use of grooving coupled with simple bosses and a limited range of stamps point to a fifth-century date rather than later.

Pots, Nos 6–8, may hang together as a group employing only vertical, diagonal or horizontal ornament equated by Myres (1977, 38) with the Angles. A fifth-century date would suit their simplicity. However, if No. 8 was part of No. 5, the faint indications of a boss, consisting of an apparent gap in the slashing coupled with a slight thickening beneath, would point to a scheme starting with horizontal grooving at the top, then a line of stamps above a slashed band interrupted for a row of bosses of indeterminate form. There is nothing which must be late fifth century or later.

The exceptional urn is No. 12 which, if it does not have pendant triangles between the bosses, and if the pot is shown the correct way up, indicates at least stamped panels lying between stamped lugs. So far, the definitions used by Myres to segregate groups of pots have only served to exclude the material here and, following the same line, there is no *Buckelurn* which might be a pointer to a date earlier than the second half of the fifth century, but the sample provided by the group may not be big enough to be truly representative of the time when the deposit was closed. The disadvantage of No. 12 is the absence of good parallels for the lugs. The suggestion from the profile, as illustrated, is that all the ornament is likely to have been on the shoulder above what may have been a fairly sharp change in the profile not far short of a carination, and this may reflect a fairly early style.

In short, there is no unambiguous evidence for the length of time represented by the deposit. Its beginning is tied to the earlier part of the fifth century. Compared with Group 2 from F254, Nos 21–40, the assemblage is marked by individual pots which can be isolated, leaving a much smaller proportion of unassociated rims and body sherds: there should be a very low residual factor here and this would relate to little previous Anglo-Saxon pottery on this site. But, apart from the mortarium, most of the plain pots, including the biconical urn, are in large reconstructable pieces while the decorated series is, by comparison, in small sherds. In standard arguments, the latter would represent the residual element. But here, the biconical and globular urns are the principal pots and suit the comb while the decorated sherds are indeterminate and should be tied to these. Hence it is argued that the maximum date range for the group is likely to be 400–475, that the minimum is probably 425–450 and that its beginning is more likely to go back rather than its end move forward: the rest of the Anglo-Saxon pottery from Orton Hall Farm presents similar problems in dating and, if the metalwork indicates an end to the site no earlier than the early sixth century, and Group 1 is the earliest Anglo-Saxon pottery on the site, it places a strain on the rest of the evidence to close that as late as 475.

Catalogue Nos 1–20
(Figs 118–19)

Plain urns

1 Biconical, complete profile with slightly out-turned rim, fairly well marked carination and foot-ring. Well smoothed, surface originally fumed near black, with a thin oxidised zone under it. The fracture shows hardly visible coils and the hand-made character is only properly revealed by the slight traces of fingering on the rim and base. Fabric: mid grey, dense fine to small rounded, or partially so, grits.

2 Complete profile, reconstructable from non-joining sections in the lower part, with slightly everted rim, the profile more globular than slack biconical. Once smoothed, lower exterior pale grey-brown shading to dark grey, interior, dark grey. Fabric: dense medium to large sharp grits mixed with possibly burnt oolitic limestone with a sparse amount of finely crushed shell.

3 Globular, more or less complete profile with upright rim finished with a slight external beading. Burnished in and out, near black outside, near black shading to a greyish buff inside. Patchy discolouration suggests burning. Fabric: moderate small to medium sharp grits.

4 Probably globular, only the guaranteed profile is illustrated, with an upright rim. Dark grey in and out, there is an oxidised zone near the outer surface. Fabric: dense mainly small rounded and sharp grits mixed with burnt oolitic limestone and a higher proportion of finely crushed shell than in No. 2.

Decorated urns

5 Fragments of the upper part of a steep shoulder with a slightly everted rim. At least four horizontal grooves with traces of stamps below, and one other sherd has the edges of two more stamps. Thickening of the sherds with the stamps suggests that there may have been bosses with the stamps lying between. Grass-wiped, dark brown/black. Fabric: moderate fine to small sharp grits.

6 Three joining fragments of shoulder, apparently plain, with diagonal grooving on widest part of profile: scheme not reconstructable. Dark grey/brown in and out, paler fracture, chopped grass marks inside, grass-wiped outside. Fabric: sparse small to medium sharp grits.

7 Fragments from lower shoulder with at least six horizontal grooves above a slashed zone. Two sherds belonging to this zone have different thicknesses, but there is no other sign of a boss. Less hard-fired, near black in and out, well burnished outside, grass-wiped inside. Fabric: sparse small sharp grit.

8 Body sherd with part of a line of vertical slashes with a break where the wall thickens. Possibly part of No. 5: firing, colour and finish closely similar, but the fabric is, perhaps, more densely tempered although this may not be significant.

9 Two sherds just joining, a trace of a horizontal groove above a narrow boss with a groove down each side. A thickening in the wall on one side may indicate alternating wide and narrow bosses. Fired moderately hard, dark brown in and out, smoothed. Fabric: sparse to moderate crushed shell, some burnt limestone and small grits.

10 One sherd with remains of three wide grooves with poorly defined edges and the edge of a circular stamp below. Soft-fired, dark-grey with brown tinge, smoothed. Fabric: sparse to moderate small to medium grits, the latter sharp.

11 Three non-joining sherds, two with stamps: double annulus. One sherd has a sharp bend along one edge. Grey-brown outside, paler inside, fracture grey, with an oxidised band near outer surface, grass-wiped. Fabric: dense medium to large sharp grits.

12 Five sherds from an urn decorated with a series of vertical lugs. Two survive, both with circular stamps, with a raised dot in the bottom, along the crest. The better preserved lug has two surviving grooves next to it. The other sherds join to show a groove down one side, a repeat of the first stamp in a line across the top and a group, in at least two rows, of another stamp below. The latter stamp has a raised cross in a circle. The curve down on the other side of this arrangement may mark the edge of another groove. Very dark grey smoothed surfaces with marks of chopped grass. Fabric: moderate to dense small to medium grits, mainly sharp, especially the larger pieces.

Other forms

13 Pedestal-footed bowl. Plain apart from three grooves around stem of foot. Profile not complete, but the distinctive fabric unites both rim and foot. Fairly hard-fired, near black, browner on foot, burnished, except on under-surface of bowl. Fabric: moderate to dense small to medium mostly rounded grits, the larger ones tending to be sharp. There is a carbonised residue on the inner and outer surface of the rim. Although the vessel could be described as a lid similar to some from Spong Hill (Hills and Penn 1981, fig. 82, no. 1963, fig. 101, no. 2246), the lack of wear on the rim or the inside at the top contrasts with the abrasion on the foot-ring.

14 The upper part of a vessel like the last, but with no trace of a base. Fairly hard-fired, dark grey with some discolouration, grass-wiped in and out. Fabric: sparse to moderate small to medium sharp grits.

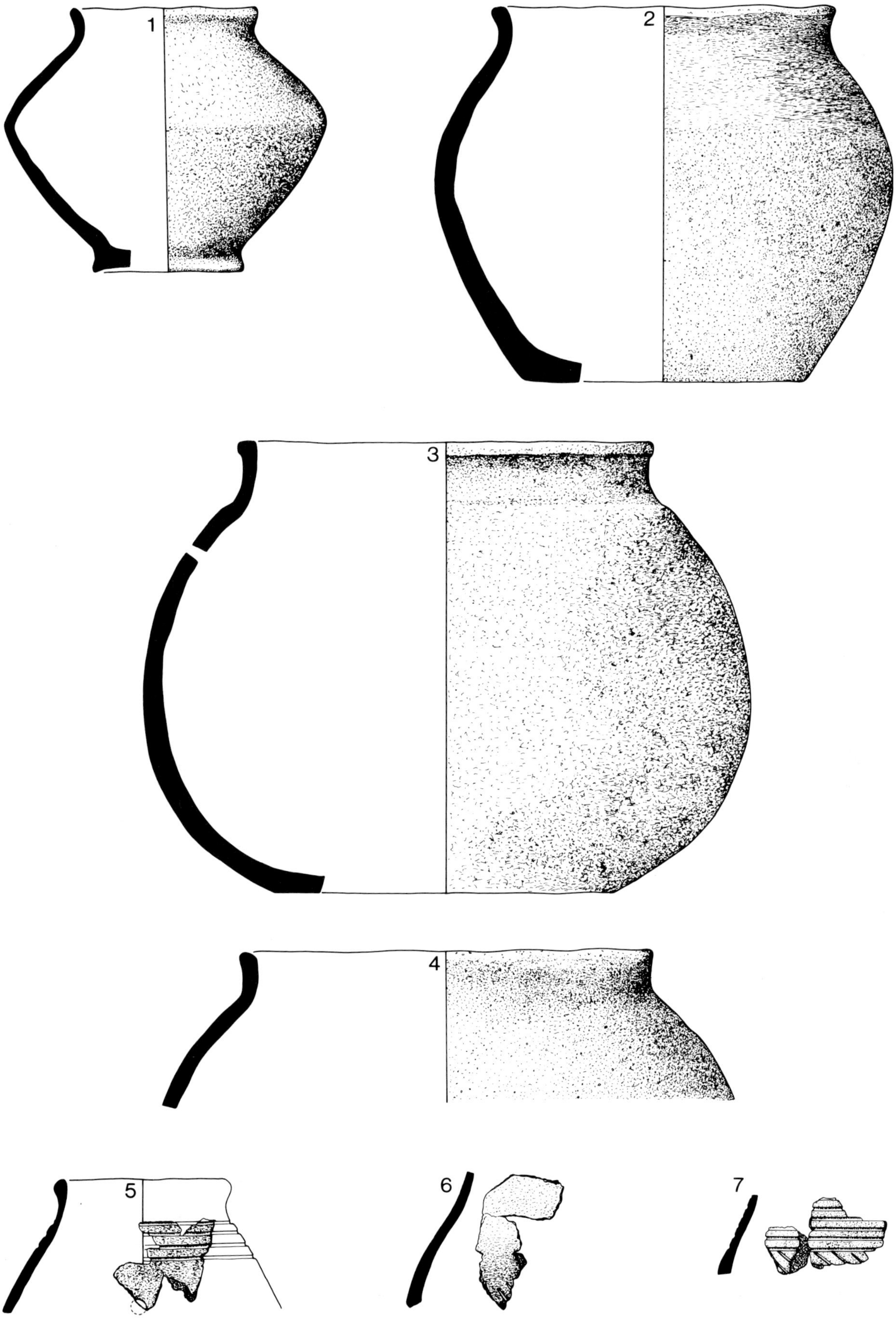

Figure 118 Anglo-Saxon pottery. Nos 1–7, Group 1, F1114. Scale 1:3.

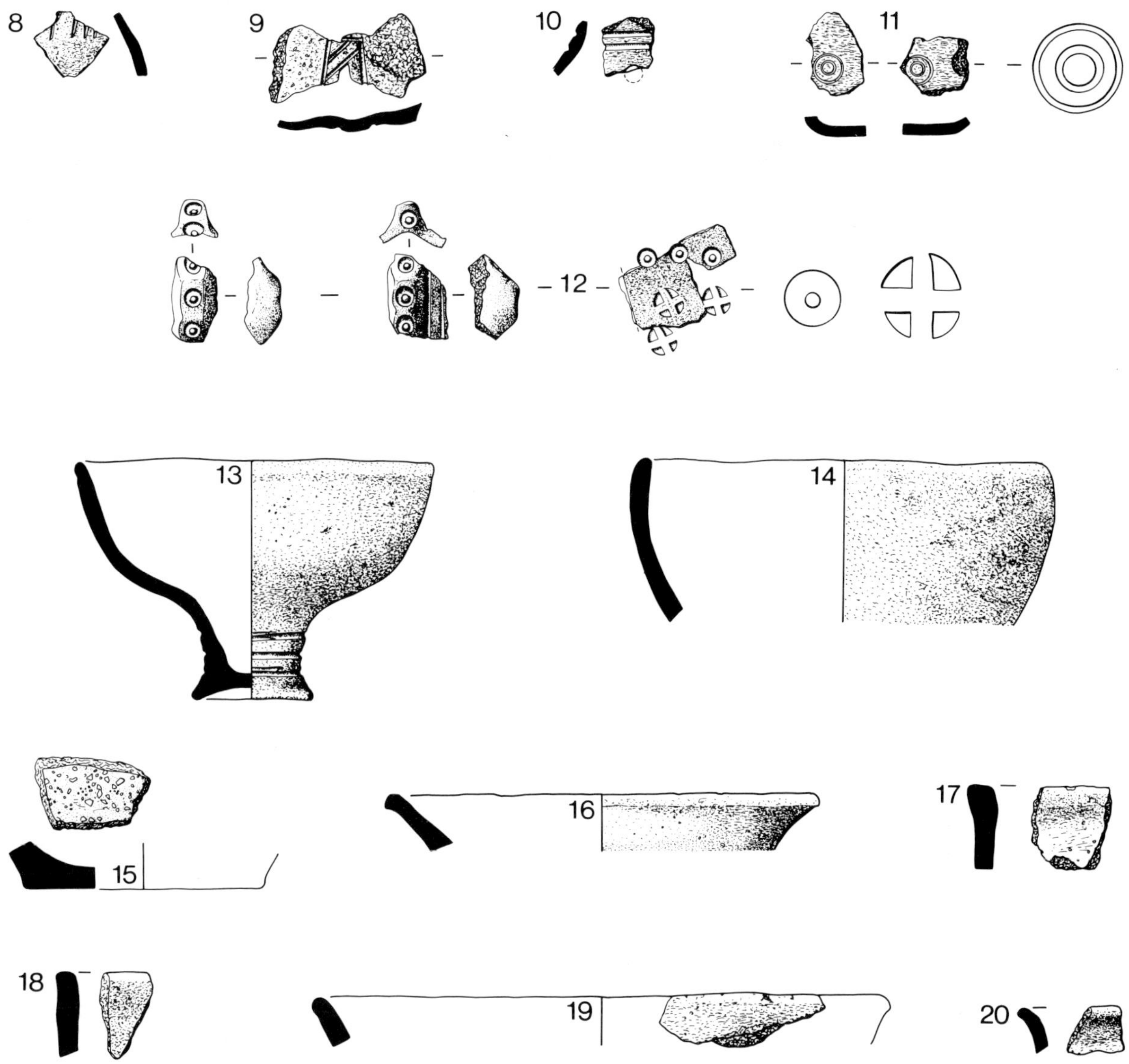

Figure 119 Anglo-Saxon pottery. Nos 8–20, Group 1, F1114. Scale 1:3, stamps at 1:1.

Again there is a carbonised residue and no sign of wear on the rim or inside.

15 Mortarium base sherd. K. Hartley has kindly examined the sherd and confirms that it is, by Romano-British standards, a mortarium, and in no Roman fabric known to her. All that survives is a short section at the junction of the wall with the flat base which has a slight vertical face before turning out to form the wall. The sherd is thick, hand-made. Nearly black in the fracture, the outer surface is partly oxidised to a dirty grey buff, the inner one worn to a dark grey. Pressed into the inner surface are large trituration grits of what appears to be crushed and partly burnt limestone, now worn flush with the face of the pot. Similar grits are to be seen in the body, but are less frequent and do not appear on the outer face. Basic fabric: moderate medium sharp grits.

Miscellaneous rims

16 Flaring, surfaces rough and greyish brown, fracture paler grey. Fabric: sparse very fine sand with finely crushed shell and crushed limestone. Before firing, the pot had been inverted on a flat surface so producing a level facet.

17 Nearly vertical, burnt reddish-brown to dirty grey buff right through, top and inner edge smoothed, the rest grass-wiped. Fabric: moderate to dense small to very large grits with, possibly, some crushed oolitic limestone. Two other sherds from this pot were equally burnt.

18 Vertical, dark brown outside, dark grey inside, smoothed. Fabric: moderate to dense small to large sharp grits.

19 Everted, dark brown outside, dark brown to pale reddish-brown inside, burnished. Fabric: moderate medium sharp grits.

20 Cavetto, dark grey, burnished. Fabric: dense very fine to small sparkling grits.

III. Group 2, (280), F254

(Fig. 120)

The sherds had been thrown into an informal pit created by the consolidation of the destruction fills of the Period 3 well beneath. There is little to suggest that the collection was as closed a group as Group 1. There were obviously more pots represented by fewer sherds for each and there were fewer joining pieces and profiles for the better represented vessels. On the other hand, the density of the sherds in the deposit was much higher than in any other context, including Group 1, and this should show that it was a genuine rubbish collection rather than a representative sample of what had occurred in the top-soil of the site before destruction by ploughing. The sherds are ordered basically in the same way as in Group 1.

Most of the sherds used by John Walker in his work came from this Group.

Despite the greater number of pots represented by few sherds, there appears to be a greater homogeneity of form than in Group 1, except for the Great Pot, No. 21, for which no parallel has been found either in size or rim form. A pot of this size and quality of form, finish and consistency in firing could only have been made by a professional potter. Not only is it a technical achievement to have made it, but its firing would not have been as easy as the more ordinary pottery of the period. As only pieces from the upper part occurred in the deposit, it may be that it had been damaged and trimmed down so that the lower part could be re-used.

Only one handled pot was found, No. 34, presumably imitating a metal cauldron (Myres 1977, 9).

The range of decoration is very limited but consistent. No. 29, with its bands of grooves and stamps, is a common scheme and would fit with a general fifth to early sixth-century date. The other two pots, Nos 30 and 31, both belong to the pendant triangle group, although it is fairly unusual to find a group of stamps without grooves to define the panel. Myres notes that most have no bosses and can be dated to the sixth century (Myres 1977, 54) while those with bosses he places later (*ibid.*, 55). If this conclusion is correct, then No. 30 may be one of the latest pots on the site. However, the fabric is distinctive and fits in well with the early wares in being unlike that usually found in sixth and seventh-century cemeteries. The profile of the pot seems to have a fairly marked change, with the ornament essentially above the greatest diameter: there is no reason to place No. 30 later than the early sixth century. The very limited ornament present in the group, coupled with an acknowledged early style on No. 29 may mean that any association with groups of pottery characterised by multiple stamps, or simple stamping on the more elongated and bag-shaped pots of the end of the stamping tradition in the seventh century, should be discounted. There is too little of No. 31 for proper discussion, but there is nothing that does not suit an end-date for the site in the earlier sixth century.

Perhaps the chief difference in the decoration between the F1114 and F254 groups is that the former contains much more in the way of linear decoration than the latter. If there is a chronological significance in this, then F254 should be later. Whether the use of more complex stamps like those on No. 29 has any bearing on date is beyond the evidence from this site.

Amongst the miscellaneous sherds are five from a pot with no rim, or reconstructable profile, having a soapy feel which contrasts with the bulk of the pottery from the site. There is also one sherd from near the base of a wall in a fabric remarkably reminiscent of No. 1.

Catalogue Nos 21–40
(Fig. 120)

Plain urns

21 The Great Pot, an enormous urn probably globular in form. The former should indicate that it was used for storage and, in that sense, is analogous to similarly large Romano-British pots. The rim form can be described as lid-seating, possibly for a round of wood fitted with a handle. The rim is wide, flared on the outside, the inner edge sweeping away in a smooth curve down the inside of the vessel. The outer edge of the rim has close-set finger-tipped decoration, but that this did not run all the way round is shown by a single short rim sherd. The higher, outer, part of the upper surface has two concentric grooves, the lower being plain apart from a groove defining the change in level. Darkish grey in and out, paler fracture, both surfaces grass-wiped, but this may have been the final stage as the surfaces betray a rhythmical unevenness suggesting a preliminary stage. Fabric: moderate small and large sharp grits with a few medium.

22 Slack-shouldered urn, upright rim with an irregular external bead. Very hard-fired, near black, grass-wiped surfaces uneven due to larger grits in the thin wall; fracture pale grey. Fabric: moderate medium to large sharp grits.

23 Three joining sherds with a short section of rim showing an inturned rim with only a slight curve becoming more upright at the very top. Near black, some discolouration suggests burning, smoothed in and out, but with more care outside. Fabric: friable, moderate to dense small to medium sharp grits.

24 At least twenty-one sherds of the same pot, of which only the rim is reconstructable; no trace of the base. The full form may have been more like a necked bowl than an urn. Dark grey shading to near black inside, mottled exterior ranging from dirty buff through dirty brown to variable dark grey suggesting burning, smoothed. Fabric: moderate to dense small to medium sharp grits.

25 Sinuous rim possibly from a bowl rather than an urn. Dark grey/brown, traces of organic inclusions in the surfaces. Fabric: sparse to moderate medium sharp grits.

26 Out-turned rim whose curve would suit a bowl more than an urn. Dark brown outside, reddish-brown inside, exterior smoothed, interior surface damaged. Fabric: moderate fine sharp grits.

27 Inturned rim with an external bead. Grey-brown, probably grass-wiped in and out. Fabric: sparse mainly medium sharp grits.

28 Two joining sherds forming most of the profile of, probably, a globular urn, the out-turn for the rim being present. Thick, grey-brown, smoothed. Fabric: moderate small to medium rounded and sharp grits.

Decorated urns

29 Out-turned rim and upper part of shoulder. Minimum of two horizontal grooves at top and bottom, three in the middle, a row of annular stamps, with a rayed pattern, in the upper zone, a row of unusual stamps in the lower. The impression is a round-cornered square with a cross formed by round depressions in the corners, each with a boss in it, and with a shallow round boss in the centre. Burnt, now a dark buff with a grey tinge here and there, smoothed, possibly grass-wiped. Fabric: dense fine to small rounded grits with a few medium sharp ones.

30 Body sherd only, the design dictated the orientation of the drawing, decorated with apparently plain bosses alternating with pendant triangles of stamps of at least three rows. There is no defining line to the panels. The stamps are circular, have a central boss in high relief with a lower annular ridge around. Very hard-fired, grey inside, dark grey outside. Fabric: dense small to medium sharp grits.

31 Body sherd with the lower part of a pendant triangle of stamps defined by three grooves on one side and with two surviving ones on the other. The stamp used is a version of that on No. 30, but with a less precisely defined annular ridge. Variable grey to dark grey in and out, signs of wiping inside, outside damaged. Fabric: moderate to dense mostly fine rounded grits, but with some medium to large sharp grits.

Other forms

32 Bowl with a definite neck finished with a slight out-turned rim. The base seems to have been rounded. Soft-fired and flaky, near black, generous grass-wiping. Carbonised residue on the outside. Fabric: sparse variable grits mixed with finely crushed shell.
There are fifteen sherds from a similar vessel in a different fabric, without rim or base.

33 Bowl, a single rim sherd, possibly more upright than shown. Near black throughout, inside grass-wiped, outside probably so. Fabric: dense fine/small sharp grains.

34 A single rim sherd which can only be reconciled with a pot if twisted to form part of a handle (*e.g.*, Myres 1977, fig. 64, no. 3994). Dark brown, smoothed. Fabric: dense fine to medium grits, also some red, powdery looking: burnt ?ironstone.

Miscellaneous rims

35 Sharply inturned rim with a mouth of *c.*100mm diameter. Dark brown, smoothed outside, grass-wiped inside. Fabric: dense fine to small grits with some medium/large sharp grains.

36 Less sharply inturned rim with a very slight bead and a mouth *c.*80mm in diameter. Dark grey to near black, outside smooth with some carbonised residue, inside possibly grass-wiped. Fabric: dense fine to small sharp grits.

37 Nearly upright rim. Near black, grey fracture, grass-wiped in and out. Fabric: moderate small sharp grits.

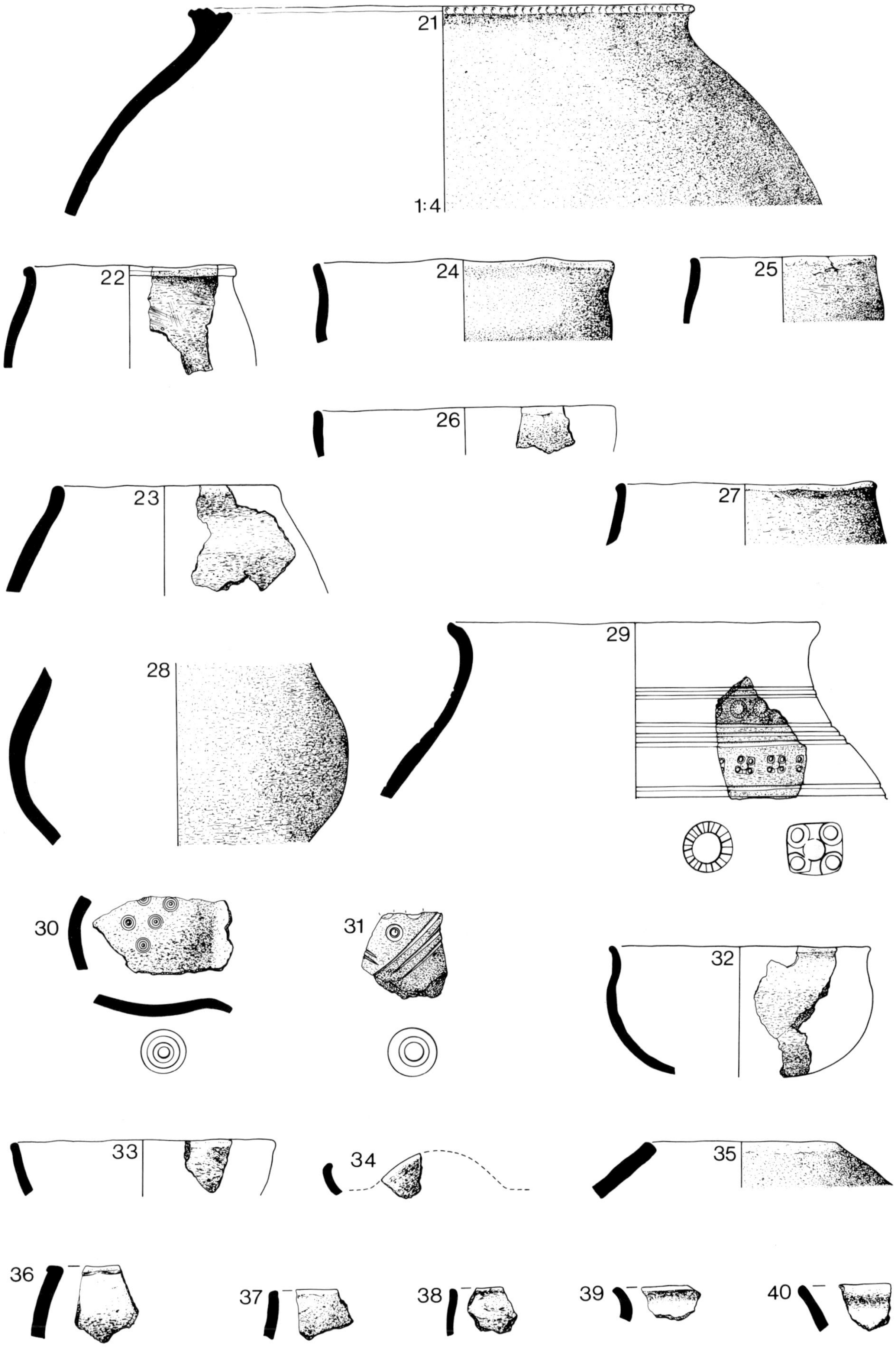

Figure 120 Anglo-Saxon pottery. Nos 21–40, Group 2, F254. Scale 1:3, except for No. 21, 1:4.

38 Nearly upright rim. Dark brown, chopped grass marks in and out. Fabric: moderate to dense fine sharp grits with a few small.
39 Cavetto rim. Dark-brown, but reddish-brown on top. Fabric: moderate fine grits and a few red pieces of possibly burnt limestone.
40 Flared rim. Dirty palish brown, very pale fracture. Fabric: moderate to dense fine to small sharp grits, red inclusions possibly burnt ironstone.

IV. Other stratified pottery

(Fig. 121, Nos 41–65)

This is ordered according to the context group in which the sherds occur and some indication of site location is given. Some comment is made about some sherds in the unillustrated material where relevant.

There are elements present here which are lacking in Groups 1 and 2. No. 55 may have come from a pot with a chevron design, as may one from (243), L121. No. 64 could have come from a combined chevron/stamped triangle pattern mixed with bosses, and No. 61 has part of a *stehende bogen* layout. Apart from these, there are repeats of bosses outlined with grooves, Nos 60 and 64, as well as a repeat of a pendant triangle marked out by grooves, No. 54, although not enough survived to show whether or not there had been stamps as well. The stamped ornament adds little to the range, if it is accepted that the definite rustication of No. 50 and the less certain case of No. 43 are removed as not being truly ornamental. The incomplete stamp on No. 56 is not identifiable.

The fabrics are very much in the same style as before, although No. 44 and a sherd from (309), L1803, are unusual. What is striking, however, are sherds which are either in the same fabric as, or remarkably close to, the biconical urn No. 1. The two sherds in (286), L512, cannot have come either from the same pot there or be parts of No. 1 itself. The developing scatter of sherds of this fabric across the site should mean that there were several pots from the same production centre and this means that No. 1 cannot have been a chance acquisition.

In summary, there is nothing in this collection which must be later than the earlier sixth century, and it could be argued that every piece is fifth century.

The eastern enclosures

41 (213), L144 Body sherd with most of a small boss. On the complete side is a single stab mark. Near black, grass-wiped. Fabric: moderate to dense fine/small sharp grits.
Also from this layer, a sherd in a fabric which is in the same style as No. 1.
42 (213), L475 A short upright rim little larger than a bead. Near black, exterior grass-wiped. Fabric: moderate small to medium sharp grits.
Also from this layer, a minute sherd with two grooves forming a sharp V. Fabric: sparse variable grits with small to large pieces of burnt limestone. Another sherd in the same fabric as No. 1.
43 (213), L322 Tiny body sherd with impressions of a stamp made up of two opposed Ds. The spacing suggests that the pot may have been rusticated using this stamp. Pale brown in and out. Fabric: moderate medium sharp grits.
44 (213), L326 Inturned straight rim rising from high shoulder. Near black, smoothed, possibly grass-wiped. Fabric: moderate mixed oolitic limestone and finely crushed shell. There is a large body sherd in the same fabric from L345.
45 (213), OHF71 VIII L+ Body sherd, orientation more or less assured, having two grooves at the top with partial combing in a narrow band underneath. The same tool was used to form a diagonal band on the left and another on the right across the end of the horizontal combing. The tool was then used to criss-cross the second diagonal band. Black inside, dirty dark grey outside, grass-wiped inside, fairly rough outside. Fabric: sparse to moderate fine to small mainly rounded, some medium sharp grits.
46 (214), OHF71 III A(2) Everted rim with a slight external bead and upper body of a thin-walled vessel reminiscent of No. 1. Black in and out, pale grey fracture, high gloss outside with grass-wiping showing under the bead. Fabric: moderate to dense fine to medium sharp grits.
47 (214), OHF71, III A(2) Slightly everted rim and upper body. The uneven rim and lumpy character of the body may be more related to function than to lack of quality in the potting: well fired and thin-walled. Very hard-fired, black outside, discoloured interior, glossed outside, generously grass-wiped, inside rough and marked by chopped grass. Fabric: sparse to moderate medium/large sharp grits.
48 (214), OHF71 III A(2) Rim very slightly everted. Near black in and out, burnt on rim, inside rough, outside grass-wiped with marked near-vertical impressions of grass stems immediately under rim. Fabric: moderate small sharp grits.
49 (214), OHF71 III A(2) Sherd from upper body, at least three horizontal grooves at the top and one at the bottom with a line of stamps between. Each stamp is an annulus with rays. Black inside, burnt a bright pink outside, inside smoothed, outside fire damaged. Fabric: dense fine to small sharp grits.
50 (243), L121 Body sherd with three rows of a vesica-shaped stamp. Near black throughout except for oxidised pale reddish-brown exterior. Fabric: dense fine to medium sharp grits.
Another sherd, possibly near a carination, with a wide groove with poorly defined edges which may have formed part of a chevron design. Pale grey fracture with surfaces fumed to a dark grey; fine and medium grits, the latter sharp.

By Barn 4

n.ill. (262), L8 contained three very thick sherds from a very large pot. Near black except for oxidised reddish-brown exterior, grass-marked, especially the interior. Fabric: sparse medium to large sharp grits as well as organic material.
n.ill. (265), L476 A sherd with a fabric like that of No. 1, but showing more laminations, thin potted, fine finish. Black with oxidised surface, burnt?
51 (274), L86: pit. Single sherd from an urn with a slight upright rim thickened on the outside, slack shouldered. Very hard-fired, burnt rim, darker grey/brown exterior, grass-wiped exterior. Fabric: dense mainly fine to small with medium to large additions, some mica, sparkling grits.
(274), AF31=L86 pit. One of the seven sherds has what looks like the edge of a stamp.
52 (274), AF22 (1) Rim inturned with only a slight evertion, very narrow mouth, only 60mm diameter. Black, grass-wiped outside, grass-marked, large pieces, inside. Fabric: moderate small to medium rounded and sharp grits.

F200 Pond

53 (277), L313 Upright rim with thickening outside at the top. Hard-fired, black throughout, finely burnished on upper rim, carbonised residues on exterior. Fabric: moderate to dense tempering of fine sharp grits.
Also a sherd of a similar fabric, but not the same, with a groove on it.
(277), L1363 One sherd grass-marked in and out, very hard fired; another large body sherd, thin walled, generously grass-wiped, both with sparkling grits.

Large pit, F263, in Main Yard

54 (281), L523 Body sherd with the lowest part of a triangle formed by two pairs of grooves. Dark grey throughout, carefully smoothed surfaces worked to regular curves. Fabric: moderate fine to large sharp grits with a little finely crushed shell.

Sunken-featured Building

n.ill. One sherd from (282), L409, bears one end of a very narrow boss. Hard-fired, pale to medium grey, damaged surfaces. Fabric: dense tempering of fine grits reminiscent of No. 1.
n.ill. (281), L993 Body sherd only, not certainly Anglo-Saxon, but unlike all Roman fabrics, fairly hard, brown/black inside, oxidised orange-brown outside. Generously grass-marked in and out and an organic content in the body. Fabric: moderate small to medium rounded and sharp grits.

Sag infill of Period 2 ditch

55 (286), L344 Body sherd from lower shoulder with the remains of one horizontal groove with three diagonal ones above: possibly part of a chevron design. Discoloured grey, damaged surfaces. Fabric: dense fine to small with a few large sharp grits.

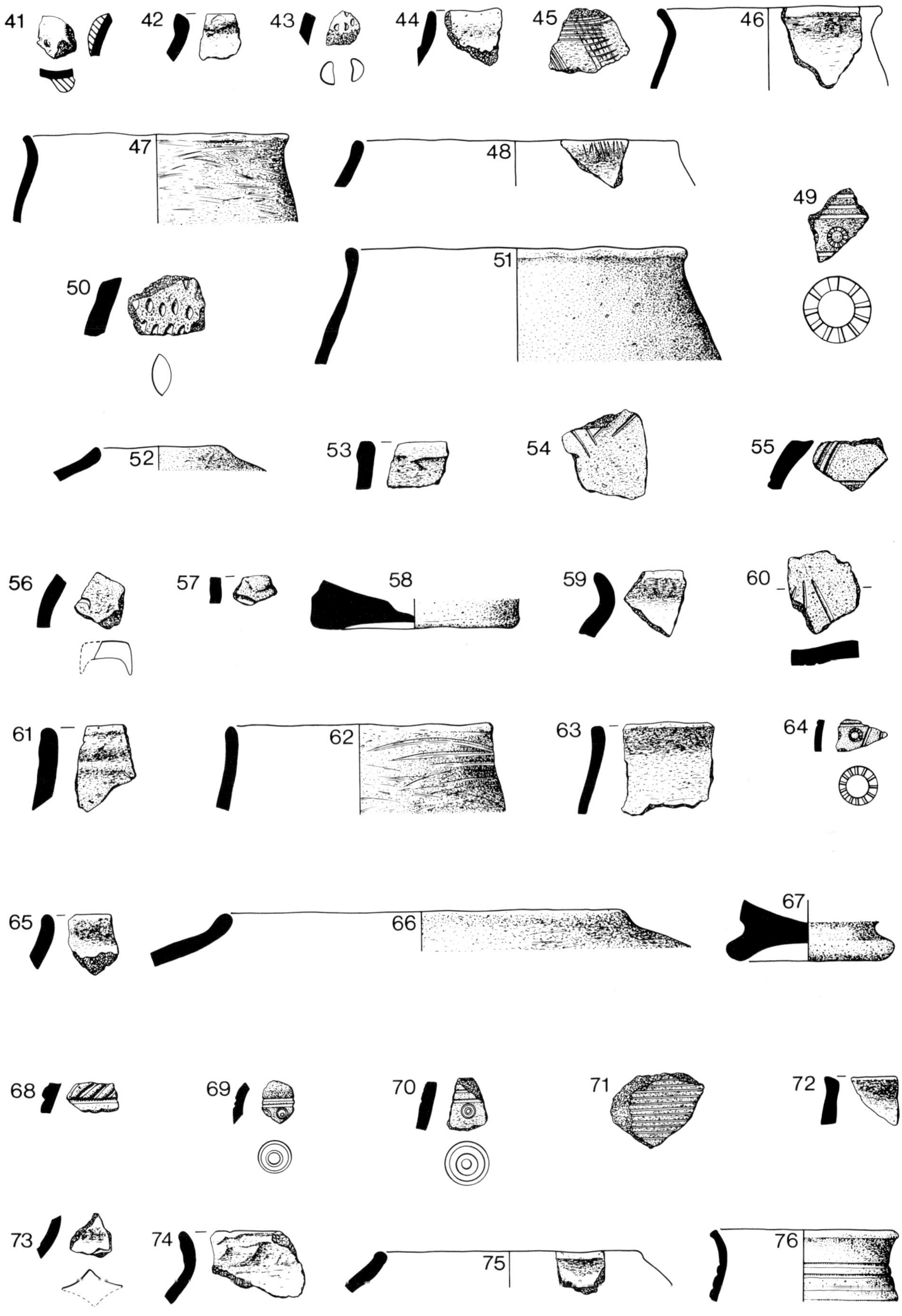

Figure 121 Anglo-Saxon pottery. Nos 41–65, other stratified; Nos 66–75, unstratified; No. 76, previous excavation. Scale 1:3, Stamps 1:1.

56 (286), L344 Body sherd with part of a stamp on it. The latter consists of a deep slight curve turning sharply to end in a shallow pointed tail. Near black. Fabric: moderate to dense fine to small/medium sharp grits.

57 (286), L344 Tiny fragment of upright rim with a levelled top and a trace of a sharp junction with a shoulder. At this point there is a sign that there had been a curved groove arranged *stehende bogen* fashion. Dark grey-brown, smoothed. Fabric: dense fine grits. The outer face of the rim itself appears to have had other decoration formed either by a curved motion or implement, but the poor condition of the sherd makes this uncertain.
Also in this context group, in (286), L512, a sherd with the same firing and finishing characteristics as No. 1, but much thicker, and the same fabric. There is another just like No. 1 and thinner than the last. Also two joining sherds, from a well-made pot with a smooth exterior, with a very worn internal surface.

The Rectangular Building

58 (292), L575 Base sherd. Dished underneath and with abrasions on the foot-ring, vertical wall before turning out. Dark grey, smoothed. Fabric: dense fine to small sharp grits.
Also a small fragment of an upright rim finely burnished and a body sherd with generous grass-wiping on the outside.

The Small Yard and environs

59 (309), L1803 Two pieces of rim, slightly everted, one sherd including part of the shoulder. Grey fracture with pale to dark brown surfaces, smoothed rim, grass-wiped shoulder. Fabric: sparse to moderate fine and medium/large sharp grits with a small amount of crushed oolitic limestone.
Also another sherd, very hard-fired, grass-wiped exterior with signs that part had been treated with sharp grits which had been wiped off, leaving short lengths of scoring. Fabric: dense small to medium sharp grits. Another sherd, thin-walled, and with large and small inclusions of what looks like burnt limestone, with very little grit.

60 (315), L1466 Body sherd with part of a boss defined on each side by a groove and with another down the middle. Burnt, thick, smoothed. Fabric: moderate medium sharp grits.

61 (316), L2007 Rim, thick slight out-turning, no neck. Near black. Fabric: dense fine and small/medium sharp grits.

62 (318), L1922 and L1923 Joining sherds of a thin rim slightly out-turned and having part of a *stehende bogen* formed of two grooves high up. Moderately hard-fired, black exterior, dirty brown fracture and interior, inside rough, outside grass-wiped. Fabric: moderate fine to small with medium sharp grits.

63 (318), L2275 Rim of the same form as the last, but from a thinner pot. Black with a dark grey fracture, traces of grass- wiping at bottom of sherd inside. Fabric: sparse to moderate small and medium sharp grits.

Secondary Anglo-Saxon ditch in south-west corner of Main Yard

64 (320), L2599 Small sherd from a decorated thin-walled pot. Only part of a pendant triangle of stamps defined by at least two grooves above and another on one side at an angle down the side of a thin boss which has another groove on the other side. The stamp used is a smaller version of that in the upper row of No. 29, but with thirteen rays. Near black, smoothed. Fabric: dense fine sharp grits.

Gulley along south side of Barn 3

65 (322), L2901 Slightly out-turned rim. Near black, reddish tinge inside, rough surfaces. Fabric: sparse to moderate fine and small/medium sharp grits.

V. Unstratified

(Fig. 121, Nos 66–75)

All these layers represent areas of initial site-stripping. The comments and catalogue entries are arranged in layer number order. All that need be noted here is that L37 covered the eastern end of the site, the rest of the material is too miscellaneous in its distribution for close analysis of its location to be justified.

The illustrated sherds have been chosen to give as complete a picture as possible of the decorative styles occurring on the site. There has been a greater degree of selection from amongst the plain forms and only the better or more interesting rims have been drawn. The only new decorative element to appear here is the slashed cordon No. 68, the only one found on the site. The base No. 67 is the only example recovered of a proper foot-ring. The rims conform with what has gone before except for No. 75 where there is not enough surviving for it to be entirely sure whether the effect is adventitious or was designed to have a *stehende bogen* pattern.

n.ill. A sherd from L1970 has an oxidised exterior, but is otherwise grey, and has a line of either rustication or vesica-shaped stamps (see No. 50). The piece is too abraded for there to be certainty.

n.ill. L1 A sherd with a roughened surface formed by impressing grits into the still damp clay. There is a clear edge between this treatment and a burnished zone which is assumed to be at the top of the sherd: fairly hard, dark grey in and out with a brown tinge. Fabric: dense fine with occasional medium sharp grits. There is a sherd from L37 from another pot on which the whole surface is roughened and with some of the grits still in position: near black in and out, dark grey fracture. Fabric: moderate small to medium sharp grits.

66 L37 Rim from neckless vessel whose shoulder is so flat that the opening could be described as being only a hole. Grey in and out, grass-wiped outside and at the top inside. Fabric: moderate fine to small sharp grits.

67 L37 From the top fill of F200 pond. Base with a well-formed flaring foot-ring deeply dished underneath. The inside face gives the impression of having been damaged by wear rather than by any other agency and this may imply that the pot had been more of a bowl than an urn. Standing edge of foot-ring abraded. Near black, except where oxidised a dirty buff by burning on the outside. Fabric: dense fine to medium sparkling grits.

68 L37 Body sherd with slashed cordon defined by a groove above and below. The rest of the scheme is lost except for a hint of the edge of a circular stamp below. The cordon can be seen in the fracture to have been formed from an applied strip of clay. Dark grey. Fabric: moderate to dense fine to small mixed round and sharp grits.

69 L37 Small abraded sherd from the shoulder with two horizontal grooves above a line of circular stamps each with a boss in the centre of an annular ridge. Grey. Fabric: moderate fine sharp grits mixed with a few small/medium ones.

70 L37 Body sherd with a circular stamp of the same type as the last lying between three horizontal grooves on one side and a single one on the other. Dark grey, grass-marked inside. Fabric: dense fine sharp grits with medium/large inclusions of a pinkish colour.
Also from this layer a sherd in a fabric very similar to No. 1.

71 L295 Body sherd with no less than eleven narrow horizontal grooves. Moderately hard-fired, dark grey inside, burnt to a pale grey outside, the heat has tended to separate the surface of the pot from the body. Fabric: moderate small to medium sharp grits with a little mica.

72 L295 Rim. Pale grey fracture, surfaces fumed to black, smoothed. Fabric: sparse fine grits. The fabric could be described as being 'romanising'. There is another sherd in the same fabric from the layer.

73 L347 Body sherd from upper shoulder with part of a stamp which, if geometrically reconstructed, would have been a lozenge with concave sides. Black, grass-wiped outside. Fabric: moderate fine sharp grits.

74 L1027 Rim, fairly sharply out-turned with marked fingering beneath the suggestion of a bead. Black, grass-wiped. Fabric: sparse to moderate fine and medium sharp grits with a little crushed shell. This could have been a bowl like one from West Stow (Myres 1977, fig. 68, no. 4000).

75 L1743 Rim. Black exterior, oxidised to brown inside. The outside has been grass-wiped in such a way that there may have been a deliberate attempt to create arcs in the *stehende bogen* manner, but there is not enough left of the pot to see if this was part of an intentional repetition. Fabric: moderate fine to medium sharp grits.
Also from this layer a sherd with what look like parallel lines of grooves. Inspection, however, shows that these had been made by a narrow tool with a nick in the end producing a slight ridge in the bottom of the groove.

n.ill. L2825 Rim from a pot whose opening is more of a hole than a conventional mouth. Dark grey, smoothed. Fabric: dense fine to small sharp grits.

VI. Previous excavations

(Fig. 121, No. 76)

The relevant parts of the 1971 excavations which could be directly related to the main excavation have been incorporated in the main catalogue. There is only one layer which possibly yields a little additional information: OHF71 IV L+. This lay within the area of the Period 4 north-east barn. Two non-joining sherds from the same pot have well-formed horizontal grooves of which there had been not less than five. Black, rough burnished. Fabric: dense fine sharp grits. These might have come from a corrugated pot.

Mr G.F. Dakin's excavations produced the first identified Anglo-Saxon pot, recognised because of its decoration. In 1973, the bulk of the pottery recovered by him was examined so that an estimate could be made of the incidence of Anglo-Saxon pottery. Most of this came from a small group of layers whose finds are now missing. One or more of the relevant deposits came from F991, the Period 5 ditch which ran along the north side of Barn 1 and which was extensively dug by Mr Dakin. The layer designations, in case the bags reappear, are:

HF A 3 WX SEa
HF A 3 WX NEa
HF A 3 WX SE ditch a
HF A III wall W a & b level

The one surviving collection of finds with Anglo-Saxon pottery is HF A 3 WX SE ditch b. Its coding is similar to one of the missing layers which had a ticket identifying the deposit as having been under the stones in the ditch. This should mean that, if F991 was the ditch concerned, the Anglo-Saxon pottery would have been sealed by the destruction deposits of the barn. The use of 'a', 'b' *etc.* in the layer designations, especially 'a & b level', is almost certainly an indication of depth and this may mean that the two sherds from the surviving layer were also sealed by the destruction of the Great Barn.

76 Everted rim and upper shoulder from a thin-walled urn with horizontal grooves, three surviving with the very edge of a fourth. There is a slight internal bead. Fairly hard-fired, near black surfaces, paler fracture. Fabric: dense to moderate fine to medium sharp grits.

Endnote

1. I am very grateful to Miss Barbara Green for reading this text. Any errors which remain are my own.

Chapter 8. The Bones

I. The Animal Bones

by Joan King
(MFTable 80; MF9)

Introduction

The bones from the site were numerous and most were well preserved. The report deals with 12,153 identifiable fragments of domestic and wild mammals. Some mollusca, fish, bird and human bones were found and reports on these by other specialists are given elsewhere.

The animal bones were sorted in individual layers which were then divided into six groups, one of which was divided into the east and west parts of the site:

pre-Period 1
Period 1 east
Period 1 west
Period 2
Period 3
Period 4
Period 5

These were examined for any difference in animal husbandry during the occupation of the site. Some comparison has been made with other local and British sites. Orton Hall Farm produced enough material for a detailed study to be undertaken on some aspects of animal husbandry.

The tables and graphs have been kept uniform with those produced for reports I have prepared for other sites in the Lower Nene Valley. The measurements are included to help other workers in this field who may need them for comparative purposes.

Method

All bones from individual layers were sorted and modern breaks united. Identifiable fragments were grouped into species, anatomical position determined, and measurements of mature bones were made. Age of animals at death was assessed by epiphyseal unity into the following ages:

A Juvenile, no epiphyseal unity, age under three months.
B Immature animals, one epiphyseal end united or immature dentition. Age varying from three months to three or four years depending on bones and species.
C Mature animal, epiphysis united or permanent dentition completed. Age, over three or four years depending on species.

Signs of butchering, cuts, chopping, burning, and animal gnawing were noted, as well as traces of any disease or signs of use.

Species found

The common domestic animals predominated: Cattle, Sheep, Pig, Horse; Dog and Cat were present in that order of importance. Few wild species were found: Hare, Deer and Rodent.

The total number of each species is shown in Table 80, with totals for each Period (see MF9 Bone 1). Also shown are the percentages of 'meat animals': Cattle, Sheep and Pig. Horse may also have been used for meat, but, as 80 per cent were of mature animals, they were probably used as beasts of burden and traction rather than primarily for meat production.

Cattle (MF9 Bone 2–9)

Cattle are numerically the largest group of bones found in each period except in Period 1 at the west end of the site (1B)(MF9 Bone 16), where a large collection of Sheep in one layer changed the balance. This makes Cattle the prime meat producers because, apart from their number, their size and the quantity of meat are greater than the other main meat producers. The live weight of the average beast at Orton Hall Farm would have been about 200kg, just over half the weight of present day Jersey cattle (Noddle 1980, 389).

Pre-Period 1 has the highest percentage of Cattle (MF9 Bone 1), but as only a small number of bones were found, this might not be a true picture.

The size of the Cattle conformed in the main with the Short-Horned small breed (Armitage and Clutton-Brock 1976) found in the Nene Valley. They were similar in size (MF9 Bone 2) to those found at Monument 97 (King in Mackreth forthcoming), but were slightly larger than the majority at Longthorpe (King 1987) and North Bretton (excavations, R.F.J. Jones, see MF9 Bone 3, 4) or those of Vindolanda (Hodgson 1977), but compared favourably with those from the Middle Saxon Period 1 at North Elmham Park (Noddle 1980). Metatarsal and metacarpal scatters showed little variation in size (see MF9 Bone 5, 6) during the life of the site. Horn cores showed some variation (MF9 Bone 7) and the majority, two thirds, came within the small- or short-horned group of cattle. The remainder came into the medium-horned group (MF9 Bone 8). This is unusual for the Roman period in this area. One specimen from Period 2 was of a long-horned breed, perhaps a long-horned bull brought in to improve the Cattle in Period 2. However, there was little evidence of this from the metatarsal and metacarpal scatters: Period 1 animals were in the centre of the scatter and horn cores did not increase dramatically (MF9 Bone 5, 6). There were many short- and small-horned animals in the later periods (MF9 Bone 7). Horn core circumferences fell into two main groups, the smallest being probably the cows and the medium group the steers (MF9 Bone 8). Most of the horns were forward-curving and smooth with many broken at the tip and cut from the skull (see below: Butchering and use).

The age of Cattle when slaughtered showed little change in the six main periods (MF9 Bone 9). No newborn animals appeared until Period 2 showing that veal production was minimal and that it was customary to over-winter immature animals until they were eighteen months to four years of age when about one-third of the animals, probably castrated males, were killed for meat: this would be the normal quantity in a herd of Cattle and indicates home consumption of prime meat. The majority of Cattle were slaughtered when mature, over four years of age, and probably older, having been used for breeding and milk production. This would require a good winter store of fodder. Anatomical distribution of bones showed that all parts of the animal were utilised on the site and little butchered meat was exported (MF9 Bone 1).

Sheep (MF9 Bone 1, 10–12)

Sheep were the second largest group of bones accounting, as they do, for about 30 per cent of all the groups, except pre-Period 1 which had only one Sheep fragment, and the west end of the site (1 west) in Period 1 where there was 68 per cent. This high percentage was entirely due to the bone content of L1467 (10) which had many complete bones of both mature and immature animals. None of the skulls, horn cores or metacarpal bones examined (MF9 Bone 10), using Payne's criteria (Payne 1969), showed any variations betraying the presence of Goat.

The Sheep were a small slender-horned animal showing little variation in size during the life of the site or from other sheep in the area in the Roman period (MF9 Bone 10, 11). They were slightly smaller than those from North Elmham Park which is later in date, essentially Middle Saxon to Early Medieval (see MF9 Bone 10.11). However, the animals represented by those in L1467 were smaller and slighter then the others on the site, being between 0.515m–0.58m in height to the shoulder, as opposed to 0.62m–0.719m in the rest. The aberrant group could just be a 'foreign' flock brought in for slaughter or a flock killed off when a new breed was brought in.

Few Sheep were killed when juvenile, only 1% in Period 2, increasing to 7 per cent in Period 5. The number of immature animals (six months to three years) varied slightly over the periods from 35 per cent in Periods 1 east, 3 and 5 to around 50 per cent in Periods 1 west, 2 and 4 (MF9 Bone 12). Period 1 west is abnormal because of the single large deposit already mentioned which had 652 bones most of which were complete and represented a minimum number of fifteen mature and twelve immature animals. The bones came from all parts of the body and the site records mention at least six articulated skeletons with articulated sections of vertebrae and limbs as well. While the whole deposit might represent the slaughter or death of a flock due to disease, drought or after winter snow, the condition of the skeletons suggests that alternative interpretations such as either ritual or an unusually large single meal might be appropriate. As for the former, the presence of mature animals would be unusual. The horn cores of the adult Sheep showed that they were ewes.

Periods 1 east, 3 and 5 showed a normal 'kill-off' pattern when wethers (male castrates) are killed off before maturity for meat, between the age of one and a half to three years; this showed well in the many tibias found fused at the distal end only. That ewes were kept for lambs and wool before being used for meat at an older age was indicated by the well-worn teeth of Sheep of five or more years of age. Tooth wear is dependent on the nature of the available pasture: short gritty pasture causes greater tooth wear than lush grass.

There seems to have been a greater slaughter of immature Sheep in Periods 2 and 4. However this may have been because mature animals were driven elsewhere to supply mutton, leaving the best lamb to be eaten on the site. That the mature animals would have been sold on the hoof is shown by the detail that there was no increase in numbers of heads and feet in the anatomical distribution (see MF9 Bone 12).

Pig (MF9 Bone 13, 14)

Pig made up only a small proportion of the bones recovered: 3.3 per cent of the overall 'meat' animals (MF9 Bone 16). No Pig were found in pre-Period 1. Nearly half of the surviving bones were from the head including teeth, very few complete long bones were found, therefore measurement comparisons are few. This is in common with Pig on other sites in the area indicating small herds only. In Periods 1 and 4, more mature animals were found, which would be more consistent with keeping pigs for meat. There were very few newborn, suckling, pigs (MF9 Bone 13).

The size of the animals is consistent with that of others in the area and this can be seen in the figures for the scapula neck width (MF9 Bone 14). The beasts were slightly smaller in size than the domestic pigs recovered from North Elmham Park, except during the medieval period. Several large boar tusks were found which might mean that some wild pig was being hunted.

Horse (MF9 Bone 15)

Bones of Horse were present in all periods except pre-Period 1, and made up about 3 per cent of the total bone assemblage. Measurements were consistent with a small, slender animal between eleven and thirteen hands high (MF9 Bone 15). Metatarsal lengths showed a variation of only 60mm between the longest and shortest specimen found. One-fifth of the total were from immature animals under three and a half years old, but many of the teeth found were from much older animals, up to fourteen years of age.

Dogs

The dogs from the site varied greatly in size, but were few in number. There was one complete skeleton of a very small house dog, estimated shoulder height being 270mm (L2152, (22), Period 1). This is consistent with the smaller dogs in Harcourt's Roman Dog range (Harcourt 1974) and thought to have been too small to have survived without human protection. It was an elderly dog with some osteo-arthritis and small jaws which gave rise to overcrowding of its teeth: in the upper jaw the Pre-Molar 4 and Molar 1 overlap. The remainder of the skull was unmeasurable. There was a small dog, represented by its skull only, in Period 4 (L985, (194)): length 132mm and zygomatic width 46mm.

Other dog bones came from the medium and larger range of Roman dog. A dog from L351 ((107), Period 2) had several bones of a larger size than previously found. The estimated shoulder height was 600mm and the animal had osteo-arthritis of the neck which may have been caused by the dog having been chained up. In the medium range, several separate bones from Period 5 deposits were consistent with those found at Longthorpe (King 1987): shoulder height 510mm.

Cat

Cats were represented in the last three periods and they seem to have been similar in size to the present day domestic cat.

Hare

Hare bones were found in Periods 1–5 in very small numbers. They therefore do not appear to have been utilised for meat consumption. They may have been brought in by hunting dogs, or lived in the fields around the farmstead.

Deer

Deer does not seem to have been utilised on the site, except for fragments of antlers. These appear to be from Red Deer.

Rodent

Rodents probably inhabited the site, but only a few were present in the last three periods in this collection. One Mouse hind leg, tibia and femur were found, all from an adult. Their presence, however, is marked by teeth-marks caused by gnawing other animal bones.

Butchering and use

Signs of butchering by cuts and chopping were present on 30 per cent of all bones. Most of the long bones of Cattle were chopped through, probably to make usable joints of meat, and for marrow extraction. The femur and humerus were broken roughly across the shaft. The tibia and radius were chopped near the proximal end. A few bones were left whole, mostly metapodials. One scapula had a hole in the neck and many others had cuts around the articular joint, and had their necks cut through. The ankle joints had cuts around the calcaneum at the point of articulation, and around the centre of the astragalus, this would have been to remove the feet from the main usable carcase after slaughtering. L488 ((280), Period 5) contained a predominance of foot bones — perhaps this was near an area of slaughter. Most ribs from the Cattle were cut in half as is the butchery practice today when making rib and brisket joints. Many thoracic vertebrae were cut in half probably to produce two manageable rib joints of beef. The heads of Cattle and Sheep were probably removed by chopping across the proximal end of the axis vertebrae and the mandibles were often broken behind the third molar or chopped across the ascending remus below the articulation process to facilitate the removal of the tongue.

The innominate bone was often cut or chopped through at the neck of the ilium.

The skulls of Cattle do not appear to have been broken by pole-axing as a method of slaughter although one large skull had a broken area below the horn on the left frontal bone. Many of the horn cores had been cut or broken from the skull and one skull had cut marks all round the horns and many cuts on the right frontal area showing the removal of horn for working (AF35 2 (269), Period 5). Some Cattle long bones had been used for making implements and for ornamental bone work.

Sheep also showed marks of having been decapitated by chopping across the axis vertebrae and some skulls were cut in half to remove the brain, but some were left whole. Many thoracic vertebrae were cut in half and ribs were cut across mid-way as in present day butchery practice for rib joints. Most metacarpal and metatarsals were complete and, because of the slenderness of the legs, these would have been of little use as meat. The same applied to the distal end of the tibia. Most parts of the body were represented in the samples from each period, therefore whole animals were utilised on the site.

There were not enough pig bones to show any proper butchery techniques, but many of the bones were chewed as in modern households where dogs scavenge the bones when they can. Scapulae seem to be the most common bones found after mandibles, and these show that both young and old Pigs were present in small quantities.

Many bones, about 12 per cent of the whole collection, were chewed showing that they had been left exposed so that scavengers and rodents could get at them.

Few bones were burnt, perhaps indicating that boiling rather than roasting was the preferred method of cooking.

Use shown by disease

Oxen were probably used as draught animals, as was shown by disease in several areas of the body: torn epiphysis on the distal caudal side of the humerus, caused by stress when immature, was seen on several examples (*e.g.*, L473, (213), Period 5).

One femur showed wear around the top of its head (L382, (213), Period 5) and one innominate acetabulum showed wear from a dislocated hip which had been compensated for during life (L551, (144), Period 3). Several metatarsals showed signs of osteo-arthritic changes. L1318 ((190), Period 4) contained a cervical vertebra showing wear on the caudal articulation which could have been caused by using a yoke.

L2738 ((321), Period 5) contained a mandible of a young animal which had been broken causing a penetrating injury in which an abcess had formed preventing complete healing and probably causing its death. One metatarsal shaft had a swelling caused by periostitis (L993, (282), Period 5).

Disease in Sheep was shown mainly by wear around the middle of their bite, pre-molar 3 and molar 1, where in mandibles of older animals this area becomes thickened and the bone recedes from the teeth.

From the Sheep in L1467 ((10), Period 1), one horn core showed depressions which are thought to be poverty marks from poor feed, and one metatarsal was curved from rickets, and one had a greenstick fracture. Perhaps pasture was poor in Period 1.

Disease in Horse was seen in the following bones (mainly the head and hind feet were affected):

L1665, (305), Period 5. A Horse metatarsal had some swelling around the distal epiphysis.

L362, (100), Period 3. A Horse metatarsal shaft showed chronic Periostitis caused by repeated knocking from a faulty step.

L532, (288), Period 5. This contained a skull fragment of a Horse about six years of age with a damaged orbit causing osteo-arthritic changes above the eye.

L671, (236), Period 5. This had a Horse metatarsal, tarsal and phalanges 1 and 2; these were all fused together with osteo-arthritis either from injury or wear.

These signs of disease were too few to draw any conclusions as to whether Horses were used only for riding or used for carrying goods on their backs or in carts. The carcase of the Horse was probably utilised for meat after working the animals, as the long bones of femur and humerus were broken and cuts shown on innominate and scapula fragments show that the meat producing joints were butchered for human or animal use. The other long bones were mainly complete showing that these were probably discarded.

Conclusions

The overall picture which emerges from the study of these remains gives us one of a Cattle orientated farm, with Cattle kept for milk, beef, hide and horn, with young steers reared through the winter to at least eighteen months of age. Draught animals were used for heavy work on the farm. The Cattle were small animals, including many short-horns.

Sheep played a smaller role than their numbers suggest as their carcase weight was much smaller in proportion to their numbers than that of the Cattle; they were a small breed of Sheep that was probably more useful for their wool than for meat. There was a change in 'kill-off' pattern for Sheep over the life of the site: Periods 1 east, 3 and 5 being normal for a herd of Sheep kept for lambs and wool production. Periods 1 west, 2 and 4 showed an abnormally high proportion of young animals being killed and eaten on the site. However, this may be an effect produced by the elder animals having been exported on the hoof.

Pigs played little part in the domestic waste, and therefore presumably were little used in the diet of the site's inhabitants. Small herds of foraging pigs were probably kept.

Horses were present in all periods except pre-Period 1, their numbers remaining at about 3 per cent of the total in each period.

Wild animals formed an insignificant part of the assemblage, therefore, hunting contributed little to the economy, although deer and hare were present. The large dog may have been used for this purpose, and the small dog for keeping the rodent population under control. This picture is consistent with that of other rural sites in this area in the Roman period.

II. The Bird Bones

by Mary Harman,
with notes on the abnormal by Dr J.R. Baker
(Table 81)

Most of the bird bones were identifiable. Table 81 shows the numbers of bones identified from different species recognised in each period of occupation. A large proportion of the bones was from domestic fowls; the goose and duck bones were probably from domestic birds, but these can be so similar to the wild greylag goose and mallard duck that exploitation of wild birds cannot be discounted, particularly as there is evidence for it in the Anglo-Saxon period at least. Of the wild birds represented, there are a few from rook or crow: the former were eaten quite recently in this country. There are also several bones from ravens and one from a buzzard.

The most interesting deposit, from ditch F1114, Period 5, is a group of ten tibiotarsi; three left and seven right, none complete, but most represented by at least half of the bone, from a species of redshank, a little smaller than a lapwing, but larger than a snipe. It is tempting to regard these as waste from edible birds, but this does not explain the presence of tibiotarsi only; although the number of bones is relatively small, the occurrence of this bone alone must surely be due to something other than coincidence or culinary practices, though of course the bones may well have been collected from the table.

There are small cuts on the shank of a duck tibiotarsus from Period 4, probably the result of dressing or carving.

Two fowl bones are abnormal; these were submitted to Dr Baker, whose comments follow. From Period 3, there is the proximal end of a humerus: 'the distal end of this fragment, approximately one third of the way along the bone, shows a very irregular expansion and closure of the pneumatic cavity. This is a non-union fracture possibly associated with infection.' From Period 5 a femur has 'an irregular spur of new bone immediately below the head of the femur on the medial aspect, approximately 6mm x 6mm x 6mm. The cause of this is unknown.'

Period	Fowl	Goose	Duck (cf. mallard)	Other
1	3	2	2	Rook or Crow (*Corvus frugileus or Corvus corone*: 2. Rook (*Corvus fragileus*): 1.
2	2	-	1	Buzzard (*Buteo buteo*): 1.
3	9	-	-	Raven (*Corvus corax*): 1.
4	7	3	3	Swan (*Cygnus* sp.): 1.
5	14	-	3	Raven (*Corvus corax*): 2.
Anglo-Saxon	11	5	1	Redshank: 10. Rook (*Corvus frugileus* or *Corvus corone*): 2. Raven (*Corvu corax*): 1

Table 81 Numbers of birds from different species recognised in each period.

III. Human Pathology

by Frances Lee

(MFTable 82; MF10)

So little human bone was recovered that the information it provides is of very limited value. The report is on MF10. (DFM)

Chapter 9. Economy and Use of the Site

The dating of the periods:

Period 1 c.50 – *c*.175
Period 2 *c*.175 – 225/250
Period 3 225/250 – 300/325
Period 4 300/325 – *c*.375
Period 5 *c*.375 – early sixth century.

I. Introduction

The information available for assessing the economy is limited, being mainly the recovered bone and the layout of the site. Although the terrain suggests what the likely land use was, the plan of the site provides major clues for the balance of interests in managing the estate attached to it. However, the actual character of the site brings in issues which are larger than just the plain use of the farmstead.

We can only be reasonably sure that the whole of the core of the establishment lay within the excavated area after Period 2. Period 1 has all the appearance of having been a gradual development from a centre of unknown character north-west of the excavation and so imperfectly reflects the economy of the site then. Period 2 seems still to be an outlying part and, after then, only the developing Anglo-Saxon site points to there having been important activity outside the excavation. Although the unknown early site might have been a single family farmstead like the Werrington Enclosure (Mackreth 1988) or Monument 97 (Mackreth forthcoming), the particular emphasis which the site first achieved in Period 3, and then had confirmed in Period 4, may have been the outcome of some unperceived quality in the Late pre-Roman Iron Age and first-century Roman site.

No parallel sites are known. Many similar ones may have been partly excavated, but, without the extensive stripping carried out here, their similarity must remain unrevealed. The lack of parallels means that some of the arguments advanced can only be tested against evidence from similarly large-scale excavations. The principles employed in the following discussion are that what is proposed should be based on a set of practices which not only fit the evidence here but also are rooted in what should have been unchanging facts in rural economy. Some of the propositions advanced are unsupported, not so much by direct evidence, but by practical experiment and a lack of detailed knowledge of farming as it was carried out, say, before 1700, but that this was complex is shown by Loder's accounts for 1610–20 for his land at Harwell, Berkshire (Fussell 1936, ix–xxviii). The discussion moves from considering the relationship of the excavated site to the land farmed from it, through an examination of the information contained in Joan King's bone report (Chapter 8), to an assessment of the functions of discrete parts of the site.

II. The Land

The plan of the site, the features found within it such as the Droveway and the animal management systems, the barns with driers, point to a mixed arable economy from the beginning of Period 3, if not before. The bone information and the plan let us down in the developed Anglo-Saxon phases of Period 5. There is no evidence that the Anglo-Saxon Granary was in use when the site was abandoned and the plan of Period 5 (Pl. VI) shows that discrete areas which allow functional areas to be broadly defined do not really exist.

The topography of the site is uncomplicated. The farm developed at a point or line on a gentle slope down from south to north where a held water-table broke out to form a spring. This must have been a powerful factor in choosing the site, and the prehistoric enclosures around the area where the spring rose suggest that it had been an important one in earlier times. The high water-table made the creation of ponds easy and there were at least three of these before the creation of the ridge and furrow, one dating broadly to the Iron Age. In spite of problems with water which led to frequent digging of drainage ditches, its ready availability was probably a powerful incentive to maintain the site. However, once a centre had been firmly established, there would have been a degree of inertia at the thought of moving to drier quarters.

The spring-line itself resulted in a marked difference in the land usage on either side of it. This observation arises particularly from comments made by the last farmer of the land, Mr J. Hunting, who pointed out that the ground below the site was so wet that he and his father had never seriously considered ploughing it. When the excavation began, only the field in which the site lay was ploughed, largely due to a government incentive scheme of the early 1960s which paid a bounty for newly ploughed land. Below this field, all the land was pasture with well-preserved ridge and furrow. The land above the spring was, and would have been, drier, but the brick-pits immediately south of the site have been there too long for any current local knowledge to survive on the suitability of the land for crops. Therefore, it is only partly demonstrable that, in a mixed economy, the pasture would have been mainly north of the farm with arable to the south.

The Period 2 plan (Pl. III) shows this division, and a major entrance in Period 1 through the east-to-west ditch line has been argued for (Chapter 1). In Period 2, the Droveway led from the north to the same point, a route for traffic continuing through the site on the same line. The width of the Droveway, coupled with the layout of the posts in it, should be sufficient to show that it was not designed for foot or wheeled traffic alone, but primarily for cattle.

Whether there had been one point of entry to the site on the south side is unknown as the full perimeter there was not recovered for any period. However, the Period 3 ditches and Barn 3 formed a partial barrier and it should not have been accidental that Barn 2 in Period 2 was laid out to be entered from the south. Its replacement, Barn 4, shows that access was still from the south right into Period 5 and the new east ditch of the Main Yard in Period 3 shifted the approach definitely to the south-west (Pls III and IV) and should have confined access to the area between it and Barn 3.

The major changes between periods did not necessarily reflect a change in basic land management. The increase in the number of buildings, and other changes which attended them, point to the increasing economic importance of the site. To argue that animal management radically altered because of the alterations to the enclosure patterns in the Main Yard would be a mistake. The development of enclosures around the Main Yard indicate expansion and it is here that the limits of the excavation become barriers to understanding. The overall impression is that the small enclosures of Period 1 were replaced by a much larger one in Period 2 which allowed specialised areas to be laid out inside, all depending on access from the Droveway and all, therefore, related to animal management. Period 3 represents an expansion of these followed by their removal in Period 4. During these two periods, the enclosure system around the Main Yard became more complex: the areas available for use in the Main Yard had probably become too small and had to be replaced by larger enclosures on new ground. This is married to the greater number of buildings. Whether this means that the land attached to the farm was used more intensively rather than it having been augmented cannot be known.

It may not be coincidence that the new layout of Period 2 was done at a time when the next site to the west, Monument 97 (Mackreth forthcoming), was closed down. But nothing is known to suggest that the beginning of Period 3 also coincided with a similar event on a local site.

The lack of an identifiable pond in Periods 1 and 2 when one can be allocated to pre-Period 1 (Pl. I), and to Periods 3 and 4 (Pls IV and V) may be an embarrassment. However, the unknown early centre belonging to Periods 1 and 2 may have had a pond attached to it. If a pond was not apposite in Periods 1 and 2, the re-introduction of one in Period 3 could speak volumes for a basic change in the day-to-day running of the farm. The Period 4 pond probably continued in use in an unmaintained form into the early part of Period 5. But for the rest of Period 5, much would depend on how intensive the pastoral element of the farm was in Anglo-Saxon rural economy and how that was managed.

III. The Animals

The management of animals probably always had the most profound effect on the plan of a farmstead. Their introduction into the central area, presumably in numbers, was bound to create greater problems because of the needs of feeding, watering, calving or lambing, breeding and culling, than the bringing in of an inert harvested vegetable crop. Thus it is likely that a pastoral or mixed arable farm will always have a more complicated plan than a purely arable set-up, providing animals were present in quantity: it may be that the activity level of sites seen from the air may not be related so much to the length of time they have been occupied, or their 'wealth' or 'success', but more to the type of farming carried out from them.

The three main categories of recovered bones — cattle, sheep and pig — should be typical of a production site: most animals sold off presumably took their bones with them. It is the interpretation of the site-plan which argues for a sizeable animal population: cows kept for ordinary domestic use would surely not have had so much influence on how the farmstead was laid out. Pigs may have been an exception and they were only present as a small percentage. As for sheep, the chief product was undoubtedly wool, with milk on occasion, but the breeding rate should still have allowed for a sizeable excess over the actual meat requirements of the farm and, in any case, a sheep would only have a carcase value once it was deemed to be past its best as a breeding, milk and wool-producing animal.

Cattle were the preponderant element in the animal economy, both in numbers of bones and meat yield in relation to sheep. The latter may have been kept as a small relatively stable flock whose meat was largely consumed on site as a variation in diet. Much depends, of course, on the human population of the site and this may have been quite large from Period 3 at least. The meat eaten and deriving from the animals on the farm should be reflected by the finding of all parts of the animal. This is true of both cattle and sheep. However, there was very little veal which either means that next to none was eaten, or that surplus veal animals were sold off for immediate profit. Other than that, there is no particularly serious imbalance in the numbers of immature animals in relation to the fully mature. These generalisations apply to most sites; what is needed here is a look at the specific classes of beast at Orton Hall Farm.

Cattle

The cattle were generally kept over winter, most for four years or more. Three considerations arise from this: where were they kept in mid-winter?; where was their fodder grown and kept?; what was the purpose in keeping them? Perhaps the first question ought to be: are we looking at the remains from a selected breeding group with most of the animals of less than four years going to market? If this had been the case, the chief answer to the question of their purpose is given: they were mainly for meat and hides, the matter of their dairy value is considered below. The answers to the other questions depend very much on a closer look at the bone evidence from this site.

Joan King concludes that, in general, the cattle conformed with others kept on nearby sites, and Monument 97 may have been the most significant of these (Chapter 8; Mackreth forthcoming), apart from a noticeable group of medium-horned cattle which is at variance with evidence from other local sites. For the argument which follows, it is important to know that only one polled animal was found in the collection. If the horn type could vary within a herd, an emphasis on an unusual aspect might point to the whole of that group having formed an individual breeding herd. One valuable indicator that this had been the case emerges when the quantity of bones per period is looked at.

The horn cores of this particular cattle type seem to be confined to a basal circumference of 135–205mm (MF9, diagram 3) and, following this through the periods, it can be argued that it is present all the time (MF9, diagram 4) and, when quantified, gives a preponderance of 3:2 in favour of medium-horned cattle (MF9, diagram 8). The figures for each succeeding period may include a growing quantity of residual material, but it seems likely that there had been a distinctive herd present on the farm. If it could be demonstrated that this herd had persisted into the

Anglo-Saxon phases of the site it would be powerful evidence for the continuity of the whole farm, including stock, from one cultural milieu into another. This cannot be over-stressed and needs to be tested on as many Roman sites with an Early Anglo-Saxon presence as possible as the implications are more fundamental than any study of a category of artefact could reveal. Unfortunately, the evidence from Orton Hall Farm is not strong enough by itself to support the supposition.

That a single strain of cattle was kept absolutely pure for several centuries is not to be insisted on and Joan King notes that at least one bull may have been introduced to improve the stock. Here we come up against the fundamental problem in discussing the economy of a rural site when the information is primarily archaeological. The deductions which can be made are severely restricted and, while treatments *in extenso* are long in theory, they are short on fact (Applebaum 1972). The following discussion is limited to the airing of a few probabilities using, where practicable, deductions from what are essentially unchanging matters: whatever modern management systems have devised for them, the annual cycle of behaviour and basic needs of cattle will hardly have varied over the last two thousand years beyond what highly selective breeding may have been able to achieve.

We start with the idea that there had been a specialised breeding herd present on the farm.[1] It seems that a viable herd could have had no fewer than 20 to 40 beasts and what follows assumes a minimum of 30. Three points arise from this: what area of pasture would have been needed?; how was the herd over-wintered?; how was it managed? It is assumed that ploughing was done by using trained oxen, but not that they formed part of a breeding herd.

Modern cows are largely reared on artificially fertilised grass, and also receive supplementary feeds, therefore adjustments have to be made for likely conditions in the past. A fairly pessimistic view has been taken, the pasture being assumed to have been poor to average, any fertiliser coming from the beasts themselves. The soil around the site was not grade a in modern terms and, given the lack of modern aids, probably equivalent to grade b/c. The size of the animals should reflect the amount of pasture needed and Joan King considers that each mature beast probably had a live weight of *c*.200kg. This is half or less than a modern animal, which would need about 0.6 of a hectare. Allowing the same area to compensate for what may have been poorer grass, 18 hectares, or *c*.44 acres, in all would have been needed. This is, at present, a minimum requirement, much depending on how the herd was managed.

That the beasts were over-wintered has been established, but it is far from certain how this was carried out. The assumption behind Applebaum's arguments was that cattle would have been housed as a matter of course (Applebaum 1972, 144–8), yet a look at the buildings at Orton Hall Farm does not support this, and definitely not at all in Periods 3 and 4.

We start off with two barns in Period 2. Barn 1, it is argued, was given over to fire-using activities which, themselves, probably needed fairly large working areas around them. Barn 2 was, certainly in Period 3, fitted with a drier and other fire-using features (see below) and may have been so used in Period 2 as far as the evidence actually goes. In Period 3, Barn 1 may have been available and the drain in the south wall may have had something to do with collecting urine, but that use did not continue into Period 4. Barn 3 was built in Period 3, possibly of timber to start with. The use of the area west of the partition wall would not have suited the stalling of cattle. Barn 4 was a direct replacement in every sense of Barn 2 and so cannot have been used (see below).

In short, the changes between the barns do not add up convincingly to an image of automatic provision of stalling, although Barn 1 may have been adapted in Period 3 for such a purpose which *could* have continued into Period 5. The building of the yard walls in Period 4 may have led to the disuse of the drain rather than being due to a change of use of the barn itself. Using Applebaum's determinations, a space of about 1.5m by 2.25m (*ibid.*, 145) would have been needed for each beast and, on this basis, Barn 1 could have housed 28 beasts, very close to the assumed 30 beast herd. This is not necessarily the total at any one time and is by no means close to the probable real cattle population of the site in and after Period 3 (see below). As Barn 1 was the largest on the site and was not big enough when it was first available, the thought that cattle had to be over-wintered under a roof must, for this site, be set aside. The next largest, Barn 3, would only have taken 24 beasts and is also Period 3 at earliest. There is no good reason to bring cattle in every winter in lowland Britain, a point made by Ryder in relation to the Iron Age (Ryder 1981, 332). However, plough oxen *may* have received special treatment.

The actual size of the herd, other than the assessed minimum of 30 beasts, depends on the way it would have been run. There was a noticeable element of animals younger than four years (MF9, diagram 9) and their place in the economics of a herd should be discussed before other aspects of over-wintering cattle. Not all calves born in a single season belong to one sex. It is also a fact that a dairy herd relies on there being cows to produce milk. As a breeding herd only needs the services of one bull, what happened to surplus males? Apart from replacing plough oxen, these could have been killed off as veal or fattened and then killed off before they needed to be fed all through a winter. Either way, they are slaughtered, probably all at the same time. This could mean that, allowing an approximate 75 per cent calving rate and the modern determination of the average birth-rate of 55 males to 45 females in every hundred, some 12 or 13 bullocks would need to be dealt with in a short time. They were almost certainly sent off on hoof to market. However, some may have been run, along perhaps with some of the cows, as meat animals for the site population and, along with young cows, there could have been a persistent element in the herd younger than four years. This would have the advantage that any of the stock members of the herd could be replaced if necessary and, assuming that four to five years was a likely lifetime, up to a quarter of a well-managed herd would need to be replaced in any one year. In any case, some seven or eight mature cows would also become available for marketing. On this basis, meat and hide production should have been an important element in the farming economy; it also follows that a basic herd of 30 could have been nearly a third greater at any one time. This makes the compulsory winter stalling of animals in the buildings at Orton Hall Farm hard to swallow. Cattle do not like to be kept in close-quarters with others with horns, and the almost complete absence of polled cattle on the site should be a valuable sign that beasts were not housed as a matter of course.

Shelter would only be a consideration in the foulest of weather, and only if that persisted. What is more important was that there should be winter feed available in case there was a covering of snow. It should be assumed that an adequate supply was provided every year to cope with all but the very worst conditions. As a rude guide, a maximum of sixteen weeks' feed is envisaged. One point in the running of a herd which may not be immediately apparent is that, in wet weather, there is a danger that pasture can be seriously down-graded because of repeated and intensive treading. This would have been watched for and the beasts may have been gathered in crew yards when there was need. The identification of crew yards is not easy, and the one in the Werrington Enclosure (Mackreth 1988, 77–80) was only identified because it was partly preserved under a medieval headland and phosphate levels were established. If relatively close confinement of this nature was a regular feature of winter life, then fodder would have been essential.

Scaling down from modern allowances to a probably generous range, 0.5–0.75 tons of hay may have been needed for each animal during the sixteen weeks. The basic herd of 30 would need, therefore, *c.*15–22.5 tons and taking 18 tons as an arbitrary figure, this may have been raised to *c.*21–24 tons if extra stock had been run in with the herd in the way suggested. Manuring of permanent pastures may have made it possible to recover nearly 5 tons of hay from one hectare. Therefore, in addition to the plain pasture needed, there may have been a further 4–5 hectares: on the assessed fairly low quality of ground available to the farm, the area needed to run its cattle could have been between 26 and 30 hectares.

To summarise the economics of the meat and hide side of the herd, a herd of 30, with a quarter being replaced every year from reared heifers allows the following product: 7 mature cows, 12 bullocks and 4 heifers, a total of 23 beasts. Once the system was running, there could well have been a significant proportion of one- to two-year-old bullocks and heifers, kept to improve their meat content. The turn-over represents about 50 per cent of the animals in any one year on the system outlined.

To modern eyes, however, the dairy aspect might seem more important. With milk available the year round today, it is often forgotten that this does not occur in nature. A cow milked regularly will have a shorter dry period than a plain cow in a meat herd. Looking at this conservatively and bearing in mind the marketing aspect only, the maximum lactation period is obviously during calving which would have begun, it is assumed, in April with the peak milk production in May. Figures for modern yields vary according to various factors only some of which will have applied in the past. For instance, the size of beast, the quality of feed and, perhaps, the breed itself are relevant. If a modern cow gives, say, 19 to 23 litres a day during the peak period lasting about four weeks, it may be legitimate to think that the typical Roman cow on this site may have produced between 10 and 12 litres of which only about 3 to 4 litres may have been surplus after the calf had had its fill. If the total bulk spare milk yield in a year is reduced to the equivalent of eight weeks at maximum lactation, and only the 75 per cent of successfully bred cows is considered, there would have been a maximum of 5,040 litres. Of course, this would not have been confined to eight weeks, but it is an indication of the product which might have been expected from the herd. It is hard to conceive what this means but, if one had a litre a day, it would take 90 people to get rid of it during those eight weeks. Over a year, it would need about 14 people. Such a bald approach is out of place and it should be obvious that almost all milk products were prepared. Butter and cheese are the principal ones and, in theory, each would have produced buttermilk and whey respectively as by-products as well as cream. None of these is discussed by Applebaum and there is only one mention of cheese in connection with sheep (Applebaum 1972, 215).

Butter was sometimes stored underground and several finds have been made (Ryder 1981, 314), but none is dated to Roman times. Using modern figures as a basis, 1,000 litres of milk could have produced 40.9kg of cheese or 90kg of butter (*ibid.*, fig. 61). The site produced what are called cheese-presses, but these were probably used to prepare a soft cheese akin to a cream or Colwick type. This is intended to be eaten fresh and could only be made in quantity when there was a lot of milk available. What is unknown is whether any 'hard' cheese of, say, Cheddar type was made. If there was any wish to convert extra milk into something which could be stored for the lean winter months, it would have to be a cheese of this type. Any cheese stock created would have been supplemented by using sheep's milk and this was probable as there was a healthy percentage of sheep, going up in Period 3.

There is no satisfactory evidence for large-scale production of cheese. Yet the evidence from Italy is that it should be expected, although there mainly sheep's milk was used (White 1970, 310–11) and in sufficient quantity for the whey from the milk of ten sheep to be enough to feed one pig (*ibid.*, 508, n.144). And enough cheese was made for recommendations to be put forward for how it was to be produced and stored for keeping (*ibid.*, 315). It is hard to see that what was practical in Italy with sheep would not have been doubly so for cattle on the richer pastures of Britain. However, although cheese made from cow's milk is the preferred type in Britain today, the role of sheep's milk should not be underestimated. In Italy, cheese for storing was prepared using salt and frequent pressings (*ibid.*) and baskets may have been used in Britain to allow excess liquid to be expressed: the pottery cheese presses are all small and are never found in sufficient quantity to suggest that they had formed an important part of cheese-making.

It is instructive to look at the actual numbers of bones found expressed as numbers for every twenty-five-year term, the point being that the periods were of variable lengths with Period 5 being the longest. This makes it hard to arrive at a quick assessment of what the sheer quantities may mean. Table 83 below gives the result and is based on Table 80 (MF9). The figure in parentheses after each period is the number of twenty-five-year terms in the period. Those in parentheses after the Period 5 bone counts represent a 20 per cent reduction to allow for some residual content.

What the Table shows is coarse in detail and discussion is confined to the major economic animals. The high sheep content in Period 1 is, as Joan King says, the result of a single large bone deposit possibly having a special significance. The surprising feature is the three-fold increase in both cattle and sheep between Periods 2 and 3, all the more marked as it is not repeated for pig, horse or

Species	Period 1 (5)	Period 2 (3)	Period 3 (3)	Period 4 (2)	Period 5 (6)
Cattle	142 35%	148 63.5%	432 65.8%	495 64.2%	597 (478) 61.9%
Sheep	232 57.6%	63 27.0%	197 30.0%	228 29.5%	285 (228) 29.5%
Pig	10 2.5%	13 5.6%	16 2.4%	22 2.8%	34 (27) 3.5%
Horse	6 1.5%	8 3.4%	9 1.4%	21 2.7%	41 (33) 4.3%
Dog	13 3.2%	1 0.5%	2 0.4%	6 0.8%	8 (6) 0.8%
Totals	403	233	656	772	965

Table 83 Number and percentages of bone per twenty-five-year term within Periods 1–5.

dog and there was no other major fluctuation, Periods 4 and 5 remaining at the same relative level. There is also a great increase in the number of vessels in Period 3 over that in Period 2 (Chapter 6, Table 77). Only Period 1 shows signs of coming close, but that lasted for nearly twice as long. The sudden increase cannot be the result of a jump in the residual content in the deposits. It must be related to a change in the economic value of the site. Although refinements can be introduced to allow for an increase in residuality, such as grading the figures from 10 per cent for Period 3 to 30 per cent for Period 5 *etc.*, this would not disguise the sharp difference between Periods 2 and 3. The allowance of a residual quantity for Period 5 was based upon the undoubted Roman residual pottery content in many layers; the same could not be said to have been true to the same extent for any previous period.

A study of the horn cores, again, is revealing (MF9, diagrams 3 and 4): the medium-horned cattle carry on through the rest of the site's history, even if the details for Period 5 are a little obscure, but the short-horned element only arrives in Period 3. The conclusion seems unavoidable: another herd was added to the animal population and this would have had a dramatic effect on meat and milk production, especially if the figures are evidence for a three-fold increase. It is unlikely that the new short-horned cattle element was twice the size of the old, but the greater number of medium-horned horn-cores probably shows that the new herd was the size of the old which was itself then doubled.

Period 3 was marked by the building of the House, the stone walling of the Small Yard and Barn 1, the construction of Barn 3 and the formal definition of the south side of the Main Yard. Problems in Period 2 in defining where any domestic focus may have been and what formed the western edge of the site are now resolved. The leap in the figures for both cattle and sheep must reflect the numbers of animals present and this helps to define why the facilities at the farm had been improved. It accounts for the addition of more enclosures along the east side of the Main Yard and it was now that a pond was provided inside that. There is in all this no sign of gradual growth, it smacks of a single concerted scheme. All-in-all, there must have been a change in the status of the site even if the basic economy remained the same.

A consequence of having more animals was either a reduction in the amount of arable or an increase in the amount of land managed from this centre. If, however, the drier in Barn 2 belongs exclusively to Period 3, and there was an increase in buildings in Period 4 with more driers and a Mill-house, the probability is that the farming unit increased by a factor of at least three. The driers, however, should not in themselves be taken as evidence that the arable content had altered, but perhaps the Mill-house should reflect the persistence of arable in the overall economy, although it may not be wise to argue that that had been augmented as well.

Sheep

The sheep on the site were of the normal type found in the lower Nene Valley. Table 83 shows that there was a three-fold increase in Period 3. There appears to have been nothing abnormal in the kill-off rate overall and there can be no doubt that the animals were kept firstly for their wool and secondly for their meat and milk. The amount of ground needed for a minimum viable flock has not been examined as sheep seem to be less fussy than cattle and able to cope with poorer conditions and were probably run over stubble between harvest and ploughing (Fowler 1981, 160; Ryder 1981, 306, 335). This would vitiate any exercise whose main purpose would have been to estimate the overall area of land attached to the farm.

Joan King, in looking at the percentage of immature animals slaughtered in Periods 2, 3 and 5, considers that they represent the keeping of wethers for one or two shearings before disposing of them for meat, presumably when fattened. The variations which can be seen (MF9, diagrams 10 and 11) suggest that Periods 3 and 5 differed from the others, but if these represent normal practice, then it is Periods 1 and 4 which should be looked at. In Period 1 there was a single very large and abnormal group of animals in a pit. The metacarpals indicated that they had come from shorter and more slender animals than the others recovered from the site. Joan King surmises that this element in the sheep population was either a 'foreign' flock or a 'native' one which was deliberately killed off. In terms of the chronology of the site, the second suggestion is the more attractive. Rather than a different strain being deliberately introduced, the flock from which these animals came perhaps became diseased to the extent that their owner was forced to buy a replacement flock of what was now the preferred breed which became the standard for the rest of the history of the site. As for Period 4, the oddity is the rise in the number of slaughtered immature animals. This could well be a reflection of a greater consumption of sheep on the site and perhaps suits what seems to have been an increase in the population of the farmstead.

Pig

The pig population, expressed as a percentage of the other meat producers, was small, smaller than on many other sites (*e.g.*, Mackreth 1988, 147; Mackreth forthcoming) which generally had a more marked Iron Age presence than was the case with the *excavated* area of Orton Hall Farm. There was no increase in Period 3. There was, however, a change, if the small number of bones can be taken as a true sign, in Period 4 and it went up again in Period 5.

The percentages for all five periods are very low (Ryder 1981, table 3), which could mean that there was less woodland attached to the site than to others, as it is generally assumed that pigs would have been run on the pannage system. There is little evidence that this was the common practice all over Roman Britain and any deductions about woodland based on the pig population should be treated with some care. A better indicator of woodland may actually be percentages of deer, except that one of the principles in managing woodland is that they were excluded.

An analysis of the pig bones found shows that nearly half came from the head. In other words, the best cuts of pig were not consumed on the site. Even if the pork joints had been preserved by salting or smoking for consumption in the winter, their bones should have been found unless they had all been reduced to ashes in fires. However, calcined bone was noticeably uncommon on the site. Pig may only have been reared for export either to market or to a site of higher status. After Period 1, only Period 4 has a greater proportion of slaughtered immature animals so matching sheep in the same period.

Horse

The percentage of horse was only just below that of pig, but actual percentages of bone are not a good guide to population. The animals were relatively slight and it seems that they would not have been used for ploughing or for pulling carts. If horses were a normal part of the management of animals, it should mean that more were needed in Period 3, yet the figures do not show this. Indeed, the percentages could be said to be distinctly low (Ryder 1981, table 3) and this is borne out by the actual numbers of bones per quarter century (see Table 83). There is, however, an increase in Period 4, but the increase in Period 3 is very marked and will not be a reflection in this case of an increasing degree of residuality. They were presumably ridden or used as pack animals. In the context of the time, and bearing in mind the isolated nature of most rural sites, the horse would have been the only efficient means of getting around the countryside.

Most of the horses on the site were mature and, as has been demonstrated elsewhere (*e.g.*, Mackreth 1988, 148), there was no squeamishness about eating them once they had ceased to be useful.

Deer

The tiny amount of deer, only 17 bones for all periods (MF9, Table 80) and these mainly antler, shows that they played no serious part in either the domestic or main economy of the site: deer management would have produced more waste no matter how far away they were confined. Deer were a preferred meat source during the winter in the Middle Ages and were reserved only for the highest classes. The same might have applied in the Roman period. As our knowledge of the general use of the landscape round Durobrivae increases, it becomes less easy to argue that there had been considerable areas of waste in which deer could have roamed at will. It looks increasingly as though they would have been kept in special reservations. This implies that venison was only for those who could afford to keep land out of cultivation. In other words, deer were kept essentially as they are in England today, in parks.

IV. Cereals and Mills

There were more millstones than querns on the site. A mill was not necessary for the plain production of flour or meal for domestic purposes, the only reason for having one would be to produce flour or meal in quantity. Little practical experiment on mills seems to have been done since the work carried out by Jacobi (1914) and the following discussion is more an exploration than a positive statement of likely conditions.

We start with the form of the mill and how it was driven. The first has been discussed by R.J. Spain (Chapter 5) and the probable system for mounting has been discussed in Chapter 2 but some consideration of the Saalburg experiment is helpful here. Although parts of geared mills have been found in Britain (Manning 1964; Neville 1856, 9–10, pl. 3, 28) their identification rests primarily on parts recovered at Zugmantel (Jacobi 1914, *Abb.* 43–4, *Taf.* XVII). Mills as such are usually equated with urban populations as the hub of a water-mill from Lincoln (Wacher 1975, 136) as well as a Pompeian-style stone from London (Wheeler 1930, 109, fig.34) demonstrate, or with the military (Richmond 1966, 161, 82–3; Simpson 1976, 26–50). However, rural mills are known at Fullerton (Wilson 1965, 217) and Ickham, Kent (Young 1981).

Jacobi, using Vitruvius' description of a water-driven mill, reconstructed a geared system driven by men by means of a crank (Jacobi 1914, 94, *Abb.* 46). The experiments showed that one man could not work the mill, but with four to six, it was easy. We are not told whether six men were needed to overcome the inertia of the system and then four only to keep it going, or whether there were relays of two, once a flywheel effect in the large cog-wheel had been achieved. However, they were able to produce 100kg of flour or meal in an hour.

At Orton Hall Farm, the simple mechanism proposed is that of a main spindle, with radial arms fitted directly to it, seated in the mill on the floor and in a beam overhead. In other words, no gearing: there would have had to be direct evidence for this in the form of parts of the 'small wheel' found at Zugmantel (*ibid.*, *Abb.* 44). As the purpose of gearing was to increase the speed of the mill beyond that possible by direct drive, a straight transfer of the results of the Saalburg experiment cannot be made. The gear ratio adopted by Jacobi allowed the stones to turn nine times to every complete revolution of the crank and the rotation of a mill at Orton Hall Farm must have been much slower than the 160 times per minute at Saalburg. Much would depend on the mechanical efficiency of each design and a direct drive may have lost less energy than would have been dissipated in turning a geared mill. At the very least, the Orton Hall Farm mill would have used two men in order to balance out the forces acting laterally on the spindle (St John Hope and Fox 1898, fig. 1).

Even if we assume that all things were equal and that the grain used in both types of mill had been the same and ground between stones set the same distance apart, we have no knowledge of the quality of flour each would have produced or how many times the 'overtails' were sieved out and reground. Even beyond that, we have no certain knowledge of how often or for how long mills would have been in operation. The Saalburg case would perhaps be easier to deal with as the mill or mills there were presumably to serve the garrison of the fort and the size of that would set a limit. The crux of the problem at Orton Hall Farm is that we do not know why a mill was needed in the first place, still less why three. If the mills were not for the purpose of supplying just those on the farm, resident or non-resident, were they for private gain or public requirements? Was the flour or meal exported in all or in part? Without some examination, however crude, of these points, the implications arising from one or more mills cannot be brought out.

What is at issue here, however, is not so much the quality of any flour or for whom it was destined, but what kind of through-put should be expected and, therefore, what kind of storage facilities may have been needed. We start by looking at the relationship of a given amount of flour, the amount of grain required to produce it and the space needed to store it. A detailed analysis has already been undertaken by Moritz and, at the level of argument possible, it is only necessary to take his assumed best weight of wheat and a volume of one cubic metre: 48lb/ft^3 (Moritz 1958, 190) or 21.77kg/ft^3 gives 769kg/m^3, or a weight of just over three-quarters of a ton, and is the equivalent of *c.*114.6 *modii.*

The flour produced from this, if wholemeal, would have been of the order of 480kg (*ibid.*, 191, 197–8, 204, 206). The Saalburg experiment was able to produce 100kg/hour, but of what quality is unknown. In other words, using four to six people, it would have taken nearly five hours to deal with a cubic metre or 115 *modii* of grain. Removing the increase due to gearing, 18 turns of a directly driven mill per minute would, theoretically, have taken about 45 hours to deal with 115 *modii* or one cubic metre of grain. But this speed is unrealistic and a similarly large amount of ground grain could only be achieved by doubling or trebling the number of mills. In these terms, three of them in the Mill-house could have coped with 115 *modii* a week without much difficulty.

Following this line of argument through, and modifying to allow for 100 *modii* per week for six months of the year only, the storage capacity for grain on the site would have needed to be about 23 cubic metres of grain. The estimated 93m floor area of granary in an upper level of the Mill-house would have provided plenty of room: if only 50 per cent of the floor had been used and the grain stored loose in bins, it would have been 0.5m deep, well within the constraints of any reasonable structure. There would have been no thrust on the side walls to speak of, the load on the floor would have been about 384kg/m and inspection would have been relatively easy. Grain may have been stored in sacks and this should have reduced the weight on each square metre of floor, but only if a greater area was taken up. If 20 per cent of the floor had to be set aside for access and feeding the mills from the upper level, the weight would have been reduced. In these terms, it should have been possible to keep enough grain in store to have kept the mills working the whole year round.

The Mill-house was put up in Period 4 when the farm reached its maximum capacity as an economic unit, a time when the animal bones point to a change in part of the animal management system. These details suggest that there had been a real change in the way the farm was run. It may be questioned whether ordinary landowners would have needed to change their farming policy so dramatically, as the means available to them for disposing of the produce from their lands should not have changed significantly from what they had been in earlier times. The emphasis on the greater production of meat from immature animals and the installation of mills could imply that the produce was being directed at a single consumer, and quantity may point to a specialised one at that (see Chapter 10, p. 237).

V. Aspects of the Use of the Site

Animals

The chief activities which need to be considered are the care of animals, the processing and storage of produce from the arable part of the farm, and the housing and provisioning of the inhabitants themselves. The evidence is generally weak or non-existent, and the division between what can and cannot be acceptably deduced is not always clear. The discussion is confined to the excavated area and the plans of the various periods.

To begin with animals, there seems to have been no special provision for housing cattle. If flocks of sheep can be left out in all weathers in the highlands of Britain today, it should have been possible to do precisely the same in lowland Britain in Roman times. Sheep would have been more hardy than the average cow which, in any case, was more valuable, dead or alive. As both cows and sheep are thought to have been present in some quantity, somewhere between 90 and 120 cattle in Periods 3 and 4 on the basis of the type of herd described, they could only have been allowed into the core of the farmstead under special circumstances.

The same is not true of pigs as they were a much smaller part of the animal population. They may have been kept close to the farm, but no sties can be identified. However, if they were run in woodland, their permanent quarters *could* have been some distance away, but this seems unlikely. Because horses were almost certainly for the use of the people running the establishment, they would always have been kept close at hand. As the relationship between person and horse should be assumed to be more personal than with any other animal apart from dogs, more care would probably have been taken of horses during foul weather and it may be that proper stabling arrangements were provided. If the living accommodation was mainly on an upper floor in the House, there could have been room on the ground floor. Horses tended to have a high value and they may have been brought into the security of the Small Yard virtually every night; if so, stables are definitely implied.

The most important period of the year in the management of both cattle and sheep, one during which it was essential that as little as possible could go wrong, was during calving and lambing. It has been argued that, as part of the over-wintering arrangements, cattle were brought in close to the farmstead. It may have been that, for calving itself, the beasts were brought into the farm and may even have received temporary housing until the health of the offspring was assured. There is no need to look for accommodation for all the cows in calf, for they would not all give birth at the same time and the signs of imminent birth are distinctive. Perhaps only cows with a history of difficult births were given special care along with those which looked as though they were in trouble. In these cases, the east part of Barn 3 may have served as temporary accommodation, although use of Barn 1 may not be out of the question. In Period 2, before the increase in the animal population, Barn 1 was the only building which could have served these functions unless Barn 2 was reserved for it. A marked feature of the Main Yard in Periods 2 and 3 is the collection of what look like specialised enclosures along the northern edge, all apparently designed to be easily approached from the Droveway. These could have been part of an elaborate system of sorting and separation tied in with marketing arrangements and, possibly, with the castration of unwanted potential rams and bulls.

The provision of a pond in Period 3, coupled with the enlargement of the minor enclosures to its north probably reflects a greater population of animals, especially cattle. In Period 4, when the buildings reached their greatest extent and the enclosures in the Main Yard were reduced to their minimum, the shift of the pond to the east might be due to a transfer of functions from the Main Yard to the eastern enclosures, especially if some of the features which can only be assigned to the Roman part of Period 5 really should be moved back to Period 4. The trouble is that we do not know if these enclosures were connected with the lands outside the core of the farmstead.

Before Period 4, the Droveway was the main route for herds or flocks, but the building of a yard wall across the line of the Droveway in that period appears to have cut off the Main Yard. This should indicate a major alteration in the routing of animals. In other words, while there is definitely no evidence that there had been a cut in their numbers, there was in a way a reduction in the facilities available for their management. If the main route was moved, it should have been to the east side of the site because of the walling in of the centre and west side. A sign that this may have been so lies in traces of a track being laid in to the north end of the reduced Barn 4.

Over-wintered beasts need fodder and straw for bedding. Until relatively modern times these were made into properly thatched stacks. If covered storage was available, presumably hay would have been the first to have been moved into it. There is no direct evidence that such stacks had been built in Roman Britain, although it is assumed (Morris 1979, 38). Applebaum only mentions hayricks once and then has them under a roof (Applebaum 1972, 147). However, there is the hazard of spontaneous combustion which applies both to straw and hay, but providing both crops were dried off and covered there was little danger. For this reason, stacks set away from buildings would seem to be the more practical proposition. In Periods 2 and 3, the area east of the Main Yard may have provided the best siting for them. They would have had the advantage of being both separate from buildings and general activities in the Main Yard, under easy supervision and readily accessible. With Period 4, however, things change. If the enclosures here were taken over for use by animals, fodder and bedding would need to be elsewhere. The apparent emptying of the Main Yard may have been accompanied by siting the necessary stacks there, especially along the south side.

Crops

Storage of a crop comes at the end of its processing and there is even less of a hint as to where threshing of grain might have been done. In Period 1 the whole of any cereal crop was probably handled in the area north-west of the excavation. This may have been true in Period 2, but Barn 2 is a likely candidate, especially as it was now that the layout of the site probably reflected the fundamental division of the land into arable and pasture. The barn lay outside the Main Yard suggesting that its use was not compatible with that which was surely for animal management. Threshing need not be carried out under a roof and no Romano-British barn has been demonstrated to have the opposed doors so typical of medieval and later ones. It should be imagined that arrangements were flexible enough to allow threshing to be moved under cover if the weather was against it taking place outside. The features inside Barn 1 in Period 2 place that out of court.

In Period 3 the position becomes complicated. Barn 2 was now essentially part of the Main Yard, but acquired fire-using features. Barn 1 seemed to lose these and was, as far as could be seen, empty of any impediment, but now had a drain. Its purpose is not certain and the barn may have been used as part of the animal management system. Barn 3 might have played some part in the handling of crops: its main door, certainly when it had stone walls, seems to have been at the east end facing traffic coming in from the south and, like Barn 2 in Period 2, could well have been laid out so that there was a large space for threshing between it and any entry into the Main Yard here. In Period 4 the arrangements become more complex, but for the processing of corn, more simple: if the granary had been over the mills, threshing may well have been carried out on the north side of the Main Yard nearby. But all these generalisations assume that there was only one type of cereal being grown. The mills imply that wheat was the main crop, but if barley was also grown, is it wise to assume that both crops would have been handled and stored in the same areas?

As for what could have happened in Period 5, there were too many changes both in the plan and character of the site for simple propositions to be made. For instance, the earlier part of the Period may be assumed to have been Roman and, apart from the loss of a good deal of the large rectangular building and part of Barn 4, the essential structural unity of the Period 4 plan was probably retained. The next stage may have been entirely Anglo-Saxon and can be reduced to the proposition that the Small Yard was kept, although the Roman House may have been replaced by the Hall. This would allow the Granary to have been put up in conjunction with it. In that case, the original granary no longer served or existed, or a supplementary store was needed. There is no guarantee that the new Granary survived to the end of Period 5.

It will not have escaped notice that, after Period 2, it has been a tacit assumption that the first and second north-east barns do not enter into any argument concerning storage or grain processing, yet it is here that the driers were located. Reynolds determined, when dealing with the reconstructed form of the Foxholes Farm drier, that such an installation would not have been capable of processing more than a small crop (Reynolds 1979, 38–9). Yet the implication of the general layout of, and tenor of the remains at, Orton Hall Farm is that animal management and the handling of grain was large-scale and two driers at most does not immediately seem commensurate with this. The large number of driers at both Hambledon and West Blatchington may point in another direction and neither site provides a good parallel for Orton Hall Farm.

Using Reynolds' results, the free atmosphere at Orton Hall Farm taking moisture away would have improved the efficiency of his best management system (Reynolds

1979, 32). His experiments suggested that the grain would probably have been no more than 0.025m deep and about 3.5 hours would have been needed to dry the batch. Two or three sessions could have been fitted into a day at harvest time.

The governing factor should have been the amount of grain to be treated. Using the same quantities as those arrived at when discussing mills and granaries (see above), $1m^3$ would have taken 19.9 hours in Period 3, 7.56 hours in Period 4 and 22.4 hours in Period 5 taking the actual area of the driers in each period into account. In other words, the amount of grain which could be handled by the triple mill arrangement throughout the year, and at two sessions a day, means that 120 days would have been needed in Period 3, 46 in Period 4 and 135 in Period 5. As the maximum operation of the farm was clearly only during Period 4, it could be argued that there may well have been a direct correlation between the size of the drying area and the expected yield. At three drying sessions a day, only 33 days would have been needed in Period 4.

Whether this is an acceptable figure rests on the criteria adopted in harvesting and storing grain. Reynolds concluded that the 1,200 hours needed for what he called a modest harvest of ten tonnes put corn drying out of court at Foxholes Farm (*ibid.*, 38). However, that was smaller than the smallest at Orton Hall Farm and surely the variation in the size of driers must have been related to the expected through-put of whatever was to be processed. Perhaps more telling is his comment that country practice in relatively modern times did not consider that dry grain storage was necessary (*ibid.*, 40–1).

Equipment

So far we have been concerned with the essential economic aspects of a farm. An abiding impression of modern farms is the large amount of equipment which has to be housed and maintained. In this respect, farms of medieval or earlier times were primitive, but the same needs had to be catered for and how it was managed is actually the hardest to detect on the site. One would assume that there had been carts and ploughs at least to take care of. One would also take it that a large establishment, and Orton Hall Farm should be counted as one, would need the services of a blacksmith. The only evidence for the presence of one consisted of an Anglo-Saxon pit with a large amount of slag in it, otherwise, there was very little slag on the site and no concentration to suggest that a hearth had been nearby. There were really too few buildings before Period 4 to consider that one had been exclusively allocated for semi-industrial use and the housing of equipment and tackle. Barn 1 in Period 2 possibly, but the best candidate is perhaps Barn 3. Its division into two functional areas, admittedly done at an unknown time, would allow cart storage to the east and more storage to the west where a blacksmith *may* have used F1135 (Chapter 3.II, No. **22**), but without leaving any characteristic residues. This barn could have been so used from the beginning of Period 3, but there is an alternative interpretation for Barn 3 (see below).

VI. The Inhabitants

There is no satisfactory way in which the numbers of farm-workers can be assessed. Estimates would have to be based on guessing the numbers of people needed to manage set numbers of animals. We have no knowledge of how specialised some of the tasks would have been, or whether the people employed doubled up their tasks according to the needs of the farming year, or were supplemented by casual or customary labour when the need arose. There was possibly a division between those who basically worked out in the fields and those who worked in the farmstead. There is also the fluctuating fortune of the site. For instance, if there had been a triple mill system running through the year, did it need an increase in personnel to run it? Is it fair to assume that a large farm of the third and fourth centuries would have had a work-force of at least twenty men and five to ten women? The agricultural revolution of the last thirty to fifty years has hidden from our eyes the fact that agriculture used to be very labour intensive.

If, for instance, we have to accept that the basic site population from the beginning of Period 3 was twenty to thirty people, and was based within the farmstead's centre as the great increase in the amount of pots suggests (Chapter 6, Table 77), where did they live? Or, at the very least, where did they sleep? Is it necessary to think that the status of most was so low that they would have slept where they could amongst the farm buildings? Or do we have to look at the House and see the upper level of that having been arranged as the kind of double-bunked and crowded barracks the writer remembers seeing in South Africa for black workers, or was the rectangular building given over to them?

Even if the total population on the site was limited to twenty people, and surely it could hardly have been less after Period 2, the cooking of their food would have been no light undertaking, or did they have to cope as best they might where they could? The brutal answer to all these questions may be that living conditions were as poor as can be conceived. But the quality of organisation of the farm here somehow goes against this. As for accommodation, the reality may well have been somewhere between the extremes. If the House was reserved for the farmer, or bailiff, or tenant and his immediate family, extended or not, the rest of the farm-hands should have had quarters according to the bias of their main work. Thus, the Mill-house may have had its staff close by. Those whose work was largely confined to the barns may have had partitioned-off sections and some traces which might have belonged to such were found in Barn 4. There is also the room added to the west end of Barn 3.

In such terms, the amount of space generally available may even have been generous. However, Period 4, when it is possible that there had been a strong official influence present, is awkward. The one building which has defied any functional interpretation is the large rectangular one in the middle of the north range east of the Droveway. Could this have been arranged as some kind of communal living quarters?

Food

The life of the lesser mortals on the site was almost certainly communal. No matter where they slept, worked or rested, the supply of their food and drink was almost certainly centralised. The consequence of that is that the preparation of both should also have been, but not necessarily, together. The question of where the cooking arrangements lay is the harder one to answer. Discussion in Chapter 3 suggested that some of the 'furnace'-like features may have been all that survived of cooking hearths or ranges. The headman and his family may have had separate facilities from his underlings, as well as a better diet. Assuming that the ordinary workers ate communally, a fairly sizeable room would have been needed, but if they collected their food and took it to their living quarters to eat, a large mess area does not have to be looked for. The absence of marked concentrations of pottery assemblages of 'table-wares' in any one period could point to the latter having been the normal practice.

However, in Period 4 the pottery from F434, which lay outside the entrance to the rectangular building, was domestic in character (Chapter 6, p. 155) and may be a pointer to the nature of the occupation there. The greatest population on the farm would have been during Periods 3 and 4 and it was at the start of 3 that Barn 3 was put up. This was divided into two parts, the western one having a stone-lined feature which might have been the base of a cooking range (Chapter 3.II, No.**22**). This part was also big enough to have sat all the numbers of workers and to have housed the stores essential for a kitchen. Although various functions have been discussed for Barn 3, this division and the addition of an extra room, along with the pentice-style structures attached to it, mark it out from the other barns. The simplest explanation for its real purpose may be that it was the communal centre for the farm-hands.

Drink

The original function of driers, where they occur in ones or twos in any period on any site, may be very different from the commonly held view. The interpretation offered here is that they were an essential part of a brewery. In common with medieval times, the standard drink of ordinary people in Roman Britain was probably beer, or, more properly, ale, which requires no cultivated hops.

The evidence for beer having been both drunk and produced on a suitably large scale is summed up neatly both by the drinking vessels found, especially in the third century, and the presence of British beer in Diocletian's price edict (Frere 1987, 285). If the site population was as large as envisaged, all that was necessary could hardly have been brought in from outside and, in common with large establishments of the Middle Ages, it should have been brewed on the spot. And surely the staple drink of the work-force would not have been prepared in a haphazard manner and by hole-in-the-corner methods. A proper centralised system should be expected, yet apart from one at Woodchester (Applebaum 1972, 112, 165–7, fig. 33), not one Romano-British brewery has been adequately identified.

The one reasonably common feature of large sites, and particularly of those belonging to the third and fourth centuries, is the drier. That this may have been used as part of the brewing process has been commented on (Morris 1979, 7; Reynolds 1979, 41–2). The identification of the Woodchester brewery (Applebaum above) is not useful as its layout and design is highly individual and cannot represent what was common throughout Roman Britain, otherwise others of a similar character would surely have been found.

The basic brewing process is both simple and unlikely to have altered very much over the centuries. The only major alterations would have been governed by changing tastes, refinements designed to produce cleaner or purer brews of differing types, and by the need to mass-produce to satisfy large urban communities. Only three materials, at base, are involved: grain — usually barley — water and yeast. There are only three basic types of equipment needed: a device to finish the malt, a vat or vats, and vessels for fermentation. The first stage is to prepare the malt. This is done by steeping the grain to soften it and begin the germination process; the grain is then heaped or spread to allow germination to continue. When this has been judged to have gone far enough, the grain is heated to stop further growth. The grain is transferred to a vat which is then filled with water and heated so that the sugars converted from the starch during malting can dissolve into the water. At the same time the flavour of the malt is taken up by the water to form the basis of the taste of the drink. The water was probably heated to boiling point to aid the process. When this stage has finished, the cooled liquor is transferred to a fermentation vessel and yeast added. After four to six days, the beer is ready. There may have been refinements such as the rootlets of the sprouting grain being removed, the malt being slightly crushed or 'cracked', and fining of the finished product (Boston 1976, 43–53).

Applying these different activities to what survived at Orton Hall Farm, Barns 2 and 4 were provided, in essence, with three different facilities. Each contained driers and these, in Periods 3 and 5, were associated with what are described as vat bases; both barns were given a water supply from a well placed as close to the main entrance as possible. The absence of a vat base in Period 4 is neither here nor there as the area in which the driers of that period lay was very heavily cut about and degraded in Period 5.

The simplest brewing process using these elements could have been to steep the grain in the vat in the first place, put it on the floor of the drier for germination to continue and then to stop it by the simple expedient of raising the heat of the drier. The quality of the malt would have been tested by empirical means, like every other stage of the process. The malt would then be put back into the vat, water added and the whole of that heated up. The liquor would then be drawn off and fermented with yeast. The spent malt would almost certainly have been used as a supplementary animal feed. No evidence for fermentation vessels was found, but these were almost certainly barrels and there is no good reason to suppose that they would have left any trace beyond, perhaps, fragments of hoops, if any had been in iron.

The important point is that beer could be produced all the year round and it was probably consumed in a fair quantity. At least one medieval monastic regulation was to the effect that new beer should not be served until it was at least four days old (Clark 1897, 185). This means that there had to be an overlap arrangement involving at least two sets of barrels, but not necessarily twice the number needed for one brewing. Beer drinkers can be fussy, and different strengths of brew can be prepared, but it may be ill-advised to think that winter-warmers or Best October (Swift 1909, 287, 308–10) were regularly made, if at all.

But high days and holidays may have called for extra effort.

That a brewing plant was a fixed feature of the barns once the first drier was built is quite conceivable. It may seem that there is a lot of unaccounted-for space in each barn, even allowing that most or all of one aisle may have been filled with barrels. If wheat was the grain stored in a granary belonging to the Mill-house, it is possible that barley, the preferred beer grain, was kept close to where it was to be used. It must be significant that, although Barn 2 was put up in Period 2, the features in it are best dated to Period 3, the time when the farm was greatly increased in size, with an influx of extra people to work it. The evidence of the site is that the drier formed part of a major economic unit into Period 5, and there was always a need to have at least one. If their size was related to the amount of malt to be prepared, then Period 3 capacity was, at 7.02m, roughly the same as that of Period 5 at 6.24m. But Period 4's capacity was at most 18.5m, more than doubling the requirements at a time when the buildings had their greatest extent.

The only large new building was the great rectangular one and, therefore, it could have been the living quarters for the implied extra hands. When that was reduced in size in Period 5, the north-east barn was cut down and the drier-area was reduced to its smallest area. The correlation has some attractions, but it may not be wise to build on it.

The frequency of driers and their sturdy construction suits an activity carried out through the year. A brewery would need to be housed in water-tight and weatherproof buildings. The temperature and moisture of stored grain need to be regulated and a steady germination rate depends upon the right temperatures. In winter, the drier was probably heated to encourage germination and careful temperature control would require the vents which are found in the better preserved examples. Temperature control is also vital in the fermentation stage: in at least one monastic house there were directions for keeping watch during the early stages, laying straw around the barrels and the lighting of fires if need be in winter, and closing windows in summer (Clark 1897, 185).

Much of what has been said is probably true for rural sites, but it is in towns that one should expect to find fairly large breweries: individual households may have catered for themselves, but the crowds coming to market from the surrounding countryside would expect a plentiful supply of refreshments. There is an enigmatic building on the south-east corner of *insula* III at Silchester. It has a colonnade or arcade along the main front overlooking the open area in which the *forum* stands. When excavated, the presence of one heated room, and what was described as a hypocaust in a large rear room, led the excavator to identify the whole building as a bath-house (Fox 1892, 281–3, pl. XXII). The hypocaust conforms in character with other driers and there is no evidence that it had heated a room as such. There was a water supply laid, not only into the building, but also directly into the large room containing the drier. The quality of the excavation carried out at the time was not such that all other features in the same area would necessarily have been identified and there is nothing in the whole plan of the building against it having been a *taberna*. Its prominent siting coupled with a covered area open to the street unusual in Silchester is in favour of the interpretation.

A fully functioning brewery at Orton Hall Farm would only have been of a substantial nature if there was a big enough call for its output. The remains do not allow a satisfactory estimate of brewing capacity, but a rough idea of the degree of likely demand can be given. As a guide, we can look at what was the standard issue to servants and boarders in that other great ale-drinking period, the Middle Ages. Whether there was a basic uniformity in practice is uncertain, but it appears that one loaf of bread and a gallon of beer per day was fairly average (Clark 1897, 217; Salzmann 1967, 477; Hart and Lyons 1884, 57–85; Hart and Lyons 1886, 134). This quantity given to twenty people at Orton Hall Farm would have needed nearly three hogsheads a week. If brewing took place once a week, six barrels at the least is not unreasonable as there would be the three being drunk and three being fermented; there may have been more if spares were kept or if there was an obligatory space of time before reuse. One brewing of this quantity would need one vat of 630 litres capacity which, when looking at the only large metal vessels known from Roman Britain, lead vats, is a sizeable amount. The capacity of those which can be measured is between 114 and 360 litres (Guy 1981, 275). They have no known function but if their relative quantity matches the number of breweries represented by the majority of driers, they *could* have been used as brewing vats, unless it is thought that the Christian symbols found on some renders them unfit for such a use. There is no objection to the use of lead as that was also commonly used for brewing and other industrial uses in the Middle Ages (*e.g.*, Salzmann 1967, 552).

VII. Conclusion

The image which emerges from the analysis of the economy and use of the complex of buildings and enclosures on the site is of a large establishment having many features in common with a medieval manor. The social order was different and therefore the architectural expression represented by the plan is not the same. However, both tended to act as major collection and storage points for a much larger holding than that enjoyed by the average peasant or yeoman farmer. Both had large staffs and these had the same basic needs in each period. The unknown element in Roman times is what part was played by customary rights and dues in the economy of such a unit. The strict image of a manor house should perhaps be replaced by that of a monastic grange: there was no person of high social status living permanently at Orton Hall Farm. But, like a grange, the Roman site may have acted as a collection point for the produce from a lot of separate holdings coming here at the will of a landowner or as a result of the rights which pertained to him. The most likely part of the economy subject to such a practice is the cereal element and, because of this, no attempt at assessing the area sown has been made.

The abiding impression is one of tight organisation and management, the controlling hand being able to alter at will the capacities of the different activities designed to take place. At some time not far short of the middle of the third century the site was radically replanned so that it could cater for a greatly enhanced number of animals and probably also a greater yield of grain. The change was dramatic, but it grew out of what was already there. The greater intensity of activity in Period 4, shown in the shift in the management of animals, the introduction of what seems to have been a mill-house and what could have been an increased beer production, does not easily fit the kind of establishment usually associated with a private landowner. The site may have been in official hands in Period 4 and had probably been so from at least the beginning of Period 3. If so, it probably remained imperial property right up to the time Anglo-Saxons took it over.

Endnote

1. I am grateful to Mr and Mrs D. Meyer of 'The Kennels', Ightfield, Whitchurch, Shropshire, for introducing me to some of the complexities of the subject.

Chapter 10. The Site in its Setting: Romans and Anglo-Saxons[1]

I. Settlement in the Nene and Welland Valleys

The basic distribution of known Roman sites in the area between the Welland and the Nene is shown in Figure 1. The picture is distorted by the uneven quality of the evidence: at the very least, modern Peterborough masks a large area along the Fen edge. The blank in the north-east corner of the figure is Borough Fen, an area too wet to have been settled. The other major blank area lies north-west of Durobrivae, shown along with its suburbs in solid black, and is now partly covered by a very large aerodrome. The whole of the Soke of Peterborough, as it was from Henry VIII's time until 1964, was the subject of a field survey carried out to a uniform standard by David Hall and his colleagues and this revealed a general lack of material along the southern edge of the Welland Valley. This may be particularly significant as the gravels in the valley bottom are very clean and air photographs reveal the archaeology in remarkable detail. The Nene Valley gravels, frequently masked by alluvial deposits, are 'dirty' in comparison. This makes the contrast in the distribution of sites between the two valleys the more striking: there can be no denying that settlement was thick in the Nene Valley compared with that in the Welland Valley.

Much of the development west of Durobrivae could have been linked to an industrial element in its economy. Iron-working is the most obtrusive on the ground, while pottery production is best known from the masses of locally made wares known both in the local area as well as further afield. Other industries either do not appear as such because their evidence is confined to specific sites, or were too specialised to affect rural settlement. The best illustration of this is the use of Barnack Rag for building as well as sculpture (Hartshorne 1847; Toynbee 1964, 211); there is no evidence for Roman activity in the area of the medieval quarries at Barnack and there must have been other outcrops which could have been exploited. The probable quarry at Jubilee Copse, Upton, and the certain one at Water Newton (excavations, B.R. Hartley) point to small-scale operations.

The figure also shows a broad band of settlement running north-east from Durobrivae to the edge of the Fens, broadly defined by the Car Dyke, and this kind of density almost certainly ran south under modern Peterborough to the Nene. The lack of knowledge of the minor roads and tracks which would have pulled the whole together leaves a rather shapeless pattern. This is true, of course, of most of Roman Britain, areas such as the Fens themselves being exceptional. However, elsewhere in the upland area there are signs of a whole series of minor routes as yet too disjointed to be mapped at such a small scale. There are strong signs that parts of the system in use under the Romans had begun much earlier and, in places, continued much longer. An extreme example is a route along the south side of the Welland, essentially south of the string of sites shown there. It was linked up with what appears to have been Bronze Age nuclei of the type excavated at Fengate (Pryor 1980) and was connected to or immediately next to centres of Iron Age occupation marked by enclosures similar to those at Werrington (Mackreth 1988) and Monument 97 (Mackreth forthcoming). The track continued in use throughout the Roman period, some of the sites along it acquiring Anglo-Saxon pottery and, between Helpston and Glinton, it only ceased to be used when Etton was enclosed in 1809.

Apart from tracks such as these, there are long ditch-lines with slightly irregular courses tied in with strings of sites dating, it appears, from Iron Age times in the first instance. This can be seen in the Welland Valley where recent excavation has revealed part of the enclosure system attached to such a line (Pryor *et al.* 1985, fig. 3). Similar lines of sites and ditches occur on the relatively high land between the Nene and Welland. Although the ground is too poor for good air photographs, it is tempting to see the long ditches laid out in the latter part of the first century AD in Period 1 at Orton Hall Farm as having run through to Monument 97, the next site to the west, to link with the field system developing there at approximately the same time. Such long ditch-lines represent, in one sense, a major land division of some form, possibly to segregate major economic areas in the local landscape, perhaps areas of predominantly pastoral character from arable.

These ditches running for kilometres across the countryside were almost certainly track lines of a less well defined type than the ones referred to in the Welland Valley. If so, one ran east-to-west through Orton Hall Farm, tying together the line of sites shown on the map and there may have been another running north-east through the sites west of the Car Dyke north of Peterborough. The Roman main roads emphasised the crossing point of the Nene at Durobrivae. There is no evidence that this represented a pre-Roman feature, but it almost certainly represented a rationalisation of an earlier arrangement much closer to medieval Peterborough probably serving the Westwood settlement.

The impression given by the archaeology of the area is that there was, from at least the third century BC, increasing exploitation of the landscape at the expense of natural wood and waste which itself was probably increasingly managed. The areas in which this development cannot be seen are anomalous, especially when this has been subjected to a uniform and concerted fieldwork programme. The generally blank area west of Ermine Street between the Welland and the Nene stands out here; the lack of discovered sites should not be taken to indicate lack of use, when managed woodland is as likely an explanation. The large-scale iron-smelting industry sited west of Durobrivae needed a regular supply of fuel and this, in a generally developed landscape, would have been supplied from coppiced woodland (*cf.* Hooke 1981, 140–2). A ready source of fuel would also have been attractive to potters, especially as there were suitable clay deposits near at hand as well as a developing market. However, one should not forget the possibility of there having been turbaries out in the Fens.

The excavations and fieldwork carried out in the fifteen or so years after 1970 around Peterborough have demonstrated that the settlement pattern was dynamic: many sites had their foci of activity moved from time to time and many were abandoned. But the processes are unknown. However, the history of Monument 97 may prove instructive. Founded sometime before the middle of the first century BC, the site was redesigned in the latter part of the first century AD at a time when Orton Hall Farm, only 0.9km away, was beginning to develop. Monument 97 seems to have come to an end between 150 and 175, when the domestic centre at Orton Hall Farm was rebuilt near the end of Period 1 and the Period 2 layout was created. The new layout was designed for a large animal component in the site's economy. The demise of Monument 97 at this time, hardly a coincidence, may have marked a reorganisation of its land so that it could be run from Orton Hall Farm.

The abandoned site was subsumed into a field system, and a cemetery, dated by its burial custom to the Late Roman period (Mackreth forthcoming), was established in the corner of what had been the largest enclosure and might have represented an element in the local population which felt an association with the old site. To some extent the same picture emerges from what can be seen of the development of the Roman sites at Werrington (Mackreth 1988). The Iron Age enclosure there became part of a larger land division system and a new focus sprang up next to the old one. In time, a major Late Roman centre was created a few hundred metres to the north and its field system developed across the almost flattened earthworks of the Iron Age enclosure. Similar clusterings elsewhere in the area covered by Figure 1 may point in the same direction: an increase in the size of holdings run from fewer centres.

If this was a general case, the larger amount of land attached to Orton Hall Farm from the beginning of Period 3, indicated by the evidence of the bones (Chapter 9, p. 225) may have been part of a rationalisation in land-holdings in the area. As Orton Hall Farm was never a major domestic centre, if mosaics and bath-houses are the signs of such, it might be asked from where the site was run. The nearest candidate for a villa centre lay in the western part of the park attached to Orton Hall itself (RCHM 1969, 29–31, (6)(14)). This seems to have developed from the third century and also was succeeded by Anglo-Saxon settlement (*ibid.*, 31, (16)). The evidence points to a major change in the organisation of the uplands generally in the second half of the second century AD and there is some reason for thinking that this was a reflection of what was happening in the Fens.

II. The Fens and Durobrivae in the Second Century

It is hard to see the foundation of sites in the Fens in the second half of the second century as an isolated phenomenon, if the Fens are thought to have been an imperial estate. The area is separated from the uplands by the Car Dyke, and it may aid understanding of the essentially different character of the two areas if the true purpose of the Car Dyke could be established. For a major earthwork now generally accepted as being Roman in date, the Car Dyke is perhaps the least explored and understood in Britain. The only other earthwork comparable in size and date is the Vallum along the south side of Hadrian's Wall, but that can hardly be anything other than military in some form while the Car Dyke should have a civil function, unless it had been a canal to supply the Roman army in the North. This was, of course, Stukeley's explanation and of such elegance that it has only recently been upset.

The work of Brian Simmons in Lincolnshire has shown incontrovertibly that the Dyke is neither level nor continuous (Simmons 1979). He also, following earlier observers, notes the difference between the physical character of what is called the Car Dyke north of the River Slea and the sections south of that (Simmons 1980, 65). In short, the northern part is most likely to have been a natural watercourse. S.J. Hallam points to a basic topographical difference between the same two stretches of country: north of the Slea, there is a marked drop along the west edge of the Dyke or river while to the south the countryside shelves almost imperceptibly from upland to fen without a marked natural physical feature (Hallam 1970, 32). This is a condition which applies right down to the Nene and in the region of the Cambridgeshire Car Dyke.

The Dyke is now generally accepted as having been a catch-water drain first and foremost. Simmons proposes that the Dyke was one essential element in the drainage of the Fens, being linked with the *Midfendic* to ensure that water on the higher lands to the east and west should not drain into the trough between (Simmons 1979, 192–6).

It is possible that a truncated version of this system was used to deal with Borough Fen, the large blank area already noted in the settlement pattern in the Fens. The course of the Nene in Roman times lay south of Whittlesey (Evans 1979), the line previously favoured being the Cat's Water (Hallam 1970, 35). This ran, as near as may be estimated, from the junction of the Car Dyke with the Nene north to join the Old South Eau, possibly the Roman Welland, an outflow from the branch of the Welland north of Maxey. The south branch was continued by Old Pepper Lake joining the Cat's Water at St Vincent's Cross.

The Car Dyke, deep and wide with a bank on each side, was generally set out in straight lengths. Whether there had been a berm between the Dyke and its banks is not well-established, but it has been seen often enough for it to be considered a consistent feature (Clark 1949, 148, pl. XVI; still visible at the watershed at Eye), and in this the whole is remarkably like the Vallum. However, in this instance, the berm may have been used for periodic cleaning out of the Dyke (Clark 1949, 150). It is odd that it should have been thought necessary to put a bank on the upland side. It is to be imagined, at the very least, that surface water could have been carried into the Dyke by way of field ditches, yet, patently, this was not allowed for.

While Simmons' argument is attractive and his demonstration convincing (Simmons 1979, fig. 7), there is a flaw: his system does not run south of Bourne. It may be that the effective coast-line of the Wash ran too close to the Dyke for the scheme to have been used (Simmons 1980, 65, fig. 31). If so, why was the Car Dyke necessary on such a grand scale, or needed at all?: the water from the uplands would drain naturally into a limited marshy area whose channels could have been improved to take the water away more quickly.

The crux of the perceived problem of the Dyke's uniform size, irrespective of the kind of land through which it ran, is represented by the section between the Welland and the Nene. Despite its sinuous course, the Dyke's artificial character is assured as it runs counter to all the natural drainage of the area. Its effectiveness as a drain can be seen from what we know of its base levels, as these drop away from the watershed between the two river systems (Pryor 1978). As a catchwater drain along the whole length it seems ineffective, as there is only one stream, Werrington Brook, which is captured. It is, of course, possible that problems with ground-water may have been a reason for its digging and this may have formed part of its function.

Between the Welland and the Nene, the Dyke respects the 25ft contour except at the watershed at Eye where a move of over a kilometre to the east would have been needed (*ibid.*, fig. 15). It was clearly expedient here to take a short cut, but its size was in no way diminished, even though, as the ground slopes away fairly markedly to the east, a shift of one or two hundred metres would have reduced its base level and, presumably, increased its effectiveness. If the Cat's Water, or the *Midfendic*, are equal parts of a drainage scheme, why is the Car Dyke so massive between these two rivers where it does not seem to have been very efficient?

The answer to the overt purpose of the Dyke is encapsulated in a comment made by Salway (1970, 10): 'the whole of these greater imperial estates lay outside the normal system of *civitates*.' If so, they lay outside ordinary civil administration and justice. Just as the Vallum is argued to have been the demarcation of an area which was strictly military (Frere 1987, 118–19), so the Car Dyke could have represented a significant change in the character of land administration, a possibility tentatively alluded to by Simmons (1980, 64). The Dyke, therefore, would be the boundary of the imperial estate in the Fens.

That there was such an estate is generally accepted, but its existence has been argued from the nature of the settlement coupled with the artificiality of some drains, droves, and roads there (Salway 1970, 10). The large labour force needed to create what is in effect a barrier and its uncompromising character when it could have been modified with advantage, as at Eye, should reflect some form of directive from the highest level in the empire.

There was no need for such a work north of the Slea, and if the true purpose of the Dyke has been established, there is no need to look for a dominant man-made feature between the Nene and the Old West Ouse as the major watercourses running north would form just as effective a marker (Clark 1949, fig. 9). Only in the land of meres south of Whittlesey may difficulties have arisen and it is here that what is thought to be a marker of a boundary of public land was found (RIB 230).

There can be no doubt that such a large tract of land would have needed to be administered on a large scale. The produce would need to be tallied and directed to its consumers by a procurator of standing (Crawford 1976, 52) who, even if he worked through middlemen (Salway 1970, 10; Crawford 1976, 47–50), would have needed considerable clerical assistance. There should have been an *entrepôt* to handle major classes of produce at least and one feature of the Fens is that there is no hint of a town or port. There are only two candidates which could have served as the major centre of both administration and exchange: Lincoln and Durobrivae. Lincoln had the higher status, but was, and is, distinctly peripheral to the Fens. Durobrivae, on the other hand, was ideally sited and has the distinction of being the largest site of its kind known in Britain. As a centre of major communications it was unrivalled along the Fen edge and, more importantly, the Fen Causeway, the only major road across the wetlands, led directly to it.

Durobrivae could not have monopolised all the trade of, or supply to, such an extensive area, and contacts with the nearer parts of East Anglia and the fringe areas of Lincolnshire should be expected. But only Durobrivae stands out as the natural 'centre' for the Fens and it is here that the major administrator should have carried out his business. Attention has been drawn to the presence of two major buildings in the walled area of Durobrivae itself. One may have been a *mansio*, but the other is not so easy to explain as it has a temple enclosure attached (Mackreth 1979). The suggestion that it may have had something to do with the running of the Fens (*ibid.*, 21) seems less than likely in view of its great area. The simplest answer would be for the whole to have been a religious enclave within the town. A better candidate under Castor village offers itself.

The vast Late Roman structure there replaced an earlier building with a bath-house which was large, even by the standard of those found in large villas (Green and Green 1987, fig. 6). As the late building was surely put up at government expense, so the earlier structure may have been in government use and it would be natural to see in it the residence and offices of the procurator in charge of the Fens who could have administered any other lands belonging to the *patrimonium* in the neighbourhood of the town (Crawford 1976, 52). There is no good reason to suppose that the whole of the Fens would have been run from the enigmatic structure at Stonea which seems anyway to have failed by the end of the second century (Frere 1985, 287).

Little is known of the industries based on Durobrivae except for pottery and iron-smelting. Despite the proven existence of the latter, no extensive study has been undertaken in modern times. Recent observations at Laxton Lodge north of Bulwick, Northamptonshire, have revealed, for a distance up to half a kilometre, an extensive iron-working complex complete with houses and workshops (Jackson and Tylecote 1988). There can be no doubt that it was a settlement site and that its main activities were directed to the extraction of iron, if not to the production of specific items. Work began in the first century and the site continued to the end of the Roman period (*ibid.*, 293). Whether the site was alone in its nature is not known, but there can be no doubt that its fuel needs were large, but not necessarily beyond the power of private individuals to have provided from their resources. However, the large blank area lying north-west of Durobrivae could have been occupied largely by managed woodland, much as the Weald on a larger scale is thought to have been reserved for imperial use (Cleere 1974), and its size might be better suited to similar ownership.

The date, apparently, of major development in and around Durobrivae and in the Fens is the second half of the second century and is provided, above all, by pottery produced around Durobrivae and found in profusion in the Fens (Hartley and Hartley 1970, 167). Even if the earlier mortaria found were not made in the Nene Valley but came from the Mancetter/Hartshill kilns (*ibid.*, 165), these would have been channelled naturally through Durobrivae and, when Nene Valley mortaria were made in quantity, they predominated (*ibid.*, 165–7).

That the full-blown Nene Valley pottery industry cannot be shown to have become firmly established before 150 is surely significant, and the bulk of the early dating suits the succeeding twenty-five years remarkably well. The best known Nene Valley product is the colour-coat beaker, but the main range of wares produced was of fumed Nene Valley Grey Ware (Chapter 6). It was the commonest ware in the second and third centuries in the Fens. The dating given to its development would suit that given to the early colour-coats, and the forms used appear not to have emerged directly from those which had been made in local kilns in the Nene Valley before.

The suburbs of Durobrivae have only been sampled, despite the relatively extensive excavations carried out in them, and the work of both Graham Webster and Ernest Greenfield (both to be published) shows that the efflorescence of activity belonged again to the period following *c.*150. The Antonine expansion is not an unknown phenomenon, but is usually seen in isolation. When read, as here, against the simultaneous development in the Fens, it is hard to escape the conclusion that the two may have been related: there was the use in one area of resettled native farmers and, in the other, the establishment of craftspeople seeking to take advantage of the new population.

Any interaction between the Fens and Durobrivae which improved the economic life of each would undoubtedly have had an effect on the immediate environs of the town. It is against this background that the expansion of Orton Hall Farm, possibly at the expense of Monument 97, should be seen. The growth of Durobrivae would have been accompanied by an increase of the market centred on it and Orton Hall Farm, less than 6km away, was well placed to take full advantage. There would have been no difficulty in driving stock there inside a day and the strength of demand may have been all that was required to reorganise the production side of the farms in the neighbourhood. There is no need to see an official hand in this. However, the changes which mark Period 3 were not necessarily free of official coercion.

III. The Third and Fourth Centuries

If there had been an intimate association between the Fens and the uplands, the third-century fresh-water flooding in the Fens (Holmes 1970; Churchill 1970, 139–41; Bromwich 1970, 116–20) should have had an effect firstly on the town and then directly on the country around it. If the settlements in the Fens were, overall, reduced in number (*ibid.*, fig. 4c), there would presumably have been a displacement of population and a reduction in taxes or dues from the area. Even if it is not necessary to think in terms of the profits from the Fens to the *patrimonium*, in ordinary commercial terms there may have been a loss of trade to the town itself and any decline in its economic vitality should have been felt in its hinterland. There was a regeneration of settlement in the Fens in the fourth century, if not from towards the end of the third. Hallam detected the development of nucleated settlements (Hallam 1970, 53–4) possibly the result of the different distribution of usable land once the worst of the flooding had gone. However, they could have reflected a change in the management of the countryside as a whole, it being more detectable in the Fens because of the relatively intense study carried out there (Phillips 1970).

With only one large scale excavation to discuss, it is not easy to see clearly what the effects of the flooding in the Fens may have had on the upland zone. At the beginning of Period 3 at Orton Hall Farm, not far short of the middle of the third century, there had been a major change in the layout of the site. Not only were there more buildings, but there was a great increase in the number of livestock on the farm. Two explanations offer themselves. Firstly, a purely local one, leading to more land being controlled from the farmstead. This would imply that at least one other site had been closed down in favour of Orton Hall Farm. Secondly, we may actually be looking at a direct result of the withdrawal of population and stock from the Fens in the face of rising waters. As many sites in the Fens would have become virtually unusable, and their farm-hands and animals would presumably have needed relocation, there might have been an effect on the immediate upland zone. However, it appears that the degree of flooding was less, if not largely absent, in the fens north of the broad Welland-Nene zone (Smith 1970, 157). The flooding there presumably took time to take its toll on the fen farms and there may not have been a need to deal with a displacement of people and stock in a very short time.

The number of excavated sites which might offer information on this aspect is pitifully small. At Werrington, there was an absence of third-century activity (Mackreth 1988), Period 3 there closing at *c.*175/200. There was no domestic focus within the excavation and the pottery recovered may not be a true reflection of the actual date of the end of the period. At Maxey, there was a complete break in occupation and this has been attributed to wetter conditions although signs of actual flooding were not found (Pryor *et al.* 1985, 309). It is possible that there had also been a third-century replanning of the farmstead in North Bretton (excavations, R.F.J. Jones) and there seems to have been a shift in occupation in the large rural site at Lynch Farm.[2] The development of the site at Werrington is known in broad terms and there is no question of a lack of a major third-century presence there. Rather, the evidence can be read as a regrouping at that time. The same may have happened at Maxey, but there is less definition of the shifting fortunes of that site.

Period 3 at Orton Hall Farm stands out both for its replanning and for its vastly greater animal population. There is evidence that the nearest known late site, that in Orton Park, Huntingdonshire, also expanded in the third century (RCHM 1969, 29–30 (6); Dakin 1961) so Period 3 cannot represent the abandonment of that site. The middle and later part of the third century is commonly thought to have been a time of inflation, public strife and general economic debility, but one during which Britain remained remarkably unaffected (Frere 1987, 172–3, 285; Salway 1981, 242–4). But even so, Period 3 is markedly against any tendency to remain static and it is tempting to link it with a retreat from the Fens. This would presumably mean that there had been some link in ownership between the farm and the Fens and any displacement from there would have been to another, similarly owned, property. Taking a wider view, there is no need to suppose that all resettlement would have been along the Fen edge; some could have been many miles inland and well away from Durobrivae.

In Period 4 the development of the farmstead was further away from the standard image of a Romanised rural site and, in its milling capacity, had a feature which could suggest an official use of the site in the fourth century rather than have been the simple provision of a private landowner. The difficulty is in seeing what kind of official requirement would have needed supplying from such a site. There needs to have been some detectable element in the neighbourhood pointing to a powerful official presence. The same argument applied to the water-mill at Ickham (Young 1981) where there was artefactual evidence for the presence of soldiers, or civil servants, and the forts at Reculver and Richborough provided recognisable bodies of men who needed to be fed.

Around Durobrivae the signs are less obvious. However, there is the very large building around Castor church for which there is no good explanation. Analysis of the remains and the records of it show that it was built on a palatial scale and was of such architectural pretension that it could have been a *palatium*, an official residence for some personage whose rank and function merited such a structure (Mackreth 1984). The problem, however, is to determine who may have lived there and conducted his official duties from it.[3] From what is suspected of the areas covered by the various provinces of Britain, it would seem that Durobrivae, would have been peripheral to any province in which it lay, unless it served as the centre for the hypothesised short-lived province of Caesariensis (Frere 1987, 198–9).[4] But as London is thought to have been the 'Caesarea' implied by the name Caesariensis, Durobrivae seems inappropriate.

The chief civil functionary in Britain was the vicar. His centre of operations can hardly have been anywhere else than in London and the chance that he had a residence at Castor seems very remote. If one looks to the army, there are two possible candidates, the duke of the Britains and the count of Britain or of the Saxon Shore. However, our knowledge of the development of their offices and the relationship between them is too problematical to be summarised in an essay of this character. Briefly, if it could be shown that southern Britain had an essentially unified military command with the navy forming a major element, then the commander could have operated out of the lands bordering the Wash, and Durobrivae would have been pre-eminently suitable as his chief urban centre. Much would depend on the date of the Castor building. At present this is ill-defined being only later than 275– 325.

A position either as *dux* or *comes* would have carried with it considerable privilege and power, and a *palatium* to act both as a headquarters and a proper expression of his rank would have been appropriate. Mann points out that the *comes Britanniae* should have had his headquarters 'within easy reach of both the northern frontier and the Saxon Shore system' and that it could have been anywhere between York and London, with Lincoln and Leicester being candidates (Mann 1977, 14). Durobrivae was better placed than either, lying as it did on the Roman equivalents of the A1 and the A47, with good water communications out of the Wash and north and south along the coasts. This is, of course, geographical determinism, but with the major building at Castor and the still large reserves of the *res privata* in the Fens, the chances that here had been some kind of headquarters are greatly strengthened.

Any locally stationed troops would have needed not only housing, but feeding and clothing. Their food, as well as fodder for horses, would have come largely through the *annona* and *capitus* (Jones 1964, 626–7). At the very least, their bread would have been baked from flour which those supplying the *annona* would have had to have ground themselves, be they townspeople or landowners (*ibid.*, 629).

The proposed Mill-house at Orton Hall Farm may be significant in this context for its estimated production could hardly have been consumed on the site itself. It seems unlikely that flour was being supplied to the townsfolk of Durobrivae from so far away, but it is hard to think of another body of people who would have needed a special supply if it was not the military, army or navy. While it could be conjectured that troops were stationed in every large town (*ibid.*, 631; Tomlin 1976, 189; Mann 1977, 14), that this was actually the case in Britain is not proven.

The water-mill at Ickham, Kent,[5] raised similar problems when it came to considering who was being supplied. However, the presence of a large masonry building associated with lead sealings, covering, as it happens, the same period as that of Period 4 at Orton Hall Farm, led to the opinion that the site had been official. The metal-working at Ickham as well as some of the small finds pointed strongly to the army (Young 1981, 35). It was argued that it was the army based at Reculver and Richborough being catered for (*ibid.*, 32). Metal-working is not evidenced at Orton Hall Farm and the quantity of copper-alloy fittings of non-civilian character was much smaller (Nos 23, 24 and 27 in the finds catalogue) than at Ickham, which could actually have been staffed by the military. At Orton Hall Farm there could have been a civilian population under the control of officials who did not themselves *have* to be members of the army as such.

The marked emphasis on extra accommodation, storage and what is taken to be brewing as well as milling, could be the best evidence that, from the beginning of Period 4, *c.*300/325, the farm had passed into the hands of either the army or the *res privata*. The vague hint that cult figures (Catalogue No. 106) were being disposed of at the beginning, or during, Period 4 may mean that, here, overt paganism was being frowned upon. Without pressing confiscation or purchase as mechanisms by which it changed hands, the possibility remains that Orton Hall Farm had ceased to be in private ownership since the beginning of Period 2 when the Main Yard was first laid out.

However, if a connection with state supply is to be upheld, it should be possible to argue that the fortunes of the site reflected those of the army itself. Precise correlation cannot be expected, but the known and deduced history of the later fourth century points firmly to there having been a series of reductions in the armed forces which, in turn, will have had an impact on the sources supplying them. In historical terms, the rebellion of Magnus Maximus is taken to mark a watershed in the disposition of the army: from this time onwards it never reached the same order of magnitude as before (Reece 1973, 241–4). The terminal date for Period 4 is given as *c*.375 and, within the bounds of the dating evidence available to us, it could have been a decade later.

An assessment by J.R. Perrin of the excavations in the south-western suburbs of Durobrivae carried out by Ernest Greenfield shows that the late fourth-century occupation was less intense than it had been up to at least the middle of the century. In Normangate Field, Castor, although there is much fourth-century pottery, it is not linked to structures.[6] The impression that there was a general lessening of activity in the suburbs of Durobrivae is a reasonably strong one, backed by a change in the character of the coinage in use which makes Britain more comparable with the continent (*ibid.*, 244). In such a context the changes marking the beginning of Period 5 at Orton Hall Farm are not exceptional, but one feature in its archaeological record seems unusual: the general absence of any of the distinctive pots made, for example, at Stibbington which are the latest known pottery made in the Nene Valley. The ordinary fourth-century pottery could run on, but, had it done so, it would normally be expected to occur with Stibbington-style pots (see below). The nature of the archaeology (described in Chapter 1) and the presence of Anglo-Saxon pottery shows that the site continued through the fifth century and into the sixth. Abandonment of the site is not the explanation for the absence of a strong element of the recognised latest Roman pottery. The temptation is to equate the absence of one stratum of pottery with the introduction of another, in this instance Anglo-Saxon.

IV. Roman and Anglo-Saxon (late fourth century–early sixth century)

The description of Period 5 will have made the state of the evidence plain. There was no good sign for a complete break between the Roman and Anglo-Saxon occupations. Equally there was none which insisted that people brought up under very different traditions lived side by side. However circumstantial the evidence for structures having continued from Period 4 into the Anglo-Saxon phases of the site, to insist that none did continue makes nonsense of the Anglo-Saxon layout. The outcome would seem to be that a working farm passed, largely intact, to a new type of occupant. Until there is direct evidence to the contrary, it should be assumed that the root economy of the countryside suffered little during what was either the swift transition of political control at one extreme or the gradual introduction of a new people under some constraints at the other. The gradually increasing number of Late Roman sites with Anglo-Saxon pottery on them (Fig. 1) suggests that the basic forms of land management at the end of the fourth century were maintained for a while, and their existence lends some support to the inferences drawn from the excavation at Orton Hall Farm. It should also mean that enough of the pre-existing organisation behind the basic economy also continued and this, again, suggests that the indigenous population remained largely in place carrying out its customary functions. The crux of the problem lies in the relationship between the two classes of the most common dating material found: the Roman and Anglo-Saxon pottery.

It is now many years since the initial discussion of what was called Romano-Saxon pottery (Myres 1956). There have been many shifts of opinion as to the precise significance of this style (Myres 1969, 66–70; Hurst 1976, 290– 2; Gillam 1979; Roberts 1982) but there seems not to have been any particular study of the reverse side of the coin: if Romans were producing pottery to suit Anglo-Saxon taste, why is it not found on Anglo-Saxon sites in sufficient quantity to be reported more often (Myres 1968, 222; West 1985, 128) and where is the Anglo-Saxon pottery which copied Roman forms? The answer is the two 'types' of pottery are found equally infrequently in mixed Late Roman and Anglo-Saxon deposits. Orton Hall Farm produced one sherd of a mortarium in an Anglo-Saxon fabric (Anglo-Saxon Pottery, No. 15) and more than one biconical pot in a Romanising fabric (Anglo-Saxon Pottery, No. 1; and other stratified pottery, L409, L476, L512). The latter is hand-made and seems, therefore, unlikely to be a Roman product. Two copies of a Roman pie dish are recorded from Mucking (Wilson 1969, 231). The large collection of cinerary urns from Spong Hill included one lid which bears an uncanny resemblance to Nene Valley colour-coated flanged lids (Hills 1977, pot 1360) belonging to the middle and late fourth century (Chapter 6, Fig. 36). But perhaps the best examples of copying are of a bat's-head spout from West Stow (West 1985, 27, fig. 92, 7) as well as a form obviously derived from a Roman 'dog' dish (*ibid.*, 52, fig. 217, 14).

The Orton Hall Farm sherd was identified as a mortarium because of the grits pressed into the inner surface just as they were into the inner wall of the West Stow example (*ibid.*, 27). This implies that it was not just the form which was borrowed, but also its function. It would be as well to bear in mind that when recognisably Roman goods ceased to be made, the indigenous population becomes invisible. The occurrence of such a pot might be considered appropriate for a one-time Roman site like Orton Hall Farm, but cannot be easily explained away on exclusively Anglo-Saxon sites like West Stow.

The emphasis placed on the so-called Romano-Saxon pottery has been entirely on what are held to be early Germanic decorative traits in Roman wares. While the kilns in and around Much Hadham remain unpublished, it is not possible to confirm an impression that the direction of the influence may actually have been the other way round, a point made by J.P. Gillam (1979, 107–12).

It cannot be demonstrated that Roman pottery was in use alongside Anglo-Saxon at Orton Hall Farm other than as still usable survivals. Therefore, large-scale Roman pottery industries had probably collapsed by the time Anglo-Saxon pottery began to be made in Britain. As it could be argued that the industries carried on into the first quarter of the fifth century, the start date for Period 5 should be moved forward from *c*.375. At Linford, however, enough reconstructable Roman vessels were recovered along with Anglo-Saxon pottery to suggest that the two were being made almost side by side, or that the Anglo-Saxon pots were imported and used along with newly acquired Roman material (Barton 1962). Such instances are few indeed and we wait for the results of the excavation at Mucking to shed further light on this.

Even if a date could be fixed for the introduction of Anglo-Saxons into the farmstead, it should not automatically be taken to have been the case at other sites of the same apparent type. Late Roman sites without Anglo-Saxon pottery may still have existed, the native inhabitants remaining, but be undetectable because of a drastic change in their material culture. The vague hints of a peaceful contact between the two cultural groups at Orton Hall Farm need to be set into some kind of chronological framework and, before any other considerations can be explored, that must arise from the archaeological evidence present on this site.

In theory, the earliest time that Anglo-Saxons could have arrived is the beginning of Period 5. That moment depends on the assessed date of the Roman pottery. A fair amount of Roman pottery was available at the beginning of the fifth century, as the fill of the late drier at the Great Casterton villa showed (Corder 1961, 64–9; Perrin 1981b) and there was no known Anglo-Saxon occupation there even if a cemetery immediately outside the walls of the town (Wilson, D.M. 1967, 268) suggested that Anglo-Saxons had arrived while Romans were still in residence. The drier deposit was marked by a recognisable element in the style of that made at Stibbington and represented in the late fourth-century well there (Guide, pots 68, 74 and 78). This material was absent from Orton Hall Farm. As this site lay only 9.5km from the production centre as opposed to the 16.5km of the other, the absence may have been a function of either date or status. The picture of decline in the late fourth-century pottery industry presented by Fulford may have applied and the change of relative quantities outlined by him (Fulford 1979, 125–6) is perhaps underlined by the greater occurrence of Oxford pottery and the appearance of 'odd' elements in the ceramic make-up of Period 5 (Chapter 6, p. 176–7).

As the Roman pottery offers no chronological help, we must turn to the Anglo-Saxon material and to one particular deposit. In it were the biconical and mortarium sherds as well as the Barred Zoomorphic Comb (Catalogue No. 251). The first two items offer only the crudest indicators, as they could only have been made at a stage of a relationship not yet dated. The comb, however, suggests that the optimum date for the deposit was the first quarter of the fifth century. The site in North Bretton produced a Mahndorf-type brooch (Böhme 1986, 530–1, *Abb.* 52, 5) which lies within the first half of the same century. Most of the dating available for objects of this sort is derived from their disuse: they could have been made and used for at least ten to twenty years before they entered the ground. Thus it is possible to set the earliest date for the deposition of the comb at, say, 410 and argue that it could have arrived *c*.390. However, at either of those dates, Anglo-Saxon occupation of the site should have been at the will of the local leaders of the Roman community, if not with the consent of a higher authority.

There are two philosophies governing the character of the end of Roman Britain, one pessimistic and the other the reverse. Dr Reece, representing the first, sees Roman Britain as doomed to collapse once the artificial props provided by a relatively strong central government were removed (Reece 1980, 85). This view coincides to some measure with that of J.N.L. Myres who sees a central government, unable to man defences from its own resources, resorting to the wholesale employment of barbarians (Myres 1969, 65–6). According to both, Anglo-Saxon occupation of Orton Hall Farm at the beginning of Period 5 would be reasonable.

The optimists deal with the matter differently. The quality of life, and therefore the condition of the surviving archaeology is of less importance to them than the political integrity of the lowest effective level of the Roman state, the *civitas*. The two visits of St Germanus are produced as evidence for the essential cohesiveness of a society which can only be called Roman (Frere 1987, 361–4; Salway 1981, 451–60; Wacher 1978, 263–6; Johnson 1980, 115–17). In this instance, it was the lack of troops supplied as a matter of course by the central government which made the surviving local powers in Britain use a well-known mechanism to enrol the aid of barbarians (Frere 1987, 369; Salway 1981, 418, 440; Wacher 1978, 265) although the direct inference is sometimes skirted round (Johnson 1980, 133–43). The optimistic view would certainly favour a date not before *c*.425 for their introduction as settlers and this is indicated by the comb and brooch from Orton Hall Farm and North Bretton, although the dating of these is conditioned by their deposition and not by their period of use.

The only two levels of tolerated barbarian settlers in the empire, outside properly constituted army units, are *laeti* and *foederati*. The first had certainly been introduced into the empire by the end of the third century (Jones 1964, 60) and it seems to have been the custom to settle them under the control of either a Roman officer or a landowner (*ibid.*, 157). Their military role is less well established and, as they could be assigned to a private landowner, it looks as though they would have been unarmed. By the middle of the fourth century at least, it appears that it was expected that their offspring would be taken as conscripts (*ibid.*, 614). That they were settled as tribal groups, but dispersed in set territories, is clear (*ibid.*, 620). None is known to have been in Britain (*ibid.*) and, in any case, they should not have had enough social cohesion to have produced household goods for themselves. Had this been so, and had there been any in Britain, a recognisable non-Roman element on fourth-century sites would surely have been identified by now.

The ceramic evidence from Orton Hall Farm and elsewhere is that, if cultural interaction produced Anglo-Saxon copies of Roman forms, the newcomers were able to rule themselves within a Roman framework of some form: they were able to make goods for themselves and market them. This argues for a degree of freedom at variance with the impression of control by either officers or individual landowners. Hence these foreign people would not have been classified as *laeti* (Wightman 1985, 256).[7]

Turning to *foederati*, in the context of the fifth century, the significant event here was the treaty with the Visigoths in 382. The terms allowed the tribe to settle within the empire and maintain its identity, receiving land in return for supplying recruits. It was a precedent which led to further trouble: the Visigoths did not settle down, yet other tribes sought to be given land on the same terms (Jones 1964, 157–8). The Visigoths were presumably given lands already partly occupied, and some form of agreement to cover the relationship between the two communities would have been necessary. Not all *foederati* were of this kind, however. Many formed contingents of the Roman army; some were captives, others were under contract (*ibid.*, 200), including large groups under their leaders who received subsidies (*ibid.*, 201).

With Roman respect for the law, it would be surprising if land allocations to barbarians were not based on some form of established practice. If the prime function of the tribal group, even if it was made up of family units who expected to carry on their normal peaceful way of life, was military in Roman eyes, it is likely that the customs associated with *hospitalitas* were used. Regulations governing the billeting of soldiers would have been framed by Augustus's time, but possibly needed thorough revision at the creation of the *comitatenses* and the *palatini*, who, if there were no specially built barracks, would have had to be billeted out. It should be assumed that the relevant laws would be known to every major landowner at least, since the practices associated with the regulations were open to abuse.

In essence, the householder made a third of his house over to his 'guest'; the abuses and extra demands which arose were over disputable matters of whether furniture was included, a meal provided and, for senior officers, a bath (*ibid.*, 631–2). The application of the general principles to the land settlement of other unwelcome 'guests' would yield a partition of the land, and probably stock, equipment and farm-buildings, of one third, more or less, according to their degree of menace (Wightman 1985, 249–53). This happened in Italy in the 440s and almost certainly followed a precedent set in parts of Gaul from 418, if not in 382 in relation to the Visigoths, or at an earlier period in North Gaul (*ibid.*, 209, 253–6).

Orton Hall Farm either belonged to a large estate and would have been suitable for sub-division, the landlord not having to divide his own residence, or, if in official hands in Period 4 and also probably Period 3, it could have been handed over lock, stock and barrel, no private interest having to be placated. If the arrival of the Anglo-Saxons was near the beginning of Period 5, it could only have been at the level of the Count and the Vicar that orders for their settlement came. If, however, the date has to be moved into the fifth century, the chances are that any land allotment was made after the departure of Constantine III and it would have been the residual authorities who would have been responsible: at worst, the *civitates*, at best, a Provincial or Diocesan assembly (Jones 1964, 765). At such a time there may have been little compunction in disposing of land belonging to the *res privata*, there being no one present to object.

The one thing which stands out from the archaeological evidence is that there was a time when the Anglo-Saxons were able to produce goods to satisfy traditional needs when enough Roman goods were in use to have influenced the new manufacturies. This argues for a high level of social cohesion in the general body of barbarians and this should reflect a type of overall settlement remarkably akin to federate plantations. The study of the fabrics of Early Saxon pottery has only just begun, but John Walker has been able to establish that a significant proportion of the wares from Orton Hall Farm had been made at least 60km away (Walker 1978). This would argue for a large area under one authority. Unfortunately, it is not certain that this condition belonged to the early fifth century. If federate settlement is the answer to the circumstances on this particular site, many others may have been involved, but not all new settlements need have been directly attached to an established centre such as Orton Hall Farm.

It is worth more than a passing thought that the apparent degradation of Orton Hall Farm from Periods 4 to 5 could have been the by-product of such an act. The arrival of new people could have lowered the economic potential of the farm to such a level that the costs of maintenance were thought too high in relation to the now reduced income. Indeed, if it had been government land, there may have been no funds forthcoming. If the land had still been in the hands of a private landowner, he may still have been required to pay taxes on the land while the barbarians were exempt (Jones 1964, 252). Such conditions would provide a nearly perfect background for the devolution of the site through phase *a* of Period 5, if not through part of phase *b* as well. It should be acceptable that what is called phase *c* was entirely under the domination of the Anglo-Saxons, or people who owed nothing to Rome and are not separately distinguishable in the archaeological record.

Apart from surmising that the earliest date for controlled settlement may have been under Magnus Maximus, followed by the years after the death of Constantine III, the evidence can be used to support any view except complete destruction of all that was Roman. But the final abandonment of the site may have been due to the faint echoes of the foundation of Mercia in secondary historical sources. There is no particular reason why any site, taken in isolation, should cease to exist at a particular moment and it is the misfortune of Orton Hall Farm that it is the only one locally which has any adequate evidence for a termination in the sixth century. Dr Davies, in discussing late notices of what seem to be early annals, came to the conclusion that there had been a source preserving traditions related to events in the Midlands and eastern districts in the sixth century (Davies 1977). These early annals point to a major invasion of East Anglia in the early sixth century and from there the earliest members of the Mercian royal house in Britain (or is it England?) moved west in the sixth century. Mercia became an identifiable entity later in the same century (*ibid.*, 22–4). Does Orton Hall Farm cease to exist during a period of political change and were these new Anglo-Saxons, products of a fresh migration, fighting against others settled over a century before in a once recognisable Roman society?

Endnotes

1. I am very grateful to Dr J.P. Wild for reading the draft and for making many helpful suggestions. I am, however, responsible for the general tenor of the essay and for any errors.
2. I give my thanks to Dr J.P. Wild for the information in advance of his own publication.
3. I am grateful to Professor Mann, Dr J.P. Wild and Mr M.W.C. Hassall for comments on various possibilities which, in the event, proved to take up more space than could be justified here.
4. I am indebted to Mr M.W.C. Hassall for pointing out this possibility.
5. My thanks are due to Mr M.W.C. Hassall for drawing my attention to this site.
6. My thanks go to Dr J.P. Wild for providing details in advance of the publication of the Nene Valley Research Committee's excavations there.
7. I am grateful to Dr J.P. Wild for drawing my attention to this reference.

Bibliography

Entries for printed text and for microfiche.

Addyman, P.V. and Leigh, D., 1973 — 'The Anglo-Saxon village at Chalton, Hampshire: second interim report', *Medieval Archaeol.* 17, 1–25

Addyman, P.V., Leigh, D. and Hughes, M.F., 1972 — 'Anglo-Saxon houses at Chalton, Hampshire', *Medieval Archaeol.* 16, 13–31

Alexander, J, and Ozanne, A., 1960 — 'Report on the investigation of a round barrow on Arreton Down, Isle of Wight', *Proc. Prehist. Soc.* 26, 263–302

Alarcão, J., 1975 — 'Bouteilles Carrées à fond decoré du Portugal Romain', *J. Glass Stud.* 17, 47–53

Allason-Jones, L., 1983 — 'The Finds', in Miket 1983, 108–42

Allason-Jones, L., 1989 — *Ear-rings in Roman Britain,* Brit. Archaeol. Rep. Brit. Ser. 201 (Oxford)

Allason-Jones, L. and Miket, R., 1984 — *Catalogue of Small Finds from South Shields Roman Fort,* (Gloucester)

Anderson, A.C., 1980 — *A Guide to Roman Fine Wares*, VORDA Res. Ser.1

Anderson, A.C. and Anderson, A.S. (eds), 1981 — *Roman Pottery Research in Britain and North-West Europe*, Brit. Archaeol. Rep. Int. Ser.123 (Oxford)

Applebaum, S., 1972 — 'Roman Britain', in Finberg 1972, 1–277

Armitage, P.L. and Clutton-Brock, J., 1976 — 'A system for classification and description of horn cores of cattle from archaeological sites', *J. Archaeol. Sci.* 3, 329–48

Arthur, P. and Marsh, G.D. (eds), 1978 — *Early fine wares in Roman Britain*, Brit. Archaeol. Rep.57

Ashby, T., 1905 — 'Excavations at Caerwent, Monmouthshire, on the site of the Romano-British city of Venta Silurum, in the year 1904', *Archaeologia* 59, pt 2, 289–310

Ashby, T., 1906 — 'Excavations at Caerwent, Monmouthshire, on the site of the Romano-British City of Venta Silurum, in the year 1905', *Archaeologia* 60, pt 2, 111–30

Ashby, T., Hudd, A.E. and Martin, A.J., 1904 — 'Excavations at Caerwent, Monmouthshire, on the site of the Romano-British city of Venta Silurum, in the years 1901–3', *Archaeologia* 59, pt 1, 87–124

Atkinson, D., 1942 — *Report on Excavations at Wroxeter (The Roman City of Viroconium) in the County of Salop, 1923–1927,* (Oxford)

Barton, K.J., 1962 — 'Settlements of the Iron Age and pagan Saxon periods at Linford, Essex', *Trans. Essex Archaeol. Soc.* 3 ser. 1, 57–104

Bass, W., 1981 — *Human Osteology: A Laboratory and Field-Manual of the Human Skeleton*, Missouri Archaeol. Soc.

Bedwin, O., 1978 — 'The excavation of a Romano-British site at Ranscombe Hill, South Malling, East Sussex, 1976', *Sussex Archaeol. Collect.* 116, 241–55

Bell, M., 1977 — 'Excavations at Bishopstone', *Sussex Archaeol. Collect.* 115

Berry, A.C. and Berry, R.J., 1967 — 'Epigenetic variations in the human cranium', *J. Anatomy* 101, 361–79

Bidwell, P.T., 1979 — *The Legionary Bath-House and Basilica and Forum at Exeter*, Exeter Archaeol. Rep., Vol.I (Exeter)

Böhme, A., 1972 — *Die Fibeln der Kastelle Saalburg und Zugmantel*, Saalburg Jahrbuch, Bericht des Saalburg Museums 29

Böhme, H.W., 1986 — 'Das Ende der Römerherrschaft in Britannien und die Angel-sächsische Besiedlung Englands im 5. Jahrhundert', *Jahrbuch des Römisch-Germanischen Zentral-museums Mainz* 33, pt 2, 469–574

Bolton, E.G., 1968 — 'Romano-British kiln at Greetham, Rutland', *Trans. Leicestershire Archaeol. Hist. Soc.* 43, 1–3

Boon, G.C., 1974 — *Silchester, the Roman Town of Calleva* (Newton Abbot)

Boston, R., 1976 — *Beer and Skittles* (London)

Bowman, A.K. and Thomas, J.D., 1983 — *Vindolanda, The Latin Writing-Tablets*, Britannia Monograph Ser.4 (London)

Bradley, R., 1970 — 'The excavation of a Beaker settlement at Belle Tout, East Sussex, England', *Proc. Prehist. Soc.* 36, 312–79

Brailsford, J.W., 1962 — *Hod Hill, volume one, Antiquities from Hod Hill in the Durden Collection* (London)

Brassington, M., 1971 — 'A Trajanic kiln complex near Little Chester, Derby', *Antiq. J.* 51, 36–69

Brassington, M., 1980 — 'Derby Racecourse kiln excavations, 1972–3', *Antiq. J.* 60, 8–47

Bridgewater, N.P., 1962 — 'The Huntsham Romano-British villa, first report', *Trans. Woolhope Natur. Fld Club* 37, 179–91

Brinson, J.G.S., 1963 — 'Rivenhall', in Hull 1963a

Brodribb, A.C.C., Hands, A.R. and Walker, D.R., 1968 — *Excavations at Shakenoak Farm, near Wilcote, Oxfordshire, Part I: Sites A & D,* (privately printed)

Brodribb, A.C.C., Hands, A.R. and Walker, D.R., 1971 — *Excavations at Shakenoak Farm, near Wilcote, Oxfordshire, Part II: Sites B & H,* (privately printed)

Brodribb, A.C.C., Hands, A.R. and Walker, D.R., 1972 — *Excavations at Shakenoak Farm, near Wilcote, Oxfordshire: Part III, Site F,* (privately printed)

Brodribb, A.C.C., Hands, A.R. and Walker, D.R., 1973 — *Excavations at Shakenoak Farm, near Wilcote, Oxfordshire, Part IV: Site C,* (privately printed)

Bromwich, J., 1970 — 'Freshwater flooding along the Fen margins south of the Isle of Ely during the Roman period', in Phillips 1970, 114–26

Brothwell, D.R., 1972 — *Digging up Bones* (London)

Brown, A.E. and Alexander, J.A., 1982 — 'Excavations at Towcester 1954: the Grammar School site', *Northamptonshire Archaeol.* 17, 24–59

Brown, A.E. and Woodfield, C., with Mynard, D.C., 1983 — 'Excavations at Towcester, Northamptonshire, the Alchester Road suburb', *Northamptonshire Archaeol.* 18, 43–140

Brown, D., Campbell, J. and Hawkes, S.C. (eds), 1981 — *Anglo-Saxon Studies in Archaeology and History*, Brit. Archaeol. Rep. Brit. Ser.92 (Oxford)

Bunch, B. and Corder, P., 1954 — 'A Romano-British pottery kiln at Weston Favell, near Northampton', *Antiq. J.* 34, 218–24

Burstow, G.P. and Wilson, A.E., 1939 — 'A Roman Bath, Highdown Hill, Sussex', *Sussex Archaeol. Collect.* 80, 63–87

Casey, P.J. (ed), 1979 — *The End of Roman Britain*, Brit. Archaeol. Rep. Brit. Ser.71 (Oxford)

Challands, A., 1974 — 'A Roman industrial site and villa at Sacrewell, Thornhaugh', *Durobrivae, Rev. Nene Valley Archaeol.* 2, 13–16

Champion, T., 1977 — 'Chalton', *Curr. Archaeol.* 5, 12 (Nov.), 364–9

Churchill, D.M., 1970 — 'Post Neolithic to Romano-British sedimentation in the South Fenlands of Cambridgeshire and Norfolk', in Phillips 1970, 132–46

Clark, J.W. (ed), 1897 — *The Observances in use at the Augustinian Priory of S. Giles and S. Andrew at Barnwell, Cambridgeshire*, (Cambridge)

Clark, J.G.D., 1949 — 'Report on excavations on the Cambridgeshire Car Dyke, 1947', *Antiq. J.* 29, 145–63

Clark, J.G.D., Price, E. S. and Longworth, I. H., 1960 — 'Excavations at the Neolithic site at Hurst Fen, Mildenhall, Suffolk', *Proc. Prehist. Soc.* 26, 202–45

Clarke, G., 1979 — *The Roman Cemetery at Lankhills*, Winchester Studies 3, Pre-Roman and Roman Winchester, Part II (Oxford)

Cleere, H., 1974 — 'The Roman iron industry of the Weald and its connexions with the *Classis Britannica*', *Archaeol. J.* 131, 171–99

Clough, T.H.McK. and Cummins, W.A., 1979 — *Stone Axe Studies*, Counc. Brit. Archaeol. Res. Rep.23 (London)

Cocks, A.H., 1921 — 'A Romano-British homestead in the Hambledon Valley, Buckinghamshire', *Archaeologia* 71, 141–98

Collis, J., 1978 — *Winchester Excavations, Excavations in the Suburbs and Western Parts of the Town, Vol.II* (Winchester)

Corder, P., 1932 — *The Roman Pottery at Throlam, Holme-on-Spalding Moor, East Yorkshire*, Roman Malton and Dist. Rep.3 (Hull)

Corder, P. (ed), 1951 — *The Roman Town and Villa at Great Casterton, Rutland,* (Nottingham)

Corder, P. (ed), 1954 — *The Roman Town and Villa at Great Casterton, Rutland, Second Interim Report for the years 1951–53* (Nottingham)

Corder, P. (ed), 1961 — *The Roman Town and Villa at Great Casterton, Rutland, Third Report for the years 1954–58* (Nottingham)

Corder, P. and Birley, M., 1937 — 'A pair of fourth-century Romano-British pottery kilns near Crambeck; with a note on the distribution of Crambeck ware', *Antiq. J.* 17, 392–413

Crawford, D.J., 1976 — 'Imperial Estates', in Finley 1976, 35–70

Crossley, D.W. (ed), 1981 — *Medieval Industry*, Counc. Brit. Archaeol. Res. Rep.40 (London)

Crummy, N., 1979 — 'A chronology of Romano-British bone pins', *Britannia* 10, 157–63

Crummy, N., 1983 — *The Roman Small Finds from Excavations in Colchester 1971–9*, Colchester Archaeol. Rep.2 (Colchester)

Cummins, W.A., 1979 — 'Neolithic stone axes: distribution and trade in England and Wales', in Clough and Cummins 1979, 5–12

Cunliffe, B. (ed), 1968 — *Fifth Report on the Excavation of the Roman Fort at Richborough, Kent*, Rep. Res. Comm. Soc. Antiq. London, 23 (Oxford)

Cunliffe, B., 1971 — *Excavations at Fishbourne, 1961–1969, Volume II, The Finds*, Rep. Res. Comm. Soc. Antiq. London, 27 (Leeds)

Cunliffe, B., 1975 — *Excavations at Porchester Castle, Volume I, Roman*, Rep. Res. Comm. Soc. Antiq. London, 32 (Dorking)

Cunnington, W., 1867 — 'On the examination of the Roman station at Baydon', *Wiltshire Archaeol. Nat. Hist. Mag.* 10, 104–9

Curle, A.O., 1923 — *The Treasure of Traprain, A Scottish Hoard of Roman Silver Plate,* (Glasgow)

Curwen, E.C., 1933 — 'Excavations on Thundersbarrow Hill, Sussex', *Antiq. J.* 13, 109–51

Dakin, G.F., 1961 — 'A Romano-British site at Orton Longueville, Huntingdonshire', *Proc. Cambridge Antiq. Soc.* 54, 50–67

Dannell, G.B., 1973 — 'The Potter Indixivixus', in Detsicas 1973, 139–42

Dannell, G.B. and Wild, J.P., 1987 — *Longthorpe II, The Military Works-Depot: an episode in Landscape History*, Britannia Monograph Ser.8 (London)

Dannell, G.B., Hartley, B.R., Wild, J.P. and Perrin, J.R., 1993 — 'Excavations on a Romano-British pottery production site at Park Farm, Stanground, Peterborough, 1965–7', *J. Roman Pottery Stud.* 6, 51–93

Darling, M.J., 1977a — *A Group of Late Roman Pottery from Lincoln*, Lincoln Archaeol. Trust Monograph Ser. 16–1, (London)

Darling, M.J., 1977b — 'Pottery from early military sites in Western Britain', in Dore and Greene 1977, 57–100

Davies, W., 1977 — 'Annals and the origin of Mercia', in Dornier 1977, 17–29

Déchelette, J., 1904 — *Les Vases Céramiques Ornés de la Gaule Romaine,* (Paris)

Dennis, G., 1978 — '1–7 St Thomas Street', in *Southwark, Excavations 1972–74, II*, Joint Publication No.1, London and Middlesex Archaeol. Soc. and Surrey Archaeol. Soc., 291–422

Detsicas, A.P. (ed), 1973 — *Current Research in Romano-British Pottery*, Counc. Brit. Archaeol. Res. Rep.10 (London)

Detsicas, A. (ed), 1981 *Collectanea Historica, Essays in Memory of Stuart Rigold* (Maidstone)

Dool, J. and Wheeler, H., *et al.*, 1985 'Roman Derby, Excavations 1968–1983', *Derbyshire Archaeol. J.* 105

Dore, J. and Greene, K. (eds), 1977 *Roman Pottery Studies in Britain and Beyond*, Brit. Archaeol. Rep. Supplementary Ser.30 (Oxford)

Dornier, A. (ed), 1977 *Mercian Studies*, (Leicester)

Down, A., 1974 *Chichester Excavations II*, (Chichester)

Down, A., 1979 *Chichester Excavations IV*, (Chichester)

Down, A. and Rule, M., 1971 *Chichester Excavations I*, (Chichester)

Drury, P.J., 1976a 'Braintree: excavations and research, 1971–76', *Essex Archaeol. Hist.* 3 ser, 8, 1–143

Drury, P.J., 1976b 'Rettendon Ware kiln-debris and other material from Sandon', *Essex Archaeol. Hist.* 8, 253–8

Dudley, D., 1967 'Excavations on Nor'Nour in the Isles of Scilly, 1962– 6', *Archaeol. J.* 124, 1–64

Evans, J., 1894 'On some iron tools and other articles formed of iron found at Silchester in the year 1890', *Archaeologia* 54, pt 1, 139–56

Ellis, P. (ed), forthcoming *The Roman Baths and Macellum at Wroxeter, a Report on the Excavations by Graham Webster, 1955–1985*, Eng. Heritage Arch. Rep.

Evans, R., 1979 'The early courses of the River Nene', *Durobrivae, Rev. Nene Valley Archaeol.* 7, 8–10

Evison, V.I. (ed), 1981 *Angles, Saxons, and Jutes, Essays presented to J.N.L. Myres*, (Oxford)

Farrar, R.A.H., 1973 'The techniques and sources of Romano-British Black-burnished Ware', in Detsicas 1973, 67–103

Field, K.J. and Mynard, D.C., 1979 'The excavation of an Iron Age and first century AD site at Stoke Goldington', *J. Northampton Mus. Art Gall.* 13, 177–98

Finberg, H.P.R. (ed), 1972 *The Agrarian History of England and Wales, Volume I:ii, A.D. 43–1042*, (Cambridge)

Finley, M.I. (ed), 1976 *Studies in Roman Property*, (Cambridge)

Finnegan, M., 1973 'Non metric variation of the infracranial skeleton', *J. Anatomy* 125, 23–37

Foster, P.J., 1976 'Romano-British finds at Kettering', *Northamptonshire Archaeol.* 11, 170–7

Foster, P.J., Harper, R. and Watkins, S., 1977 'An Iron-Age and Romano-British settlement at Hardwick Park, Wellingborough, Northamptonshire', *Northamptonshire Archaeol.* 12, 55–96

Fowler, P.J., 1981 'Later Prehistory', in Piggott 1981, 63–298

Fowler, P.J. and Walthew, C.V., 1971 'Archaeology and the M5 Motorway', *Trans. Bristol Gloucestershire Archaeol. Soc.* 90, 22–63

Fox, G.E., 1892 'Excavations on the site of the Roman city at Silchester, Hants, in 1891', *Archaeologia* 53, pt 1, 263–85

Frend, W.H.C., 1968 'A Roman farm-settlement at Godmanchester', *Proc. Cambridge Antiq. Soc.* 61, 19–44

Frere, S.S., 1972 *Verulamium Excavations, Vol.I*, Rep. Res. Comm. Soc. Antiq. London, 28 (Oxford)

Frere, S.S., 1977 'Roman Britain in 1976, I, Sites Explored', *Britannia* 8, 356–425

Frere, S.S., 1983 *Verulamium Excavations, Vol.II*, Rep. Res. Comm. Soc. Antiq. London, 41 (London)

Frere, S.S., 1984 *Verulamium Excavations, Vol.III*, Oxford Comm. Archaeol. Monograph I (Oxford)

Frere, S.S., 1985 'Roman Britain in 1984, I, Sites Explored', *Britannia* 16, 252–316

Frere, S.S., 1987 *Britannia: a History of Roman Britain*, (London)

Frere, S.S. and St Joseph, J.K., 1974 'The Roman Fortress at Longthorpe', *Britannia* 5, 1–129

Frere, S.S., Stow, S. and Bennet, P., 1982 *Excavations on the Roman and Mediaeval Defences at Canterbury*, Archaeol. Canterbury, Vol.II (Maidstone)

Friendship-Taylor, R.M., 1979 'The excavation of the Belgic and Romano-British settlement at Quinton, Northamptonshire. Site B, 1973–1977', *J. Northampton Mus. Art Gall.* 13, 3–176

Fulford, M.G., 1975 *New Forest Roman Pottery*, British Archaeol. Rep. Brit. Ser.17 (Oxford)

Fulford, M., 1979 'Pottery production and trade at the end of Roman Britain: the case against continuity', in Casey 1979, 120–32

Fussell, G.E. (ed), 1936 *Robert Loder's Farm Accounts, 1610–1620*, Camden Soc. 3 ser. 53

Garrod, P. and Atkin, M., forthcoming *Kingsholm II*, Eng. Heritage Arch. Rep.

Gascoigne, P.E., 1969 'The Roman villa at Cold Cupboard Farm, Farmington', *Trans. Bristol Gloucestershire Archaeol. Soc.* 88, 34–67

Gelling, P.S., 1957 'Report on Excavations at Bays Meadow, Droitwich', *Trans. Proc. Birmingham Archaeol. Soc.* 75, 1–23

Gentry, A.P., 1976 *Roman Military Stone-Built Granaries in Britain*, Brit. Archaeol. Rep. Brit. Ser.32

Gillam, J.P., 1951 'Notes on pottery from the destruction layer', in Corder 1951, 24–40

Gillam, J.P., 1970 *Types of Roman Coarse Pottery Vessels in Northern Britain*, (3rd edn, Newcastle)

Gillam, J.P., 1973 'Sources of pottery in northern sites', in Detsicas 1973, 53–62

Gillam, J.P., 1976 'Coarse fumed ware in North Britain and beyond', *Glasgow Archaeol. J.* 4, 57–80

Gillam, J.P., 1979 'Romano-Saxon pottery: an alternative interpretation', in Casey 1979, 103–18

Goethert-Polaschek, K., 1977 *Katalog der Römischen Gläser des Rheinischen Landesmuseums Trier*, Trier

Going, C.J., 1987 *The Mansio and Other Sites in the south-eastern sector of Caesaromagus: the Roman Pottery*, Chelmsford Archaeol. Trust Rep.3.2, Counc. Brit. Archaeol. Res. Rep.62 (London)

Goodburn, R., 1976 'Roman Britain in 1975, I, Sites Explored', *Britannia* 7, 291–377

Goodburn, R., 1984 'The non-ferrous metal objects', in Frere 1984, 83–108

Goodburn, R. and Bartholemew, P. (eds), 1976 *Aspects of the Notitia Dignitatum*, Brit. Archaeol. Rep. Supplementary Ser.15 (Oxford)

Goodburn, R. and Grew, F., 1984 'Objects of bone, and objects of stone', in Frere 1984, 69–82

Goodchild, R.G., 1943 'T-shaped corn-drying ovens in Roman Britain', *Antiq. J.* 23, 148–53

Gordon 1792 'Extract of a Letter to Mr. Pownall from the Rev. Dr. Gordon, in Account of some sepulchral Antiquities discovered at Lincoln, John Pownall', *Archaeologia* 10, 345–9

Gracie, H.S., 1970 'Frocester Roman villa, first interim report, 1961–1970', *Trans. Bristol Gloucestershire Archaeol. Soc.* 89, 20–86

Gracie, H.S. and Price, E. G., 1979 'Frocester Court Roman villa, second interim report, 1968–1977: the Courtyard',*Trans. Bristol Gloucestershire Archaeol. Soc.* 97, 9–64

Graham, A., 1978 'The Bonded Warehouse, Montague Close', in *Southwark, Excavations 1972–74, I,* Joint Publication No.1, London and Middlesex Archaeol. Soc. and Surrey Archaeol. Soc., 237–90

Green, B., 1973 'The grave goods from the cremations', in Myres and Green 1973, 77–113

Green, C., 1977 *Excavations in the Roman kiln field at Brampton, 1973–4,* E. Anglian Archaeol. 5, 31–95

Green, C. and Draper, J., 1978 'The Milcoak Roman Villa, Handley, Towcester, Northamptonshire, report on the excavations of 1955 and 1956', *Northamptonshire Archaeol.* 13, 28–66

Green, C. and Green, I., 1987 'Excavations in 1957–8', in Green *et al.* 1987, 112–28

Green, C., Green, I. and Dallas, C., with Wild, J.P., 1987 'Excavations at Castor, Cambridgeshire in 1957–8 and 1973', *Northamptonshire Archaeol.* 21, 109–148

Green, H.M.J., 1960 'Roman Godmanchester: Part I. An Early Dwelling', *Proc. Cambridge Antiq. Soc.* 53, 8–22

Green, H.M.J., 1961 'Roman Godmanchester: Part II. The town defences', *Proc. Cambridge Antiq. Soc.* 54, 68–82

Green, H.S., 1980 *The Flint Arrowheads of the British Isles*, Brit. Archaeol. Rep. Brit. Ser.75 (Oxford)

Greene, K., 1978 'Imported fine wares in Britain to AD250: a guide to identification', in Arthur and Marsh 1978, 15–30

Gregory, T., 1976 'A hoard of Late Roman metalwork from Weeting, Norfolk, *Norfolk Archaeol.* 36, pt 3, 265–72

Gregory, T., 1977 *Enclosure at Ashill*, E. Anglian Archaeol. 5, 9–30

Guide, (nd) *Roman Pottery from the Nene Valley: a Guide*, M.D. Howe, J.R. Perrin and D.F. Mackreth, Peterborough City Mus. Occ. Paper 2, (Peterborough)

Guido, M., 1978 *The Glass Beads of the Prehistoric and Roman Periods in Britain and Ireland*, Rep. Res. Comm. Soc. Antiq. London 35 (London)

Gurney, D., 1985 'The Roman Pottery', in Pryor *et al.* 1985, 121–51

Guy, J.G., 1981 'Roman circular lead tanks in Britain', *Britannia* 12, 271–6

Hadman, J. and Upex, S., 1975 'A Roman pottery kiln at Sulehay, near Yarwell', *Durobrivae: Rev. Nene Valley Archaeol.* 3, 16–18

Hadman, J. and Upex, S., 1979 'Ashton 1977–8', *Durobrivae: Rev. Nene Valley Archaeol.* 7, 28–30

Hallam, S.J., 1970 'Settlement round the Wash', in Phillips 1970, 22–113

Hammerson, M.J. and Murray, C., 1978a 'Other Roman pottery', in Sheldon 1978, 440–59

Hammerson, M.J. and Murray, C., 1978b 'Other Roman pottery', in Graham 1978, 265–80

Hammerson, M.J. and Murray, C., 1978c 'Other Roman pottery', in Dennis 1978, 337–76

Hanworth, R., 1968 'The Roman villa at Rapsley, Ewhurst', *Surrey Archaeol. Collect.* 65, 1–70

Harcourt, R.A., 1974 'The dog in prehistoric and early historic Britain', *J. Archaeol. Sciences* 1, 151–75

Harden, D.B. (ed), 1956 *Dark Age Britain, Studies presented to E.T. Leeds with a bibliography of his work,* (London)

Harden, D.B., 1962 'Glass in Roman York', in RCHM 1962, 136–41

Harden, D.B., 1973 'The Glass', in Brodribb *et al.* 1973, 98–107

Hart, W.H. and Lyons, P.A. (eds), 1884 *Cartularium Monasterii Rameseia*, I (London)

Hart, W.H. and Lyons, P.A. (eds), 1886 *Cartularium Monasterii Rameseia*, II (London)

Hartley, B.R., 1960 *Notes on the Roman Pottery Industry in the Nene Valley*, Peterborough Mus. Soc. Occ. Papers No.2 (Peterborough)

Hartley, B.R. and Dickinson, B. forthcoming *Index of Potters' Stamps on Samian Ware*

Hartley, K.F., 1961 'The stamped mortaria', in Hemsley 1961, 8–13

Hartley, K.F., 1971 'Mortaria from Mancetter 1964', in Mahany 1971, 28–34

Hartley, K.F., 1973 'The marketing and distribution of mortaria', in Detsicas 1973, 39–51

Hartley, K.F. and Hartley, B.R., 1970 'Pottery in the Romano-British Fenland', in Phillips 1970, 165–9

Hartshorne, C.H., 1847 'Description of a Statue of Minerva Custos, and other Roman Antiquities, recently discovered on the Estate of the Duke of Bedford at Sibson and Bedford Purlieus, in the County of Northampton', *Archaeologia* 32, 1–15

Haupt, D., 1984 'Römischer Töpfereibezirk bei Soller, Kreis Düren', *Rheinische Ausgrabungen* 23, 1984: Beiträge zur Archäologie des Römischen Rheinlands 4, Bonn, 391–476

Hawkes, C.F.C., 1947 'Britons, Romans and Saxons round Salisbury and in Cranborne Chase', *Archaeol. J.* 104, 27–81

Hawkes, C.F.C. and Hull, M.R., 1947 *Camulodunum, First Report on the Excavations at Colchester, 1930–1939*, Rep. Res. Comm. Soc. Antiq. London 14 (Oxford)

Hayes, J.W., 1984 'The Roman pottery from the Cat's Water Subsite, Fengate', in Pryor 1984, 179–93

Hayes, R.H. and Whitley, E., 1950 *The Roman Pottery at Norton, East Yorkshire*, Roman Malton Dist. Rep.7 (Leeds)

Hemsley, R., 1961 'A Romano-British Pottery Kiln at Manduessedum', *Trans Birmingham Warwickshire Archaeol. Soc.* 77, 5–17

Henig, M., 1974 'An intaglio from Hall Farm, Orton Longueville', *Durobrivae: Rev. Nene Valley Archaeol.* 2, 20

Henig, M., 1978 *A Corpus of Roman Engraved Gemstones from British Sites*, Brit. Archaeol. Rep. Brit. Ser. 8 (Oxford)

Hermet, F., 1934 *La Graufesenque (Condatomago)*, (Paris)

Hills, C., 1977 *The Anglo-Saxon Cemetery at Spong Hill, North Elmham, Part I, Catalogue of Cremations*, E. Anglian Archaeol. 6

Hills, C.M., 1981 'Barred zoomorphic combs of the migration period', in Evison 1981, 96–125

Hills, C. and Penn, K., 1981 *The Anglo-Saxon Cemetery at Spong Hill, North Elmham, Part II, Catalogue of Cremations*, E. Anglian Archaeol. 11

Hines, J., 1984 *The Scandinavian Character of Anglian England in the pre-Viking Period*, Brit. Archaeol. Rep. Brit. Ser.124 (Oxford)

Hodgson, G.W.I., 1977 *The Animal Remains from Excavations at Vindolanda, 1970–1975*, Vindolanda II (Hexham)

Holmes, S.C.A., 1970 'Outline geology of the Roman Fenland and stratification deposits near the Old West River, Cambridgeshire', in Phillips 1970, 127–31

Holwerda, J., 1941 *De Belgische waar in Nijmegen*, (The Hague)

Hooke, D., 1981 'The Droitwich salt industry: an examination of the West Midland charter evidence', in Brown *et al.* 1981, 123–53

Houlder, C.H., 1979 'The Langdale and Scafell Pike axe factory sites: a field survey', in Clough and Cummins 1979, 87–9

Huggins, P.J., 1978 'Excavation of a Belgic and Romano-British farm with Middle Saxon cemetery and churches at Nazeingbury, Essex', *Essex Archaeol. Hist.* 10, 29–117

Hughes, T.McK., 1902 'On the potter's field at Horningsea, and a comparative notice of the kilns and furnaces found in the neighbourhood', *Proc. Cambridge Antiq. Soc.* 10, 174–94

Hull, M.R., 1963a 'Roman Gazetteer', *A History of Essex, Volume III*, Victoria Hist. Counties of England (London), 35–203

Hull, M.R., 1963b *The Roman Potters' Kilns of Colchester*, Rep. Res. Comm. Soc. Antiq. London, 21 (Oxford)

Hunter, A.G. and Kirk, J.R., 1953 'Excavations at Campsfield, Kidlington, Oxon, 1949', *Oxoniensia* 17/18, 36–62

Hunter, R. and Mynard, D., 1977 'Excavations at Thorplands near Northampton, 1970 and 1974', *Northamptonshire Archaeol.* 12, 97–154

Hurst, J.G., 1976 'The Pottery', in Wilson 1976, 283–348

Isings, C., 1957 *Roman Glass from Dated Finds*, (Groningen)

Isings, C., 1971 *Roman Glass in Limberg*, (Groningen)

Jackson, D.A., 1977 'Further excavations at Aldwincle, Northamptonshire, 1969–71', *Northamptonshire Archaeol.* 12, 9–54

Jackson, D.A. and Ambrose, T.M., 1978 'Excavations at Wakerley, Northants., 1972–5', *Britannia* 9, 115–242

Jackson, D.A. and Tylecote, R.F., 1988 'Two new Romano-British iron-working sites in Northamptonshire — a new type of furnace?', *Britannia* 19, 275–98

Jackson, D.A., Harding, D.W. and Myres, J.N.L., 1969 'The Iron Age and Anglo-Saxon site at Upton, Northants.', *Antiq. J.* 49, 202–21

Jackson, R. P. J. and Potter, T. W., forthcoming *Excavations at Stonea, Cambs 1980–1985* (British Museum Publications)

Jacobi, H., 1914 'Römische Getreidemühlen', *Saalburg Jahrbuch, Bericht des Saalburg Museums* 3, 75–95

James, S., Marshall, A. and Millett, M., 1984 'An early medieval building tradition', *Archaeol. J.* 141, 182–215

Jenkins, F., 1967 'Romano-British clay figurines', in Dudley 1967, 19–21

Johnson, S., 1980 *Late Roman Britain* (London)

Johnson, S., 1983 *Burgh Castle: Excavations by Charles Green, 1958–61*, E. Anglian Archaeol. 20

Johnston, D.E., 1969 'Romano-British pottery kilns near Northampton', *Antiq. J.* 49, 75– 97

Johnston, D.E. (ed), 1977 *The Saxon Shore*, Counc. Brit. Archaeol. Res. Rep.18 (London)

Jones, A.H.M., 1964 *The Later Roman Empire 284–602, a Social, Economic and Administrative Survey*, (Oxford)

Jones, M.U., 1974 'Excavations at Mucking, Essex, a second interim report', *Antiq. J.* 54, 183–99

Kent, J.P.C. and Painter, K.S., 1977 *Wealth of the Roman World, AD 300–700* (London)

Kenyon, K.M., 1948 *Excavations at the Jewry Wall Site, Leicester*, Rep. Res. Comm. Soc. Antiq. London 15 (Oxford)

Keppie, L.J.F., 1975 'Finds, stone to clay', in Robertson *et al.* 1975, 31–126

King, J.M., 1987 'The animal bones', in Dannell and Wild 1987, 185–8

Knorr, R., 1919 *Töpfer und Fabriken verzierter Terra-Sigillata des ersten Jahrhunderts*, (Stuttgart)

Knorr, R., 1952 *Terra-Sigillata-Gefässe des ersten Jahrhunderts mit Töpfernamen*, (Stuttgart)

Krogman, W.M., 1978 *The Human Skeleton in Forensic Medicine*, (Springfield, Illinois)

Krug, A., 1980 'Antike Gemmen im Römisch-Germanischen Museum Köln', *Bericht der Römisch-Germanischen Kommission* 61, 152–260

Kunow, J., 1980 'Der Leithorizont der Augenfibel und die Stufe Eggers B1', *Archäologisches Korrespondenzblatt* 10, 157–61

Lancel, S., 1967 *Verrerie antique de Tipasa* (Paris)

Large, E.H. and Lucas, R., 1970 'Third interim report on the Roman villa at Halstock', *Proc. Dorset Natur. Hist. Archaeol. Soc.* 92, 134

Leach, P., 1982 *Ilchester, Volume I, Excavations 1974–75*, Western Archaeol. Trust, Monograph 3 (Bristol)

Leech, R., 1982 *Excavations at Catsgore, 1970–73, A Romano-British Village*, Western Archaeol. Trust, Monograph 2 (Bristol)

Leeds, E.T., 1945 'The distribution of the Angles and Saxons archaeologically considered', *Archaeologia* 91, 1–106

Leeds, E.T. and Atkinson, R.J.C., 1944 'An Anglo-Saxon cemetery at Nassington, Northamptonshire', *Antiq. J.* 24, 100–28

Lethbridge, T.C., 1931 *Recent Excavations in Anglo-Saxon Cemeteries in Cambridgeshire and Suffolk, A Report*, Cambridge Antiq. Soc., Quarto Pub. n. ser.III (Cambridge)

LMC, 1954 *Medieval Catalogue*, London Mus. Cat.7 (London, 1954)

Losco-Bradley, S., 1977 'Catholme', *Curr. Archaeol.* 5, No.12 , 358–64

Loughlin, N., 1977 'Dales Ware: a contribution to the study of Roman coarse pottery, in Peacock 1977, 85–146

LRBC, 1960 *Late Roman Bronze Coinage, A.D. 324–498*, R.A.G. Carson, P.V. Hill and J.P.C. Kent (London, 1960)

Ludowici, W., 1927 *Katalog V: Stempel, Namen und Bilder römischer Töpfer aus meinen Ausgrabungen in Rheinzabern, 1901–1914*

MacDonald, G. and Curle, A.O., 1929 'The Roman Fort at Mumrills, near Falkirk', *Proc. Soc. Antiq. Scotland* 63, 396–575

MacGregor, A., 1976 *Finds from a Roman Sewer System and an Adjacent Building in Church Street*, Archaeol. of York, The Small Finds, 17/1, (London)

MacGregor, A., 1978 *Roman Finds from Skeldergate and Bishophill*, Archaeol. of York, The Small Finds, 17/2, (London)

Mackreth, D.F., 1979 'Durobrivae', *Durobrivae, Rev. Nene Valley Archaeol.* 7, 19–21

Mackreth, D.F., 1984 'Castor', *Durobrivae, Rev. Nene Valley Archaeol.* 9, 22–5

Mackreth, D.F., 1988 'Excavation of an Iron Age and Roman enclosure at Werrington, Cambridgeshire', *Britannia* 19, 59–151

Mackreth, D.F. forthcoming *Monument 97, Orton Longueville, Cambridgeshire, a Late Pre-Roman Iron Age and Early Roman Farmstead*, E. Anglian Archaeol.

Mackreth, D.F., forthcoming b 'The Brooches', in Jackson and Potter, forthcoming

Mackreth, D.F., forthcoming c 'The Brooches', in Timby, forthcoming

Mackreth, D.F., forthcoming d 'The Brooches', in Garrod and Atkin, forthcoming

Mackreth, D.F., forthcoming e 'The Brooches', in Wilson, forthcoming

Mackreth, D.F., forthcoming f 'The Brooches', in Ellis (ed), forthcoming

Mahany, C., 1971 'Excavations at Manduessedum, 1964', *Trans. Birmingham Warwickshire Archaeol. Soc.* 84, 18–44

Manby, T.G., 1979 'Typology, materials and distribution of flint and stone axes in Yorkshire', in Clough and Cummins 1979, 65–81

Mann, J.C., 1977 '*Duces* and *comites* in the 4th century', in Johnston 1977, 11–15

Manning, W.H., 1964 'A mill pivot from Silchester', *Antiq. J.* 44, 38–40

Manning, W.H., 1972 'The iron objects', in Frere 1972, 163–95

Manning, W.H., 1974 'Objects of iron', in Neal 1974, 157–87

Manning, W.H., 1976 *Catalogue of Romano-British Ironwork in the Museum of Antiquities, Newcastle-upon-Tyne*, (Newcastle)

Manning, W.H., 1984 'The iron objects', in Frere 1984, 83–108

Martin, A., Ashby, T. and Hudd, A.E., 1901 'Excavations at Caerwent, Monmouthshire, on the site of the Roman city of Venta Silurum, in 1899 and 1900', *Archaeologia* 57, pt.2, 295–316

McWhirr, A., Viner, L. and Wells, C., 1982 *Romano-British Cemeteries at Cirencester*, Cirencester Excavations II (Cirencester)

Meates, G.W., 1979 *The Roman Villa at Lullingstone, Kent, Volume I, The Site*, Kent Archaeol. Soc. Monograph Ser.1 (Chichester)

Mellor, A.S. and Goodchild, R., 1942 'The Roman villa at Atworth, Wilts.', *Wiltshire Archaeol. Natur. Hist. Mag.* 49, 46–95

Middleton, J.H., 1890 'On a Roman villa in Spoonley Wood, Gloucestershire and on Romano-British houses generally', *Archaeologia* 52, pt 2, 651–68

Miket, R., 1983 *The Roman Fort at South Shields: Excavation of the Defences, 1977–81*, (Gateshead)

Miller, L., Schofield, J. and Rhodes, M., 1986 *The Roman Quay at St. Magnus House, London, Excavations at New Fresh Wharf, Lower Thames Street, London, 1974–78*, London and Middlesex Archaeol. Soc., Spec. Paper 8

Millett, M. and James, S., 1983 'Excavations at Cowdery's Down, Basingstoke, Hampshire, 1978–81', *Archaeol. J.* 140, 151–279

Moorhouse, S.A., 1981 'The medieval pottery industry and its markets', in Crossley 1981, 96–125

Moritz, L.A., 1958 *Grain-Mills and Flour in Classical Antiquity*, (Oxford)

Morley Hewitt, A.T., 1969 *Roman Villa, West Park, Rockbourne, near Fordingbridge, Hants.*, (Fordingbridge)

Morris, P., 1979 *Agricultural Buildings in Roman Britain*, Brit. Archaeol. Rep. Brit. Ser.70 (Oxford)

Mossop, J.C., 1960 'Potters' stamps on Castor ware platters', *Antiq. J.* 40, 226–8

Munby, J. and Henig, M. (eds), 1977 *Roman Life and Art in Britain*, Brit. Archaeol. Rep.41 (Oxford)

Mynott, E., 1977 'The Roman villa site at Keston', *Kent Archaeol. Rev.* 45–53, 215–18

Myres, J.N.L., 1956 'Romano-Saxon pottery', in Harden 1956, 16–39

Myres, J.N.L., 1968 'The Anglo-Saxon pottery from Mucking', *Antiq. J.* 48, 222–8

Myres, J.N.L., 1969 *Anglo-Saxon Pottery and the Settlement of England*, (Oxford)

Myres, J.N.L., 1977 *A Corpus of Anglo-Saxon Pottery of the Pagan Period*, (Cambridge)

Myres, J.N.L. and Green, B., 1973 *The Anglo-Saxon Cemeteries of Caistor-by-Norwich and Markshall, Norfolk*, Rep. Res. Comm. Soc. Antiq. London 30 (Oxford)

Neal, D.S., 1974 *The Excavation of the Roman Villa in Gadebridge Park, Hemel Hempstead, 1963–8*, Rep. Res. Comm. Soc. Antiq. London 31 (Leeds)

Neal, D.S., 1976 *Northchurch, Boxmoor and Hemel Hempstead Station: The Excavation of Three Roman Buildings in the Bulbourne Valley*, Hertfordshire Archaeol. 4, 1–135

Neal, D.S. and Butcher, S.A., 1974 'Miscellaneous objects of bronze', in Neal 1974, 128–50

Neville, R.C., 1852 *Saxon Obsequies illustrated by ornaments and weapons discovered by the Hon. R.C. Neville during the Autumn of 1851*, (London)

Neville, R.C., 1856 'Description of a Remarkable Deposit of Roman Antiquities of Iron, Discovered at Great Chesterford, Essex, in 1854', *Archaeol. J.* 13, 1–13

Noddle, B., 1980 'The animal bones', in Wade-Martins 1980, 375–409

Norris, N.E.S., 1956 'Miscellaneous researches, 1949–56', *Sussex Archaeol. Collect.* 94, 1–12

Norris, N.E.S. and Burstow, G.P., 1950 'A prehistoric and Romano-British site at West Blatchington, Hove', *Sussex Archaeol. Collect.* 89, 1–56

O'Neil, H., 1947 'The Roman villa at Park Street, near St. Albans, Hertfordshire, report on the excavations of 1943–45', *Archaeol. J.* 102, 21–110

Oswald, F., 1936–7 *Index of Figure-Types on Terra Sigillata*, Univ. Liverpool Annals Archaeol. Anthropol. Supplement (Liverpool)

Partridge, C., 1981 *Skeleton Green, a Late Iron Age and Romano-British Site*, Britannia Monograph Ser.2

Payne, S., 1969 'A metrical distinction between sheep and goat metacarpals', in Ucko and Dimbleby 1969, 295–305

Peacock, D. P. S. (ed), 1977 *Pottery and Early Commerce* (London)

Peers, C. and Radford, C.A.R., 1943 'The Saxon monastery of Whitby', *Archaeologia* 89, 27–88

Perrin, J.R., 1980 'Pottery of "London Ware" type from the Nene Valley', *Durobrivae: Rev. Nene Valley Archaeol.* 8, 9–10

Perrin, J.R., 1981a *Roman Pottery from the Colonia: Skeldergate and Bishophill*, Archaeol. of York, The Pottery 16/2, (London)

Perrin, J.R., 1981b 'The Late Roman pottery of Great Casterton — thirty years on', in Anderson and Anderson 1981, 447–63

Perrin, J.R., 1988 'The grey wares', in Mackreth 1988, 12–121

Perrin, J.R. and Webster, G., 1990 'Roman Pottery from Excavations in Normangate Field, Castor, Peterborough, 1962–3', *J. Roman Pottery Stud.* 3, 35–62

Perring, D., 1977 'Aspects of art in Romano-British pottery', in Munby and Henig 1977, 253–82

Phillips, C.W. (ed), 1970 *The Fenland in Roman Times, Studies of a major area of peasant colonization with a Gazetteer covering all known sites and finds*, Royal Geog. Soc. Res. Ser.5 (London)

Philp, B., 1973 'Site 21. The Roman villa at Darenth, Kent', *Excav. W. Kent, 1960– 1970*, (Dover), 119–54

Piggott, C.M., 1946 'The Late Bronze Age razors of the British Isles', *Proc. Prehist. Soc.* 12, 121–41

Piggott, S. (ed), 1981 *The Agrarian History of England and Wales, Vol.I.i, Prehistory*, (Cambridge)

Pitt Rivers, A.H.L.F., 1887 *Excavations in Cranborne Chase near Rushmore on the borders of Dorset and Wiltshire*, Vol.I, (privately printed)

Pitts, M.W., 1978 'On the shape of waste flakes as an index of technological change in lithic industries', *J. Archaeol. Sci.* 5, 17–37

Pollard, S., 1974 'A Late Iron-Age settlement and a Romano-British villa at Holcombe near Uplyme, Devon', *Proc. Devon Archaeol. Soc.*32, 59–161

Potter, T.W., 1965 'The Roman pottery from Coldham Clamp and its affinities', *Proc. Camb. Antiq. Soc.* 58, 12–37

Potter, T.W. and Potter, C.F., 1982 *A Romano-British village at Grandford, March, Cambridgeshire*, Brit. Mus. Occ. Paper 35 (London)

Prashma, M.G., 1980 'Recommendations for age and sex diagnoses of skeletons', *J. Human Evol.* 9, 517–49

Price, J., 1979 'The Glass', in Gracie and Price 1979, 37–46

Price, J.E. and Price, F.G.H., 1884 *A Guide to the Roman Villa recently discovered at Morton between Sandown and Brading, Isle of Wight*, (11th edn, Ventnor)

Pretty, E., 1852 'On the discovery of Roman remains, near Towcester, Northamptonshire', *J. Brit. Archaeol. Ass.* 7, 107–14

Pryor, A., 1978 'The Car Dyke', *Durobrivae, Rev. Nene Valley Archaeol.* 6, 24–5

Pryor, F., 1980 *Excavation at Fengate, Peterborough, England, The Third Report*, Northamptonshire Archaeol. Soc. Monograph 1, Royal Ontario Mus. Archaeol. Monograph 6 (Leicester)

Pryor, F., 1984 *Excavations at Fengate, Peterborough, England: The Fourth Report*, Northamptonshire Archaeol. Soc. Monograph 2, Royal Ontario Mus. Archaeol. Monograph 7

Pryor, F., French, C., Crowther, D., Gurney, D., Simpson, G. and Taylor, M., 1985 *Fenland Project No. 1, Archaeology and Environment in the Lower Welland Valley*, E. Anglian Archaeol. 27

Rahtz, P.A., 1963 'A Roman villa at Downton', *Wiltshire Archaeol. Natur. Hist. Mag.* 58, 303–41

Rahtz, P.A., 1976 'Buildings and rural settlement', in Wilson 1976, 49–98

Rahtz, P.A. and Greenfield, E., 1977 *Excavations at Chew Valley Lake*, Dept. Environment Archaeol. Rep. 8 (London)

Reece, R., 1973 'Roman coinage in the Western Empire', *Britannia* 4, 227–51

Reece, R., 1980 'Town and country: the end of Roman Britain', *World Archaeol.* 12.1, 77–92

Rees, S.E., 1979 *Agricultural Implements of Prehistoric and Roman Britain*, Brit. Archaeol. Rep. Brit. Ser.69 (Oxford)

Reynolds, P.J., 1979 'Romano-British corn-drying oven: an experiment', *Archaeol. J.* 136, 27–42

RIB *The Roman Inscriptions of Britain, I, Inscriptions on Stone*, R.G. Collingwood and R.P. Wright (Oxford, 1965)

RIC *Roman Imperial Coinage,* H. Mattingley, E. A. Sydenham, C. H. V. Sutherland and R. A. G. Carson (eds)

Richardson, B., 1986 'Pottery', in Miller *et al.* 1986, 96–138

Richmond, I. (ed), 1966 *Handbook to the Roman Wall* (Newcastle)

Roach Smith, C., 1857 *Collectanea Antiqua* 4, pt 2 (London)

Robertson, A. R., Scott, M. and Keppie, L., 1975 *Bar Hill: A Roman Fort and its Finds*, Brit. Archaeol. Rep. 16 (Oxford)

Roberts, W.I., 1982 *Romano-Saxon Pottery*, Brit. Archaeol. Rep. Brit. Ser.106 (Oxford)

Rodwell, W., 1978 'Stamp-decorated pottery of the early Roman period in eastern England', in Arthur and Marsh 1978, 225–92

Rogers, G.B., 1974 *Poteries Sigillées de la Gaule Centrale I*, Gallia Supplément 28

Rogerson, A., 1977 *Excavations at Scole, 1973,* E. Anglian Archaeol. 5, 97–122

Rollo, L., 1988 'The shell-gritted wares', in Mackreth 1988, 107–20

Rollo, L., forthcoming 'The Pottery', in Mackreth forthcoming

Rowley, T. (ed), 1974 *Anglo-Saxon Settlement and Landscape, Papers presented to a Symposium, Oxford 1973*, Brit. Archaeol. Rep. Brit. Ser.6

Royal Commission on Historical Monuments, 1962 *Eburacum, Roman York, Volume I, An inventory of the Historical Monuments in the City of York,* (London)

Royal Commission on Historical Monuments, 1969 *Peterborough New Town, a Survey of the Antiquities in the Areas of Development,* (London)

Rutter, J.G. and Duke, G., 1958 *Excavations at Crossgates near Scarborough, 1947–56,* Scarborough Dist. Archaeol. Soc. Res. Rep.1 (Scarborough)

Ryder, M.L., 1981 'Livestock', in Piggott 1981, 301–410

St John Hope, W.H., 1903 'Excavations on the site of the Roman city at Silchester, Hants, 1902', *Archaeologia* 58, pt 2, 413–28

St John Hope, W.H. and Fox, G.E., 1898 'Excavations on the site of the Roman city at Silchester, Hants, in 1897', *Archaeologia* 56, pt 1, 103–26

St John Hope, W.H. and Fox, G.E., 1905 'Excavations on the site of the Roman city at Silchester, Hants, in 1903 and 1904', *Archaeologia* 59, pt 2, 333–70

Salway, P., 1970 'The Roman Fenland', in Phillips 1970, 1–21

Salway, P., 1981 *Roman Britain* (Oxford)

Salzmann, L.F., 1967 *Building in England down to 1540, a Documentary History* (Oxford)

Saville, A., 1981 *Grimes Graves, Norfolk, Excavations 1971–72, Vol. II, The Flint Assemblages*, Dept. Environment Archaeol. Rep.11 (London)

Schiffer, M.B., 1976 *Behavioural Archaeology*, (New York)

Sheldon, H., 1978 '93–95 Borough High Street', in *Southwark, Excavations 1972–74, II*, Joint Publication No.1, London and Middlesex Archaeol. Soc. and Surrey Archaeol. Soc., 423–72

Shepard, A.O., 1974 *Ceramics for the Archaeologist*, Carnegie Inst. Washington, Pub. 609 (5th edn, Washington)

Simmons, B., 1979 'The Lincolnshire Car Dyke: navigation or drainage?', *Britannia* 10, 183–96

Simmons, B., 1980 'Iron Age and Roman coasts around the Wash', in Thompson 1980, 56–73

Simpson, G. (ed), 1976 *Watermills and Military Works on Hadrian's Wall, Excavations in Northumberland, 1907–1913, by F. Gerald Simpson,* (Kendal)

Simpson, W.G., 1981 'Excavations in Field OS 124, Maxey, Cambridgeshire', *Northamptonshire Archaeol.* 16, 34–64

Smedley, N. and Owles, E., 1961 'Some Suffolk kilns II: two kilns making colour-coated ware at Grimstone End, Pakenham', *Proc. Suffolk Inst. Archaeol.* 28, 203–25

Smith, A.G., 1970 'The stratigraphy of the northern Fenland', in Phillips 1970, 147–64

Smith, I.F., 1965 *Windmill Hill and Avebury: Excavations by Alexander Keiller, 1925–1939,* (Oxford)

Smith, I.F., 1979 'The chronology of British stone implements', in Clough and Cummins 1979, 13–22

Smith, J.T., 1964 'The Roman villa at Denton', *Archit. Archaeol. Soc. County of Lincoln, Rep, Papers* 10, pt 1, 75–104

Stanfield, J.A. and Simpson, G., 1958 *Central Gaulish Potters,* (Durham)

Stanford, S.C., 1974 *Croft Ambrey,* (Hereford)

Stead, I.M., 1976 *Excavations at Winterton Roman Villa, and other Roman sites in north Lincolnshire, 1958–1969,* Dept. Environment Archaeol. Rep.9 (London)

Stead, I.M., 1980 *Rudston Roman Villa* (Leeds)

Stead, I.M. and Rigby, V., 1986 *Baldock, The Excavation of a Roman and Pre-Roman Settlement, 1968–72,* Britannia Monograph Ser.7 (London)

Stow, S., 1982 'Non-ferrous objects', in Frere *et al.* 1982, 121–6

Sumner, H., 1914 *Excavations on Rockbourne Down, Hampshire* (London)

Swan, V.G., 1984 *The Pottery Kilns of Roman Britain*, Royal Comm. Hist. Monuments Supplementary Ser.5 (London)

Swift, J., 1909 *A Complete Collection of Genteel and Ingenious Conversation, according to the most Polite Mode and Method, now used at Court, and the Best Companies of England*, Everyman's Library 347 (London), 235–325

Taylor, M.V., 1959 'Roman Britain in 1958, I, Sites Explored', *J. Roman Stud.* 49, 102–35

Taylor, M.V., 1962 'Roman Britain in 1961, I, Sites Explored', *J. Roman Stud.* 52, 160–90

Taylor, M.V. and Wilson, D.R., 1961 'Roman Britain in 1960, I, Sites Explored', *J. Roman Stud.* 51, 157–91

Thompson, F.H. (ed), 1980 *Archaeology and Coastal Change*, Soc. Antiq. London Occ. Papers n. ser.1 (London)

Thompson, I., 1982 *Grog-tempered 'Belgic' Pottery of South-Eastern England*, Brit. Archaeol. Rep. Brit. Ser.108 (Oxford)

Tilson, P., 1973 'A Belgic and Romano-British site at Bromham', *Beds. Archaeol. J.* 8, 23–66

Timby, J., forthcoming *Kingscote: a Roman Estate Centre in the Cotswolds*, Cotswold Archaeol. Trust Monogr. 2

Todd, M., 1968a 'Trent Valley Ware', a Roman coarse ware of the Middle and Lower Trent Valley, M. Todd, *Trans. Thoroton Soc. Notts.* 72, 38–41

Todd, M. (ed), 1968b *The Roman Fort at Great Casterton, Rutland, Excavations of 1960 and 1962, directed by the late Sir Ian Richmond and the late Dr. Philip Corder* (Nottingham)

Todd, M. (ed), 1978 *Studies in the Romano-British Villa* (Leicester)

Tomlin, R.S.O., 1976 *'Notitia Dignitatum Omnium tam Civilium quam Militarium'*, in Goodburn and Bartholemew 1976, 189–209

Toynbee, J.M.C., 1964 *Art in Britain under the Romans* (Oxford)

Trotter, M. and Gleser, G.C., 1952 'Estimation of stature from long bones of American whites and negroes', *American J. Phys. Anth.* n. ser. 10 (4), 463–514

Ubelaker, D.H., 1974 *Reconstruction of Demographic Profiles from Ossuary Skeletal Samples: A Case Study from Tideswater Potomac*, Smithsonian Contrib. Anthropol. 18 (Washington)

Ucko, P.J. and Dimbleby, G.W., 1969 *The Domestication and Exploitation of Plants and Animals,* (London)

Vanderhoeven, M., 1962 *De Romeinse Glasverzamelung in het Provinciaal Gallo-Romeinse Museum,* (Tongeren)

Vogt, E., 1960 'Interpretation und museale Answertung almannischer Grab-funde', *Zeitschrift für Schweizerische Archäologie und Kunstgeschicte* 20, 70–90

Wacher, J.S., 1969 *Excavations at Brough-on-Humber, 1958–1961*, Rep. Res. Comm. Soc. Antiq. London 25 (Leeds)

Wacher, J.S., 1975 *The Towns of Roman Britain* (London)

Wacher, J., 1978 *Roman Britain* (London)

Wade-Martins, P., 1980 *Excavations in North Elmham Park, 1967–1972*, E. Anglian Archaeol. 9

Wainwright, G.J., 1971 'The excavation of prehistoric and Romano-British settlements near Durrington Walls, Wiltshire, 1970', *Wiltshire Archaeol. Natur. Hist. Mag.* 66, 76–128

Wainwright, G.J., 1972 'The excavation of a Neolithic settlement on Broome Heath, Ditchingham, Norfolk, England', *Proc. Prehist. Soc.* 38, 1–97

Walker, T.J., 1899 'The traces of the Roman occupation left in Peterborough and the surrounding district', *J. Brit. Archaeol. Ass.* 55, 51–62

Walker, F.G., 1912 'Roman pottery kilns at Horningsea, Cambridgeshire', *Proc. Cambridge Antiq. Soc.* 17, 14–69

Walker, J., 1978 'Anglo-Saxon traded pottery', in Todd 1978, 224–8

Waugh, H. and Goodburn, R., 1972 'The non-ferrous objects', in Frere 1972, 114–62

Webster, G. (ed), 1976 *Romano-British Coarse Pottery: A Student's Guide*, Counc. Brit. Archaeol. Res. Rep. 6 (2nd edn, London)

Webster, G. and Booth, N., 1947 'A Romano-British pottery kiln at Swanpool, near Lincoln', *Antiq. J.* 27, 61–79

Webster, G. and Smith, L., 1982 'The excavations of a Romano-British rural establishment at Barnsley Park, Gloucestershire, 1961–1979, Part II, *c.*AD360–400+', *Trans. Bristol Gloucestershire Archaeol. Soc.* 100, 65–189

Webster, J., 1975 'Objects of bronze and silver', and 'Objects of bone and antler', in Cunliffe 1975, 198–225

Wedlake, W.J., 1958 *Excavations at Camerton, Somerset*, privately printed

Wedlake, W.J. 1982 *The Excavation of the Shrine of Apollo at Nettleton, Wiltshire, 1956–1971*, Rep. Res. Comm. Soc. Antiq. London 40 (Dorking)

West, S., 1985 *West Stow, The Anglo-Saxon Village*, E. Anglian Archaeol. 24

Wheeler, R.E.M., 1930 *London in Roman Times*, London Mus. Cat.3 (London)

Wheeler, R.E.M., 1943 *Maiden Castle, Dorset*, Rep. Res. Comm. Soc. Antiq. London 12 (Oxford)

Wheeler, R.E.M. and Wheeler, T.V., 1932 *Report on the Excavation of the Prehistoric, Roman, and Post- Roman Site in Lydney Park, Gloucestershire*, Rep. Res. Comm. Soc. Antiq. London 9 (Oxford)

Wheeler, R.E.M. and Wheeler, T.V., 1936 *Verulamium, a Belgic and two Roman Cities*, Rep. Res. Comm. Soc. Antiq. London 11 (Oxford)

White, D.A., 1967 'Excavations at a Romano-British Settlement on the Fenland Edge at Earith, Hunts', *Proc. Cambridge Antiq. Soc.* 60, 7–71

White, K.D., 1967 *Agricultural Implements of the Roman World* (Cambridge)

White, K.D., 1970 *Roman Farming* (London)

Whiting, W., Hawley, W. and May, T., 1921 *Report on the Excavation of the Roman Cemetery at Ospringe, Kent*, Rep. Res. Comm. Soc. Antiq. London 8 (Oxford)

Wightman, E.M., 1985 *Gallia Belgica*, (London)

Wild, J.P., 1974 'Roman settlement in the Lower Nene Valley', *Archaeol. J.* 131, 140–70

Williams, A. M., 1909 'The Romano-British Buildings at Stroud, near Petersfield, Hants', *The Antiquary* 45, 375–81

Williams, D.F., 1977 'The Romano-British Black-Burnished industry: an essay on characterization by heavy mineral analysis', in Peacock 1977, 163–220

Williams, J.H. (ed), 1974 *Two Iron Age Sites in Northampton*, (Northampton)

Wilson, D.M., 1967 'Medieval Britain in 1966, I, Pre-Conquest', *Medieval Archaeol.* 11, 262–72

Wilson, D.M., 1969 'Medieval Britain in 1968, I, Pre-Conquest', *Medieval Archaeol.* 13, 230–43

Wilson, D.M. (ed), 1976 *The Archaeology of Anglo-Saxon England*, (London)

Wilson, D.R., 1965 'Roman Britain in 1964, I, Sites Explored', *J. Roman Stud.* 55, 199– 220

Wilson, D.R., 1968 'Roman Britain in 1967, I, Sites Explored', *J. Roman Stud.* 58, 176–206

Wilson, D.R., 1969 'Roman Britain in 1968, I, Sites Explored', *J. Roman Stud.* 59, 198–234

Wilson, M.G., 1968 'Other objects ...', in Cunliffe 1968, 93–110

Wilson, P., forthcoming *Excavations at the Roman Town of Cataractonium, 1958–1993*, Eng. Heritage Arch. Rep.

Woodfield, C., 1983 'The remainder of the Roman pottery', in Brown *et al.* 1983, 74–100

Woods, P.J., 1970 'Excavations at Brixworth, Northants, 1965–70. The Romano-British Villa. Part 1, The Roman Coarse Pottery and Decorated Samian Ware', *J. Northampton Mus. Art Gall.* 8

Woods, P.J. and Hastings, B.C., 1984 *Rushden: The Early Fine Wares*, (Northampton)

Wyman-Abbott, G., 1920 'Further discoveries in Anglo-Saxon cemeteries at Woodston, Hunts.', *Peterborough Natur. Hist., Sci. Archaeol. Soc.* 49, 34–40

Young, C.J., 1977 *The Roman Pottery Industry of the Oxford Region*, Brit. Archaeol. Rep. Brit. Ser. 43 (Oxford)

Young, C.J., 1981 'The Late Roman water-mill at Ickham, Kent, and the Saxon Shore', in Detsicas 1981, 32–41

Zimmermann, W.H., 1974 'A Roman Iron Age and Early Migration settlement at Flögeln, Kr. Wesermünde, Lower Saxony', in Rowley 1974, 56–73

Index

aerial photography, 232
agriculture
land use, 220–1, 224
medieval period, xi, 19, 75
medieval/post-medieval period, 27
modern development, xvii, 42–3, 228
Period 1 pasture, 218
ridge and furrow, xi, 27, 42, 43, 220
see also animals; cereals; crops
ale, 229, 230
Anglo-Saxon period
Early, 40, 89, 239
Middle, 89
brooches, 102
buildings, 5, 36–7, 85-91
small rectangular, 39, 85, 91
western (over Roman House), 26, 70, 85, 87–9 (Fig. 56)
see also granaries (The Granary); Hall; House, Anglo-Saxon; Sunken-featured Buildings
evidence for dating, 27
finds, 92, 102–5 (Fig. 74), 106–7 (Pl. IX; Fig. 75)
metalwork, 42
and Period 5, xv, 28, 29 (Fig. 12), 30, 31, 32, 33
and Roman settlement/culture, 33, 41, 178, 182, 189–90, 237–9
rural economy, 221–2
Saxon Shore system, 236
settlement, 27, 35, 39, 42, 164, 165, 177, 231, 233
arrival of settlers, xiv
eastern limit of, xi
and population, 188
site development, 220, 227
site features, 15, 36–7, 39, 40, 41, 84, 228
see also under pottery
animals, 8, 23, 39, 236
barriers against, 4, 7
bone, 31, 216–18, 221, 223–4 (Table 83), 225, 226
see also bird bones
in economy, 233
increase in, xv, 235
management of, 2, 4, 9, 15, 17, 36, 216, 220, 221–5, 226–7, 230–1
and Mill-house, 72
and milling, 108, 113
and Rectangular Building, 22, 72
stalling/pens, 10, 23, 222
annex (to Anglo-Saxon Hall), 88–9 (Fig. 56), 173 (Fig. 105)
see also Anglo-Saxon period, buildings (western)
antler, 225
Antonine period, 121, 139, 182, 190, 235
Hadrianic-Antonine period, 191
Ashton, 119, 121, 179
Atworth: drier, 80
Augustus, Emperor, 239
axe, greenstone, 1

Bancroft villa: barn at, 69
barley, 227, 229, 230
Barnack rag, 72, 232
Barns, 55–70, 215
analysis/structure, 65–7, 70
construction and reconstruction, 6, 67–70 (Tables 1–2)
conversion to brewery (Period 3), xv
driers, 220
Barn 1, xi, 8, 25–6, 38, 39, 41, 42, 55–8 (Figs 35–7)
analysis, 66, 67
animal management, 222, 226–7
construction and reconstruction, 67, 68, 70
crop handling, 227
dating, 9
equipment storage, 228
pottery, 139, 181, 182, 189, 215
site features, 13–14, 18, 20, 39–40, 83, 84, 91, 143
structure, 5, 10–11, 17, 18, 66, 67, 224
survival of, 89
Barn 2, 15, 19, 58–60 (Figs 38–40), 63, 65, 74
analysis, 66, 67
animal management, 227
construction, 9, 67, 68, 69, 70, 230
and crops, 227
driers, 75–6 (Figs 48 & 50), 224, 229
passes out of use, 14
pottery, 139, 141 (Table 23), 143 (Table 31), 181, 182, 189
refurbishment, 15
replaced by Barn 4, 21, 66, 67, 155, 220, 222
site features, 12–14, 179
structure of, 18, 66, 67
use, 222
Barn 3, 19–20, 26–7, 42, 55, 60–3 (Figs 41–2), 83
analysis and structure, 66–7
animal management, 226
construction, 14, 18, 68, 69, 222, 224, 229
crop handling, 227
equipment storage, 228
pottery, 143, 181, 182
site features, 20, 36, 39–40
and site population, 228
siting of, 220, 221
Barn 4, 20, 22, 41, 55, 63–5 (Figs 43–4), 72, 74
access to, 23
analysis, 67
animal management, 227
construction and reconstruction, 67, 68, 69
driers, xi, 75–8 (Figs 49–52), 229
features in vicinity, 36, 37
Period 4, 13, 19, 75–6
Period 5, 34–5, 75–8
reduction of, 41
replaces Barn 2, 20–2, 66, 67, 155, 220, 222
Roman pottery, 171 (Fig. 104)
and site changes, 227
and site population, 228
structure, 15, 18, 67
barrels, 230
beads, 98, 99 (Fig. 63)
'beaker-jug', 178·
beakers, 114, 117, 177, 178, 181, 182, 188, 189
colour-coated, 122, 139, 235
demise of, 164, 178, 179
'fine' ware, 118
Period 1 distribution, 8
shape/decoration, 142
Stanground, 116
beer, 229, 230, 231
belt fittings, 40, 95, 97 (Fig. 62)
bird bones, 40, 216, 218–19 (Table 81)
Bishopstone, Sussex, 85, 90
blacksmiths, 83, 86, 228
bone, 2, 3, 30, 188, 220, 233
animal, 31, 216–18, 221, 223–4 (Table 83), 225, 226
bird, 40, 216, 218-19 (Table 81)
human, 33, 216, 219
working, 216
bootplates, 100, 101 (Fig. 64)
Borough Fen, Cambridgeshire, 232, 233
bottles, 189
Bourne, 119, 120, 233
bowls, 8, 117, 122, 139, 143, 155, 165, 177–8, 179, 180, 188, 206
red colour-coated, 118
shell-gritted, 119, 179, 181, 182, 189
thick-walled shell-tempered, 27
types, 120
'boxes', 118, 164, 180
bracelets, 96–9 (Figs 62–3)
Brading, Isle of Wight: drier, 80
bread, 230, 236
brewing, xv, 77, 179, 180, 229–30, 231, 236
brick pits, 43
brick-lined drain, 18
Brixworth, Northamptonshire, 121
Bronze Age, xv, 1, 232
brooches
Anglo-Saxon, 102
Roman, 2, 92–5 (Fig. 61)
Mahndorf-type, 238
buckles, 40, 97 (Fig. 62)
buildings, xv, 5, 20, 182, 221, 224, 228, 235
added stone, 25
for brewing, 230
in Durobrivae, 234
nine-post: Period 5, 2
Period 1 phase *f*, 7–8
Roman period, 236
stone, 23–4
see also Mill-house
timber *see under* timber
unidentified, 3–4
see also House (Roman & Anglo-Saxon); round houses *and under* Roman period *and* Anglo-Saxon period
Bulwick, Northamptonshire, 234
burials, 21, 33 (Fig. 13)
burnt clay, 3, 7, 9, 35
butchering, 216, 217–18
butter, 223
buzzards, 218

Caesariensis, Roman province of, 236
candlesticks, 179, 189
Car Dyke, 232, 233–4
carbonised grain, 80
carpentry, 85
Castleford, Yorkshire: pottery, 189
'Castor box', 179
Castor, Cambridgeshire, 191, 234, 236
see also Normangate Field
Catholme, 90
cats, 216
Cat's Water, Fengate, Peterborough, 181, 233, 234
Catsgore, Avon: drier, 79–80
cattle, 216, 217, 218, 220, 221–4, 226, 227
cauldron, metal, 210
causeway, 12, 13
cemeteries, 206, 210, 233
see also burials
cereals, 74, 77, 225–6, 227–8
and brewing, 229, 230
drying, 79, 80
storage, 91
see also driers; granaries; Mill-house; mills; millstones
chalk, 3, 78, 79
Chalton, Hampshire, 85, 87, 89, 90

charcoal, 9, 33, 34, 80, 83
cheese, 223
cheese presses, 122, 155, 180, 223
Chesterton, Cambridgeshire, 114, 117, 119, 121, 191, 192
Christian symbols, 230
cistern/tank, 23
cladding, fragments of, 8
clay, 1, 12, 13
as backfill, 38
and Barn 1, 57
burnt, 3, 7, 9, 35
flooring, 23, 91
green, 3
lining of features, 16, 25
pit: London Brick Company, xi
plug, 4
for pottery, 116, 117, 118, 119, 120, 180, 233
coins, 17, 92, 93, 142, 189, 237
of Arcadius, 40, 42
and dating/phasing, xv
first/second century, 92
fourth century, 24, 30
House of Theodosius, 42
colanders, 122, 139, 155, 165, 177, 179, 180
combs, 100, 101 (Fig. 64), 207
Barred Zoomorphic, 40, 106–7 (Pl. IX; Fig. 75), 206, 238
computer program: pottery, xiv
condiment set, 180
Constantine III, 239
cooking, 2, 83, 228, 229
of birds/fowl, 218
and bones, 218
casserole-type vessels, 179, 180
hearths, 83, 229
pottery, 15, 189
see also food; kitchens; recipes
copper alloy
brooches, 92–5 (Fig. 61)
objects, 102 (Fig. 65), 236
razor, 1
coppiced woodland, 233
Corbridge, Northumberland: pottery, 189
corn *see* cereals
cosmetic items, 100, 101 (Fig. 64)
'cottage' houses, 70
counters, 100, 101 (Fig. 64)
Cowdery's Down, Basingstoke, Hampshire, 85, 89, 90
Crambeck, East Yorkshire, 118
cremation urns, 205
crops, 220, 227–8
vegetable, 77, 221, 229
see also cereals
crows, 218
crucibles, 179
cruck-framing, in buildings, 85–6
cult objects, 102 (Fig. 65), 236
cups, 122, 179

Dakin, Mr G. F., xi, 2, 5, 18, 26, 39, 215
and the barns, 55
and granary, 89
Darenth: drier, 79, 80
dating
Anglo-Saxon features, 31, 40
of artefacts, 238
Barn 1, 9, 55, 84
Barn 2, 12–13
Barn 3, 14, 19
Barn 4, 21, 22, 34
coins, 92
driers, 15
granary, 90
House, 18
Main Yard, 23, 25
Period 1 features, 2, 3, 5, 6, 8
Period 2 features, 10, 11, 12, 16
Period 3 features, 11, 12, 17, 18–19, 20
Period 4 features, 21, 25, 26, 27, 237
Period 5 features, 35, 36, 37, 38, 40
and periodisation, 27
pond (F200), 36
post-Period 5 features, 43
of pottery, 8, 29 (Fig. 12), 34, 92, 114, 122, 139, 142, 192, 235
pottery and phasing, xv
and pre-Period 1, 1
Rectangular Building, 22
ridge and furrow, 42
sites, 232
stone building, 24
decoration, of pottery, 117, 119, 120, 210
deer, 216, 225
Denton, Lincolnshire: barn at, 68 (Table 2)
Diocletian, 229
disc-type millstones, 108, 113
disease, animal, 218
dishes, 122, 139, 143, 155, 165, 177–8, 179, 180, 188
shell-gritted, 119
dogs, 216, 217, 218, 224, 226
Downton, Wiltshire: drier, 80
Dragonby, Lincolnshire, 181
drains, 18, 19, 21, 24
driers, 75–80, 227, 228
Barn 2: Period 3, 222, 224
F156, 12, 15, 21, 60, 75–6 (Figs 48 & 50), 79, 143, 144 (Fig. 90)
Barn 4, xi, 63, 65, 78
Period 4, 21
F168 (No. 2), 21, 35, 75–6 (Figs 49–50), 79
F169 (No. 3), 21, 75–6 (Figs 49–50), 79
Period 5, 41
F159/F160 (Nos 4 & 5), 34, 75-8 (Figs 51–2), 79
and brewing, 229-30
'developed', 15
excavation of, xiii
floor areas, 77 (Table 3)
Foxholes Farm, 227
Great Casterton villa, 238
positioning in barns, 69
drink, 229–30
milk, 218, 221, 222, 223, 224
see also brewing; food
drinking vessels, 229
drought, 1
Droveway, 13, 17, 18, 19, 24–5, 220, 221, 227, 228
and Main Yard, 23
old, 20, 26, 37
Period 2, 2, 8, 9 (Fig. 7), 10, 11, 16 (Fig. 9)
surrounding features, 20
ducks, 218
Durobrivae, 117, 232
development of, 191, 236
excavations, 204, 237
and the Fens, 234–5
landscape around, 225
as main local market, 177
pottery industry, 116, 120, 121, 122
settlement, 232
see also Water Newton
Durrington Walls, Wiltshire: drier, 79, 80
Dutch medieval pottery, 118

East Anglia, invasion of, 239
economy and site use, 220–31
Ecton, Northamptonshire, 121
Enclosure Award (1728), 43
Ermine Street, 232
estimated vessel equivalent (EVE), 114–15
Etton, Cambridgeshire, 232
Eye, Cambridgeshire, 233, 234

Fen Causeway, 234
Fen/Fen-edge sites xii (Fig. 1), 114, 232, 236
fences, 4, 10, 11, 15, 22, 23, 24, 39, 43, 90
Fengate, Peterborough, 119, 191, 232
Cat's Water, 181, 233, 234
Fens
drainage of, 233
economy of, 235–6
markets in, 117
Roman period, 232
second century development, 233–5
fermentation/distillation, 3
field systems, 232, 233
finger rings, 95–6 (Pl. VIII), 97 (Fig. 62)
fire
features associated with, 15, 75–84
see also driers; furnaces; hearths; ovens
fish, 216
flagons, 116, 139, 177, 178, 179, 189
cream ware, 117
'fine' ware, 118
pewter, 32
flask, unguent, 189
Flavian period, 121, 122, 182
Fletton Parkway, xi
flint, 1, 92
flint-working, 1
Flögeln, Germany, 91
flooding, xiii, 21, 235
floors
of Anglo-Saxon buildings, 91
Barn 3, 19, 60–2, 66
Barn 4, 21
clay, 23
of driers, 77 (Table 3), 78, 79, 80
of granaries, 74, 90
gravel, 37
Period 4 building, 13
of rectangular buildings, 22, 85
stone, 24, 37
tiles: of barn, 58
flour, 74, 225, 226, 236
foederati, 238
food
for animals, 223, 227, 229, 236
bread, 230, 236
consumption, 3
meat, 221, 225
and pastoral farming, 223
preparation, 20, 83, 178, 179–80, 182, 191, 228–9
Roman military needs, 236
storage, 178, 179
see also cooking; drink
footwear, 100, 101 (Fig. 64)
fossil coprolites, 180
fossil shell, 116
fowl, 218
Foxholes Farm, 79, 227, 228
Fullerton, Hampshire, 225
furnaces, 83, 229
F999, 36, 37, 81, 91
F1081–1082, 5, 37, 91
'furnace' base (F1135): Barn 3, 19, 83
location in barns, 69
Small Yard, 18
see also hearths

Gadebridge, Hertfordshire: drier, 80
gates/gateways, 4, 5, 8, 24, 25 (Fig. 10), 37
Anglo-Saxon, 28-30 (Fig. 11)
for animal control, 9
into eastern central enclosure, 14 (Fig. 8)
Period 1, 6 (Fig. 5), 7
Period 3, 11
Small Yard, 18
Gaul, 239
geology, xv-xvii
glass, 105
dating, 8
vessels, 13, 177, 179, 181, 189
window, 7

Glinton, Cambridgeshire, 232
Godmanchester, Cambridgeshire: pottery, 114
grain *see* cereals
granaries, 74, 228, 230
The Granary, xv, 85, 89–91 (Fig. 57), 220, 227
Great Casterton, Rutland
drier, 78
pottery, 114, 116, 117, 122, 155, 189
villa, 118, 165, 238
Great Pot, 206, 210
Greenfield, Ernest, 237
Greetham, Leicestershire: pottery products, 119, 120
gypsum, 116

Hadham, Hertfordshire: pottery, 118, 142, 155, 164, 165
Hadrian, Emperor, 117
Hadrianic-Antonine period, 191
Hadrian's Wall, 114, 233
Hall, David, 232
Hall, Anglo-Saxon, xv, 88–9 (Fig. 56), 90, 91
annex to, 173 (Fig. 105)
and replacement of House, 227
see also Anglo-Saxon period (buildings, western)
Hambledon, Buckinghamshire: driers, 78, 79, 227
hare, 216
Harrold, Bedfordshire, 119
harvesting, 228
Harwell, Berkshire, 220
hearths, 12, 16, 33, 35, 39, 81–2 (Figs 53–4), 83, 86, 228
Anglo-Saxon, 41
Barn 2, 75
central (Barn 1), 10–11
cooking, 229
sunken (F1103), 40
see also furnaces
hearths/ovens, 10, 15
hedges, xi, 4, 7, 10, 15, 42
Helpston, Cambridgeshire, 232
Henry VIII, King, 232
hides, cattle, 221, 222, 223
Highdown Hill: drier, 79
hops, 229
horn
of cattle, 218, 221, 222
cores, 224
vessels, 177, 179, 189
for working, 216
horses, 23, 39, 216, 218, 223, 225, 226, 236
House, Anglo-Saxon (eastern building), 28, 30, 31–2, 37, 38, 39, 41, 85–6 (Fig. 55), 87, 89, 91
House, Roman, 20, 24, 25–6, 34, 37, 38–9, 41–2, 143, 155
Anglo-Saxon structure over, 70, 85, 87–9
animal management, 226
building of, 224
function, 18
pottery, 182
replaced by Hall, 227
and site population, 228
and site status, 181
Hunting, Mr John, xi, xiv, 220
hypocausts, 79, 80, 230

Ickham, Kent, 225, 236
inhumations *see* burials
inkwells, 179
Intaglio, 96 (Pl. VIII)
Irchester, Northamptonshire, 181
iron, 2
iron-pan mass (F687), 2
iron-smelting, 232, 234
objects, 180
reaping hook, 40
slag, 83
Iron Age, xiv, 1, 177
animal management, 222, 224
cultivation, 220
pond, 43
pottery, 1, 119, 120, 121, 122
settlement and economy, 220, 232, 233
site location, 181
ironstone, 119, 192
ironworking, 192, 232, 234
Iwerne, Dorset: granary, 74

jars, 122, 139, 143, 165, 177–8, 179, 180, 181, 188
grey ware, 117
Horningsea, 114, 180
Iron Age, 120
red colour-coated, 118
shell-gritted, 119
storage, 180, 189
Jubilee Copse, Upton, 232
jugs, 177, 179

kilns, pottery, 114, 116, 119, 121, 192, 203, 204, 235, 237
see also Lower Nene Valley
kitchens, 3, 133 (Fig. 87), 135 (Fig. 88), 179, 229

labour, farm, xv, 228–9, 235
laeti, 238
lamps, 179, 180, 189
land
economy of, 220–1
management of, 224, 237
lapwing, 218
Laxton Lodge (Bulwick), Northamptonshire, 234
lead, 84, 180, 230
lean-to shack, 23
legumes, 77
Leicester, 119, 236
lids, 177, 179, 180
Limes frontier (Taunus mountain range), 108
limestone
Anglo-Saxon buildings, 91
of floors, 87
lining of drier, 75
in metalling, 19
in Mill-house, 72
post fills, 9, 10, 25
post-hole packing, 6, 15, 16, 17, 26, 27
in pottery, 119
limestone rag, 7, 15, 16, 24
and Anglo-Saxon gateway, 30
burnt, 34
in driers, 77
on flooring, 37
as packing, 21, 37
as post-packing, 38, 42
Lincoln, 225, 234, 236
Linford, Essex, 237
lintel, timber, 18
Loder, Robert: farm accounts, 220
London, 189, 225, 236
London Brick Company, xi, xiv, xvii
Longthorpe, Cambridgeshire, 122, 206
Longthorpe fortress, 117, 121, 191
Lower Nene Valley, 216, 224
mortaria, 199
pottery industry, 114, 115, 116–19, 120, 121, 122, 155, 177, 178, 179, 192
Lower Rhineland pottery, 117, 122, 192
Lullingstone, Kent: granary, 74
Lynch Farm, 235

Magnus Maximus, 236, 239
Main Yard, 20, 35–7, 38, 40, 41, 42, 155, 182, 224
access to, 26
Anglo-Saxon occupation, 27
buildings, 85, 89
barns, 19, 58, 63
Mill-house, 72
Rectangular Building, 70
Sunken-featured Building, 86
and crop handling, 227
development, 236
eastern side, 14–15, 20–2
enclosures, 17
features, 10, 13, 20, 22, 220, 221, 227
metalling, 18, 143
north side, 15–17
north-east corner, 23
north-west corner, 23–5
Period 2: glass and samian, 8
Period 3, 5
Period 4, 34
Period 5, 41
pottery, 28, 170 (Fig. 103)
Roman (Period 3), 145 (Fig. 91), 147–8 (Figs 92–3), 150 (Fig. 94)
south side, 26–7
malt, 77, 229, 230
Mancetter-Hartshill pottery kilns, 192, 235
manuring, 42, 223
Maxey, Cambridgeshire, 233, 235
medieval period
agriculture, 228
plough damage, xi, 17, 19, 27, 75, 188
ale drinking, 230
barn use, 227
brewing, 229
building, 85
deer as meat source, 225
objects, 92, 103 (Fig. 74), 105
Peterborough, 232
pottery, 3, 42, 118
quarrying, 232
recipes, 179–180
site features, 42
vernacular architecture, 68
Mercia, foundation of, 239
metalling
of Droveway, 24
Main Yard, 14, 19
Main Yard/Small Yard, 18–19
metalwork, 42, 207
metalworking, 236
mica: in pottery, 116, 119
Midfendic, 233, 234
Migration period: Germany, 91
milk, 218, 221, 222, 223, 224
Mill-house, xv, 34, 72–4 (Figs 46–7), 155, 180, 224, 231, 236
construction, 226
granary, 230
pottery, 182
remains of, 23–4
and site population, 228
and site status, 181
mills/milling, 74, 77, 108, 110, 225–6, 227, 228, 236
millstones, 72, 92, 105–13 (Figs 77–9), 225
mollusca, 216
monasteries, medieval, 229, 230
Monument 97, Orton Longueville, Cambridgeshire, xvii, 220, 221
Iron Age enclosures, 232
pottery, 121, 181
site development, 233, 235
mosaics, 79, 181, 233
Much Hadham pottery kilns *see* Hadham
Mucking, Essex, 87, 237
'Munsell Soil Color Charts', 115

nails, 69, 80
Nene, River, xvii, 116, 233, 234, 235
Nene Valley
pottery, 121, 235, 237
settlement in, 232-3

well, 18
see also Lower Nene Valley
Nene Valley Research Committee, xi, xiv, 105, 115, 203, 239
Neolithic period, xv, 1
New Forest pottery industry, 118
Normangate Field, Castor, Cambridgeshire, 114, 117, 121, 191, 192, 237
North Bretton, 235, 238
North Elmham, Norfolk, 90, 205
Northchurch, Hertfordshire, 179, 180

oils, use of, 179, 180
ointments, 180
Old Pepper Lake, 233
Old Sleaford, Lincolnshire, 78, 181
Old South Eau, 233
Old West Ouse, 234
Orton Longueville parish, Cambridgeshire, xi
see also Monument 97
Orton Park, Huntingdonshire, 235
Orton Township, xi
osteo-arthritis, 218
ovens, 2, 15, 33, 83, 84, 179
F901, 8, 10
oxen, 218, 222
Oxford Clay, xv, xvii, 116

Park Street, St Albans: drier, 80
pathology, 219
perfumes, 179
periglacial action, xvii
Period 1
animal bone, 218
animal management, 224, 225
Barn 1, 55
cereal crops, 227
continuity with Period 2, 11
eastern round house, 37
features, 5–6, 7–8, 9, 10, 13, 18, 23, 24, 26, 35, 37, 39, 81, 87, 91
north-south traffic route, 14
parameters, xiv
pasture, 218
phases of, xv, 1, 2–7, 8, 10, 18
pottery, 4, 7, 8, 10, 121–38, 139, 143, 165, 177, 179, 191, 192, 194, 224
and population size, 182, 188
types, 181–2, 183
vessels, 178, 179
Pre-Period 1, 1, 221
Roman House, 70
site description, 1–8 (Figs 4–6)
site development, xv, 221, 233
site occupation, 189
site status, 181
Period 2, 1
animal bone, 218, 223
animal management, 224, 226–7
barns, 228, 230
and Barn 1, 55, 58
and Barn 4, 63
construction, 69, 70
first barn built, 6
two barns laid out, xv
use, 222
cereal crops, 227
Droveway, 2, 8, 9 (Fig. 7), 16 (Fig. 9)
features, 15, 17, 18, 20, 21, 23, 27, 37, 39, 81, 84, 90
first stage of farmyard creation, xiv
'furnaces', 83
Main Yard: glass and samian, 8
north-south traffic route, 14
phases, xv, 8, 9–14
pottery, 122, 138–42, 165, 177, 191, 194
and population size, 182, 188
types, 181–2, 184
vessels, 179, 189
Roman House, 70
site description, 3, 4, 8–14 (Fig. 7)
site development, 221, 224, 233, 236
site population, 228
Period 3, xv, 14–20 (Figs 8–9)
animal bone, 218, 223
animal management, 222, 224, 225, 226, 227
barns, 228, 229, 230
and Barn 1, 55, 58
and Barn 2, 58
and Barn 4, 63
construction, 69, 70
and consolidation of west end, xiv
crop handling, 227, 228
drier, 75–6
features, 9 (Fig. 7), 10, 11, 12, 21, 23, 28, 30, 34, 37, 39, 40, 41, 81, 55, 224
gateway, 11
land use, 220
Main Yard, 5
pond, 1, 5, 20
pottery, 12, 138, 139, 142–55, 177, 189, 192, 194, 195
and population size, 182, 188
types, 181–2, 185
vessels, 178, 179, 189
settlement, 239
site development, xv, 164, 221, 224, 233, 235
site population, 228, 229
site use, 231
Small Yard, 8, 26
vat base (F105), 179
walls, 8, 11
well, 5, 13, 42, 209
yard deposits, 2
Period 4
animal bone, 218, 224
animal management, 222, 224, 225, 226, 227, 230–1
barns
Barn 1, 58
Barn 4, 13, 19, 65
construction, 69
brewing, 230
buildings, 15, 224
crop handling, 227, 228
driers, 75–6, 78
features, 6, 9, 12, 17, 18, 19, 20–7 (Fig. 10), 30, 34, 36, 37, 39, 40, 41, 165, 229
gravel surfacing, 10
Mill-house, 72, 226
periodisation, 15, 20
pond (F200), 1, 11, 36
pottery, 28, 142, 155–64, 177, 192, 195, 197, 215
and population size, 182, 188
types, 181–2, 186
vessels, 178, 179, 189
rectangular building, 8, 12, 13, 14, 20, 70
settlement, 237, 239
site description, 3
site development, xiv, xv, 221, 236, 237
site population, 228, 229
Period 5
Anglo-Saxon occupation, xiv, xv, 239
animal bone, 218, 223, 224
animal management, 222, 224
barns, 229
Barn 1, 58
Barn 3, 19
Barn 4, 65
coins, 92
crop handling, 227, 228
driers, 75–8, 229
features, 1, 5, 9, 11, 13, 14, 15, 17, 18, 20, 21, 22, 23, 24, 27, 43, 81, 89, 91, 215, 227, 230
grave, 21
land use, 220
and Mill-house, 74
and periodisation, 15, 41
phases, 41, 42
pottery, 162, 164–77, 189, 197, 198, 205, 206, 224
and population size, 182, 188
types, 181–2, 187
vessels, 178, 179, 180
rectangular building, 22, 70, 72
settlement patterns, 237, 238
site damage during, 26
site description, 27–42 (Figs 11–13)
site developments, 221, 237
structures, 2, 85
Sunken-featured Building, 13
post-Period 5 activity, 42–3
Peterborough
medieval period, 232
modern, xi, 232, 233
see also Fengate
Peterborough Development Corporation, xiv
Peterborough Museum, xi, xiv, 205–6
pewter flagon, 32
Piddington, Northamptonshire, 121
pigs, 216, 217, 218, 221, 223, 224–5, 226
pilae: of driers, 80
pins, 98–100 (Fig. 63)
pipe-clay figurine, 102 (Fig. 65)
plaster coating, 8
Pompeii
hour-glass type millstones, 108
raised hearth, 83
style of stone, 225
ponds, 14, 22
and animal management, 224, 227
creation of, 220
F200, 1, 11, 34, 36, 157 (Table 48), 166 (Table 63), 168 (Fig. 102)
replaces pond F500, 23
F225, 1
F500, 1, 17, 23
Iron Age, 43
Period 3, 5, 20
relocation of, 155
Roman, 43
population, 221, 228–30
changes, 178
in the Fens, 235
increase in, 182, 224
and Roman/Anglo-Saxon settlement, 237–8
1size of, and pottery, 182-8 (Table 77)
pottery
ANGLO-SAXON, xiii, 27–8, 31–42 *passim*, 70, 81, 83, 87, 89, 90, 91, 164, 165, 181, 182, 189, 192, 205–15, 232, 237, 238
Early, 239
Great Pot, 206, 210
Group 1, 206–9 (Figs 118–19), 212
Group 2, 207, 209–12 (Fig. 120)
and Roman pottery, 27–8, 189–90
stamped, 205–6
stratified, 212–14 (Fig. 121)
unstratified, 213 (Fig. 121), 214
Barn 1, 26
Barn 3, 19
colour ranges, 115
computer program, xiv
for cooking *see under* cooking
dating, 11, 12, 23, 25, 29 (Fig. 12), 92
Period 2, 8–9
and phasing, xv
IRON AGE 119, 120, 121, 122
shell-gritted, 84, 119–20, 121, 134, 155, 177
'belgic' types, 120
'transitional', 119, 120
MEDIEVAL, 3, 42, 118
and population, 182–8 (Table 77), 228, 229
PREHISTORIC, 1
Rectangular Building, 72

residual, 27, 28, 114, 139, 142, 155, 165, 188–9, 190, 224
ROMAN, 27–8, 32, 33, 34, 36, 40, 41, 42, 114–204, 206, 224, 237, 238
and Anglo-Saxon period, 189–90, 237
forms, 177–9
Late, 28, 34, 35, 40–1
Period 1, 121–38 (Tables 5–17; Figs 80–8), 181–2, 183 (Fig. 108)
Period 2, 138–42 (Tables 18–28; Fig. 89), 181–2, 184 (Fig. 109)
Period 3, 142–56 (Tables 29–44; Figs 90–6), 181–2, 185 (Fig. 110)
Period 4, 155–64 (Tables 45–57; Figs 96–100), 181–2, 186 (Fig. 111)
Period 5, 162 (Fig. 100), 164–77 (Tables 58–75; Figs 101–6), 181–2, 187 (Fig. 112), 206
and trade, 117, 121, 177, 192, 233
Small Yard, 18
'survival', 189
TYPES
amphorae, 122, 139, 142, 155, 165, 177, 179
'belgic' wares, 115, 121, 122, 181
Bourne products, 119, 120
burnt pot bases, 2
calcite-gritted wares, 120
colour-coated wares, 116, 117, 118–19, 122, 137 (Table 16), 139, 142 (Table 27), 153 (Table 43), 155, 164 (Table 57), 165, 176 (Table 75), 177, 178, 179, 192, 235, 237
red, 118, 119
cream wares, 114, 116, 117, 142, 165
cream/white wares, 117, 138 (Table 17), 142 (Table 28), 154 (Table 44), 165
Dales ware, 120, 181
Durobrivae industry, 232, 234, 235
East Midlands products, 120
East Yorkshire products, 118, 120
fumed wares, 116, 118
Gallo-Belgic type, 117
Greetham products, 119, 120
grey wares, 114–121 [U]passim[u], 136–7 (Tables 14–15), 139, 141 (Tables 25–6), 152–3 (Tables 41–2), 155, 163 (Tables 55–6), 175–6 (Table 73–4), 177, 178–9, 189, 235
grog-tempered wares, 121
Hadham region wares, 118, 142, 155, 164, 165
Hofheim type vessel, 117
Horningsea ware, 114, 142, 155, 165, 180
Knapton ware, 120
Lincolnshire products, 120
'local' wares, 116–21, 177, 181
London ware, 165
London ware type, 115, 121, 122, 139, 142
Lower Nene Valley industry, 114, 115, 116–19, 120, 121, 122, 155, 177, 178, 179, 192
Lower Rhineland wares, 117, 122, 155, 192
Mancetter-Hartshill wares, 192, 235
mortaria, xv, 8, 13, 20, 22, 27, 28, 37, 83, 114, 117, 122, 139, 142, 177–82, 188, 189, 191–203 (Table 79; Figs 114–17), 206, 235, 237, 238
Nene Valley industry, 235, 237
New Forest industry, 118
and Northamptonshire, 114, 121
Oxfordshire wares, 118, 122, 142, 192, 238
colour-coated, 164, 165, 177
mortaria, 155, 192, 199
'rhenish' types, 142
shell-gritted (RSG), 27, 33, 115, 117, 119–20, 134 (Table 13), 141 (Table 24), 152 (Table 40), 163 (Table 54), 165, 174 (Table 72), 177–82, 189
samian ware, xv, 8, 13, 117, 122, 139, 142, 155, 165, 177–82, 188 (Table 78), 189, 190–1 (Fig. 113)
Flavian, 121, 122
imitation, 118, 121, 177, 178
red, 119
South Gaulish, 190
slipped, 120
see also colour-coated wares
stamped, 116, 121
see also samian; mortaria
Stibbington-style, 237, 238
Swanpool products, 118, 120
'table-wares', 229
Trent Valley products, 34, 118, 120
Trent Valley type, 155, 165
Verulamium region, 192
ROMANO-SAXON, 237
uses of, 3
see also individual vessels
prehistoric period, 1, 22, 220
see also Bronze Age; Iron Age; Neolithic period
pulses, 77

quarrying, 40, 232
quartz, in pottery, 116, 119, 120, 121
querns, xiv, 110, 225
Quinton, Northamptonshire, 121, 179, 180

Rapsley, Surrey: barn at, 68 (Table 2), 69
ravens, 218
razor, copper alloy, 1
reaping hook, iron, 40
recipes, 3
Rectangular Building (Roman), 8, 12, 13, 14, 20, 22–3, 35, 41, 70–2 (Figs 45–6), 74, 227
creation of, 155
features in vicinity, 37, 38
and labour force, 230
pottery, 156 (Fig. 96), 168 (Fig. 102), 170 (Fig. 103), 182, 229
use of, 228
rectangular buildings (Anglo-Saxon), 39, 85, 91
Reculver, Kent: Roman fort, 236
redshank species, 40, 218
Richborough, Kent: Roman fort, 236
ridge and furrow cultivation, xi, 27, 42, 43, 220
rims, 114, 115, 117, 120, 180, 206, 207, 214
'late cornice', 122
mortaria, 191
ritual, in burial, 2
Rockbourne Down, Hampshire: driers, 79, 80
rodents, 216, 218
Roman period
Late Iron Age/early Roman: pottery, 119
and Anglo-Saxons, 28, 33, 41, 178, 182, 237–9
animal management, 226, 227
beads, 98, 99 (Fig. 63)
belt fittings, 95, 97 (Fig. 62)
bracelets, 96–9 (Figs 62-3)
brewing, 229, 230
brooches, 92–5 (Fig. 61)
buildings, 55–74
layout, 90
see also Barns; House, Roman; Mill-house; Rectangular Building (Roman)
and Car Dyke, 233
coins, 92
combs/cosmetic items, 100, 101 (Fig. 64)
Conquest, 1, 120, 121, 122, 181
cooking, 83
counters, 100, 101 (Fig. 64)
cult objects, 102 (Fig. 65)
culture, 191
and dating, 26, 36
degradation of site: Period 5, xv
disc-type millstones, 108
driers, 79, 80
and Fens, 234, 236–7
finger rings, 95–6 (Pl. VIII), 97 (Fig. 62)
footwear, 100, 101 (Fig. 64)
major road, 117
millstones, 110
Nene/Welland Valleys, 232, 233
Period 2, 35
Period 5, 39, 41, 42
pins, 98–100 (Fig. 63)
ponds, 43
rural economy, 222
settlement, 30
site economy, 220
and woodland, 225
Late, 5, 31, 89, 188
building, 234
evidence for dating, 27
Nene/Welland Valleys, 233
pewter flagon, 32
structures in Germany, 91
see also under pottery
roofing
of barns, 63, 66, 67, 68, 69
of driers, 79
of granary, 90
Rectangular Building, 70, 72
temporary, 8
rooks, 218
Rothwell, 206
round houses, 1
eastern, 2, 3, 7, 8, 37, 135 (Fig. 88), 182
western, 2, 3, 8, 135 (Fig. 88)
rubbish, 20, 138, 209
disposal, 7, 13, 89, 142, 188
from the House, 18
organic, 17
pits, 16, 28, 30, 32, 36, 38
Rushden, Northamptonshire: pottery manufacture, 121

Saalburg, Germany, 108, 225, 226
St Albans: Park Street, 80
St Germanus, 238
St Vincent Cross, 233
salt, 223, 225
sculpture, 232
Shakenoak, Oxfordshire: granary, 74
sheep, 216, 217, 218, 221, 223, 224, 226
Sibson, Cambridgeshire: pottery kiln, 114, 116
Silchester, Hampshire, 180, 230
skin vessels, 177, 179, 180, 189
slag, xiv, 32, 192, 228
iron, 83
Slea, River, 233, 234
Small Yard, 17–19, 20, 38–9, 41–2, 90
and animal management, 226
buildings, 87, 89
Period 3, 8, 26
pottery, 159 (Fig. 98), 161 (Fig. 99), 181, 182
Roman House, 70
site changes, 227
and site features, 14, 91
stone walling, 224
use of, 22
smithy, Anglo-Saxon, 32
snipe, 218
Soke Parkway, Peterborough, xi
Soke of Peterborough, 232
South Malling, Sussex: drier, 80
'spit-roast' beast, 31
Spong Hill, North Elmham, Norfolk, 205, 237
Spoonley Wood, Gloucestershire: barn, 69
stamps, of pottery, 205-6
Stanground, Cambridgeshire: pottery, 114, 116, 117, 118, 121

Stibbington, Cambridgeshire, 114, 116, 118, 237, 238
Stilton, Cambridgeshire, 116, 117, 121, 192, 203, 204
stoke-holes, 143, 179
  of Barn 2, 60
  of driers, 15, 34, 75, 77, 78, 79, 80
stone
  barns, 9, 12, 65–6, 227
    Barn 1, 10, 17, 18, 55, 57, 58, 66
    Barn 2, 58
    Barn 3, 19, 60
    Barn 4 walls, 21, 67
    barn construction, 67, 68, 69, 70
  building, 14, 23–4, 25, 37, 83
    see also Mill-house
  burnt, 33
  in driers, 75, 79, 80
  fill, 38, 40
  flooring, 37
  and the granary, 89, 90
  hardcore in ditches, 37–8
  in metalling, 14, 18
  packing, 7, 22
  as post-packing, 11, 43, 87
  and Rectangular Building, 70
  and Roman House, 70
  rubble, xi
  sole-plate, 37
  walling of Small Yard, 224
  walls, 8, 11, 23, 26, 39
  of wells, 18, 19, 21, 34
  see also millstones
Stonea, Cambridgeshire, 234
storage
  of food, 178, 179
  of grain *see* granaries
  pit, 30
strainers, 179, 180
stratigraphy, xv
Stroud, 68 (Table 2), 74
structures *see* buildings
Sulehay, Yarwell, Northamptonshire: pottery kiln, 117
sunken hearth (F1103), 40
Sunken-featured Buildings, 13, 36, 41, 42, 85, 86–7 (Fig. 55), 206
Swanpool (near Lincoln) pottery products, 118, 120

tank bases (F529/F539), 72
tank (F426), 22
tank/cistern, 23
Taunus mountain range, 108
taxes, 235, 239
Theodosius, House of: coins, 42
Third Terrace river gravels, xv–xvii
Thistleton, Leicestershire: pottery, 119
Thundersbarrow Hill: drier, 80
tiles, 58, 79, 181
timber
  barns
    Barn 1, 10, 55, 58, 66
    Barn 2, 58, 67
    Barn 3, 19, 222
    Barn 4, 67
    construction, 9, 68, 69
  drier floors, 79, 80
  gateway lintel, 18
  granary floors, 74
  and post-holes, 16
  Roman House, 70
  steyning, 11, 13
  structures/buildings, 12, 26, 30, 38, 39, 42, 90, 182
    Anglo-Saxon (over House), 70, 85, 87–9 (Fig. 56)
    framed structure, 37
    north of boundary ditch, 20
    replaced by stone building, 24
    timber-framing, 5
    wall of building, 22
  vessels, 177, 179, 180, 189
  well-frame, xiv, 18, 19, 21
tools, Anglo-Saxon, 102
topography, xvii, 220
Towcester, Northamptonshire, 119
trackways, 7, 8, 12, 23, 34–5
  see also Droveway
trade
  Anglo-Saxon period, 239
  in animals, 225
  and Durobrivae, 235
  in the Fens, 234, 235
  in pottery, 117, 121, 177, 192, 233
Trent Valley pottery, 34, 118, 120
turbaries, 233
tureens, 179

Upper Estuarine series (clay), 116
Upper Jurassic age, xv
Upton, Northamptonshire
  Jubilee Copse, 232
  Sunken-featured Building, 87
urns, 206, 207
  biconical, 206, 207, 212
  cinerary, 237
  cremation, 205
  decorated, 206
  globular, 206, 207

Vallum (Hadrian's Wall), 233
vases, 180
vat bases, 83, 84, 229
  Barn 2, 12, 60
  Barn 4, 41, 65
  F105, 21, 143, 144 (Fig. 90), 179
  F161, 34
vats, 15, 229, 230
  lead, 230
vegetable crops, 77, 221, 229
vegetation: site of Barn 4, 63
Verulamium, 78, 114, 122, 180, 192
vessels, 13
  functions, 179-80
  lead, 84
  see also individual vessels
Visigoths, 238, 239
Vitruvius, 225

Wakerley, Northamptonshire, 68 (Table 2), 69, 121
wall-plates, of barns, 66, 67, 69
walls, 8, 14
  Barn 1, 17, 18
  Barn 3, 19
  Barn 4, 21
Wash, the, 233, 236
Water Newton, Cambridgeshire, 114, 117, 232
  see also Durobrivae
water supply, 13, 17, 23, 229, 230
water tanks, 16, 23, 25, 36, 38
water-table, 1, 4, 11, 17, 21, 22, 30, 35, 38, 87, 220
  geology of, xvii
  and well, 13
  winter rise of, xiii
watermills, 108
  Ickham, Kent, 225, 236
Weald, the, 234
well house, 20
Welland, River, 233, 234, 235
Welland Valley, 232–3
wells, 5, 11, 18, 36, 41, 42, 67, 72, 229
  F172, 21, 34, 35, 41, 63, 171 (Fig. 104), 172 (Table 69)
  F218, 10, 13, 15
  F254, 17, 18, 19, 20, 143, 151 (Fig. 95), 155, 207
    disappearance of, 23
    pottery, 209–12 (Fig. 120)
  'F969', 18, 143, 151–2 (Fig. 95; Table 39)
  F1016, 18, 26, 143, 146 (Table 34), 148 (Fig. 93), 155
  F1052, 18, 26, 157 (Table 49), 158 (Fig. 97), 182
  in Small Yard, 18, 26, 39
  Stibbington, 118
  successive, 18, 165
  timber base-frames of, xiv
  wicker-lined, 38
Welton Wold, 79
Werrington Brook, 234
Werrington, Cambridgeshire, 232, 233, 235
Werrington Enclosure, 119, 120, 121, 179, 220, 223
West Blatchington, Sussex: driers, 227
West Stow, Suffolk, 85, 87, 90, 205, 206, 237
Weston Favell, Northamptonshire, 121
Westwood, Cambridgeshire, 232
wheat, 226, 227, 230
Whittlebury, Northamptonshire, 18
Whittlesey, Cambridgeshire, 233, 234
wicker, 36
wicker-lined well, 38
wickerwork, of fencing, 90
windows
  of barns, 69
  glass, 7
Winterton: barns and granary, 74
wood *see* timber
Woodchester brewery, 229
woodland, 225, 226
  clearance of, xvii
  management of, 232–3, 234
Woodston, Cambridgeshire, 206
wool, 218, 221, 224
wrist-clasp, 36
Wroxeter, Shropshire: pottery, 189

Yarwell, Northamptonshire: Sulehay, 117
Yaxley, Cambridgeshire, 116, 117, 121
yeast, 229
York, 190, 236

Zugmantel, Germany, 108, 225